Electrical, Computers, and Communication Networks

WileyPLUS is built around the activities you perform

Prepare & Present

Create outstanding class presentations using a wealth of resources, such as PowerPoint™ slides, image galleries, interactive simulations, and more. Plus you can easily upload any materials you have created into your course, and combine them with the resources Wiley provides you with.

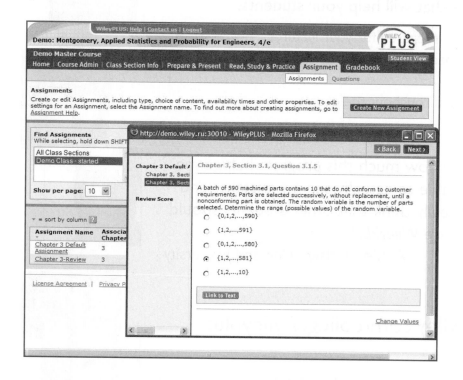

Create Assignments

Automate the assigning and grading of homework or quizzes by using the provided question banks, or by writing your own. Student results will be automatically graded and recorded in your gradebook. *WileyPLUS* also links homework problems to relevant sections of the online text, hints, or solutions— context-sensitive help where students need it most!

*Based on a spring 2005 survey of 972 student users of *WileyPLUS*

TO THE STUDENT

You have the potential to make a difference!

Will you be the first person to land on Mars? Will you invent a car that runs on water? But, first and foremost, will you get through this course?

WileyPLUS is a powerful online system packed with features to help you make the most of your potential, and get the best grade you can!

With Wiley**PLUS** you get:

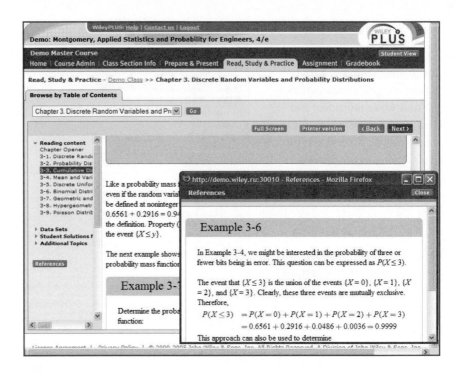

A complete online version of your text and other study resources

Study more effectively and get instant feedback when you practice on your own. Resources like self-assessment quizzes, tutorials, and animations bring the subject matter to life, and help you master the material.

Problem-solving help, instant grading, and feedback on your homework and quizzes

You can keep all of your assigned work in one location, making it easy for you to stay on task. Plus, many homework problems contain direct links to the relevant portion of your text to help you deal with problem-solving obstacles at the moment they come up.

The ability to track your progress and grades throughout the term.

A personal gradebook allows you to monitor your results from past assignments at any time. You'll always know exactly where you stand.

If your instructor uses *WileyPLUS*, you will receive a URL for your class. If not, your instructor can get more information about *WileyPLUS* by visiting www.wiley.com/college/wileyplus

---

"It has been a great help, and I believe it has helped me to achieve a better grade."

Michael Morris, *Columbia Basin College*

69% of students surveyed said it helped them get a better grade.*

Engineering
Statistics

Fourth Edition

Student Study Edition

Engineering Statistics

Fourth Edition

Student Study Edition

Douglas C. Montgomery
George C. Runger
Norma Faris Hubele
Arizona State University

BICENTENNIAL
1807
WILEY
2007
BICENTENNIAL

John Wiley & Sons, Inc.

ACQUISITIONS EDITOR *Jenny Welter*
SENIOR PRODUCTION EDITOR *Sujin Hong*
MARKETING MANAGER *Chris Ruel*
SENIOR DESIGNER *Kevin Murphy*
COVER PHOTO *Norm Christiansen*
ANNIVERSARY LOGO DESIGN *Richard Pacifico*
EDITORIAL ASSISTANT *Mark Owens*

This book was set in Times New Roman by TechBooks and printed and bound by R. R. Donnelley. The cover was printed by Phoenix Colors.

This book is printed on acid-free paper. ∞

To order books or for customer service, please call 1-800-CALL WILEY (225-5945).

Library of Congress Cataloging-in-Publication Data

Montgomery, Douglas C.
  Engineering statistics/Douglas C. Montgomery, George C. Runger, Norma Faris Hubele.—4th ed.
     p.   cm.
  Includes index.
  ISBN-13: 978-0-470-52694-1 (cloth)

     1. Statistics.   2. Probabilities.   I. Runger, George C.   II. Hubele, Norma Faris, 1953-   III. Title.
  QA276.12.M6453 2006b
  519.5′024′62—dc22

2006030759

Printed in the United States of America
10 9 8 7 6 5

To

Meredith, Neil, Colin, and Cheryl

Rebecca, Elisa, George, and Taylor

Yvonne and Joseph Faris, My Parents

Other Wiley books by Douglas C. Montgomery

Website: www.wiley.com/college/montgomery

Introduction to Statistical Quality Control, Sixth Edition
by Douglas C. Montgomery
For a first course in statistical quality control. Statistical techniques are emphasized throughout, with a strong engineering and management orientation.

Design and Analysis of Experiments, Seventh Edition
By Douglas C. Montgomery
An introduction to design and analysis of experiments, with the modest prerequisite of a first course in statistical methods. For senior and graduate students, and practitioners, to design and analyze experiments for improving the quality and efficiency of working systems.

Applied Statistics and Probability for Engineers, Fourth Edition
By Douglas C. Montgomery and George C. Runger
Introduction to engineering statistics, with topical coverage appropriate for either a one- or two-semester course. An applied approach to solving real-world engineering problems.

Probability and Statistics in Engineering, Fourth Edition
By William W. Hines, Douglas C. Montgomery, David M. Goldsman, and Connie M. Borror
Website: www.wiley.com/college/hines
For a first two-semester course in applied probability and statistics for undergraduate students, or a one-semester refresher for graduate students, covering probability from the start.

Introduction to Linear Regression Analysis, Fourth Edition
By Douglas C. Montgomery, Elizabeth A. Peck, and G. Geoffrey Vining
A comprehensive and thoroughly up-to-date look at regression analysis, still the most widely used technique in statistics today.

Response Surface Methodology: Process and Product Optimization Using Designed Experiments, Second Edition
By Raymond H. Myers and Douglas C. Montgomery
Website: www.wiley.com/college/myers
The exploration and optimization of response surfaces, for graduate courses in experimental design, and for applied statisticians, engineers, and chemical and physical scientists.

Generalized Linear Models: With Applications in Engineering and the Sciences
By Raymond H. Myers, Douglas C. Montgomery, and G. Geoffrey Vining
Website: www.wiley.com/college/myers
An introductory text or reference on Generalized Linear Models (GLMs). The range of theoretical topics and applications appeals both to students and practicing professionals.

Introduction to Time Series Analysis and Forecasting
By Douglas C. Montgomery, Cheryl L. Jennings, and Murat Kulahci
Methods for modeling and analyzing time series data, to draw inferences about the data and generate forecasts useful to the decision maker. Minitab and SAS are used to illustrate how the methods are implemented in practice. For advanced undergrad/first-year graduate, with a prerequisite of basic statistical methods. Portions of the book require calculus and matrix algebra.

About the Authors

Douglas C. Montgomery, Regents' Professor of Industrial Engineering and Statistics at Arizona State University, received his B.S., M.S., and Ph.D. degrees in engineering from Virginia Polytechnic Institute. He has been a faculty member of the School of Industrial and Systems Engineering at the Georgia Institute of Technology and a professor of mechanical engineering and director of the Program in Industrial Engineering at the University of Washington, where he held the John M. Fluke Distinguished Chair of Manufacturing Engineering. The recipient of numerous awards including the Stewart Medal of the American Society for Quality, two Brumbaugh Awards, the William G. Hunter Award, and two Shewell Awards from the ASQ. He is the editor of *Quality and Reliability Engineering International* and a former editor of the *Journal of Quality Technology*.

George C. Runger, Ph.D., is a Professor in the department of Industrial Engineering at Arizona State University. His research is on data mining, real-time monitoring and control, and other data-analysis methods with a focus on large, complex, multivariate data streams. His work is funded by grants from the National Science Foundation and corporations. In addition to academic work, he was a senior engineer at IBM. He holds degrees in industrial engineering and statistics.

Norma Faris Hubele, Professor of Engineering and Statistics at Arizona State University, and Director of Strategic Initiatives for the Ira A. Fulton School of Engineering, holds degrees in mathematics, operations research, statistics and computer and systems engineering. She is co-owner and serves as vice-president for Quality Control and Statistical Analysis at the Phoenix-based Refrac System. She is on the editorial board of the *Journal of Quality Technology* and *Quality Technology & Quantity Management*, as a founding member. Her specializations include capability analysis, transportation safety and statistics in litigation.

Preface

INTENDED AUDIENCE

Engineers play a significant role in the modern world. They are responsible for the design and development of most of the products that our society uses, as well as the manufacturing processes that make these products. Engineers are also involved in many aspects of the management of both industrial enterprises and business and service organizations. Fundamental training in engineering develops skills in problem formulation, analysis, and solution that are valuable in a wide range of settings.

Solving many types of engineering problems requires an appreciation of variability and some understanding of how to use both descriptive and analytical tools in dealing with variability. Statistics is the branch of applied mathematics that is concerned with variability and its impact on decision making. This is an introductory textbook for a first course in engineering statistics. Although many of the topics we present are fundamental to the use of statistics in other disciplines, we have elected to focus on meeting the needs of engineering students by allowing them to concentrate on the applications of statistics to their disciplines. Consequently, our examples and exercises are engineering based, and in almost all cases, we have used a real problem setting or the data either from a published source or from our own consulting experience.

Engineers in all disciplines should take at least one course in statistics. Indeed, the Accreditation Board on Engineering and Technology is requiring that engineers learn about statistics and how to use statistical methodology effectively as part of their formal undergraduate training. Because of other program requirements, most engineering students will take only one statistics course. This book has been designed to serve as a text for the one-term statistics course for all engineering students.

The Fourth edition has been extensively revised and includes some new examples and many new problems. In this revision we have focused on rewriting topics that our own teaching experience or feedback from others indicated that students found difficult.

ORGANIZATION OF THE BOOK

The book is based on a more comprehensive text (Montgomery, D. C., and Runger, G. C., *Applied Statistics and Probability for Engineers*, Fourth Edition, Hoboken, NJ: John Wiley & Sons, 2006) that has been used by instructors in a one- or two-semester course. We have taken the key topics for a one-semester course from that book as the basis of this text. As a result of this condensation and revision, this book has a modest mathematical level. Engineering students who have completed one semester of calculus should have no difficulty reading nearly all of the text. Our intent is to give the student an understanding of statistical methodology and how it may be applied in the solution of engineering problems, rather than the mathematical theory of statistics. Margin notes help to guide the student in this interpretation and understanding. Throughout the book, we provide guidance on how statistical methodology is a key part of the problem-solving process.

Chapter 1 introduces the role of statistics and probability in engineering problem solving. Statistical thinking and the associated methods are illustrated and contrasted with other engineering modeling approaches within the context of the engineering problem-solving method. Highlights of the value of statistical methodologies are discussed using simple examples. Simple summary statistics are introduced.

Chapter 2 illustrates the useful information provided by simple summary and graphical displays. Computer procedures for analyzing large data sets are given. Data analysis methods such as histograms, stem-and-leaf plots, and frequency distributions are illustrated. Using these displays to obtain insight into the behavior of the data or underlying system is emphasized.

Chapter 3 introduces the concepts of a random variable and the probability distribution that describes the behavior of that random variable. We introduce a simple 3-step procedure for structuring a solution to probability problems. We concentrate on the normal distribution, because of its fundamental role in the statistical tools that are frequently applied in engineering. We have tried to avoid using sophisticated mathematics and the event–sample space orientation traditionally used to present this material to engineering students. An in-depth understanding of probability is not necessary to understand how to use statistics for effective engineering problem solving. Other topics in this chapter include expected values, variances, probability plotting, and the central limit theorem.

Chapters 4 and 5 present the basic tools of statistical inference: point estimation, confidence intervals, and hypothesis testing. Techniques for a single sample are in Chapter 4, and two-sample inference techniques are in Chapter 5. Our presentation is distinctly applications oriented and stresses the simple comparative-experiment nature of these procedures. We want engineering students to become interested in how these methods can be used to solve real-world problems and to learn some aspects of the concepts behind them so that they can see how to apply them in other settings. We give a logical, heuristic development of the techniques, rather than a mathematically rigorous one. In this edition, we have focused more extensively on the *P*-value approach to hypothesis testing because it is relatively easy to understand and is consistent with how modern computer software presents the concepts.

Empirical model building is introduced in *Chapter 6*. Both simple and multiple linear regression models are presented, and the use of these models as approximations to mechanistic models is discussed. We show the student how to find the least squares estimates of the regression coefficients, perform the standard statistical tests and confidence intervals, and use the model residuals for assessing model adequacy. Throughout the chapter, we emphasize the use of the computer for regression model fitting and analysis.

Chapter 7 formally introduces the design of engineering experiments, although much of Chapters 4 and 5 was the foundation for this topic. We emphasize the factorial design and, in particular, the case in which all of the experimental factors are at two levels. Our practical experience indicates that if engineers know how to set up a factorial experiment with all factors at two levels, conduct the experiment properly, and correctly analyze the resulting data, they can successfully attack most of the engineering experiments that they will encounter in the real world. Consequently, we have written this chapter to accomplish these objectives. We also introduce fractional factorial designs and response surface methods.

Statistical quality control is introduced in *Chapter 8*. The important topic of Shewhart control charts is emphasized. The $\overline{X}$ and R charts are presented, along with some simple control charting techniques for individuals and attribute data. We also discuss some aspects of estimating the capability of a process.

The students should be encouraged to work problems to master the subject matter. The book contains an ample number of problems of different levels of difficulty. The end-of-section exercises are intended to reinforce the concepts and techniques introduced in that section. These exercises are more structured than the end-of-chapter supplemental exercises, which generally require more formulation or conceptual thinking. We use the supplemental exercises as integrating problems to reinforce mastery of concepts, as opposed to analytical technique. The team exercises challenge the student to apply chapter methods and concepts to problems requiring data collection. As noted below, the use of statistics software in problem solution should be an integral part of the course.

USING THE BOOK

We strongly believe that an introductory course in statistics for undergraduate engineering students should be, first and foremost, an *applied course.* The primary emphasis should be on data description, inference (confidence intervals and tests), model building, designing engineering experiments, and statistical quality control *because these are the techniques that they as practicing engineers will need to know how to use.* There is a tendency in teaching these courses to spend a great deal of time on probability and random variables (and, indeed, some engineers, such as industrial and electrical engineers, do need to know more about these subjects than students in other disciplines) and to emphasize the mathematically oriented aspects of the subject. This can turn an engineering statistics course into a "baby math-stat" course. This type of course can be fun to teach and much easier on the instructor because it is almost always easier to teach theory than application, but it does not prepare the student for professional practice.

In our course taught at Arizona State University, students meet twice weekly, once in a large classroom and once in a small computer laboratory. Students are responsible for reading assignments, individual homework problems, and team projects. In-class team activities include designing experiments, generating data, and performing analyses. The supplemental problems and team exercises in this text are a good source for these activities. The intent is to provide an active learning environment with challenging problems that foster the development of skills for analysis and synthesis.

STUDENT STUDY EDITION

The *Student Study Edition* has been developed to better help students and instructors identify when resources outside the text are available and relevant to support student understanding.
Icons in the *Student Study Edition* pinpoint:

 Exercises included in the *Student Solutions Manual*

 Animations on the book website or *WileyPLUS*

Exercises (and specifically GO Tutorial problems) available for instructors to assign in *WileyPLUS*

Exercises for which it is recommended computer software be used

 Exercises for which summary statistics are given, and the complete sample of data is available on the book website

USING THE COMPUTER

In practice, engineers use computers to apply statistical methods in solving problems. Therefore, we strongly recommend that the computer be integrated into the course. Throughout the book, we have presented output from Minitab as typical examples of what can

FEATURED IN THIS BOOK

Learning Objectives
Learning Objectives at the start of each chapter guide the student in what they are intended to take away from this chapter, and serve as a study reference.

→

LEARNING OBJECTIVES

After careful study of this chapter, you should be able to do the following:

1. Use simple linear or multiple linear regression for building empirical models of engineering and scientific data.
2. Analyze residuals to determine if the regression model is an adequate fit to the data or to see if any underlying assumptions are violated.
3. Test statistical hypotheses and construct confidence intervals on regression model parameters.
4. Use the regression model either to compute a mean or to make a prediction of a future observation.
5. Use confidence intervals or prediction intervals to describe the error in estimation from a regression model.
6. Comment on the strengths and weaknesses of your empirical model.

← **Definitions, Key Concepts, and Equations**
Throughout the text, definitions, and key concepts and equations are highlighted by a box to emphasize their importance.

than $\hat{\Theta}_2$, the estimator $\hat{\Theta}_1$ is more likely to produce an estimate close to the true value θ. A logical principle of estimation, when selecting among several estimators, is to choose the estimator that has minimum variance.

Minimum Variance Unbiased Estimator

> If we consider all unbiased estimators of θ, the one with the smallest variance is called the **minimum variance unbiased estimator** (MVUE).

The concepts of an unbiased estimator and an estimator with minimum variance are extremely important. There are methods for formally deriving estimates of the parameters of a probability distribution. One of these methods, the **method of maximum likelihood,** produces point estimators that are approximately unbiased and very close to the minimum variance estimator. For further information on the method of maximum likelihood, see Montgomery and Runger (2006).

In practice, one must occasionally use a biased estimator (such as S for σ). In such cases, the mean square error of the estimator can be important. The **mean square error** of an estimator $\hat{\Theta}$ is the expected squared difference between $\hat{\Theta}$ and θ.

Mean Square Error of an Estimator

> The **mean square error** of an estimator $\hat{\Theta}$ of the parameter θ is defined as
>
> $$\text{MSE}(\hat{\Theta}) = E(\hat{\Theta} - \theta)^2 \qquad (4\text{-}3)$$

Margin Notes
Margin notes help to guide the student in interpreting and understanding statistics.

→

| Normal Probability Plot Interpretation | A normal probability plot of the residuals from the paper tensile strength experiment is shown in Fig. 5-9. Figures 5-10 and 5-11 present the residuals plotted against the factor levels and the fitted value $\bar{y}_i$, respectively. These plots do not reveal any model inadequacy or unusual problem with the assumptions. |

Figures

Numerous Figures throughout the text illustrate statistical concepts in multiple formats.

Table 6-1 Salt Concentration in Surface Streams and Roadway Area

Observation	Salt concentration (y)	Roadway area (x)
1	3.8	0.19
2	5.9	0.15
3	14.1	0.57
4	10.4	0.40
5	14.6	0.70
6	14.5	0.67
7	15.1	0.63
8	11.9	0.47
9	15.5	0.75
10	9.3	0.60
11	15.6	0.78
12	20.8	0.81
13	14.6	0.78
14	16.6	0.69
15	25.6	1.30
16	20.9	1.05
17	29.9	1.52
18	19.6	1.06
19	31.3	1.74
20	32.7	1.62

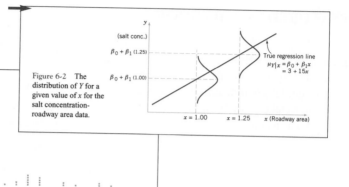

Figure 6-2 The distribution of Y for a given value of x for the salt concentration-roadway area data.

Figure 6-1 Scatter diagram of the salt concentration in surface streams and roadway area data in Table 6-1.

Minitab Output

Throughout the book, we have presented output from Minitab as typical examples of what can be done with modern statistical software, and have also included analysis of Minitab output in the homework exercises.

Table 7-13 Analysis for Example 7-4 Process Yield from Minitab

Factorial Design

Full Factorial Design

Factors:	2	Base Design:	2, 4
Runs:	9	Replicates:	1
Blocks:	none	Center pts (total):	5

All terms are free from aliasing

Fractional Factorial Fit

Estimated Effects and Coefficients for y

Term	Effect	Coef	StDev Coef	T	P
Constant		40.4444	0.06231	649.07	0.000
A	1.5500	0.7750	0.09347	8.29	0.000
B	0.6500	0.3250	0.09347	3.48	0.018
A*B	−0.0500	−0.0250	0.09347	−0.27	0.800

Analysis of Variance for y

Source	DF	Seq SS	Adj SS	Adj MS	F
Main Effects	2	2.82500	2.82500	1.41250	40.42
2-Way Interactions	1	0.00250	0.00250	0.00250	0.07
Residual Error	5	0.17472	0.17472	0.03494	
Curvature	1	0.00272	0.00272	0.00272	0.06
Pure Error	4	0.17200	0.17200	0.04300	
Total	8	3.00222			

6-13. Use the partially complete Minitab output below to answer the following questions.

(a) Find all of the missing values.
(b) Find the estimate of σ^2.
(c) Test for significance of regression. Test for significance of β_1. Comment on the two results. Use $\alpha = 0.05$.
(d) Construct a 95% CI on β_1. Use this CI to test for significance of regression.
(e) Comment on results found in parts (c) and (d).
(f) Write the regression model and use it to compute the residual when $x = 2.18$ and $y = 2.8$.
(g) Use the regression model to compute the mean and predicted future response when $x = 1.5$. Given that $\bar{x} = 1.76$ and $S_{xx} = 5.326191$, construct a 95% CI on the mean response and a 95% PI on the future response. Which interval in larger? Why?

Predictor	Coef	SE Coef	T	P
Constant	0.6649	0.1594	4.17	0.001
X	0.83075	0.08552	?	?

S = ?   R − Sq = 88.7%   R − Sq(adj) = ?

Analysis of Variance

Source	DF	SS	MS	F	P
Regression	1	3.6631	3.6631	?	?
Residual Error	12	0.4658	?		
Total	13	?			

Problem-Solving Procedures

The text introduces a sequence of steps in solving probability problems, and applying hypothesis-testing methodology, and explicitly exhibits these procedures in examples.

3-Step Procedure for Probability ➡️

To determine a probability for a random variable, it can be helpful to apply three steps:

1. Determine the random variable and distribution of the random variable.
2. Write the probability statement in terms of the random variable.
3. Compute the probability using the probability statement and the distribution.

These steps are shown in the solutions of some examples in this chapter. In other examples and exercises you might use these steps on your own.

EXAMPLE 3-1
Current in a Wire

Let the continuous random variable X denote the current measured in a thin copper wire in milliamperes. Assume that the range of X is [0, 20 mA], and assume that the probability density function of X is $f(x) = 0.05$ for $0 \le x \le 20$. What is the probability that a current measurement is less than 10 milliamperes?

Define the random variable and distribution.

Solution. The random variable is the current measurement with distribution given by $f(x)$. The pdf is shown in Fig. 3-8. It is assumed that $f(x) = 0$ wherever it is not specifically defined. The probability requested is indicated by the shaded area in Fig. 3-8.

Write the probability statement.

$$P(X < 10) = \int_{0}^{10} f(x)\, dx = 0.5$$

Compute the probability.

As another example,

7-Step Procedure for Hypothesis Testing ⬇️

$$P(5 < X < 10) = \int_{5}^{10} f(x)\, dx = 0.25$$ ∎

4-3.5   General Procedure for Hypothesis Testing

This chapter develops hypothesis testing procedures for many practical problems. Use of the following sequence of steps in applying hypothesis testing methodology is recommended:

1. **Parameter of interest:** From the problem context, identify the parameter of interest.
2. **Null hypothesis, H_0:** State the null hypothesis, H_0.
3. **Alternative hypothesis, H_1:** Specify an appropriate alternative hypothesis, H_1.
4. **Test statistic:** State an appropriate test statistic.
5. **Reject H_0 if:** Define the criteria that will lead to rejection of H_0.
6. **Computations:** Compute any necessary sample quantities, substitute these into the equation for the test statistic, and compute that value.
7. **Conclusions:** Decide whether or not H_0 should be rejected and report that in the problem context. This could involve computing a P-value or comparing the test statistic to a set of critical values.

Steps 1–4 should be completed prior to examination of the sample data. This sequence of steps will be illustrated in subsequent sections.

be done with modern computer software. In teaching, we have used Statgraphics, Minitab, Excel, and several other statistics packages or spreadsheets. We did not clutter the book with examples from many different packages because *how* the instructor integrates the software into the class is ultimately more important than *which* package is used. All text data and the instructor manual are available in electronic form.

In our large-class meeting times, we have access to computer software. We show the student how the technique is implemented in the software as soon as it is discussed in class. We recommend this as a teaching format. Low-cost student versions of many popular software packages are available, and many institutions have statistics software available on a local area network, so access for the students is typically not a problem.

Computer software can be used to do many exercises in this text. Some exercises, however, have small computer icons in the margin. We highly recommend using software in these instances.

The second icon is meant to represent the book website. This icon marks problems for which summary statistics are given, and the complete sample of data is available on the book website. Some instructors may wish to have the students use the data rather than the summary statistics for problem solutions.

Example Problems
A set of example problems provides the student detailed solutions and comments for interesting, real-world situations.

EXAMPLE 4-3
Propellant
Burning Rate

Aircrew escape systems are powered by a solid propellant. The burning rate of this propellant is an important product characteristic. Specifications require that the mean burning rate must be 50 cm/s. We know that the standard deviation of burning rate is $\sigma = 2$ cm/s. The experimenter decides to specify a type I error probability or significance level of $\alpha = 0.05$. He selects a random sample of $n = 25$ and obtains a sample average burning rate of $\bar{x} = 51.3$ cm/s. What conclusions should he draw?

Solution. We may solve this problem by following the seven-step procedure outlined in Section 4-3.5. This results in the following:

1. **Parameter of interest:** The parameter of interest is μ, the mean burning rate.
2. **Null hypothesis, H_0:** $\mu = 50$ cm/s
3. **Alternative hypothesis, H_1:** $\mu \neq 50$ cm/s
4. **Test statistic:** The test statistic is

$$z_0 = \frac{\bar{x} - \mu_0}{\sigma/\sqrt{n}}$$

5. **Reject H_0 if:** Reject H_0 if the P-value is smaller than 0.05. (Note that the corresponding critical region boundaries for fixed significance level testing would be $-z_{0.025} = -1.96$ and $z_{0.025} = 1.96$.)
6. **Computations:** Since $\bar{x} = 51.3$ and $\sigma = 2$,

$$z_0 = \frac{51.3 - 50}{2/\sqrt{25}} = \frac{1.3}{0.4} = 3.25$$

EXERCISES FOR SECTION 2-4

2-27. The following data are the joint temperatures of the O-rings (°F) for each test firing or actual launch of the space shuttle rocket motor (from *Presidential Commission on the Space Shuttle Challenger Accident*, Vol. 1, pp. 129–131): 84, 49, 61, 40, 83, 67, 45, 66, 70, 69, 80, 58, 68, 60, 67, 72, 73, 70, 57, 63, 70, 78, 52, 67, 53, 67, 75, 61, 70, 81, 76, 79, 75, 76, 58, 31.

(a) Calculate the sample mean.
(b) Calculate the sample variance and sample standard deviation.
(c) Construct a box plot of the data.

2-29. The "cold start ignition time" of an automobile engine is being investigated by a gasoline manufacturer. The following times (in seconds) were obtained for a test vehicle: 1.75,

End-of-Section Exercises
Exercises at the end of each section emphasize the material in that section.

Supplemental Exercises
At the end of each chapter, a set of supplemental exercises cover the scope of the chapter topics, and require the student to make a decision about the approach they will use to solve the problem.

SUPPLEMENTAL EXERCISES

2-50. The pH of a solution is measured eight times by one operator using the same instrument. She obtains the following data: 7.15, 7.20, 7.18, 7.19, 7.21, 7.20, 7.16, and 7.18.
(a) Calculate the sample mean. Suppose that the desirable value for this solution was specified to be 7.20. Do you think that the sample mean value computed here is close enough to the target value to accept the solution as con-

(b) Compute the sample variance and sample standard deviation using the definition in equation 2-3. Explain why the results from both equations are the same.
(c) Subtract 35 from each of the original resistance measurements and compute s^2 and s. Compare your results with those obtained in parts (a) and (b) and explain your findings.

TEAM EXERCISES

5-106. Construct a data set for which the paired t-test statistic is very large, indicating that when this analysis is used, the two population means are different; however, t_0 for the two-sample t-test is very small, so the incorrect

miles per gallon in urban driving than Car Type B. The standard or claim may be expressed as a mean (average), variance, standard deviation, or proportion. Collect two appropriate random samples of data and

Team Exercises
At the end of each chapter, these exercises challenge students to apply chapter methods and concepts to problems requiring data collection.

Important Terms and Concepts
At the end of each chapter is a list of important terms and concepts, for an easy self-check and study tool.

IMPORTANT TERMS AND CONCEPTS

Adjusted R^2	C_p statistic	Outliers	Sample correlation coefficient, r
All possible regressions	Empirical model	Polynomial regression	
Analysis of variance (ANOVA)	Forward selection	Population correlation coefficient, ρ	Significance of regression
Backward elimination	Indicator variables	Prediction interval	Simple linear regression
Coefficient of determi-	Influential observations	Regression analysis	Standard errors of
	Interaction		

STUDENT RESOURCES

- **Data Sets**   Data sets for all examples and exercises in the text. Visit the student section of the book website at www.wiley.com/college/montgomery to access these materials.
- **Student Solutions Manual**   Detailed solutions to all of the odd-numbered exercises in the book. The Student Solutions Manual may be purchased from the website at www.wiley.com/college/montgomery.

 This icon in the book will show which exercises are included in the accompanying *Student Solutions Manual.*

INSTRUCTOR RESOURCES

The following resources are available only to instructors who adopt the text:

- **Solutions Manual**   All solutions to exercises in the text.
- **Data Sets**   Data sets for all examples and exercises in the text.
- **Image Gallery of Text Figures**
- **PowerPoint Lecture Slides**
- **Case Studies**   In-depth case studies that illustrate an integration of several analysis methods will be posted as they are developed.

These instructor-only resources are password-protected. Visit the instructor section of the book website at www.wiley.com/college/montgomery to register for a password to access these materials.

We will use this site to communicate our latest information about innovations and recommendations for effectively using this text, and we hope to elicit your feedback.

MINITAB

A Student Version of Minitab is available as an option to purchase in a set with this text. Student versions of software often do not have all the functionality that full versions do. Consequently, student versions may not support all the concepts presented in this text. If you would like to adopt for your course the set of this text with the Student Version of Minitab, please contact your local Wiley representative at www.wiley.com/college/rep.

Alternatively, students may find information about how to purchase the professional version of the software at www.minitab.com.

WileyPLUS

Instructors have the option to adopt the WileyPLUS course management tool along with this text for the following resources:

- Your own customizable course website, populated with the textbook resources.
- All of the resources that are on the Student and Instructor Companion Sites will also be in the student and instructor sections of your WileyPLUS website, so everything will be in one location.
- The entire text in a searchable and linked HTML format. For example, the Important Terms and Concepts at the end of each chapter in the online text link directly to where those concepts are explained in the chapter, for an easy-to-use study tool.
- Online reading and homework assignments, to help you track your students' progress.

These icons in the book will show which exercises (and specifically GO Tutorial problems) available for instructors to assign in *WileyPLUS*.

For more information about WileyPLUS, visit www.wiley.com/college/wileyplus.

ACKNOWLEDGMENTS

We express our appreciation for support in developing some of this text material from the Course and Curriculum Development Program of the Undergraduate Education Division of the National Science Foundation. We are grateful to Dr. Dale Kennedy and Dr. Mary Anderson–Rowland for their generous feedback and suggestions in teaching our course at Arizona State University. We also thank Dr. Teri Reed Rhoads, Purdue University, Dr. Lora Zimmer, and Dr. Sharon Lewis for their work in the development of the course based on this text when they were graduate assistants. We are very thankful to Ms. Nuttha Lurpongpukanag and Dr. Connie Borror for their thorough work on the instructor's solutions manual, and to Dr. Sarah Street, Dr. James C. Ford, Dr. Craig Downing, and Mr. Patrick Egbunonu for their assistance in checking the accuracy and completeness of the text, solutions manual, and WileyPLUS. We appreciate the staff support and resources provided by the Department of Industrial Engineering at Arizona State University, and our department chair, Dr. Ronald Askin.

Several reviewers provided many helpful suggestions, including Dr. Thomas Willemain, Rensselaer, Dr. Hongshik Ahn, SUNY, Stony Brook; Dr. James Simpson, Florida State/FAMU; Dr. John D. O'Neil, California Polytechnic University, Pomona; Dr. Charles Donaghey, University of Houston; Professor Gus Greivel, Colorado School of Mines; Professor Arthur M. Sterling, LSU; Professor David Powers, Clarkson University; Dr. William Warde, Oklahoma State University; and Dr. David Mathiason. We are also indebted to Dr. Smiley Cheng of the University of Manitoba for permission to adapt many of the statistical tables from his excellent book (with Dr. James Fu), *Statistical Tables for Classroom and Exam Room*. John Wiley & Sons, Prentice-Hall, the Biometrika Trustees, The American Statistical Association, the Institute of Mathematical Statistics, and the editors of *Biometrics* allowed us to use copyrighted material, for which we are grateful.

**This project was supported, in part,
by the
National Science Foundation
Opinions expressed are those of the authors
and not necessarily those of the Foundation**

Douglas C. Montgomery

George C. Runger

Norma Faris Hubele

Contents

CHAPTER 5   Decision Making for Two Samples   216

CHAPTER 6   Building Empirical Models   282

1

The Role of Statistics in Engineering

LEARNING OBJECTIVES

After careful study of this chapter, you should be able to do the following:

1. Identify the role that statistics can play in the engineering problem-solving process.
2. Discuss how variability affects data collected and used in making decisions.
3. Discuss the methods that engineers use to collect data.
4. Explain the importance of random samples.
5. Identify the advantages of designed experiments in data collection.
6. Explain the difference between mechanistic and empirical models.
7. Explain the difference between enumerative and analytic studies.

1-1 THE ENGINEERING METHOD AND STATISTICAL THINKING

Engineers solve problems of interest to society by the efficient application of scientific principles. The **engineering** or **scientific method** is the approach to formulating and solving these problems. The steps in the engineering method are as follows:

1. Develop a clear and concise description of the problem.
2. Identify, at least tentatively, the important factors that affect this problem or that may play a role in its solution.

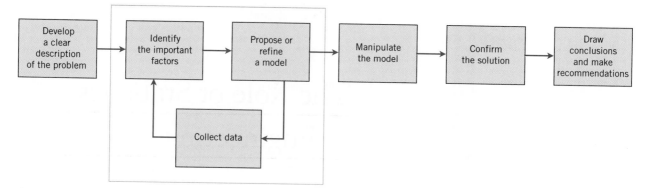

Figure 1-1   The engineering problem-solving method.

3. Propose a model for the problem, using scientific or engineering knowledge of the phenomenon being studied. State any limitations or assumptions of the model.

4. Conduct appropriate experiments and collect data to test or validate the tentative model or conclusions made in steps 2 and 3.

5. Refine the model on the basis of the observed data.

6. Manipulate the model to assist in developing a solution to the problem.

7. Conduct an appropriate experiment to confirm that the proposed solution to the problem is both effective and efficient.

8. Draw conclusions or make recommendations based on the problem solution.

The steps in the engineering method are shown in Fig. 1-1. Note that the engineering method features a strong interplay between the problem, the factors that may influence its solution, a model of the phenomenon, and experimentation to verify the adequacy of the model and the proposed solution to the problem. Steps 2–4 in Fig. 1-1 are enclosed in a box, indicating that several cycles or iterations of these steps may be required to obtain the final solution. Consequently, engineers must know how to efficiently plan experiments, collect data, analyze and interpret the data, and understand how the observed data are related to the model they have proposed for the problem under study.

The field of **statistics** deals with the collection, presentation, analysis, and use of data to make decisions and solve problems. Because many aspects of engineering practice involve working with data, obviously some knowledge of statistics is important to any engineer. Specifically, statistical techniques can be a powerful aid in designing new products and systems, improving existing designs, and designing, developing, and improving production processes.

Statistical methods are used to help us describe and understand **variability.** By variability, we mean that successive observations of a system or phenomenon do not produce exactly the same result. We all encounter variability in our everyday lives, and **statistical thinking** can give us a useful way to incorporate this variability into our decision-making processes. For example, consider the gasoline mileage performance of your car. Do you always get exactly the same mileage performance on every tank of fuel? Of course not—in fact, sometimes the mileage performance varies considerably. This observed variability in gasoline mileage depends on many factors, such as the type of driving that has occurred most recently (city versus highway), the changes in condition of the vehicle over time (which could include factors such as tire inflation, engine compression, or valve wear), the brand and/or octane number of the gasoline used, or possibly even the weather conditions that have been experienced recently. These factors represent potential **sources of variability** in the system. Statistics gives us a framework for describing this variability and for learning about which potential sources of variability are the most important or which have the greatest impact on the gasoline mileage performance.

We also encounter variability in most types of engineering problems. For example, suppose that an engineer is developing a rubber compound for use in O-rings. The O-rings are to be employed as seals in plasma etching tools used in the semiconductor industry, so their resistance to acids and other corrosive substances is an important characteristic. The engineer uses the standard rubber compound to produce eight O-rings in a development laboratory and measures the tensile strength of each specimen after immersion in a nitric acid solution at 30°C for 25 minutes [refer to the American Society for Testing and Materials (ASTM) Standard D 1414 and the associated standards for many interesting aspects of testing rubber O-rings]. The tensile strengths (in psi) of the eight O-rings are 1030, 1035, 1020, 1049, 1028, 1026, 1019, and 1010. As we should have anticipated, not all of the O-ring specimens exhibit the same measurement of tensile strength. There is **variability** in the tensile strength measurements. Because the measurements exhibit variability, we say that tensile strength is a **random variable.** A convenient way to think of a random variable, say X, which represents a measured quantity, is by using the **model**

$$X = \mu + \epsilon$$

where μ is a constant and ϵ is a random disturbance, or "noise" term. The constant remains the same, but small changes in the environmental conditions and test equipment, differences in the individual O-ring specimens, and potentially many other factors change the value of ϵ. If none of these disturbances was present, the value of ϵ would always equal zero, and X would always be equal to the constant μ. However, this never happens in engineering practice, so the actual measurements we observe exhibit variability. We often need to describe, quantify, and ultimately (because variability is often detrimental to our final objectives) reduce variability.

Figure 1-2 is a **dot diagram** of the O-ring tensile strength data. The dot diagram is a very useful plot for displaying a small body of data, say, up to about 20 observations. This plot allows us to easily see two important features of the data: the **location,** or the middle, and the **scatter** or **variability.** When the number of observations is small, it is usually difficult to see any specific pattern in the variability, although the dot diagram is a very convenient way to observe data features such as **outliers** (observations that differ considerably from the main body of the data) or **clusters** (groups of observations that occur closely together).

The need for statistical thinking arises often in the solution of engineering problems. Consider the engineer developing the rubber O-ring material. From testing the initial specimens, he knows that the average tensile strength is 1027.1 psi. However, he thinks that this may be too low for the intended application, so he decides to consider a modified formulation of the rubber in which a Teflon additive is included. Eight O-ring specimens are made from this modified rubber compound and subjected to the nitric acid emersion test described earlier. The tensile test results are 1037, 1047, 1066, 1048, 1059, 1073, 1070, and 1040.

The tensile test data for both groups of O-rings are plotted as a dot diagram in Fig. 1-3. This display gives the visual impression that adding the Teflon to the rubber compound has led to an increase in the tensile strength. However, there are some obvious questions to ask. For instance,

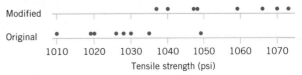

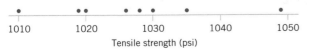

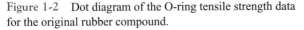

Figure 1-2   Dot diagram of the O-ring tensile strength data for the original rubber compound.

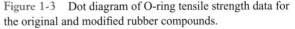

Figure 1-3   Dot diagram of O-ring tensile strength data for the original and modified rubber compounds.

how do we know that another set of O-ring specimens will not give different results? Is a sample of eight O-rings adequate to give reliable results? If we use the test results obtained thus far to conclude that adding the Teflon to the rubber formulation will increase the tensile strength after exposure to nitric acid, what risks are associated with this decision? For example, is it possible (or even likely) that the apparent increase in tensile strength observed for the modified formulation is only due to the inherent variability in the system and that including this additional ingredient (which would increase both the cost and the manufacturing complexity) really has no effect on tensile strength? Statistical thinking and methodology can help answer these questions.

Often, physical laws (such as Ohm's law and the ideal gas law) are applied to help design products and processes. We are familiar with this reasoning from general laws to specific cases. However, it is also important to reason from a specific set of measurements to more general cases to answer the previous questions. This reasoning is from a sample (such as the eight rubber O-rings) to a population (such as the O-rings that will be sold to customers) and is referred to as **statistical inference.** See Fig. 1-4. Clearly, reasoning based on measurements from some objects to measurements on all objects can result in errors (called sampling errors). However, if the sample is selected properly, these risks can be quantified and an appropriate sample size can be determined.

Engineers and scientists are also often interested in comparing two different conditions to determine whether either condition produces a significant effect on the response that is observed. These conditions are sometimes called "treatments." The rubber O-ring tensile strength problem illustrates such a situation; the two different treatments are the two formulations of the rubber compound, and the response is the tensile strength measurement. The purpose of the study is to determine whether the modified formulation results in a significant effect—increased tensile strength. We can think of each sample of eight O-rings as a random and representative sample of all parts that will ultimately be manufactured. The order in which each O-ring was tested was also randomly determined. This is an example of a **completely randomized designed experiment.**

When statistical significance is observed in a randomized experiment, the experimenter can be confident in the conclusion that it was the difference in treatments that resulted in the difference in response. That is, we can be confident that a cause-and-effect relationship has been found.

Sometimes the objects to be used in the comparison are not assigned at random to the treatments. For example, the September 1992 issue of *Circulation* (a medical journal published by the American Heart Association) reports a study linking high iron levels in the body with increased risk of heart attack. The study, done in Finland, tracked 1931 men for 5 years and showed a statistically significant effect of increasing iron levels on the incidence of heart attacks. In this study, the comparison was not performed by randomly selecting a sample of men and then assigning some to a "low iron level" treatment and the others to a "high iron level" treatment. The researchers just tracked the subjects over time. This type of study is called an **observational study.** Designed experiments and observational studies are discussed in more detail in the next section.

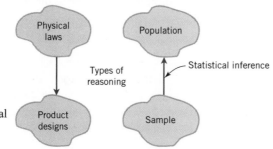

Figure 1-4   Statistical inference is one type of reasoning.

It is difficult to identify causality in observational studies because the observed statistically significant difference in response between the two groups may be due to some other underlying factor (or group of factors) that was not equalized by randomization and not due to the treatments. For example, the difference in heart attack risk could be attributable to the difference in iron levels or to other underlying factors that form a reasonable explanation for the observed results—such as cholesterol levels or hypertension.

1-2 COLLECTING ENGINEERING DATA

In the previous section, we illustrated some simple methods for summarizing and visualizing data. In the engineering environment, the data are almost always a **sample** that has been selected from some **population.** Generally, engineering data are collected in one of three ways:

- A **retrospective study** based on historical data
- An **observational study**
- A **designed experiment**

A good data collection procedure will usually result in a simplified analysis and help ensure more reliable and generally applicable conclusions. When little thought is put into the data collection procedure, serious problems for both the statistical analysis and the practical interpretation of results can occur.

Montgomery, Peck, and Vining (2006) describe an acetone-butyl alcohol distillation column. A schematic of this binary column is shown in Fig. 1-5. We will use this distillation column

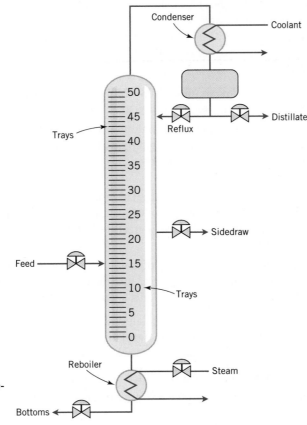

Figure 1-5 Acetone-butyl alcohol distillation column.

to illustrate the three methods of collecting engineering data identified above. There are three factors that may influence the concentration of acetone in the distillate (output product) stream from the column. These are the reboil temperature (controlled by steam flow), the condensate temperature (controlled by coolant flow), and the reflux rate. For this column, production personnel maintain and archive the following records:

- The concentration of acetone (in g/l) in a test sample taken every hour from the product stream
- The reboil temperature controller log, which is a record of the reboil temperature versus time
- The condenser temperature controller log
- The nominal reflux rate each hour

The process specifications require that the nominal reflux rate is held constant for this process. The production personnel very infrequently change this rate.

1-2.1 Retrospective Study

A **retrospective study** uses either all or a sample of the **historical** process data from some period of time. The objective of this study might be to determine the relationships among the two temperatures and the reflux rate on the acetone concentration in the output product stream. In most such studies, the engineer is interested in using the data to construct a **model** relating the variables of interest. For example, in this case the model might relate acetone concentration (the dependent variable) to the three independent variables, reboil temperature, condenser temperature, and reflux rate. These types of models are called **empirical models,** and they are illustrated in more detail in Section 1-3.

A retrospective study takes advantage of previously collected, or historical, data. It has the advantage of minimizing the cost of collecting the data for the study. However, there are several potential problems:

1. We really cannot isolate the effect of the reflux rate on concentration because it probably did not vary much over the historical period.

2. The historical data on the two temperatures and the acetone concentration do not correspond directly. Constructing an approximate correspondence would probably require making several assumptions and a great deal of effort, and it might be impossible to do reliably.

3. Production maintains both temperatures as tightly as possible to specific target values through the use of automatic controllers. Because the two temperatures do not vary very much over time, we will have a great deal of difficulty seeing their real impact on the concentration.

4. Within the narrow ranges that they do vary, the condensate temperature tends to increase with the reboil temperature. Because the temperatures vary together, it will be difficult to separate their individual effects on the acetone concentration.

Retrospective studies, although often the quickest and easiest way to collect engineering process data, often provide limited useful **information** for controlling and analyzing a process. In general, their primary disadvantages are as follows:

1. Some of the important process data often are missing.

2. The reliability and validity of the process data are often questionable.

3. The nature of the process data often may not allow us to address the problem at hand.

4. The engineer often wants to use the process data in ways that they were never intended to be used.

5. Logs, notebooks, and memories may not explain interesting phenomena identified by the data analysis.

Using historical data always involves the risk that, for whatever reason, some of the important data were not collected or were lost or were inaccurately transcribed or recorded. Consequently, historical data often suffer from problems with data quality. These errors also make historical data prone to outliers.

Just because data are convenient to collect does not mean that these data are useful. Often, there are data that are not considered essential for routine process monitoring and that are not convenient to collect, but these data have a significant impact on the process. Historical data cannot provide this information if information on some important variables was never collected. For example, the ambient temperature may affect the heat losses from the distillation column. On cold days, the column loses more heat to the environment than during very warm days. The production logs for this acetone-butyl alcohol column do not routinely record the ambient temperature. Also, the concentration of acetone in the input feed stream has an effect on the acetone concentration in the output product stream. However, this variable is not easy to measure routinely, so it is not recorded either. Consequently, the historical data do not allow the engineer to include either of these factors in the analysis even though potentially they may be important.

The purpose of many engineering data analysis efforts is to isolate the root causes underlying interesting phenomena. With historical data, these interesting phenomena may have occurred months, weeks, or even years earlier. Logs and notebooks often provide no significant insights into these root causes, and memories of the personnel involved fade over time. Analyses based on historical data often identify interesting phenomena that go unexplained.

Finally, retrospective studies often involve very large (indeed, even *massive*) data sets. The engineer will need a firm grasp of statistical principles if the analysis is going to be successful.

1-2.2   Observational Study

We could also use an observational study to collect data for this problem. As the name implies, an **observational study** simply observes the process or population during a period of routine operation. Usually, the engineer interacts or disturbs the process only as much as is required to obtain data on the system, and often a special effort is made to collect data on variables that are not routinely recorded, if it is thought that such data might be useful. With proper planning, observational studies can ensure accurate, complete, and reliable data. On the other hand, these studies still often provide only limited information about specific relationships among the variables in the system.

In the distillation column example, the engineer would set up a data collection form that would allow production personnel to record the two temperatures and the actual reflux rate at specified times corresponding to the observed concentration of acetone in the product stream. The data collection form should provide the ability to add comments to record any other interesting phenomena that may occur, such as changes in ambient temperature. It may even be possible to arrange for the input feed stream acetone concentration to be measured along with the other variables during this relatively short-term study. An observational study conducted in this manner would help ensure accurate and reliable data collection and would take care of problem 2 and possibly some aspects of problem 1 associated with the retrospective study. This approach also minimizes the chances of observing an outlier related to some error in the data. Unfortunately, an observational study cannot address problems 3 and 4. Observational studies can also involve very large data sets.

1-2.3   Designed Experiments

The third way that engineering data are collected is with a **designed experiment.** In a designed experiment, the engineer makes deliberate or purposeful changes in controllable variables (called **factors**) of the system, observes the resulting system output, and then makes a decision or an inference about which variables are responsible for the changes that he or she observes in the output performance. An important distinction between a designed experiment and either an observational or retrospective study is that the different combinations of the factors of interest are applied randomly to a set of experimental units. This allows cause-and-effect relationships to be established, something that cannot be done with observational/retrospective studies.

[margin notes: →deliberate changes →randomly applied cause/effect relationships to be established]

The O-ring example is a very simple illustration of a designed experiment. That is, a deliberate change was introduced into the formulation of the rubber compound with the objective of discovering whether or not an increase in the tensile strength could be obtained. This is an experiment with a single factor. We can view the two groups of O-rings as having the two formulations applied randomly to the individual O-rings in each group. This establishes the desired cause-and-effect relationship. The engineer can then address the tensile strength question by comparing the mean tensile strength measurements for the original formulation to the mean tensile strength measurements for the modified formulation. Statistical techniques called **hypothesis testing** and **confidence intervals** can be used to make this comparison. These techniques are introduced and illustrated extensively in Chapters 4 and 5.

A designed experiment can also be used in the distillation column problem. Suppose that we have three factors: the two temperatures and the reflux rate. The experimental design must ensure that we can separate out the effects of these three factors on the **response variable,** the concentration of acetone in the output product stream. In a designed experiment, often only two or three levels of each factor are employed. Suppose that two levels of the temperatures and the reflux rate are used and that each level is coded to a ± 1 (or low, high) level. The best experimental strategy to use when there are several factors of interest is to conduct a **factorial experiment.** In a factorial experiment, the factors are varied together in an arrangement that tests all possible combinations of factor levels.

Figure 1-6 illustrates a factorial experiment for the distillation column. Because all three factors have two levels, there are eight possible combinations of factor levels, shown geometrically as the eight corners of the cube in Fig. 1-6a. The tabular representation in Fig. 1-6b

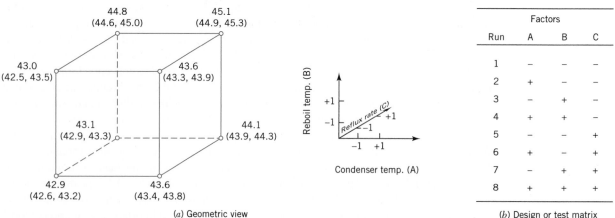

(a) Geometric view   (b) Design or test matrix

Figure 1-6   A factorial design for the distillation column.

shows the test matrix for this factorial experiment; each column of the table represents one of the three factors, and each row corresponds to one of the eight runs. The − and + signs in each row indicate the low and high settings for the factors in that run. The actual experimental runs would be conducted in **random order,** thus establishing the random assignment of factor level combinations to experimental units that is the key principle of a designed experiment. Two trials, or **replicates,** of the experiment have been performed (in random order), resulting in 16 runs (also called observations).

Some very interesting tentative conclusions can be drawn from this experiment. First, compare the average acetone concentration for the eight runs with condenser temperature at the high level with the average concentration for the eight runs with condenser temperature at the low level (these are the averages of the eight runs on the left and right faces of the cube in Fig. 1-6*a*, respectively), or 44.1 − 43.45 = 0.65. Thus, increasing the condenser temperature from the low to the high level increases the average concentration by 0.65 g/l. Next, to measure the effect of increasing the reflux rate, compare the average of the eight runs in the back face of the cube with the average of the eight runs in the front face, or 44.275 − 43.275 = 1. The effect of increasing the reflux rate from the low to the high level is to increase the average concentration by 1 g/l; that is, reflux rate apparently has an effect that is larger than the effect of condenser temperature. The reboil temperature effect can be evaluated by comparing the average of the eight runs in the top of the cube with the average of the eight runs in the bottom, or 44.125 − 43.425 = 0.7. The effect of increasing the reboil temperature is to increase the average concentration by 0.7 g/l. Thus, if the engineer's objective is to increase the concentration of acetone, there are apparently several ways to do this by making adjustments to the three process variables.

There is an interesting relationship between reflux rate and reboil temperature that can be seen by examination of the graph in Fig. 1-7. This graph was constructed by calculating the average concentration at the four different combinations of reflux rate and reboil temperature, plotting these averages versus the reflux rate, and then connecting the points representing the two temperature levels with straight lines. The slope of each of these straight lines represents the effect of reflux rate on concentration. Note that the slopes of these two lines do not appear to be the same, indicating that the reflux rate effect is *different* at the two values of reboil temperature. This is an example of an **interaction** between two factors. The interpretation of this interaction is very straightforward; if the low level of reflux rate (−1) is used, reboil temperature

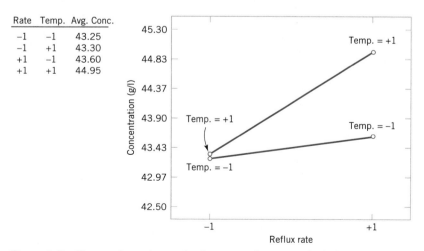

Figure 1-7 The two-factor interaction between reflux rate and reboil temperature.

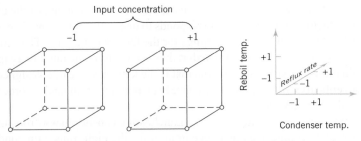

Figure 1-8   A four-factor factorial experiment for the distillation column.

has little effect, but if the high level of reflux rate (+1) is used, increasing the reboil temperature has a large effect on average concentration in the output product stream. Interactions
occur often in physical and chemical systems, and factorial experiments are the only way to
investigate their effects. In fact, if interactions are present and the factorial experimental strategy is not used, incorrect or misleading results may be obtained.

   We can easily extend the factorial strategy to more factors. Suppose that the engineer
wants to consider a fourth factor, the concentration of acetone in the input feed stream. Figure
1-8 illustrates how all four factors could be investigated in a factorial design. Because all four
factors are still at two levels, the experimental design can still be represented geometrically as
a cube (actually, it's a *hypercube*). Note that as in any factorial design, all possible combinations of the four factors are tested. The experiment requires 16 trials. If each combination of
factor levels in Fig. 1-8 is run one time, this experiment actually has the same number of runs
as the replicated three-factor factorial in Fig. 1-6.

   Generally, if there are k factors and they each have two levels, a factorial experimental
design will require 2^k runs. For example, with $k = 4$, the 2^4 design in Fig. 1-8 requires 16
tests. Clearly, as the number of factors increases, the number of trials required in a factorial
experiment increases rapidly; for instance, eight factors each at two levels would require
256 trials. This amount of testing quickly becomes unfeasible from the viewpoint of time
and other resources. Fortunately, when there are four to five or more factors, it is usually unnecessary to test all possible combinations of factor levels. A **fractional factorial experiment** is a variation of the basic factorial arrangement in which only a subset of the factor
combinations are actually tested. Figure 1-9 shows a fractional factorial experimental design for the four-factor version of the distillation column experiment. The circled test combinations in this figure are the only test combinations that need to be run. This experimental
design requires only 8 runs instead of the original 16; consequently it would be called a **one-
half fraction.** This is an excellent experimental design in which to study all four factors. It

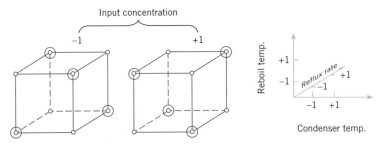

Figure 1-9   A fractional factorial experiment for the distillation column.

will provide good information about the individual effects of the four factors and some information about how these factors interact.

Factorial and fractional factorial experiments are used extensively by engineers and scientists in industrial research and development, where new technology, products, and processes are designed and developed and where existing products and processes are improved. Because so much engineering work involves testing and experimentation, it is essential that all engineers understand the basic principles of planning efficient and effective experiments. Chapter 7 focuses on these principles, concentrating on the factorial and fractional factorials that we have introduced here.

1-2.4   Random Samples

As the previous three sections have illustrated, almost all statistical analysis is based on the idea of using a **sample** of data that has been selected from some **population.** The objective is to use the sample data to make decisions or learn something about the population. In general, the population is the complete collection of items or objects from which the sample is taken. A sample is just a subset of the items in the population.

For example, suppose that we are manufacturing semiconductor wafers, and we want to learn about the resistivity of the wafers in a particular lot. In this case, the lot is the population. Our strategy for learning about wafer resistivity would be to select a sample of (say) three wafers and measure the resistivity on those specific wafers. This is an example of a **physical population;** that is, the population consists of a well-defined, often finite group of items all of which are available at the time the sample is collected.

Data are often collected as a result of an engineering experiment. For example, recall the O-ring experiment described in Section 1-1. Initially eight O-rings were produced and subjected to a nitric acid bath, following which the tensile strength of each O-ring was determined. In this case the eight O-ring tensile strengths are a sample from a population that consists of all of the measurements on tensile strength that could possibly have been observed. This type of population is called a **conceptual population.** Many engineering problems involve conceptual populations. The O-ring experiment is a simple but fairly typical example. The factorial experiment used to study the concentration in the distillation column in Section 1-2.3 also results in sample data from a conceptual population.

The way that samples are selected is also important. For example, suppose that you wanted to learn about the mathematical skills of undergraduate students at Arizona State University. Now, this involves a physical population. Suppose that we select the sample from all of the students who are currently taking an engineering statistics course. This is probably a bad idea, as this group of students will most likely have mathematical skills that are quite different than those found in the majority of the population. In general, samples that are taken because they are convenient, or that are selected through some process involving the judgment of the engineer, are unlikely to produce correct results. For example, the sample of engineering statistics students would likely lead to a biased conclusion regarding mathematical skill in the population. This usually happens with judgment or convenience samples.

In order for statistical methods to work correctly and to produce valid results, **random samples** must be used. The most basic method of random sampling is **simple random sampling**. To illustrate simple random sampling, consider the mathematical skills question discussed previously. Assign an integer number to every student in the population (all of the ASU undergraduates). These numbers range from 1 to N. Suppose that we want to select a simple random sample of 100 students. Now we could use a computer to generate 100 random integers from 1 to N where each integer has the same chance of being selected.

Choosing the students who correspond to these numbers would produce the simple random sample. Notice that every student in the population has the same chance of being chosen for the sample. Another way to say this is that all possible samples of size $n = 100$ have the same chance of being selected.

Definition

A **simple random sample** of size n is a sample that has been selected from a population in such a way that each possible sample of size n has an equally likely chance of being selected.

In Chapter 3, Section 3-13, we will provide a more mathematical definition of a simple random sample and discuss some of its properties.

It is not always easy to obtain a random sample. For example, consider the lot of semiconductor wafers. If the wafers are packaged in a container, it may be difficult to sample from the bottom, middle, or sides of the container. It is tempting to take the sample of three wafers off the top tray of the container. This is an example of a convenience sample, and it may not produce satisfactory results, because the wafers may have been packaged in time order of production and the three wafers on top may have been produced last, when something unusual may have been happening in the process.

Retrospective data collection may not always result in data that can be viewed as a random sample. Often, retrospective data are data of convenience, and they may not reflect current process performance. Data from observational studies are more likely to reflect random sampling, because a specific study is usually being conducted to collect the data. Data from a designed experiment can usually be viewed as data from a random sample, if the individual observations in the experiment are made in random order. Completely randomizing the order of the runs in an experiment helps eliminate the effects of unknown forces that may be varying while the experiment is being run, and it provides assurance that the data can be viewed as a random sample.

Collecting data retrospectively on a process or through an observational study, and even through a designed experiment, almost always involves sampling from a conceptual population. Our objective in many of these data studies is to draw conclusions about how the system or process that we are studying will perform in the future. An **analytic study** is a study or experiment where the conclusions are to be drawn relative to a **future population.** For example, in the distillation column experiment we want to make conclusions about the concentration of future production quantities of acetone that will be sold to customers. This is an analytic study involving a conceptual population that does not yet exist. Clearly in addition to random sampling there must be some additional assumption of **stability** of this process over time. For example, it might be assumed that the sources of variability currently being experienced in production are the same as will be experienced in future production. In Chapter 8 we introduce control charts, an important statistical technique to evaluate the stability of a process or system.

The problem involving sampling of wafers from a lot to determine lot resistivity is called an **enumerative study.** The sample is used to make conclusions about the population from which the sample was drawn. The study to determine the mathematical abilities of ASU undergraduates is also an enumerative study. Note that random samples are required in both enumerative and analytic studies, but the analytic study requires an additional assumption of stability. Figure 1-10 provides an illustration.

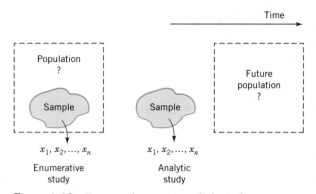

Figure 1-10   Enumerative versus analytic study.

1-3   MECHANISTIC AND EMPIRICAL MODELS

Models play an important role in the analysis of nearly all engineering problems. Much of the formal education of engineers involves learning about the models relevant to specific fields and the techniques for applying these models in problem formulation and solution. As a simple example, suppose we are measuring the current flow in a thin copper wire. Our model for this phenomenon might be Ohm's law:

$$\text{Current} = \text{voltage/resistance}$$

or

$$I = E/R \tag{1-1}$$

We call this type of model a **mechanistic model** because it is built from our underlying knowledge of the basic physical mechanism that relates these variables. However, if we performed this measurement process more than once, perhaps at different times or even on different days, the observed current could differ slightly because of small changes or variations in factors that are not completely controlled, such as changes in ambient temperature, fluctuations in performance of the gauge, small impurities present at different locations in the wire, and drifts in the voltage source. Consequently, a more realistic model of the observed current might be

$$I = E/R + \epsilon \tag{1-2}$$

where ϵ is a term added to the model to account for the fact that the observed values of current flow do not perfectly conform to the mechanistic model. We can think of ϵ as a term that includes the effects of all of the unmodeled sources of variability that affect this system.

Sometimes engineers work with problems for which there is no simple or well-understood mechanistic model that explains the phenomenon. For instance, suppose we are interested in the number average molecular weight (M_n) of a polymer. Now we know that M_n is related to the viscosity of the material (V) and that it also depends on the amount of catalyst (C) and the temperature (T) in the polymerization reactor when the material is manufactured. The relationship between M_n and these variables is

$$M_n = f(V, C, T) \tag{1-3}$$

say, where the *form* of the function *f* is unknown. Perhaps a working model could be developed from a first-order Taylor series expansion, which would produce a model of the form

$$M_n = \beta_0 + \beta_1 V + \beta_2 C + \beta_3 T \qquad (1\text{-}4)$$

where the βs are unknown parameters. As in Ohm's law, this model will not exactly describe the phenomenon, so we should account for the other sources of variability that may affect the molecular weight by adding a random disturbance term to the model; thus

$$M_n = \beta_0 + \beta_1 V + \beta_2 C + \beta_3 T + \epsilon \qquad (1\text{-}5)$$

is the model that we will use to relate molecular weight to the other three variables. This type of model is called an **empirical model;** that is, it uses our engineering and scientific knowledge of the phenomenon, but it is not directly developed from our theoretical or first-principles understanding of the underlying mechanism. Data are required to estimate the βs in equation 1-5. These data could be from a retrospective or observational study, or they could be generated by a designed experiment.

   To illustrate these ideas with a specific example, consider the data in Table 1-1. This table contains data on three variables that were collected in an observational study in a semiconductor manufacturing plant. In this plant, the finished semiconductor is wire bonded to a frame. The variables reported are pull strength (a measure of the amount of force required to break the bond), the wire length, and the height of the die. We would like to find a model relating pull strength to wire length and die height. Unfortunately, there is no physical mechanism that we can easily apply here, so it doesn't seem likely that a mechanistic modeling approach will be successful.

   Figure 1-11*a* is a **scatter diagram** of the pull strength *y* from Table 1-1 versus the wire length x_1. This graph was constructed by simply plotting the pairs of observations (y_i, x_{1i}), $i = 1, 2, \ldots, 25$ from Table 1-1. We used the computer package Minitab to construct this plot. Minitab has an option that produces a dot diagram along the right and top edges of the scatter diagram, allowing us to easily see the distribution of each variable individually. So in a sense, the scatter diagram is a two-dimensional version of a dot diagram.

Table 1-1 Wire Bond Data

Observation Number	Pull Strength y	Wire Length x_1	Die Height x_2	Observation Number	Pull Strength y	Wire Length x_1	Die Height x_2
1	9.95	2	50	14	11.66	2	360
2	24.45	8	110	15	21.65	4	205
3	31.75	11	120	16	17.89	4	400
4	35.00	10	550	17	69.00	20	600
5	25.02	8	295	18	10.30	1	585
6	16.86	4	200	19	34.93	10	540
7	14.38	2	375	20	46.59	15	250
8	9.60	2	52	21	44.88	15	290
9	24.35	9	100	22	54.12	16	510
10	27.50	8	300	23	56.63	17	590
11	17.08	4	412	24	22.13	6	100
12	37.00	11	400	25	21.15	5	400
13	41.95	12	500				

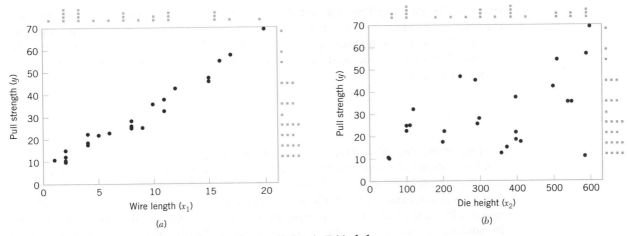

Figure 1-11   Scatter plots of the wire bond pull strength data in Table 1-1.

The scatter diagram in Figure 1-11*a* reveals that as the wire length increases, the pull strength of the wire bond also increases. Similar information is conveyed by the scatter diagram in Figure 1-11*b*, which plots pull strength y against die height x_2. Figure 1-12 is a **three-dimensional scatter diagram** of the observations on pull strength, wire length, and die height. Based on these plots, it seems reasonable to think that a model such as

$$\text{Pull strength} = \beta_0 + \beta_1(\text{wire length}) + \beta_2(\text{die height}) + \epsilon$$

would be appropriate as an empirical model for this relationship. In general, this type of empirical model is called a **regression model.** In Chapter 6 we show how to build these models and test their adequacy as approximating functions. Chapter 6 presents a method for estimating the parameters in regression models, called the method of least squares, that traces its origins to work by Karl Gauss. Essentially, this method chooses the parameters (the βs) in the empirical model to minimize the sum of the squared distances between each data point and the plane represented by the model equation. Applying this technique to the data in Table 1-1 results in

$$\widehat{\text{Pull strength}} = 2.26 + 2.74(\text{wire length}) + 0.0125(\text{die height}) \qquad (1\text{-}6)$$

where the "hat," or circumflex, over pull stength indicates that this is an estimated quantity.
Figure 1-13 is a plot of the predicted values of pull strength versus wire length and die height obtained from equation 1-6. Note that the predicted values lie on a plane above the wire

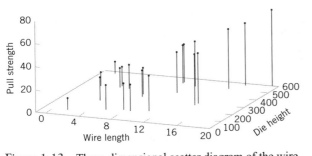

Figure 1-12   Three-dimensional scatter diagram of the wire and pull strength data.

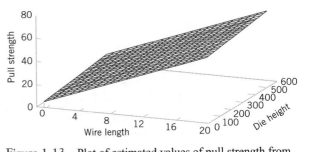

Figure 1-13   Plot of estimated values of pull strength from the empirical model in equation 1-6.

length–die height space. From the plot of the data in Fig. 1-12, this model does not appear unreasonable. The empirical model in equation 1-6 could be used to predict values of pull strength for various combinations of wire length and die height that are of interest. Essentially, the empirical model can be used by an engineer in exactly the same way that a mechanistic model can be used.

1-4   OBSERVING PROCESSES OVER TIME

Data are often collected over time. This occurs in many practical situations. Perhaps the most familiar of these is business and economic data reflecting daily stock prices, interest rates, the monthly unemployment and inflation rates, and quarterly sales volume of products. Newspapers and business publications such as the *Wall Street Journal* typically display this data in tables and graphs. In many engineering studies, the data are also collected over time. Phenomena that might affect the system or process often become more visible in a time-oriented plot of the data, and the stability of the process can be better judged. For example, the **control chart** is a technique that displays data over time and permits the engineer to assess the stability of a process.

Figure 1-14 is a **dot diagram** of acetone concentration readings taken hourly from the binary distillation column described in Section 1-2. The large variation displayed on the dot diagram indicates a possible problem, but the chart does not help explain the reason for the variation. Because the data are collected over time, they are called a **time series.** A graph of the data versus time, as shown in Fig. 1-15, is called a **time series plot.** A shift in the process mean level is visible in the plot, and an estimate of the time of the shift can be obtained.

The famous quality authority W. Edwards Deming stressed that it is important to understand the nature of variation in processes over time. He conducted an experiment in which he attempted to drop marbles as close as possible to a target on a table. He used a funnel mounted on a ring stand and the marbles were dropped into the funnel. See Fig. 1-16. The funnel was aligned as closely as possible with the center of the target. Deming then used two different strategies to operate the process. (1) He never moved the funnel. He simply dropped one marble after another and recorded the distance from the target. (2) He dropped the first marble and recorded its location relative to the target. He then moved the funnel an equal and opposite distance in an attempt to compensate for the error. He continued to make this type of adjustment after each marble was dropped.

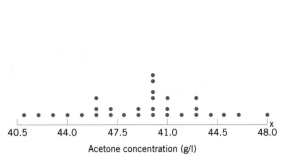

Figure 1-14   A dot diagram illustrates variation but does not identify the problem.

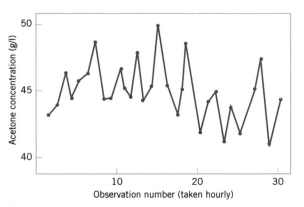

Figure 1-15   A time series plot of acetone concentration provides more information than the dot diagram.

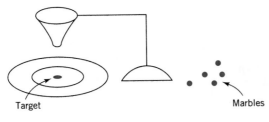

Figure 1-16
Deming's funnel
experiment.

After both strategies were completed, Deming noticed that the variability in the distance from the target for strategy 2 was approximately twice as large as for strategy 1. The adjustments to the funnel increased the deviations from the target. The explanation is that the error (the deviation of the marble's position from the target) for one marble provides no information about the error that will occur for the next marble. Consequently, adjustments to the funnel do not decrease future errors. Instead, they tend to move the funnel farther from the target.

This interesting experiment points out that adjustments to a process based on random disturbances can actually *increase* the variation of the process. This is referred to as **overcontrol** or **tampering.** Adjustments should be applied only to compensate for a nonrandom shift in the process—then they can help. A computer simulation can be used to demonstrate the lessons of the funnel experiment. Figure 1-17 displays a time plot of 100 measurements (denoted as y) from a process in which only random disturbances are present. The target value for the process is 10 units. The figure displays the data with and without adjustments that are applied to the process mean in an attempt to produce data closer to the target. Each adjustment is equal and opposite to the deviation of the previous measurement from the target. For example, when the measurement is 11 (one unit above target), the mean is reduced by one unit before the next measurement is generated. The overcontrol has increased the deviations from the target.

Figure 1-18 displays the data without adjustment from Fig. 1-17, except that the measurements after observation number 50 are increased by two units to simulate the effect of a shift in the mean of the process. When there is a true shift in the mean of a process, an adjustment can

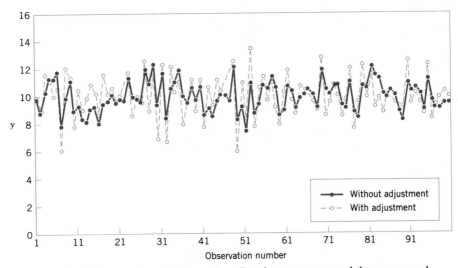

Figure 1-17   Adjustments applied to random disturbances overcontrol the process and increase the deviations from the target.

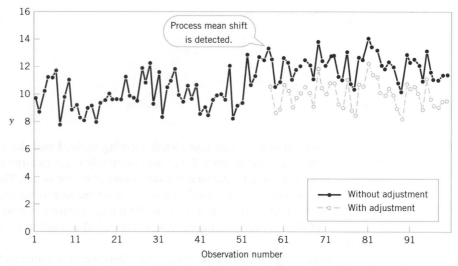

Figure 1-18   Process mean shift is detected at observation number 57, and one adjustment (a decrease of two units) reduces the deviations from the target.

be useful. Figure 1-18 also displays the data obtained when one adjustment (a decrease of two units) is applied to the mean after the shift is detected (at observation number 57). Note that this adjustment decreases the deviations from the target.

The question of when to apply adjustments (and by what amounts) begins with an understanding of the types of variation that affect a process. A **control chart** is an invaluable way to examine the variability in time-oriented data. Figure 1-19 presents a control chart for the concentration data from Fig. 1-14. The **center line** on the control chart is just the average of the concentration measurements for the first 20 samples ($\bar{x} = 51.5$ g/l) when the process is stable. The **upper control limit** and the **lower control limit** are a pair of statistically derived limits that reflect the inherent, or natural, variability in the process. These limits are located three standard deviations of the concentration values above and below the center line. If the process is operating as it should, without any external sources of variability present in the system, the concentration measurements should fluctuate randomly around the center line, and almost all of them should fall between the control limits.

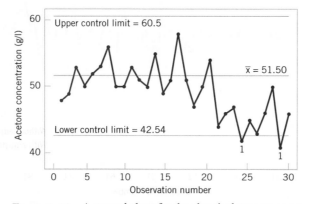

Figure 1-19   A control chart for the chemical process concentration data.

In the control chart of Fig. 1-19, the visual frame of reference provided by the center line and the control limits indicates that some upset or disturbance has affected the process around sample 20 because all of the following observations are below the center line and two of them actually fall below the lower control limit. This is a very strong signal that corrective action is required in this process. If we can find and eliminate the underlying cause of this upset, we can improve process performance considerably.

Control charts are critical to engineering analyses for the following reason. In some cases, the data in our sample are actually selected from the population of interest. The sample is a subset of the population. For example, a sample of three wafers might be selected from a production lot of wafers in semiconductor manufacturing. Based on data in the sample, we want to conclude something about the lot. For example, the average of the resistivity measurements in the sample is not expected to exactly equal the average of the resistivity measurements in the lot. However, if the sample average is high, we might be concerned that the lot average is too high.

In many other cases, we use the current data to make conclusions about the future performance of a process. For example, not only are we interested in the acetone concentration measurements produced by the binary distillation column, but we also want to make conclusions about the concentration of future production of acetone that will be sold to customers. This population of future production does not yet exist. Clearly, this is an analytic study, and the analysis requires some notion of stability as an additional assumption. A control chart is the fundamental tool to evaluate the stability of a process.

Control charts are a very important application of statistics for monitoring, controlling, and improving a process. The branch of statistics that makes use of control charts is called **statistical process control,** or **SPC.** We will discuss SPC and control charts in Chapter 8.

EXERCISES FOR CHAPTER 1

1-1. The Department of Industrial Engineering at a major university wants to develop an empirical model to predict the success of their undergraduate students in completing their degrees.

(a) What would you use as the response or outcome variable in this study?

(b) What predictor variables would you recommend using?

(c) Discuss how you would collect data to build this model. Is this a retrospective study or an observational study?

1-2. A city has 20 neighborhoods. How could the city tax appraiser select a random sample of single-family homes that could be used in developing an empirical model to predict the appraised value of a house?

(a) What characteristics of the house would you recommend that the tax appraiser consider using in the model?

(b) Is this an enumerative study or an analytic study?

1-3. Are the populations in Exercises 1-1 and 1-2 conceptual or physical populations?

1-4. How would you obtain a sample of size 20 from all of the possible distances that you might throw a baseball? Is the population that you are sampling from conceptual or physical?

1-5. A group of 100 dental patients are randomly divided into two groups. One group (the "treatment" group) receives supplemental fluoride treatments monthly for two years, while the other group (the "control" group) receives their standard semiannual dental care. At the end of the two-year study period, each patient's tooth decay is evaluated and the two groups are compared.

(a) Does this study involve a conceptual or a physical population?

(b) What is the purpose of randomly dividing the patients into two groups?

(c) Do you think that the study results would be valid if the patients elected which group they belonged to?

(d) Why are two groups needed? Couldn't valid results be obtained just by putting all 100 patients into the "treatment" group?

1-6. List two examples of conceptual populations and two examples of physical populations. For each population, describe a question that could be answered by sampling from the population. Describe how a random sample could be obtained.

1-7. An engineer wants to obtain a random sample of the output of a process manufacturing digital cameras. She samples on three different days, and on each day she selects five cameras at random between 3 PM and 4 PM from the production line output. Is this a random sample?

1-8. List three mechanistic models that you have used in your engineering studies.

1-9.   A population has four members, a, b, c, and d.
(a) How many different samples are there of size $n = 2$ from this population? Assume that the sample must consist of two different objects.
(b) How would you take a random sample of size $n = 2$ from this population?

1-10.   An automobile company analyzes warranty claim data reported by its dealer network to determine potential design problems with its vehicles.
(a) Can warranty data be viewed as a random sample of failures?
(b) Is this an enumerative study or an analytic study?
(c) Is this an observational study or a retrospective study?

IMPORTANT TERMS AND CONCEPTS FOR THIS CHAPTER

Analytic study	Empirical model	Factorial experiment	Scatter diagram
Control chart	Engineering or scientific method	Mechanistic model	Sources of variability
Designed experiment		Observational study	Statistical thinking
Dot diagram	Enumerative study	Retrospective study	Variability

IMPORTANT TERMS AND CONCEPTS DISCUSSED FURTHER IN SUBSEQUENT CHAPTERS

Confidence interval	Fractional factorial experiment	Mechanistic model	Response variable
Control chart		Model	Sample
Designed experiment	Hypothesis testing	Random variable	Scatter diagram
Empirical model	Interaction	Replication	Variability
Factorial experiment			

2
Data Summary
and Presentation

CHAPTER OUTLINE

LEARNING OBJECTIVES

After careful study of this chapter, you should be able to do the following:

1. Compute and interpret the sample mean, sample variance, sample standard deviation, sample median, and sample range.

2. Explain the concepts of sample mean, sample variance, population mean, and population variance.

3. Construct and interpret visual data displays, including the stem-and-leaf display, the histogram, and the box plot.

4. Explain how to use box plots and other data displays to visually compare two or more samples of data.

5. Know how to use simple time series plots to visually display the important features of time-oriented data.

6. Construct scatter plots and compute and interpret a sample correlation coefficient.

2-1   DATA SUMMARY AND DISPLAY

Well-constructed data summaries and displays are essential to good statistical thinking because they focus the engineer on important features of the data or provide insight about the type of model that should be used in a problem solution. The computer has become an important tool in the presentation and analysis of data. Although many statistical techniques require only a handheld calculator, much time and effort may be required by this approach, and a computer will perform the tasks much more efficiently.

Most statistical analysis is done using a prewritten library of statistical programs. The user enters the data and then selects the types of analysis and output displays that are of interest. Statistical software packages are available for both mainframe machines and personal computers. Among the most popular and widely used packages are SAS (Statistical Analysis System) for both servers and personal computers (PCs) and Minitab for the PC. We will present some examples of output from Minitab throughout the book. We will not discuss its hands-on use for entering and editing data or using commands. You will find Minitab or other very similar packages available at your institution, along with local expertise in their use.

We can describe data features **numerically.** For example, we can characterize the location or central tendency in the data by the ordinary arithmetic average or mean. Because we almost always think of our data as a sample, we will refer to the arithmetic mean as the **sample mean.**

Sample Mean

If the n observations in a sample are denoted by $x_1, x_2, \ldots, x_n$, the **sample mean** is

$$\bar{x} = \frac{x_1 + x_2 + \cdots + x_n}{n}$$

$$= \frac{\sum_{i=1}^{n} x_i}{n} \tag{2-1}$$

EXAMPLE 2-1
O-Ring Strength:
Sample Mean

Consider the O-ring tensile strength experiment described in Chapter 1. The data from the modified rubber compound are shown in the **dot diagram** (Fig. 2-1). The sample mean strength (psi) for the eight observations on strength is

$$\bar{x} = \frac{x_1 + x_2 + \cdots x_n}{n} = \frac{\sum_{i=1}^{8} x_i}{8} = \frac{1037 + 1047 + \cdots + 1040}{8}$$

$$= \frac{8440}{8} = 1055.0 \text{ psi}$$

A physical interpretation of the sample mean as a measure of location is shown in Fig. 2-1. Note that the sample mean $\bar{x} = 1055$ can be thought of as a "balance point." That is, if each observation represents 1 pound of mass placed at the point on the x-axis, a fulcrum located at $\bar{x}$ would exactly balance this system of weights. ∎

The sample mean is the average value of all the observations in the data set. Usually, these data are a **sample** of observations that have been selected from some larger **population** of observations. Here the population might consist of all the O-rings that will be sold to customers. Sometimes there is an actual physical population, such as a lot of silicon wafers produced in a semiconductor factory. We could also think of calculating the average value of all the observations in a population. This average is called the **population mean,** and it is denoted by the Greek letter μ (mu).

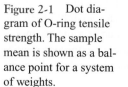

Figure 2-1   Dot diagram of O-ring tensile strength. The sample mean is shown as a balance point for a system of weights.

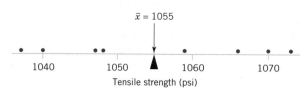

When there are a finite number of observations (say, N) in the population, the population mean is

$$\mu = \frac{\sum_{i=1}^{N} x_i}{N} \qquad (2\text{-}2)$$

The sample mean, $\bar{x}$, is a reasonable estimate of the population mean, μ. Therefore, the engineer investigating the modified rubber compound for the O-rings would conclude, on the basis of the data, that an estimate of the mean tensile strength is 1055 psi.

Although the sample mean is useful, it does not convey all of the information about a sample of data. The variability or scatter in the data may be described by the **sample variance** or the **sample standard deviation.**

Sample Variance and Sample Standard Deviation

If the n observations in a sample are denoted by $x_1, x_2, \ldots, x_n$, then the **sample variance** is

$$s^2 = \frac{\sum_{i=1}^{n} (x_i - \bar{x})^2}{n - 1} \qquad (2\text{-}3)$$

The **sample standard deviation,** s, is the positive square root of the sample variance.

The units of measurement for the sample variance are the square of the original units of the variable. Thus, if x is measured in psi, the units for the sample variance are $(\text{psi})^2$. The standard deviation has the desirable property of measuring variability in the original units of the variable of interest, x (psi).

How Does the Sample Variance Measure Variability?

To see how the sample variance measures dispersion or variability, refer to Fig. 2-2, which shows the deviations $x_i - \bar{x}$ for the O-ring tensile strength data. The greater the amount of variability in the O-ring tensile strength data, the larger in absolute magnitude some of the deviations $x_i - \bar{x}$ will be. Because the deviations $x_i - \bar{x}$ always sum to zero, we must use a measure of variability that changes the negative deviations to nonnegative quantities. Squaring the deviations is the approach used in the sample variance. Consequently, if s^2 is small, there is relatively little variability in the data, but if s^2 is large, the variability is relatively large.

EXAMPLE 2-2
O-Ring Strength:
Sample Variance

Table 2-1 displays the quantities needed to calculate the sample variance and sample standard deviation for the O-ring tensile strength data. These data are plotted in Fig. 2-2. The numerator of s^2 is

$$\sum_{i=1}^{8} (x_i - \bar{x})^2 = 1348$$

so the sample variance is

$$s^2 = \frac{1348}{8 - 1} = \frac{1348}{7} = 192.57 \text{ psi}^2$$

and the sample standard deviation is

$$s = \sqrt{192.57} = 13.9 \text{ psi}$$

■

Table 2-1   Calculation of Terms for the Sample Variance and Sample Standard Deviation

i	x_i	$x_i - \bar{x}$	$(x_i - \bar{x})^2$
1	1048	-7	49
2	1059	4	16
3	1047	-8	64
4	1066	11	121
5	1040	-15	225
6	1070	15	225
7	1037	-18	324
8	1073	18	324
	8440	0.0	1348

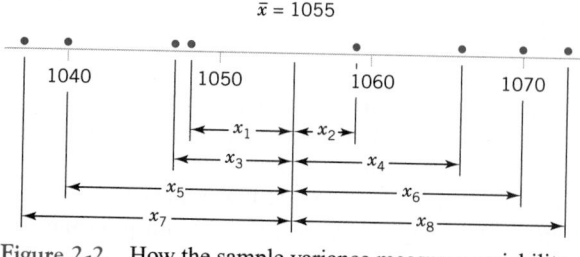

$\bar{x} = 1055$

Figure 2-2   How the sample variance measures variability through the deviations $x_i - \bar{x}$.

The computation of s^2 requires calculations of $\bar{x}$, n subtractions, and n squaring and adding operations. If the original observations or the deviations $x_i - \bar{x}$ are not integers, the deviations $x_i - \bar{x}$ may be tedious to work with, and several decimals may have to be carried to ensure numerical accuracy. A more efficient **computational formula** for the sample variance is found as follows:

$$s^2 = \frac{\sum_{i=1}^{n}(x_i - \bar{x})^2}{n-1} = \frac{\sum_{i=1}^{n}(x_i^2 + \bar{x}^2 - 2\bar{x}x_i)}{n-1} = \frac{\sum_{i=1}^{n}x_i^2 + n\bar{x}^2 - 2\bar{x}\sum_{i=1}^{n}x_i}{n-1}$$

and because $\bar{x} = (1/n)\sum_{i=1}^{n}x_i$, this simplifies to

$$s^2 = \frac{\sum_{i=1}^{n}x_i^2 - \dfrac{\left(\sum_{i=1}^{n}x_i\right)^2}{n}}{n-1} \tag{2-4}$$

Note that equation 2-4 requires squaring each individual x_i, then squaring the sum of the x_i, subtracting $(\sum x_i)^2/n$ from $\sum x_i^2$, and finally dividing by $n - 1$. Sometimes this computational formula is called the **shortcut method** for calculating s^2 (or s).

EXAMPLE 2-3
O-Ring Strength:
Alternative
Variance
Calculation

We will calculate the sample variance and standard deviation for the O-ring tensile strength data using the shortcut method, equation 2-4. The formula gives

$$s^2 = \frac{\sum_{i=1}^{n}x_i^2 - \dfrac{\left(\sum_{i=1}^{n}x_i\right)^2}{n}}{n-1} = \frac{8{,}905{,}548 - \dfrac{(8440)^2}{8}}{7} = \frac{1348}{7} = 192.57 \text{ psi}^2$$

and

$$s = \sqrt{192.57} = 13.9 \text{ psi}$$

These results agree exactly with those obtained previously.

Analogous to the sample variance s^2, there is a measure of variability in the population called the **population variance.** We will use the Greek letter σ^2 (sigma squared) to denote the population variance. The positive square root of σ^2, or σ, will denote the **population standard deviation.** When the population is finite and consists of N values, we may define the population variance as

$$\sigma^2 = \frac{\sum_{i=1}^{N} (x_i - \mu)^2}{N} \tag{2-5}$$

A more general definition of the variance σ^2 will be given later. We observed previously that the sample mean could be used as an estimate of the population mean. Similarly, the sample variance is an estimate of the population variance.

Note that the divisor for the sample variance is the sample size minus 1, $(n - 1)$, whereas for the population variance it is the population size, N. If we knew the true value of the population mean, μ, we could find the *sample* variance as the average squared deviation of the sample observations about μ. In practice, the value of μ is almost never known, and so the sum of the squared deviations about the sample average $\bar{x}$ must be used instead. However, the observations x_i tend to be closer to their average, $\bar{x}$, than to the population mean, μ. Therefore, to compensate for this we use $n - 1$ as the divisor rather than n. If we used n as the divisor in the sample variance, we would obtain a measure of variability that is, on the average, consistently smaller than the true population variance σ^2.

Another way to think about this is to consider the sample variance s^2 as being based on $n - 1$ **degrees of freedom.** The term *degrees of freedom* results from the fact that the n deviations $x_1 - \bar{x}, x_2 - \bar{x}, \ldots, x_n - \bar{x}$ always sum to zero, and so specifying the values of any $n - 1$ of these quantities automatically determines the remaining one. This was illustrated in Table 2-1. Thus, only $n - 1$ of the n deviations, $x_i - \bar{x}$, are freely determined.

EXERCISES FOR SECTION 2-1

2-1. An important quality characteristic of water is the concentration of suspended solid material in mg/l. Twelve measurements on suspended solids from a certain lake are as follows: 42.4, 65.7, 29.8, 58.7, 52.1, 55.8, 57.0, 68.7, 67.3, 67.3, 54.3, 54.0. Calculate the sample average and sample standard deviation. Construct a dot diagram of the data.

2-2. In *Applied Life Data Analysis* (Wiley, 1982), Wayne Nelson presents the breakdown time of an insulating fluid between electrodes at 34 kV. The times, in minutes, are as follows: 0.19, 0.78, 0.96, 1.31, 2.78, 3.16, 4.15, 4.67, 4.85, 6.50, 7.35, 8.01, 8.27, 12.06, 31.75, 32.52, 33.91, 36.71, and 72.89. Calculate the sample average and sample standard deviation. Construct a dot diagram of the data.

2-3. Seven oxide thickness measurements of wafers are studied to assess quality in a semiconductor manufacturing process. The data (in angstroms) are: 1264, 1280, 1301, 1300, 1292, 1307, and 1275. Calculate the sample average and sample standard deviation. Construct a dot diagram of the data.

2-4. An article in the *Journal of Structural Engineering* (Vol. 115, 1989) describes an experiment to test the yield strength of circular tubes with caps welded to the ends. The first yields (in kN) are 96, 96, 102, 102, 102, 104, 104, 108, 126, 126, 128, 128, 140, 156, 160, 160, 164, and 170. Calculate the sample average and sample standard deviation. Construct a dot diagram of the data.

2-5. An article in *Human Factors* (June 1989) presented data on visual accommodation (a function of eye movement) when recognizing a speckle pattern on a high-resolution CRT screen. The data are as follows: 36.45, 67.90, 38.77, 42.18, 26.72, 50.77, 39.30, and 49.71. Calculate the sample average and sample standard deviation. Construct a dot diagram of the data.

2-6. Preventing fatigue crack propagation in aircraft structures is an important element of aircraft safety. An engineering study to investigate fatigue crack in $n = 9$ cyclically loaded wing boxes reported the following crack lengths (in mm): 2.13, 2.96, 3.02, 1.82, 1.15, 1.37, 2.04,

2.47, 2.60. Calculate the sample average and sample standard deviation. Construct a dot diagram of the data.

2-7. The following data are direct solar intensity measurements (watts/m^2) on different days at a location in southern Spain: 562, 869, 708, 775, 775, 704, 809, 856, 655, 806, 878, 909, 918, 558, 768, 870, 918, 940, 946, 661, 820, 898, 935, 952, 957, 693, 835, 905, 939, 955, 960, 498, 653, 730, 753. Calculate the sample mean and sample standard deviation. Prepare a dot diagram of these data. Indicate where the sample mean falls on this diagram. Provide a practical interpretation of the sample mean.

2-8. An article in *Nature Genetics* (Vol. 34(1), 2003, pp. 85–90), "Treatment-Specific Changes in Gene Expression Discriminate in Vivo Drug Response in Human Leukemia Cells," reported the study of gene expression as a function of treatments for leukemia. One group received a high dose of the drug while the control group received no treatment. Expression data (measures of gene activity) from one gene are shown in the table on the right (for all of the treated subjects and some of the control subjects). Compute the sample mean and standard deviation for each group separately. Construct a dot diagram for each group separately. Comment on any differences between the groups.

2-9. For each of Exercises 2-1 through 2-8, discuss whether the data result from an observational study or a designed experiment.

High Dose	Control
16.1	297.1
134.9	491.8
52.7	1332.9
14.4	1172.0
124.3	1482.7
99.0	335.4
24.3	528.9
16.3	24.1
15.2	545.2
47.7	92.9
12.9	337.1
72.7	102.3
126.7	255.1
46.4	100.5
60.3	159.9
23.5	168.0
43.6	95.2
79.4	132.5
38.0	442.6
58.2	15.8
26.5	175.6
25.1	131.1

2-2 STEM-AND-LEAF DIAGRAM

The dot diagram is a useful data display for small samples, up to (say) about 20 observations. However, when the number of observations is moderately large, other graphical displays may be more useful.

For example, consider the data in Table 2-2. These data are the compressive strengths in pounds per square inch (psi) of 80 specimens of a new aluminum-lithium alloy undergoing evaluation as a possible material for aircraft structural elements. The data were recorded in the order of testing, and in this format they do not convey much information about compressive strengths. Questions such as What percentage of the specimens fail below 120 psi? are not

Table 2-2 Compressive Strength of 80 Aluminum-Lithium Alloy Specimens

105	221	183	186	121	181	180	143
97	154	153	174	120	168	167	141
245	228	174	199	181	158	176	110
163	131	154	115	160	208	158	133
207	180	190	193	194	133	156	123
134	178	76	167	184	135	229	146
218	157	101	171	165	172	158	169
199	151	142	163	145	171	148	158
160	175	149	87	160	237	150	135
196	201	200	176	150	170	118	149

easy to answer. Because there are many observations, constructing a dot diagram of these data would be relatively inefficient; more effective displays are available for large data sets.

A **stem-and-leaf diagram** is a good way to obtain an informative visual display of a data set $x_1, x_2, \ldots, x_n$, where each number x_i consists of at least two digits. To construct a stem-and-leaf diagram, use the following steps:

Steps for Constructing a Stem-and-Leaf Diagram	
	1. Divide each number x_i into two parts: a **stem,** consisting of one or more of the leading digits, and a **leaf,** consisting of the remaining digit.
	2. List the stem values in a vertical column.
	3. Record the leaf for each observation beside its stem.
	4. Write the units for stems and leaves on the display.

To illustrate, if the data consist of percent defective information between 0 and 100 on lots of semiconductor wafers, we can divide the value 76 into the stem 7 and the leaf 6. In general, we should choose relatively few stems in comparison with the number of observations. It is usually best to choose between 5 and 20 items. Once a set of stems has been chosen, the stems are listed along the left-hand margin of the diagram. Beside each stem all leaves corresponding to the observed data values are listed in the order in which they are encountered in the data set.

EXAMPLE 2-4
Compressive Strength

To illustrate the construction of a stem-and-leaf diagram, consider the alloy compressive strength data in Table 2-2. We will select as stem values the numbers 7, 8, 9, . . . , 24. The resulting stem-and-leaf diagram is presented in Fig. 2-3. The last column in the diagram is a frequency count of the number of leaves associated with each stem. Inspection of this display immediately reveals that most of the compressive strengths lie between 110 and 200 psi and that a central value is somewhere between 150 and 160 psi. Furthermore, the strengths are distributed approximately symmetrically about the central value. The stem-and-leaf diagram enables us to determine quickly some important features of the data that were not immediately obvious in the original display in the table. ■

Stem	Leaf	Frequency
7	6	1
8	7	1
9	7	1
10	5 1	2
11	5 8 0	3
12	1 0 3	3
13	4 1 3 5 3 5	6
14	2 9 5 8 3 1 6 9	8
15	4 7 1 3 4 0 8 8 6 8 0 8	12
16	3 0 7 3 0 5 0 8 7 9	10
17	8 5 4 4 1 6 2 1 0 6	10
18	0 3 6 1 4 1 0	7
19	9 6 0 9 3 4	6
20	7 1 0 8	4
21	8	1
22	1 8 9	3
23	7	1
24	5	1

Figure 2-3   Stem-and-leaf diagram for the compressive strength data in Table 2-2.

(a)	
Stem	Leaf
6	1 3 4 5 5 6
7	0 1 1 3 5 7 8 8 9
8	1 3 4 4 7 8 8
9	2 3 5

(b)	
Stem	Leaf
6L	1 3 4
6U	5 5 6
7L	0 1 1 3
7U	5 7 8 8 9
8L	1 3 4 4
8U	7 8 8
9L	2 3
9U	5

(c)	
Stem	Leaf
6z	1
6t	3
6f	4 5 5
6s	6
6e	
7z	0 1 1
7t	3
7f	5
7s	7
7e	8 8 9
8z	1
8t	3
8f	4 4
8s	7
8e	8 8
9z	
9t	2 3
9f	5
9s	
9e	

Figure 2-4 Stem-and-leaf displays for Example 2-5.

In some data sets, it may be desirable to provide more classes or stems. One way to do this is to modify the original stems as follows: Divide the stem 5 (say) into two new stems, 5L and 5U. The stem 5L has leaves 0, 1, 2, 3, and 4, and stem 5U has leaves 5, 6, 7, 8, and 9. This will double the number of original stems. We could increase the number of original stems by 4 by defining five new stems: 5z with leaves 0 and 1, 5t (for two and three) with leaves 2 and 3, 5f (for four and five) with leaves 4 and 5, 5s (for six and seven) with leaves 6 and 7, and 5e with leaves 8 and 9.

EXAMPLE 2-5
Batch Yield

Figure 2-4 illustrates the stem-and-leaf diagram for 25 observations on batch yields from a chemical process. In Fig. 2-4a we used 6, 7, 8, and 9 as the stems. This results in too few stems, and the stem-and-leaf diagram does not provide much information about the data. In Fig. 2-4b we divided each stem into two parts, resulting in a display that more adequately displays the data. Figure 2-4c illustrates a stem-and-leaf display with each stem divided into five parts. There are too many stems in this plot, resulting in a display that does not tell us much about the shape of the data. ■

Figure 2-5 shows a stem-and-leaf display of the compressive strength data in Table 2-2 produced by Minitab. The software uses the same stems as in Fig. 2-3. Note also that the computer orders the leaves from smallest to largest on each stem. This form of the plot is usually called an **ordered stem-and-leaf diagram.** This ordering is not usually done when the plot is constructed manually because it can be time consuming. The computer adds a column to the left of the stems that provides a count of the observations at and above each stem in the upper half of the display and a count of the observations at and below each stem in the lower half of the display. At the middle stem of 16, the column indicates the number of observations at this stem.

The ordered stem-and-leaf display makes it relatively easy to find data features such as percentiles, quartiles, and the median. The **median** is a measure of central tendency that divides the data into two equal parts, half below the median and half above. If the number of observa-

Character Stem-and-Leaf Display

Stem-and-Leaf of Strength   N = 80
Leaf Unit = 1.0

```
 1     7     6
 2     8     7
 3     9     7
 5    10     1 5
 8    11     0 5 8
11    12     0 1 3
17    13     1 3 3 4 5 5
25    14     1 2 3 5 6 8 9 9
37    15     0 0 1 3 4 4 6 7 8 8 8 8
(10)   16     0 0 0 3 3 5 7 7 8 9
33    17     0 1 1 2 4 4 5 6 6 8
23    18     0 0 1 1 3 4 6
16    19     0 3 4 6 9 9
10    20     0 1 7 8
 6    21     8
 5    22     1 8 9
 2    23     7
 1    24     5
```

Figure 2-5   A stem-and-leaf diagram from Minitab.

tions is even, the median is halfway between the two central values. From Fig. 2-5 we find the 40th and 41st values of strength as 160 and 163, so the median is $(160 + 163)/2 = 161.5$. If the number of observations is odd, the median is the central value. The **range** is a measure of variability that can be easily computed from the ordered stem-and-leaf display. It is the maximum minus the minimum measurement. From Fig. 2-5 the range is $245 - 76 = 169$.

We can also divide data into more than two parts. When an ordered set of data is divided into four equal parts, the division points are called **quartiles.** The *first* or *lower quartile, q_1,* is a value that has approximately 25% of the observations below it and approximately 75% of the observations above. The *second quartile, q_2,* has approximately 50% of the observations below its value. The second quartile is exactly equal to the median. The *third* or *upper quartile, q_3,* has approximately 75% of the observations below its value. As in the case of the median, the quartiles may not be unique. The compressive strength data in Fig. 2-5 contains $n = 80$ observations. Minitab software calculates the first and third quartiles as the $(n + 1)/4$ and $3(n + 1)/4$ ordered observations and interpolates as needed. For example, $(80 + 1)/4 = 20.25$ and $3(80 + 1)/4 = 60.75$. Therefore, Minitab interpolates between the 20th and 21st ordered observation to obtain $q_1 = 143.50$ and between the 60th and 61st observation to obtain $q_3 = 181.00$. The **interquartile range** (IQR) is the difference between the upper and lower quartiles, and it is sometimes used as a measure of variability. In general, the $100k$th **percentile** is a data value such that approximately $100k\%$ of the observations are at or below this value and approximately $100(1 - k)\%$ of them are above it. For example, to find the 95th percentile for this sample data we use the formula $0.95(80 + 1) = 76.95$ to determine that we need to interpolate between the 76th and 77th observations, 221 and 228, respectively. Hence, approximately 95% of the data is below 227.65 and 5% is above. It should be noted that when the percentile falls between two sample observations, it is common practice to use the midpoint (in contrast to the Minitab procedure of interpolation) of the two sample observations as the percentile. Using this simplified approach on this sample data, the sample first and third quartiles and the 95th percentile are 144, 181, and 224.5, respectively. In this text, we use the Minitab interpolation procedure.

Many statistics software packages provide data summaries that include these quantities. The output obtained for the compressive strength data in Table 2-2 from Minitab is shown in

Table 2-3 Summary Statistics for the Compressive Strength Data from Minitab

Variable	N	Mean	Median	StDev	SE Mean
	80	162.66	161.50	33.77	3.78
	Min	Max	Q1	Q3	
	76.00	245.00	143.50	181.00	

Table 2-3. Note that the results for the median and the quartiles agree with those given previously. SE Mean is an abbreviation for the standard error of the mean, and it will be discussed in a later chapter.

EXERCISES FOR SECTION 2-2

2-10.* The shear strengths of 100 spot welds in a titanium alloy follow. Construct a stem-and-leaf diagram for the weld strength data and comment on any important features that you notice.

5408	5431	5475	5442	5376	5388	5459	5422	5416	5435
5420	5429	5401	5446	5487	5416	5382	5357	5388	5457
5407	5469	5416	5377	5454	5375	5409	5459	5445	5429
5463	5408	5481	5453	5422	5354	5421	5406	5444	5466
5399	5391	5477	5447	5329	5473	5423	5441	5412	5384
5445	5436	5454	5453	5428	5418	5465	5427	5421	5396
5381	5425	5388	5388	5378	5481	5387	5440	5482	5406
5401	5411	5399	5431	5440	5413	5406	5342	5452	5420
5458	5485	5431	5416	5431	5390	5399	5435	5387	5462
5383	5401	5407	5385	5440	5422	5448	5366	5430	5418

Construct a stem-and-leaf display for these data.

2-11. The following data are the numbers of cycles to failure of aluminum test coupons subjected to repeated alternating stress at 21,000 psi, 18 cycles per second:

1115	1567	1223	1782	1055
1310	1883	375	1522	1764
1540	1203	2265	1792	1330
1502	1270	1910	1000	1608
1258	1015	1018	1820	1535
1315	845	1452	1940	1781
1085	1674	1890	1120	1750
798	1016	2100	910	1501
1020	1102	1594	1730	1238
865	1605	2023	1102	990
2130	706	1315	1578	1468
1421	2215	1269	758	1512
1109	785	1260	1416	1750
1481	885	1888	1560	1642

(a) Construct a stem-and-leaf display for these data.
(b) Does it appear likely that a coupon will "survive" beyond 2000 cycles? Justify your answer.

2-12. An important quality characteristic of water is the concentration of suspended solid material. Following are 60 measurements on suspended solids from a certain lake. Construct a stem-and-leaf diagram for these data and comment on any important features that you notice.

42.4	65.7	29.8	58.7	52.1	55.8	57.0	68.7	67.3	67.3
54.3	54.0	73.1	81.3	59.9	56.9	62.2	69.9	66.9	59.0
56.3	43.3	57.4	45.3	80.1	49.7	42.8	42.4	59.6	65.8
61.4	64.0	64.2	72.6	72.5	46.1	53.1	56.1	67.2	70.7
42.6	77.4	54.7	57.1	77.3	39.3	76.4	59.3	51.1	73.8
61.4	73.1	77.3	48.5	89.8	50.7	52.0	59.6	66.1	31.6

2-13. The data that follow represent the yield on 90 consecutive batches of ceramic substrate to which a metal coating has been applied by a vapor-deposition process. Construct a stem-and-leaf display for these data.

94.1	87.3	94.1	92.4	84.6	85.4
93.2	84.1	92.1	90.6	83.6	86.6
90.6	90.1	96.4	89.1	85.4	91.7
91.4	95.2	88.2	88.8	89.7	87.5
88.2	86.1	86.4	86.4	87.6	84.2
86.1	94.3	85.0	85.1	85.1	85.1
95.1	93.2	84.9	84.0	89.6	90.5
90.0	86.7	78.3	93.7	90.0	95.6
92.4	83.0	89.6	87.7	90.1	88.3
87.3	95.3	90.3	90.6	94.3	84.1
86.6	94.1	93.1	89.4	97.3	83.7
91.2	97.8	94.6	88.6	96.8	82.9
86.1	93.1	96.3	84.1	94.4	87.3
90.4	86.4	94.7	82.6	96.1	86.4
89.1	87.6	91.1	83.1	98.0	84.5

---

*Please remember that the computer icon indicates that the data are available on the book website and the problem should be solved using software.

2-14. Construct a stem-and-leaf display for the gene expression data in Exercise 2-8 for each group separately and comment on any differences.

2-15. Construct a stem-and-leaf display for the solar intensity data in Exercise 2-7 and comment on the shape.

2-16. Find the median, the quartiles, and the 5th and 95th percentiles for the weld strength data in Exercise 2-10.

2-17. Find the median, the quartiles, and the 5th and 95th percentiles for the failure data in Exercise 2-11.

2-18. Find the median and sample average for the water quality data in Exercise 2-12. Explain how these two measures of location describe different features in the data.

2-19. Find the median, the quartiles, and the 5th and 95th percentiles for the yield data in Exercise 2-13.

2-3 HISTOGRAMS

A **histogram** is a more compact summary of data than a stem-and-leaf diagram. To construct a histogram for continuous data, we must divide the range of the data into intervals, which are usually called **class intervals, cells,** or **bins.** If possible, the bins should be of equal width to enhance the visual information in the histogram. Some judgment must be used in selecting the number of bins so that a reasonable display can be developed. The number of bins depends on the number of observations and the amount of scatter or dispersion in the data. A histogram that uses either too few or too many bins will not be informative. We usually find that between 5 and 20 bins is satisfactory in most cases and that the number of bins should increase with n. Choosing the number of bins approximately equal to the square root of the number of observations often works well in practice.*

Once the number of bins and the lower and upper boundary of each bin have been determined, the data are sorted into the bins and a count is made of the number of observations in each bin. To construct the histogram, use the horizontal axis to represent the measurement scale for the data and the vertical scale to represent the counts, or **frequencies.** Sometimes the frequencies in each bin are divided by the total number of observations (n), and then the vertical scale of the histogram represents **relative frequencies.** Rectangles are drawn over each bin, and the height of each rectangle is proportional to frequency (or relative frequency). Most statistics packages construct histograms.

EXAMPLE 2-6
Golf Ball
Distance

The United States Golf Association tests golf balls to ensure that they conform to the rules of golf. Balls are tested for weight, diameter, roundness, and conformance to an overall distance standard. The overall distance test is conducted by hitting balls with a driver swung by a mechanical device nicknamed Iron Byron, after the legendary great player Byron Nelson, whose swing the machine is said to emulate. Table 2-4

Table 2-4 Golf Ball Distance Data

291.5	274.4	290.2	276.4	272.0	268.7	281.6	281.6	276.3	285.9
269.6	266.6	283.6	269.6	277.8	287.8	267.6	292.6	273.4	284.4
270.7	274.0	285.2	275.5	272.1	261.3	274.0	279.3	281.0	293.1
277.5	278.0	272.5	271.7	280.8	265.6	260.1	272.5	281.3	263.0
279.0	267.3	283.5	271.2	268.5	277.1	266.2	266.4	271.5	280.3
267.8	272.1	269.7	278.5	277.3	280.5	270.8	267.7	255.1	276.4
283.7	281.7	282.2	274.1	264.5	281.0	273.2	274.4	281.6	273.7
271.0	271.5	289.7	271.1	256.9	274.5	286.2	273.9	268.5	262.6
261.9	258.9	293.2	267.1	255.0	269.7	281.9	269.6	279.8	269.9
282.6	270.0	265.2	277.7	275.5	272.2	270.0	271.0	284.3	268.4

---

*There is no universal agreement about how to select the number of bins for a histogram. Some basic statistics textbooks suggest using Sturges' rule, which sets the number of bins $h = 1 + \log_2 n$, where n is the sample size. There are many variations of Sturges' rule. Computer software packages use many different algorithms to determine the number and width of bins, and some of them may not be based on Sturges' rule.

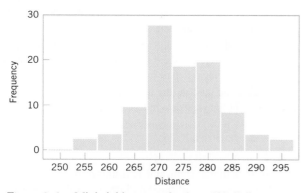

Figure 2-6   Minitab histogram for the golf ball distance data in Table 2-4.

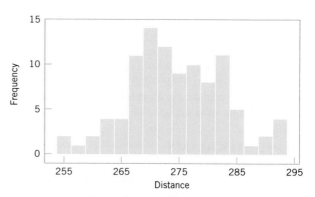

Figure 2-7   Minitab histogram with 16 bins for the golf ball distance data.

gives the distances achieved (in yards) by hitting 100 golf balls of a particular brand in the overall distance test. Because the data set contains 100 observations and $\sqrt{100} = 10$, we suspect that about 10 bins will provide a satisfactory histogram, so we selected the Minitab option that allows the user to specify the number of bins. The Minitab histogram for the golf ball distance data is shown in Figure 2-6. Notice that the midpoint of the first bin is 250 yards and that the histogram only has 9 bins that contain a nonzero frequency. A histogram, like a stem-and-leaf plot, gives a visual impression of the shape of the distribution of the measurements, as well as information about the inherent variability in the data. Note the reasonably symmetric or bell-shaped distribution of the golf ball distance data. ■

Most computer packages have a default setting for the number of bins. Figure 2-7 is the Minitab histogram obtained with the default setting, which leads to a histogram with 16 bins. Histograms can be relatively sensitive to the choice of the number and width of the bins. For small data sets, histograms may change dramatically in appearance if the number and/or width of the bins changes. For this reason, we prefer to think of the histogram as a technique best suited for **larger data sets** containing say 75 to 100 or more observations. Because the number of observations in the golf ball distance data set is moderately large ($n = 100$), the choice of the number of bins is not especially important, and the histograms in Figs. 2-6 and 2-7 convey very similar information.

Notice that in passing from the original data or a stem-and-leaf diagram to a histogram, we have in a sense lost some information because the original observations are not preserved on the display. However, this loss in information is usually small compared with the conciseness and ease of interpretation of the histogram, particularly in large samples.

Histograms are always easier to interpret if the bins are of equal width. If the bins are of unequal width, it is customary to draw rectangles whose areas (as opposed to heights) are proportional to the number of observations in the bins.

Figure 2-8 shows a variation of the histogram available in Minitab (i.e., the **cumulative frequency plot**). In this plot, the height of each bar represents the number of observations that are less than or equal to the upper limit of the bin. Cumulative frequencies are often very useful in data interpretation. For example, we can read directly from Fig. 2-8 that about 15 of the 100 balls tested traveled farther than 280 yards.

Frequency distributions and histograms can also be used with qualitative, categorical, or count (discrete) data. In some applications there will be a natural ordering of the categories (such as freshman, sophomore, junior, and senior), whereas in others the order of the

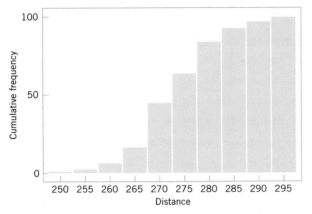

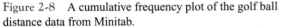

Figure 2-8   A cumulative frequency plot of the golf ball distance data from Minitab.

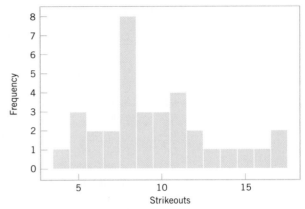

Figure 2-9   Histogram of the number of strikeouts by Randy Johnson, 2002 baseball season.

categories will be arbitrary (such as male and female). When using categorical data, the bars should be drawn to have equal width.

To construct a histogram for discrete or count data, first determine the frequency (or relative frequency) for each value of x. Each of the x values corresponds to a bin. The histogram is drawn by plotting the frequencies (or relative frequencies) on the vertical scale and the values of x on the horizontal scale. Then above each value of x, draw a rectangle whose height is the frequency (or relative frequency) corresponding to that value.

EXAMPLE 2-7
Baseball Pitcher

During the 2002 baseball season, the Arizona Diamondbacks' Randy Johnson won the National League "triple crown" for pitchers by winning 24 games, striking out 334 opposing batters, and compiling a 2.32 earned-run average. Table 2-5 contains a game-by-game summary of Johnson's performance for all 35 games in which he was the starting pitcher. Figure 2-9 is a histogram of Johnson's strikeouts. Notice that the number of strikeouts is a discrete variable. From either the histogram or the tabulated data, we can determine the following:

$$\text{Proportion of games with at least 10 strikeouts} = \frac{15}{35} = 0.4286$$

and

$$\text{Proportion of games with between 8 and 14 strikeouts} = \frac{23}{35} = 0.6571$$

These proportions are examples of **relative frequencies.**   ■

An important variation of the histogram is the **Pareto chart.** This chart is widely used in quality and process improvement studies where the data usually represent different types of defects, failure modes, or other categories of interest to the analyst. The categories are ordered so that the category with the largest number of frequencies is on the left, followed by the category with the second largest number of frequencies, and so forth. These charts are named after the Italian economist V. Pareto, and they usually exhibit **"Pareto's law"**; that is, most of the defects can usually be accounted for by a few of the categories.

Table 2-5   Pitching Performance for Randy Johnson in 2002

DATE	OPPONENT	SCORE		IP	H	R	ER	HR	BB	SO
4/1	San Diego	W,	2-0 (C)	9.0	6	0	0	0	1	8
4/6	@ Milwaukee	W,	6-3	7.0	5	1	1	1	3	12
4/11	@ Colorado	W,	8-4	7.0	3	2	2	0	2	9
4/16	St. Louis	W,	5-3	7.0	8	3	3	1	1	5
4/21	Colorado	W,	7-1(C)	9.0	2	1	0	0	1	17
4/26	@ Florida	W,	5-3	7.0	4	1	1	0	3	10
5/6	Pittsburgh	L,	2-3	7.0	7	3	2	1	0	8
5/11	@ Philadelphia	ND,	6-5 (10)	7.0	8	4	4	2	2	8
5/16	Philadelphia	W,	4-2	7.0	6	1	1	1	4	8
5/21	San Francisco	W,	9-4	7.0	6	3	3	0	3	10
5/26	Los Angeles	ND,	10-9 (10)	5.0	8	7	7	3	2	5
5/31	@ Los Angeles	W,	6-3	8.0	6	3	0	1	1	4
6/5	Houston	ND,	5-4 (13)	8.0	6	3	3	1	0	11
6/10	@ N.Y. Yankees	L,	5-7	7.2	7	5	5	2	3	8
6/15	Detroit	W,	3-1	7.0	7	1	0	0	2	13
6/20	Baltimore	W,	5-1	7.0	5	1	1	1	2	11
6/26	@ Houston	W,	9-1	8.0	3	0	0	0	3	8
7/1	Los Angeles	L,	0-4	7.0	9	4	3	0	0	6
7/6	San Francisco	ND,	2-3	7.0	7	2	2	1	2	10
7/11	@ Los Angeles	ND,	4-3	6.0	6	3	3	2	2	5
7/16	@ San Francisco	W,	5-3	7.0	5	3	3	2	3	7
7/21	@ San Diego	L,	9-11	5.0	8	8	8	1	6	9
7/26	San Diego	W,	12-0	7.0	4	0	0	0	1	8
7/31	@ Montreal	W,	5-1 (C)	9.0	8	1	1	0	3	15
8/5	@ New York	W,	2-0 (C)	9.0	2	0	0	0	2	11
8/10	Florida	W,	9-2	8.0	5	2	2	1	2	14
8/15	@ Cincinnati	W,	7-2	8.0	2	2	1	1	2	11
8/20	Cincinnati	ND,	5-3	7.0	5	2	2	1	3	12
8/25	Chicago	W,	7-0 (C)	9.0	6	0	0	0	2	16
8/30	San Francisco	L,	6-7	5.1	9	7	6	0	3	6
9/4	Los Angeles	W,	7-1 (C)	9.0	3	1	1	1	0	8
9/9	San Diego	W,	5-2	7.0	8	1	1	1	3	7
9/14	Milwaukee	W,	5-0 (C)	9.0	3	0	0	0	2	17
9/19	@ San Diego	W,	3-1	7.0	4	1	1	1	0	9
9/26	Colorado	W,	4-2 (C)	9.0	6	2	0	0	2	8
Season Totals		24-5,	2.32	260.0	197	78	67	26	71	334

Key: W = won, L = lost, ND = no decision; C = complete game, IP = innings pitched, H = hits, R = runs, ER = earned runs, HR = home runs, BB = base on balls, and SO = strikeouts.

EXAMPLE 2-8
Aircraft Accidents

Table 2-6 presents data on aircraft accident rates taken from an article in the *Wall Street Journal* ("Jet's Troubled History Raises Issues for the FAA and the Manufacturer," 19 September 2000). The table presents the total number of accidents involving hull losses between 1959 and 1999 for 22 types of aircraft and the hull loss rate expressed as the number of hull losses per million departures. Figure 2-10 shows a Pareto chart of the hull losses per million departures. Clearly the first three aircraft types account for a large percentage of the incidents on a per-million-departures basis. An interesting fact about the first three aircraft types is that the 707/720 and the DC-8 were mid-1950s designs and are not in regular

Table 2-6 Aircraft Accident Data

Aircraft Type	Actual Number of Hull Losses	Hull Losses/Million Departures
MD-11	5	6.54
707/720	115	6.46
DC-8	71	5.84
F-28	32	3.94
BAC 1-11	22	2.64
DC-10	20	2.57
747-Early	21	1.90
A310	4	1.40
A300-600	3	1.34
DC-9	75	1.29
A300-Early	7	1.29
737-1 & 2	62	1.23
727	70	0.97
A310/319/321	7	0.96
F100	3	0.80
L-1011	4	0.77
BAe 146	3	0.59
747-400	1	0.49
757	4	0.46
MD-80/90	10	0.43
767	3	0.41
737-3, 4 & 5	12	0.39

passenger service today in most of the world, whereas the MD-11 was introduced into passenger service in 1990. Between 1990 and 1999, five of the global fleet of 198 MD-11s were destroyed in crashes, leading to the high accident rate (an excellent discussion of potential **root causes** of these accidents is in the *Wall Street Journal* article). The purpose of most Pareto charts is to help the analyst separate the sources of defects or incidents into the vital few and the relatively "insignificant many." There are many variations of the Pareto chart; for some examples, see Montgomery (2005a). ∎

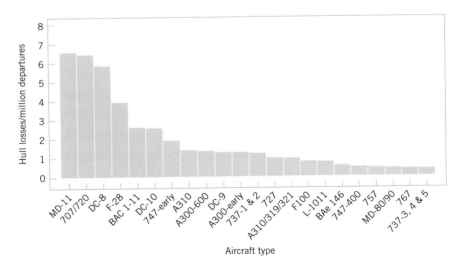

Figure 2-10   Pareto chart for the aircraft accident data.

Animation 2: Understanding Means, Standard Deviations and Histograms

EXERCISES FOR SECTION 2-3

2-20. Construct a cumulative frequency plot and histogram for the weld strength data from Exercise 2-10.

(a) Use 8 bins.

(b) Use 16 bins and compare with part (a).

2-21. Construct a cumulative frequency plot and histogram using the failure data from Exercise 2-11.

2-22. Construct a cumulative frequency plot and histogram for the water quality data in Exercise 2-12.

2-23. Construct a cumulative frequency plot and histogram for the yield data in Exercise 2-13.

2-24. Construct a cumulative frequency plot and histogram for the gene expression data from each group separately in Exercise 2-8. Comment on any differences.

2-25. Construct a cumulative frequency plot and histogram for the solar intensity data in Exercise 2-7. Use 6 bins.

2-26. The following information on structural defects in automobile doors is obtained: dents, 4; pits, 4; parts assembled out of sequence, 6; parts undertrimmed, 21; missing holes/slots, 8; parts not lubricated, 5; parts out of contour, 30; and parts not deburred, 3. Construct and interpret a Pareto chart.

2-4   BOX PLOT

The stem-and-leaf display and the histogram provide general visual impressions about a data set, whereas numerical quantities such as $\bar{x}$ or s provide information about only one feature of the data. The **box plot** is a graphical display that simultaneously describes several important features of a data set, such as center, spread, departure from symmetry, and identification of observations that lie unusually far from the bulk of the data. (These observations are called "outliers.")

A box plot displays the three quartiles on a rectangular box, aligned either horizontally or vertically. The box encloses the interquartile range (IQR) with the left (or lower) edge at the first quartile, q_1, and the right (or upper) edge at the third quartile, q_3. A line is drawn through the box at the second quartile (which is the 50th percentile, or the median). A line, or **whisker,** extends from each end of the box. The lower whisker is a line from the first quartile to the smallest data point within 1.5 interquartile ranges from the first quartile. The upper whisker is a line from the third quartile to the largest data point within 1.5 interquartile ranges from the third quartile. Data farther from the box than the whiskers are plotted as individual points. A point beyond a whisker, but less than three interquartile ranges from the box edge, is called an **outlier.** A point more than three interquartile ranges from a box edge is called an **extreme outlier.** See Fig. 2-11. Occasionally, different symbols, such as open and filled circles, are used to identify the two types of outliers. Sometimes box plots are called *box-and-whisker plots.*

Figure 2-12 presents the box plot from Minitab for the alloy compressive strength data shown in Table 2-2. This box plot indicates that the distribution of compressive strengths is fairly symmetric around the central value because the left and right whiskers and the lengths of the left and right boxes around the median are about the same. There are also outliers on either end of the data.

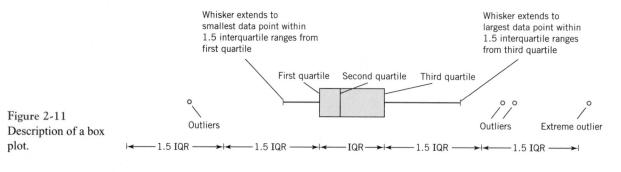

Figure 2-11
Description of a box plot.

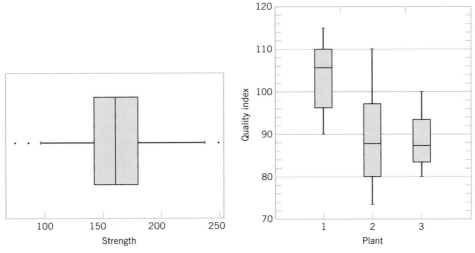

Figure 2-12   Box plot for compressive strength data in Table 2-2.

Figure 2-13   Comparative box plots of a quality index at three plants.

Box plots are very useful in graphical comparisons among data sets because they have high visual impact and are easy to understand. For example, Fig. 2-13 shows the comparative box plots for a manufacturing quality index on semiconductor devices at three manufacturing plants. Inspection of this display reveals that there is too much variability at plant 2 and that plants 2 and 3 need to raise their quality index performance.

EXERCISES FOR SECTION 2-4

2-27.   The following data are the joint temperatures of the O-rings (°F) for each test firing or actual launch of the space shuttle rocket motor (from *Presidential Commission on the Space Shuttle Challenger Accident*, Vol. 1, pp. 129–131): 84, 49, 61, 40, 83, 67, 45, 66, 70, 69, 80, 58, 68, 60, 67, 72, 73, 70, 57, 63, 70, 78, 52, 67, 53, 67, 75, 61, 70, 81, 76, 79, 75, 76, 58, 31.

(a) Compute the sample mean and sample standard deviation.
(b) Find the upper and lower quartiles of temperature.
(c) Find the median.
(d) Set aside the smallest observation (31°F) and recompute the quantities in parts (a), (b), and (c). Comment on your findings. How "different" are the other temperatures from this smallest value?
(e) Construct a box plot of the data and comment on the possible presence of outliers.

2-28.   An article in the *Transactions of the Institution of Chemical Engineers* (Vol. 34, 1956, pp. 280–293) reported data from an experiment investigating the effect of several process variables on the vapor phase oxidation of naphthalene. A sample of the percentage mole conversion of naphthalene to maleic anhydride follows: 4.2, 4.7, 4.7, 5.0, 3.8, 3.6, 3.0, 5.1, 3.1, 3.8, 4.8, 4.0, 5.2, 4.3, 2.8, 2.0, 2.8, 3.3, 4.8, 5.0.

(a) Calculate the sample mean.
(b) Calculate the sample variance and sample standard deviation.
(c) Construct a box plot of the data.

2-29.   The "cold start ignition time" of an automobile engine is being investigated by a gasoline manufacturer. The following times (in seconds) were obtained for a test vehicle: 1.75, 1.92, 2.62, 2.35, 3.09, 3.15, 2.53, 1.91.

(a) Calculate the sample mean and sample standard deviation.
(b) Construct a box plot of the data.

2-30.   The nine measurements that follow are furnace temperatures recorded on successive batches in a semiconductor manufacturing process (units are °F): 953, 950, 948, 955, 951, 949, 957, 954, 955.

(a) Calculate the sample mean, variance, and standard deviation.
(b) Find the median. How much could the largest temperature measurement increase without changing the median value?
(c) Construct a box plot of the data.

2-31.   An article in the *Journal of Aircraft* (1988) describes the computation of drag coefficients for the NASA 0012 airfoil. Different computational algorithms were used at $M_a = 0.7$ with the following results (drag coefficients are in units of drag counts; that is, 1 count is equivalent to a drag coefficient of 0.0001): 79, 100, 74, 83, 81, 85, 82, 80, and 84.

(a) Calculate the sample mean, variance, and standard deviation.
(b) Find the upper and lower quartiles of the drag coefficients.
(c) Construct a box plot of the data.
(d) Set aside the largest observation (100) and redo parts (a), (b), and (c). Comment on your findings.

2-32. The following data are the temperatures of effluent at discharge from a sewage treatment facility on consecutive days:

43	47	51	48	52	50	46	49
45	52	46	51	44	49	46	51
49	45	44	50	48	50	49	50

(a) Calculate the sample mean and median.
(b) Calculate the sample variance and sample standard deviation.
(c) Construct a box plot of the data and comment on the information in this display.
(d) Find the 5th and 95th percentiles of temperature.

2-33. The inside diameter (in inches) of 50 lightweight snaps used in assembling computer cases are measured and sorted with the following resulting data:

0.0395	0.0443	0.0450	0.0459	0.0470
0.0485	0.0486	0.0487	0.0489	0.0496
0.0499	0.0500	0.0503	0.0504	0.0504
0.0516	0.0529	0.0542	0.0550	0.0571

(a) Compute the sample mean and sample variance.
(b) Find the sample upper and lower quartiles.
(c) Find the sample median.
(d) Construct a box plot of the data.
(e) Find the 5th and 95th percentiles of the inside diameter.

2-34. Eighteen measurements of the disbursement rate (in cm³/sec) of a chemical disbursement system are recorded and sorted:

6.50	6.77	6.91	7.38	7.64	7.74	7.90	7.91	8.21
8.26	8.30	8.31	8.42	8.53	8.55	9.04	9.33	9.36

(a) Compute the sample mean and sample variance.
(b) Find the sample upper and lower quartiles.
(c) Find the sample median.
(d) Construct a box plot of the data.
(e) Find the 5th and 95th percentiles of the inside diameter.

2-35. A battery-operated pacemaker device helps the human heart to beat in regular rhythm. The activation rate is important in stimulating the heart, when necessary. Fourteen activation rates (in sec.) were collected on a newly designed device:

0.670	0.697	0.699	0.707	0.732	0.733	0.737
0.747	0.751	0.774	0.777	0.804	0.819	0.827

(a) Compute the sample mean and sample variance.
(b) Find the sample upper and lower quartiles.
(c) Find the sample median.
(d) Construct a box plot of the data.
(e) Find the 5th and 95th percentiles of the inside diameter.

2-36. Consider the solar intensity data in Exercise 2-7.
(a) Compute the sample mean, variance, and standard deviation.
(b) Find the sample upper and lower quartiles.
(c) Find the sample median.
(d) Construct a box plot of the data.
(e) Find the 5th and 95th percentiles.

2-37. Consider the gene expression data in Exercise 2-8. Use each group separately and calculate the following. Comment on any differences between the groups.
(a) Compute the sample mean, variance, and standard deviation.
(b) Find the sample upper and lower quartiles.
(c) Find the sample median.
(d) Construct a box plot of the data.
(e) Find the 5th and 95th percentiles.

2-5 TIME SERIES PLOTS

The graphical displays that we have considered thus far such as histograms, stem-and-leaf plots, and box plots are very useful visual methods for showing the variability in data. However, we noted previously in Chapter 1 that time is an important factor that contributes to variability in data, and those graphical methods do not take this into account. A **time series** or **time sequence** is a data set in which the observations are recorded in the order in which they occur. A **time series plot** is a graph in which the vertical axis denotes the observed value of the variable (say, x) and the horizontal axis denotes the time (which could be minutes, days, years, etc.). When measurements are plotted as a time series, we often see trends, cycles, or other broad features of the data that we could not see otherwise.

For example, consider Fig. 2-14a, which presents a time series plot of the annual sales of a company for the last 10 years. The general impression from this display is that sales show an upward **trend.** There is some variability about this trend, with some years' sales increasing over those of the last year and some years' sales decreasing. Figure 2-14b shows the last 3

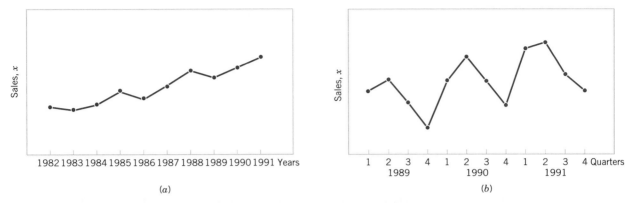

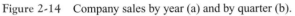

Figure 2-14   Company sales by year (a) and by quarter (b).

years of sales reported by quarter. This plot clearly shows that the annual sales in this business exhibit a **cyclic** variability by quarter, with the first- and second-quarter sales generally greater than sales during the third and fourth quarters.

Sometimes it can be very helpful to combine a time series plot with some of the other graphical displays that we have considered previously. J. Stuart Hunter (*The American Statistician,* Vol. 42, 1988, p. 54) has suggested combining the stem-and-leaf plot with a time series plot to form a **digidot plot.**

Figure 2-15 shows a digidot plot for the observations on compressive strength from Table 2-2, assuming that these observations are recorded in the order in which they occurred. This plot effectively displays the overall variability in the compressive strength data and simultaneously shows the variability in these measurements over time. The general impression is that compressive strength varies around the mean value of 162.67, and there is no strong obvious pattern in this variability over time.

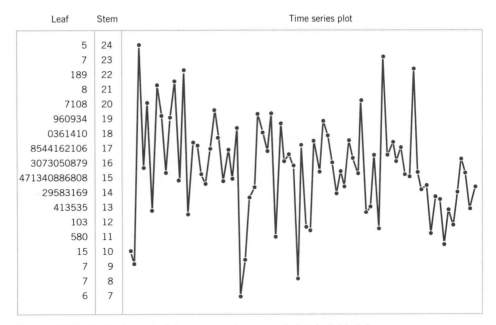

Figure 2-15   A digidot plot of the compressive strength data in Table 2-2.

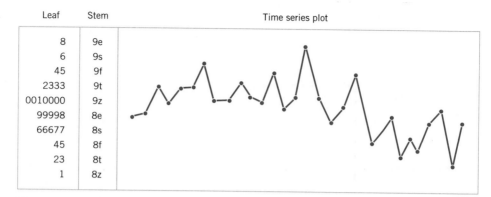

Figure 2-16   A digidot plot of chemical process concentration readings, observed hourly.

The digidot plot in Fig. 2-16 tells a different story. This plot summarizes 30 observations on concentration of the output product from a chemical process, where the observations are recorded at 1-hour time intervals. This plot indicates that during the first 20 hours of operation this process produced concentrations generally above 85 g/l, but that following sample 20 something may have occurred in the process that resulted in lower concentrations. If this variability in output product concentration can be reduced, operation of this process can be improved. The control chart, which is a special kind of time series plot, of these data was shown in Fig. 1-19 of Chapter 1.

EXERCISES FOR SECTION 2-5

2-38. Consider a server at Arizona State University. Response times for 20 consecutive jobs done on the system were recorded and are shown next in order. (Read down, then left to right.)

5.3	5.0	9.5	10.1	5.8	6.2	5.9
7.2	10.0	12.2	8.5	4.7	11.2	7.3
6.4	12.4	3.9	8.1	9.2	10.5	

Construct and interpret a time series plot of these data.

2-39. The following data are the viscosity measurements for a chemical product observed hourly. (Read down, then left to right.)

47.9	47.9	48.6	48.0	48.4	48.1	48.0	48.6
48.8	48.1	48.3	47.2	48.9	48.6	48.0	47.5
48.6	48.0	47.9	48.3	48.5	48.1	48.0	48.3
43.2	43.0	43.5	43.1	43.0	42.9	43.6	43.3
43.0	42.8	43.1	43.2	43.6	43.2	43.5	43.0

(a) Construct and interpret either a digidot plot or a separate stem-and-leaf and time series plot of these data.
(b) Specifications on product viscosity are at 48 ± 2. What conclusions can you make about process performance?

2-40. The pull-off force for a connector is measured in a laboratory test. Data for 40 test specimens are shown next. (Read down the entire column, then left to right.)

241	258	237	210	194	225	248	203
195	249	220	194	245	209	201	195
255	245	235	220	249	251	238	210
198	199	183	213	236	245	209	212
185	187	218	190	175	178	175	190

(a) Construct a time series plot of the data.
(b) Construct and interpret either a digidot plot or a stem-and-leaf and time series plot of the data.

2-41. In their book *Time Series Analysis, Forecasting, and Control* (Holden-Day, 1976), G. E. P. Box and G. M. Jenkins present chemical process concentration readings made every 2 hours. Some of these data are shown next. (Read down, then left to right.)

17.0	16.6	16.3	16.1	17.1	16.9	16.8	17.4
17.1	17.0	16.7	17.4	17.2	17.4	17.4	17.0
17.3	17.2	17.4	16.8	17.1	17.4	17.4	17.5
17.4	17.6	17.4	17.3	17.0	17.8	17.5	18.1
17.5	17.4	17.4	17.1	17.6	17.7	17.4	17.8
17.6	17.5	16.5	17.8	17.3	17.3	17.1	17.4
16.9	17.3						

Table 2-7   Annual Sunspot Numbers

1770	101	1795	21	1820	16	1845	40
1771	82	1796	16	1821	7	1846	62
1772	66	1797	6	1822	4	1847	98
1773	35	1798	4	1823	2	1848	124
1774	31	1799	7	1824	8	1849	96
1775	7	1800	14	1825	17	1850	66
1776	20	1801	34	1826	36	1851	64
1777	92	1802	45	1827	50	1852	54
1778	154	1803	43	1828	62	1853	39
1779	125	1804	48	1829	67	1854	21
1780	85	1805	42	1830	71	1855	7
1781	68	1806	28	1831	48	1856	4
1782	38	1807	10	1832	28	1857	23
1783	23	1808	8	1833	8	1858	55
1784	10	1809	2	1834	13	1859	94
1785	24	1810	0	1835	57	1860	96
1786	83	1811	1	1836	122	1861	77
1787	132	1812	5	1837	138	1862	59
1788	131	1813	12	1838	103	1863	44
1789	118	1814	14	1839	86	1864	47
1790	90	1815	35	1840	63	1865	30
1791	67	1816	46	1841	37	1866	16
1792	60	1817	41	1842	24	1867	7
1793	47	1818	30	1843	11	1868	37
1794	41	1819	24	1844	15	1869	74

Construct and interpret either a digidot plot or a stem-and-leaf and time series plot of these data.

2-42.   The annual Wolfer sunspot numbers from 1770 to 1869 are shown in Table 2-7. (For an interesting analysis and interpretation of these numbers, see the book by Box and Jenkins referenced in Exercise 2-41. The analysis requires some advanced knowledge of statistics and statistical model building.)

(a) Construct a time series plot of these data.
(b) Construct and interpret either a digidot plot or a stem-and-leaf and time series plot of these data.

2-43.   In their book *Forecasting and Time Series Analysis,* 2nd ed. (McGraw-Hill, 1990), D. C. Montgomery, L. A. Johnson, and J. S. Gardiner analyze the data in Table 2-8,

Table 2-8   United Kingdom Airline Miles Flown

	1964	1965	1966	1967	1968	1969	1970
Jan.	7.269	8.350	8.186	8.334	8.639	9.491	10.840
Feb.	6.775	7.829	7.444	7.899	8.772	8.919	10.436
Mar.	7.819	8.829	8.484	9.994	10.894	11.607	13.589
Apr.	8.371	9.948	9.864	10.078	10.455	8.852	13.402
May	9.069	10.638	10.252	10.801	11.179	12.537	13.103
June	10.248	11.253	12.282	12.953	10.588	14.759	14.933
July	11.030	11.424	11.637	12.222	10.794	13.667	14.147
Aug.	10.882	11.391	11.577	12.246	12.770	13.731	14.057
Sept.	10.333	10.665	12.417	13.281	13.812	15.110	16.234
Oct.	9.109	9.396	9.637	10.366	10.857	12.185	12.389
Nov.	7.685	7.775	8.094	8.730	9.290	10.645	11.594
Dec.	7.682	7.933	9.280	9.614	10.925	12.161	12.772

which are the monthly total passenger airline miles flown in the United Kingdom, 1964–1970 (in millions of miles).

(a) Draw a time series plot of the data and comment on any features of the data that are apparent.

(b) Construct and interpret either a digidot plot or a stem-and-leaf and time series plot of these data.

 2-44. The following table shows U.S. petroleum imports, imports as a percentage of total, and Persian Gulf imports as a percentage of all imports by year since 1973 (source: U.S. Department of Energy Web site http://www.eia.doe.gov/). Construct and interpret either a digidot plot or a separate stem-and-leaf and time series plot for each column of data.

Year	Petroleum Imports (thousand barrels per day)	Total Petroleum Imports as Percent of Petroleum Products Supplied (percent)	Petroleum Imports from Persian Gulf as Percent of Total Petroleum Imports (percent)
1973	6256.145	36.1	13.5
1974	6112.184	36.7	17.0
1975	6055.712	37.1	19.2
1976	7312.598	41.8	25.1
1977	8807.249	47.7	27.8
1978	8363.411	44.3	26.5
1979	8456.129	45.6	24.4
1980	6909.025	40.5	21.9
1981	5995.673	37.3	20.3

Year	Petroleum Imports (thousand barrels per day)	Total Petroleum Imports as Percent of Petroleum Products Supplied (percent)	Petroleum Imports from Persian Gulf as Percent of Total Petroleum Imports (percent)
1982	5113.311	33.4	13.6
1983	5051.353	33.1	8.7
1984	5436.982	34.5	9.3
1985	5067.144	32.2	6.1
1986	6223.512	38.2	14.6
1987	6677.696	40.0	16.1
1988	7402.021	42.8	20.8
1989	8060.545	46.5	23.0
1990	8017.521	47.1	24.5
1991	7626.748	45.6	24.1
1992	7887.697	46.3	22.5
1993	8620.422	50.0	20.6
1994	8996.222	50.7	19.2
1995	8834.94	49.8	17.8
1996	9478.492	51.7	16.9
1997	10161.562	54.5	17.2
1998	10708.071	56.6	19.9
1999	10852.258	55.5	22.7
2000	11459.251	58.1	21.7
2001	11871.337	60.4	23.2
2002	11530.241	58.3	19.6
2003	12264.386	61.2	20.3
2004	13145.093	63.4	18.9

2-6 MULTIVARIATE DATA

The dot diagram, stem-and-leaf diagram, histogram, and box plot are descriptive displays for **univariate** data; that is, they convey descriptive information about a single variable. Many engineering problems involve collecting and analyzing **multivariate data,** or data on several different variables. The acetone-butyl alcohol distillation column discussed in Section 1-2 and the wire bond pull strength problem in Section 1-3 are typical examples of engineering studies involving multivariate data. The wire bond pull strength data is reproduced for convenience in Table 2-9. In engineering studies involving multivariate data, often the objective is to determine the relationships among the variables or to build an empirical model, as we discussed in Section 1-3.

The scatter diagram introduced in Section 1-3 is a simple descriptive tool for multivariate data. The diagram is useful for examining the pairwise (or two variables at a time) relationships between the variables in a multivariate data set. Scatter diagrams for the wire bond pull strength data in Table 2-9 are shown in Figure 2-17. These plots were constructed using Minitab, and an option to display box plots for each individual variable in the margins of the plot was specified. As we observed previously in Section 1-3, both scatter diagrams convey the impression that there may be an approximate linear relationship between bond pull strength and wire length and between bond strength and die height. The strength of these relationships appears to be stronger for pull strength and length than it does for pull strength and die height.

Table 2-9   Wire Bond Data

Observation Number	Pull Strength y	Wire Length x_1	Die Height x_2	Observation Number	Pull Strength y	Wire Length x_1	Die Height x_2
1	9.95	2	50	14	11.66	2	360
2	24.45	8	110	15	21.65	4	205
3	31.75	11	120	16	17.89	4	400
4	35.00	10	550	17	69.00	20	600
5	25.02	8	295	18	10.30	1	585
6	16.86	4	200	19	34.93	10	540
7	14.38	2	375	20	46.59	15	250
8	9.60	2	52	21	44.88	15	290
9	24.35	9	100	22	54.12	16	510
10	27.50	8	300	23	56.63	17	590
11	17.08	4	412	24	22.13	6	100
12	37.00	11	400	25	21.15	5	400
13	41.95	12	500				

The strength of a linear relationship between two variables, y and x, can be described by the **sample correlation coefficient** r. Suppose that we have n pairs of observations on two variables $(y_1, x_1), (y_2, x_2), \ldots, (y_n, x_n)$. It would be logical to say that y and x have a positive relationship if large values of y together with large values of x and small values of y occur with small values of x. Similarly, a negative relationship between the variables is implied if large values of y occur together with small values of x and small values of y occur with large values of x. The corrected sum of cross-products given by

$$S_{xy} = \sum_{i=1}^{n} (x_i - \bar{x})(y_i - \bar{y}) = \sum_{i=1}^{n} x_i y_i - \left(\sum_{i=1}^{n} x_i \right)\left(\sum_{i=1}^{n} y_i \right) / n$$

can reflect these types of relationships. To see why this is so, suppose that the relationship between y and x is strongly positive, as in the case of pull strength and length in Figure 2-17a. In this situation, a value of x_i above the mean $\bar{x}$ will tend to occur along with a value of y_i that is above the mean $\bar{y}$, so the cross-product $(x_i - \bar{x})(y_i - \bar{y})$ will be positive. This will also occur

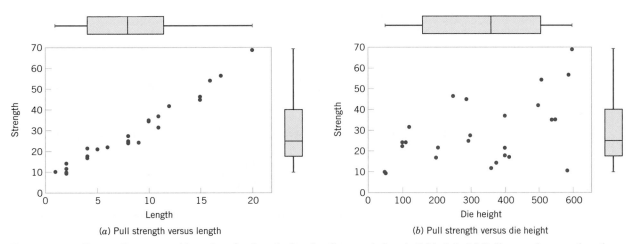

(a) Pull strength versus length   (b) Pull strength versus die height

Figure 2-17   Scatter diagrams and box plots for the wire bond pull strength data in Table 1-1. (a) Pull strength versus length. (b) Pull strength versus die height.

when both x_i and y_i are below their respective means. Consequently, a positive relationship between y and x implies that S_{xy} will be positive. A similar argument will show that when the relationship is negative, most of the cross-products $(x_i - \bar{x})(y_i - \bar{y})$ will be negative, so S_{xy} will be negative. Now the units of S_{xy} are the units of x times the units of y, so it would be difficult to interpret S_{xy} as a measure of the strength of a relationship because a change in the units of either x and/or y will affect the magnitude of S_{xy}. The sample correlation coefficient r simply scales S_{xy} to produce a dimensionless quantity.

**Sample
Correlation
Coefficient**

> Given n pairs of data $(y_1, x_1), (y_2, x_2), \ldots, (y_n, x_n)$, the **sample correlation coefficient** r is defined by
>
> $$r = \frac{S_{xy}}{\sqrt{\left(\displaystyle\sum_{i=1}^{n}(x_i - \bar{x})^2\right)\left(\displaystyle\sum_{i=1}^{n}(y_i - \bar{y})^2\right)}} = \frac{S_{xy}}{\sqrt{S_{xx}S_{yy}}} \qquad (2\text{-}6)$$
>
> with $-1 \leq r \leq +1$.

The value $r = +1$ is achieved only if all the observations lie exactly along a straight line with positive slope, and similarly, the value $r = -1$ is achieved only if all the observations lie exactly along a straight line with negative slope. Thus r measures the strength of the linear relationship between x and y. When the value of r is near zero, it may indicate either no relationship among the variables or the absence of a linear relationship. Refer to Fig. 2-18.

As an illustration, consider calculating the sample correlation coefficient between pull strength and wire length. From the data in Table 2-9, we find that

Calculating the sample correlation coefficient

$$\sum_{i=1}^{25} x_i^2 = 2396 \quad \sum_{i=1}^{25} x = 206 \quad \sum_{i=1}^{25} y_i^2 = 27179 \quad \sum_{i=1}^{25} y_i = 725.82 \quad \sum_{i=1}^{25} x_i y_i = 8008.5$$

$$S_{xx} = \sum_{i=1}^{n} x_i^2 - \frac{\left(\displaystyle\sum_{i=1}^{n} x_i\right)^2}{n} = 2396 - \frac{(206)^2}{25} = 698.56$$

$$S_{yy} = \sum_{i=1}^{n} y_i^2 - \frac{\left(\displaystyle\sum_{i=1}^{n} y_i\right)^2}{n} = 27179 - \frac{(725.82)^2}{25} = 6106.41$$

$$S_{xy} = \sum_{i=1}^{n} x_i y_i - \frac{\left(\displaystyle\sum_{i=1}^{n} x_i\right)\left(\displaystyle\sum_{i=1}^{n} y_i\right)}{n} = 8008.5 - \frac{(206)(725.82)}{25} = 2027.74$$

The sample correlation coefficient between wire bond pull strength and wire length is

$$r = \frac{S_{xy}}{\sqrt{S_{xx}S_{yy}}} = \frac{2027.74}{\sqrt{(698.56)(6106.41)}} = 0.982.$$

A similar calculation reveals that the sample correlation between wire bond pull strength and die height is $r = 0.493$.

Generally, we consider the correlation between two variables to be strong when $0.8 \leq r \leq 1$, weak when $0 \leq r \leq 0.5$, and moderate otherwise. Therefore there is a strong

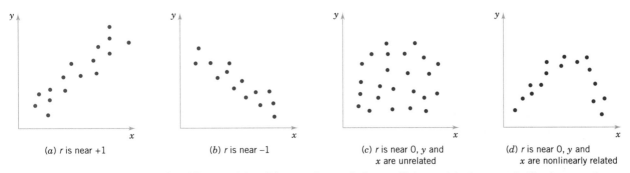

Figure 2-18   Scatter diagrams for different values of the sample correlation coefficient r. (*a*) r is near $+1$. (*b*) r is near -1. (*c*) r is near 0; y and x are unrelated. (*d*) r is near 0; y and x are nonlinearly related.

correlation between the wire bond pull strength and wire length, and a relatively weak to moderate correlation between pull strength and die height.

There are several other useful graphical methods for displaying multivariate data. To illustrate the use of these methods consider the shampoo data in Table 2-10. These data were collected during a sensory evaluation experiment conducted by a scientist. The variables foam, scent, color, and residue (a measure of the extent of the cleaning ability) are descriptive properties evaluated on a 10-point scale. Quality is a measure of overall desirability of the shampoo, and it is the nominal response variable of interest to the experimenter. Region is an indicator for a qualitative variable identifying whether the shampoo was produced in an eastern (1) or western (2) plant.

Table 2-10   Data on Shampoo

Foam	Scent	Color	Residue	Region	Quality
6.3	5.3	4.8	3.1	1	91
4.4	4.9	3.5	3.9	1	87
3.9	5.3	4.8	4.7	1	82
5.1	4.2	3.1	3.6	1	83
5.6	5.1	5.5	5.1	1	83
4.6	4.7	5.1	4.1	1	84
4.8	4.8	4.8	3.3	1	90
6.5	4.5	4.3	5.2	1	84
8.7	4.3	3.9	2.9	1	97
8.3	3.9	4.7	3.9	1	93
5.1	4.3	4.5	3.6	1	82
3.3	5.4	4.3	3.6	1	84
5.9	5.7	7.2	4.1	2	87
7.7	6.6	6.7	5.6	2	80
7.1	4.4	5.8	4.1	2	84
5.5	5.6	5.6	4.4	2	84
6.3	5.4	4.8	4.6	2	82
4.3	5.5	5.5	4.1	2	79
4.6	4.1	4.3	3.1	2	81
3.4	5.0	3.4	3.4	2	83
6.4	5.4	6.6	4.8	2	81
5.5	5.3	5.3	3.8	2	84
4.7	4.1	5.0	3.7	2	83
4.1	4.0	4.1	4.0	2	80

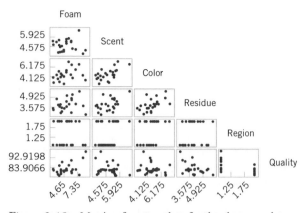

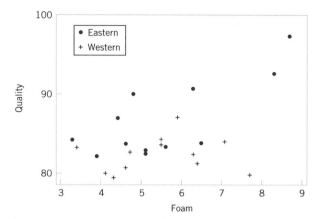

Figure 2-19   Matrix of scatter plots for the shampoo data in Table 2-10.

Figure 2-20   Scatter diagram of shampoo quality versus foam.

Figure 2-19 is a **matrix of scatter plots** for the shampoo data, produced by Minitab. This display reveals the pairwise relationships between all of the variables in Table 2-10. The individual scatter plots in this matrix reveal that there may be a positive relationship between shampoo quality and foam and negative relationships between quality and scent and between quality and the indicator variable identifying the region in which it was produced. There may also be relationships between some of the property variables, such as color and residue. Minitab will also calculate all of the pairwise correlations between these variables. The results of this are as follows:

	Foam	Scent	Color	Residue	Region
Scent	0.002				
Color	0.328	0.599			
Residue	0.193	0.500	0.524		
Region	−0.032	0.278	0.458	0.165	
Quality	0.512	−0.252	−0.194	−0.489	−0.507

Notice that none of the correlations is strong.

Figure 2-20 is a scatter diagram of shampoo quality versus foam. In this scatter diagram we have used two different plotting symbols to identify the observations associated with the two different regions, thus allowing information about *three* variables to be displayed on a two-dimensional graph. The display in Fig. 2-20 reveals that the relationship between shampoo quality and foam may be different in the two regions. Another way to say this is that there may be an **interaction** between foam and region (you may find it helpful to reread the discussion of interaction in Section 1-2.3). Obviously, this technique could be extended to more than three variables by defining several additional plotting symbols.

The variation of the scatter diagram in Fig. 2-20 works well when the third variable is **discrete** or **categorical.** When the third variable is continuous, a **coplot** may be useful. A coplot for the shampoo quality data is shown in Fig. 2-21. In this display, shampoo quality is plotted against foam, and as in Fig. 2-20, different plotting symbols are used to identify the two production regions. The descriptive variable residue in Table 2-10 is not necessarily a desirable characteristic of shampoo, and higher levels of residue would receive a rating of between 4 and 4.5 or greater in Table 2-10. The graph in Fig. 2-21*a* uses all of the observations in Table 2-10 for which the residue variable is less than or equal to 4.6, and the graph in Fig. 2-21*b* uses all of the observations for which the residue is greater than or equal to 4. Notice that there is overlap in the values of residue used in constructing the coplot; this is perfectly

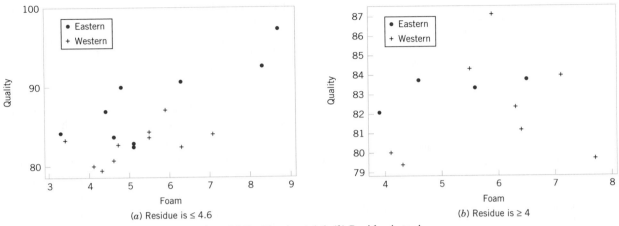

Figure 2-21   A coplot for the shampoo data. (*a*) Residue is ≤ 4.6. (*b*) Residue is ≥ 4.

acceptable. Using more than two categories may be useful as well in some problems. The coplot indicates that the positive relationship between shampoo quality and foam is much stronger for the lower range of residue, perhaps indicating that too much residue doesn't always result in a good shampoo.

Animation 14: The Correlation Game

EXERCISES FOR SECTION 2-6

2-45. An engineer at a semiconductor company wants to model the relationship between the device HFE (y) and three parameters: Emitter-RS (x_1), Base-RS (x_2), and Emitter-to-Base RS (x_3). The data are shown in the following table.

x_1 Emitter-RS	x_2 Base-RS	x_3 E-B-RS	y HFE-1M-5V
14.620	226.00	7.000	128.40
15.630	220.00	3.375	52.62
14.620	217.40	6.375	113.90
15.000	220.00	6.000	98.01
14.500	226.50	7.625	139.90
15.250	224.10	6.000	102.60
16.120	220.50	3.375	48.14
15.130	223.50	6.125	109.60
15.500	217.60	5.000	82.68
15.130	228.50	6.625	112.60
15.500	230.20	5.750	97.52
16.120	226.50	3.750	59.06
15.130	226.60	6.125	111.80
15.630	225.60	5.375	89.09
15.380	229.70	5.875	101.00
14.380	234.00	8.875	171.90
15.500	230.00	4.000	66.80
14.250	224.30	8.000	157.10
14.500	240.50	10.870	208.40
14.620	223.70	7.375	133.40

(a) Construct scatter plots of each x versus y and comment on any relationships.

(b) Compute and interpret the three sample correlation coefficients between each x and y.

2-46. Establishing the properties of materials is an important problem in identifying a suitable substitute for biodegradable materials in the fast-food packaging industry. Consider the following data on product density (g/cm³) and thermal conductivity K-factor (W/mK) published in *Materials Research and Innovation* (1999, pp. 2–8).

Thermal Conductivity y	Product Density x
0.0480	0.1750
0.0525	0.2200
0.0540	0.2250
0.0535	0.2260
0.0570	0.2500
0.0610	0.2765

(a) Create a scatter diagram of the data. What do you anticipate will be the sign of the sample correlation coefficient?

(b) Compute and interpret the sample correlation coefficient.

2-47. A study was performed on wear of a bearing y and its relationship to $x_1 =$ oil viscosity and $x_2 =$ load. The following data were obtained.

y	x_1	x_2
193	1.6	851
230	15.5	816
172	22.0	1058
91	43.0	1201
113	33.0	1357
125	40.0	1115

(a) Create two scatter diagrams of the data. What do you anticipate will be the sign of each sample correlation coefficient?
(b) Compute and interpret the two sample correlation coefficients.

2-48. To investigate fuel efficiency, the following data were
collected.

MPG, y	Weight, x_1	Horsepower x_2	MPG, y	Weight, x_1	Horsepower x_2
29.25	2464	130	17.00	4024	394
21.00	3942	235	17.00	3495	294
32.00	2604	110	18.50	4300	362
21.25	3241	260	16.00	4455	389
26.50	3283	200	10.50	3726	485
23.00	2809	240	12.50	3522	550

(a) Create two scatter diagrams of the data. What do you anticipate will be the sign of each sample correlation coefficient?

(b) Compute and interpret the two sample correlation coefficients.

2-49. The weight and systolic blood pressure of 26 randomly selected males in the age group 25 to 30 are shown in
the following table.

Subject	Weight	Systolic BP	Subject	Weight	Systolic BP
1	165	130	14	172	153
2	167	133	15	159	128
3	180	150	16	168	132
4	155	128	17	174	149
5	212	151	18	183	158
6	175	146	19	215	150
7	190	150	20	195	163
8	210	140	21	180	156
9	200	148	22	143	124
10	149	125	23	240	170
11	158	133	24	235	165
12	169	135	25	192	160
13	170	150	26	187	159

(a) Create a scatter diagram of the data. What do you anticipate will be the sign of the sample correlation coefficient?
(b) Compute and interpret the sample correlation coefficient.

SUPPLEMENTAL EXERCISES

2-50. The pH of a solution is measured eight times by one
operator using the same instrument. She obtains the following
data: 7.15, 7.20, 7.18, 7.19, 7.21, 7.20, 7.16, and 7.18.
(a) Calculate the sample mean. Suppose that the desirable
   value for this solution was specified to be 7.20. Do you
   think that the sample mean value computed here is close
   enough to the target value to accept the solution as conforming to target? Explain your reasoning.
(b) Calculate the sample variance and sample standard
   deviation. What do you think are the major sources of
   variability in this experiment? Why is it desirable to have
   a small variance of these measurements?

2-51. A sample of six resistors yielded the following resistances (ohms): $x_1 = 45$, $x_2 = 38$, $x_3 = 47$, $x_4 = 41$, $x_5 = 35$,
and $x_6 = 43$.
(a) Compute the sample variance and sample standard deviation using the method in equation 2-4.

(b) Compute the sample variance and sample standard deviation using the definition in equation 2-3. Explain why the
   results from both equations are the same.
(c) Subtract 35 from each of the original resistance measurements and compute s^2 and s. Compare your results with
   those obtained in parts (a) and (b) and explain your
   findings.
(d) If the resistances were 450, 380, 470, 410, 350, and 430
   ohms, could you use the results of previous parts of this
   problem to find s^2 and s? Explain how you would proceed.

2-52. The percentage mole conversion of naphthalene to
maleic anhydride from Exercise 2-28 follows: 4.2, 4.7, 4.7,
5.0, 3.8, 3.6, 3.0, 5.1, 3.1, 3.8, 4.8, 4.0, 5.2, 4.3, 2.8, 2.0, 2.8,
3.3, 4.8, 5.0.
(a) Calculate the sample range, variance, and standard deviation.

(b) Calculate the sample range, variance, and standard deviation again, but first subtract 1.0 from each observation. Compare your results with those obtained in part (a). Is there anything "special" about the constant 1.0, or would any other arbitrarily chosen value have produced the same results?

2-53. Suppose that we have a sample $x_1, x_2, \ldots, x_n$ and we have calculated $\bar{x}_n$ and s_n^2 for the sample. Now an $(n + 1)$st observation becomes available. Let $\bar{x}_{n+1}$ and s_{n+1}^2 be the sample mean and sample variance for the sample using all $n + 1$ observations.

(a) Show how $\bar{x}_{n+1}$ can be computed using $\bar{x}_n$ and x_{n+1}.

(b) Show that $ns_{n+1}^2 = (n - 1)s_n^2 + \dfrac{n}{n + 1}(x_{n+1} - \bar{x}_n)^2$.

(c) Use the results of parts (a) and (b) to calculate the new sample average and standard deviation for the data of Exercise 2-51, when the new observation is $x_7 = 46$.

2-54. **The Trimmed Mean.** Suppose that the data are arranged in increasing order, $T\%$ of the observations are removed from each end, and the sample mean of the remaining numbers is calculated. The resulting quantity is called a *trimmed mean*. The trimmed mean generally lies between the sample mean $\bar{x}$ and the sample median $\tilde{x}$. Why?

(a) Calculate the 10% trimmed mean for the yield data in Exercise 2-13.

(b) Calculate the 20% trimmed mean for the yield data in Exercise 2-13 and compare it with the quantity found in part (a).

(c) Compare the values calculated in parts (a) and (b) with the sample mean and median for the data in Exercise 2-13. Is there much difference in these quantities? Why?

(d) Suppose that the sample size n is such that the quantity $nT/100$ is not an integer. Develop a procedure for obtaining a trimmed mean in this case.

2-55. Consider the two samples shown here:

Sample 1: 20, 19, 18, 17, 18, 16, 20, 16

Sample 2: 20, 16, 20, 16, 18, 20, 18, 16

(a) Calculate the range for both samples. Would you conclude that both samples exhibit the same variability? Explain.

(b) Calculate the sample standard deviations for both samples. Do these quantities indicate that both samples have the same variability? Explain.

(c) Write a short statement contrasting the sample range versus the sample standard deviation as a measure of variability.

2-56. An article in *Quality Engineering* (Vol. 4, 1992, pp. 487–495) presents viscosity data from a batch chemical process. A sample of these data is presented next. (Read down the entire column, then left to right.)

(a) Draw a time series plot of all the data and comment on any features of the data that are revealed by this plot.

13.3	14.9	15.8	16.0	14.3	15.2	14.2	14.0
14.5	13.7	13.7	14.9	16.1	15.2	16.9	14.4
15.3	15.2	15.1	13.6	13.1	15.9	14.9	13.7
15.3	14.5	13.4	15.3	15.5	16.5	15.2	13.8
14.3	15.3	14.1	14.3	12.6	14.8	14.4	15.6
14.8	15.6	14.8	15.6	14.6	15.1	15.2	14.5
15.2	15.8	14.3	16.1	14.3	17.0	14.6	12.8
14.5	13.3	14.3	13.9	15.4	14.9	16.4	16.1
14.6	14.1	16.4	15.2	15.2	14.8	14.2	16.6
14.1	15.4	16.9	14.4	16.8	14.0	15.7	15.6

(b) Consider the notion that the first 40 observations were generated from a specific process, whereas the last 40 observations were generated from a different process. Does the plot indicate that the two processes generate similar results?

(c) Compute the sample mean and sample variance of the first 40 observations; then compute these values for the second 40 observations. Do these quantities indicate that both processes yield the same mean level? the same variability? Explain.

2-57. A manufacturer of coil springs is interested in implementing a quality control system to monitor his production process. As part of this quality system, they decided to record the number of nonconforming coil springs in each production batch of size 50. During production, 40 batches of data were collected and are reported here. (Read down the entire column, then left to right.)

9	13	9	8	7	7	10	9	17	17
10	4	8	9	9	9	6	11	12	19
9	11	10	4	8	10	8	4	16	16
9	8	7	9	7	11	6	8	13	15

(a) Construct a stem-and-leaf plot of the data.

(b) Find the sample average and sample standard deviation.

(c) Construct a time series plot of the data. Is there evidence that there was an increase or decrease in the average number of nonconforming springs made during the 40 days? Explain.

2-58. A communication channel is being monitored by recording the number of errors in a string of 1000 bits. Data for 20 of these strings are given here. (Read the data left to right, then down.)

3	2	4	1	3	1	3	1	0	1
3	2	0	2	0	1	1	1	2	3

(a) Construct a stem-and-leaf plot of the data.

(b) Find the sample average and sample standard deviation.

(c) Construct a time series plot of the data. Is there evidence that there was an increase or decrease in the number of errors in a string? Explain.

TEAM EXERCISES

2-59. As an engineering student you have frequently encountered data (e.g., in engineering or science laboratory courses). Choose one of these data sets or another data set of interest to you. Describe the data with appropriate numerical and graphical tools.

2-60. Select a data set that is time ordered. Describe the data with appropriate numerical and graphical tools. Discuss potential sources of variation in the data.

2-61. Consider the data on weekly waste (percent) for five suppliers of the Levi-Strauss clothing plant in Albuquerque and reported at the Web site http://lib.stat.cmu.edu/DASL/Stories/wasterunup.html. Generate box plots for the five suppliers.

2-62. Thirty-one consecutive daily carbon monoxide measurements were taken by an oil refinery northeast of San Francisco and reported at the Web site http://lib.stat.cmu.edu/DASL/Datafiles/Refinery.html. Draw a time series plot of the data and comment on features of the data that are revealed by this plot.

2-63. Consider the famous data set listing the time between eruptions of the geyser "Old Faithful" found at http://lib.stat.cmu.edu/DASL/Datafiles/differencetestdat.html from the article by A. Azzalini and A. W. Bowman, "A Look at Some Data on the Old Faithful Geyser," *Applied Statistics,* Vol. 39, 1990, pp. 357–365.

(a) Construct a time series plot of all the data.

(b) Split the data into two data sets of 100 observations each. Create two separate stem-and-leaf plots of the subsets. Is there reason to believe that the two subsets are different?

IMPORTANT TERMS AND CONCEPTS

Box plot	Median	Population variance, σ^2	Sample variance, s^2
Degrees of freedom	Multivariate data	Quartiles	Scatter diagram
Digidot plot	Ordered stem-and-leaf diagram	Range	Stem-and-leaf diagram
Dot diagram	Pareto chart	Relative frequency	Time sequence
Frequency	Percentile	Sample correlation coefficient	Time series plot
Histogram	Population mean, μ	Sample mean, $\bar{x}$	Univariate data
Interquartile range, IQR	Population standard deviation, σ	Sample standard deviation, s	
Matrix of scatter plots			

Random Variables and Probability Distributions

LEARNING OBJECTIVES

After careful study of this chapter, you should be able to do the following:

1. Determine probabilities for discrete random variables from probability mass functions and for continuous random variables from probability density functions, and use cumulative distribution functions in both cases.

2. Calculate means and variances for discrete and continuous random variables.

3. Understand the assumptions for each of the probability distributions presented.

4. Select an appropriate probability distribution to calculate probabilities in specific applications.

5. Standardize normal random variables.

6. Use the table (or software) for the cumulative distribution function of a standard normal distribution to calculate probabilities.

7. Approximate probabilities for some binomial and Poisson distributions.

8. Interpret and calculate covariances and correlations between random variables.

9. Calculate means and variances for linear combinations of random variables.

10. Approximate means and variances for general functions of several random variables.

11. Understand statistics and the central limit theorem.

Earlier in this book, numerical and graphical summaries were used to summarize data. A summary is often needed to transform the data to useful information. Furthermore, conclusions about the process that generated the data are often important; that is, we might want to draw conclusions about the long-term performance of a process based on only a relatively small sample of data. Because only a sample of data is used, there is some uncertainty in our conclusions. However, the amount of uncertainty can be quantified and sample sizes can be selected or modified to achieve a tolerable level of uncertainty if a probability model is specified for the data. The objective of this chapter is to describe these models and present some important examples.

3-1   INTRODUCTION

The measurement of current in a thin copper wire is an example of an **experiment.** However, the results might differ slightly in day-to-day replicates of the measurement because of small variations in variables that are not controlled in our experiment—changes in ambient temperatures, slight variations in gauge and small impurities in the chemical composition of the wire if different locations are selected, current source drifts, and so forth. Consequently, this experiment (as well as many we conduct) can be considered to have a **random** component. In some cases, the random variations that we experience are small enough, relative to our experimental goals, that they can be ignored. However, the variation is almost always present and its magnitude can be large enough that the important conclusions from the experiment are not obvious. In these cases, the methods presented in this book for modeling and analyzing experimental results are quite valuable.

An experiment that can result in different outcomes, even though it is repeated in the same manner every time, is called a **random experiment.** We might select one part from a day's production and very accurately measure a dimensional length. Although we hope that the manufacturing operation produces identical parts consistently, in practice there are often small variations

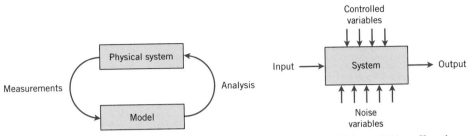

Figure 3-1   Continuous iteration between model and physical system.

Figure 3-2   Noise variables affect the transformation of inputs to outputs.

in the actual measured lengths due to many causes—vibrations, temperature fluctuations, operator differences, calibrations of equipment and gauges, cutting tool wear, bearing wear, and raw material changes. Even the measurement procedure can produce variations in the final results.

No matter how carefully our experiment is designed and conducted, variations often occur. Our goal is to understand, quantify, and model the type of variations that we often encounter. When we incorporate the variation into our thinking and analyses, we can make informed judgments from our results that are not invalidated by the variation.

Models and analyses that include variation are not different from models used in other areas of engineering and science. Figure 3-1 displays the relationship between the model and the physical system it represents. A mathematical model (or abstraction) of a physical system need not be a perfect abstraction. For example, Newton's laws are not perfect descriptions of our physical universe. Still, these are useful models that can be studied and analyzed to quantify approximately the performance of a wide range of engineered products. Given a mathematical abstraction that is validated with measurements from our system, we can use the model to understand, describe, and quantify important aspects of the physical system and predict the response of the system to inputs.

Throughout this text, we discuss models that allow for variations in the outputs of a system, even though the variables that we control are not purposely changed during our study. Figure 3-2 graphically displays a model that incorporates uncontrollable variables (noise) that combine with the controllable variables to produce the output of our system. Because of the noise, the same settings for the controllable variables do not result in identical outputs every time the system is measured.

For the example of measuring current in a copper wire, our model for the system might simply be Ohm's law,

$$\text{Current} = \text{voltage}/\text{resistance}$$

As described previously, variations in current measurements are expected. Ohm's law might be a suitable approximation. However, if the variations are large relative to the intended use of the device under study, we might need to extend our model to include the variation. See Fig. 3-3. It is often difficult to speculate on the magnitude of the variations without empirical measurements. With sufficient measurements, however, we can approximate the magnitude of the variation and consider its effect on the performance of other devices, such as amplifiers, in the circuit. We are therefore sanctioning the model in Fig. 3-2 as a more useful description of the current measurement.

As another example, in the design of a communication system, such as a computer network or a voice communication network, the information capacity available to service individuals using the network is an important design consideration. For voice communication, sufficient external lines need to be purchased to meet the requirements of a business. Assuming each line can carry

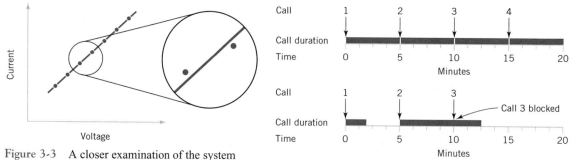

Figure 3-3   A closer examination of the system
identifies deviations from the model.

Figure 3-4   Variation causes disruptions in the system.

only a single conversation, how many lines should be purchased? If too few lines are purchased, calls can be delayed or lost. The purchase of too many lines increases costs. Increasingly, design and product development are required to meet customer requirements *at a competitive cost*.

In the design of the voice communication system, a model is needed for the number of calls and the duration of calls. Even knowing that, on average, calls occur every 5 minutes and that they last 5 minutes is not sufficient. If calls arrived precisely at 5-minute intervals and lasted for precisely 5 minutes, one phone line would be sufficient. However, the slightest variation in call number or duration would result in some calls being blocked by others. See Fig. 3-4. A system designed without considering variation will be woefully inadequate for practical use.

3-2   RANDOM VARIABLES

In an experiment, a measurement is usually denoted by a variable such as X. In a random experiment, a variable whose measured value can change (from one replicate of the experiment to another) is referred to as a **random variable.** For example, X might denote the current measurement in the copper wire experiment. A random variable is conceptually no different from any other variable in an experiment. We use the term "random" to indicate that noise disturbances can change its measured value. An **uppercase letter** is used to denote a random variable.

Random Variable

> A **random variable** is a numerical variable whose measured value can change from one replicate of the experiment to another.

After the experiment is conducted, the measured value of the random variable is denoted by a **lowercase letter** such as $x = 70$ milliamperes. We often summarize a random experiment by the measured value of a random variable.

This model can be linked to data as follows. The data are the measured values of a random variable obtained from replicates of a random experiment. For example, the first replicate might result in a current measurement of $x_1 = 70.1$, the next day $x_2 = 71.2$, the third day $x_3 = 71.1$, and so forth. These data can then be summarized by the descriptive methods discussed in Chapter 2.

Often, the measurement of interest—current in a copper wire experiment, length of a machined part—is assumed to be a real number. Then arbitrary precision in the measurement is possible. Of course, in practice, we might round off to the nearest tenth or hundredth of a unit. The random variable that represents this measurement is said to be a **continuous random variable.**

In other experiments, we might record a count such as the number of transmitted bits that are received in error. Then the measurement is limited to integers. Or we might record that a proportion such as 0.0042 of 10,000 transmitted bits were received in error. Then the measurement is fractional, but it is still limited to discrete points on the real line. Whenever the measurement is limited to discrete points on the real line, the random variable is said to be a **discrete random variable.**

Discrete and Continuous Random Variables

> A **discrete** random variable is a random variable with a finite (or countably infinite) set of real numbers for its range.
> A **continuous** random variable is a random variable with an interval (either finite or infinite) of real numbers for its range.

In some cases, the random variable X is actually discrete but, because the range of possible values is so large, it might be more convenient to analyze X as a continuous random variable. For example, suppose that current measurements are read from a digital instrument that displays the current to the nearest one-hundredth of a milliampere. Because the possible measurements are limited, the random variable is discrete. However, it might be a more convenient, simple approximation to assume that the current measurements are values of a continuous random variable.

Examples of Random Variables

> Examples of **continuous** random variables:
>   electrical current, length, pressure, temperature, time, voltage, weight
>
> Examples of **discrete** random variables:
>   number of scratches on a surface, proportion of defective parts among 1,000 tested, number of transmitted bits received in error

EXERCISES FOR SECTION 3-2

Decide whether a discrete or continuous random variable is the best model for each of the following variables.

3-1. The lifetime of a biomedical device after implant in a patient.

3-2. The number of times a transistor in a computer memory changes state in one operation.

3-3. The strength of a concrete specimen.

3-4. The number of convenience options selected by an automobile buyer.

3-5. The proportion of defective solder joints on a circuit board.

3-6. The weight of an injection-molded plastic part.

3-7. The number of molecules in a sample of gas.

3-8. The energy generated from a reaction.

3-9. The concentration of organic solids in a water sample.

3-3   PROBABILITY

A random variable is used to describe a measurement. Probability is used to quantify the likelihood, or chance, that a measurement falls within some set of values. "The chance that X, the length of a manufactured part, is between 10.8 and 11.2 millimeters is 25%" is a statement that quantifies our feeling about the possibility of part lengths. Probability statements describe the likelihood that particular values occur. The likelihood is quantified by assigning a number from the interval [0, 1] to the set of values of the random variable. Higher numbers indicate that the set of values is more likely.

The probability of a result can be interpreted as our subjective probability, or **degree of belief,** that the result will occur. Different individuals will no doubt assign different probabilities to the same result. Another interpretation of probability can be based on repeated replicates of the random experiment. The probability of a result is interpreted as the proportion of times the result will occur in repeated replicates of the random experiment. For example, if we assign probability 0.25 to the result that a part length is between 10.8 and 11.2 millimeters, we might interpret this assignment as follows. If we repeatedly manufacture parts (replicate the random experiment an infinite number of times), 25% of them will have lengths in this interval. This example provides a **relative frequency** interpretation of probability. The proportion, or relative frequency, of repeated replicates that fall in the interval will be 0.25. Note that this interpretation uses a long-run proportion, the proportion from an infinite number of replicates. With a small number of replicates, the proportion of lengths that actually fall in the interval might differ from 0.25.

To continue, if every manufactured part length will fall in the interval, the relative frequency, and therefore the probability, of the interval is 1. If no manufactured part length will fall in the interval, the relative frequency, and therefore the probability, of the interval is 0. Because probabilities are restricted to the interval [0, 1], they can be interpreted as relative frequencies.

A probability is usually expressed in terms of a random variable. For the part length example, X denotes the part length and the probability statement can be written in either of the following forms:

$$P(X \in [10.8, 11.2]) = 0.25 \qquad \text{or} \qquad P(10.8 \le X \le 11.2) = 0.25$$

Both equations state that the probability that the random variable X assumes a value in [10.8, 11.2] is 0.25.

Probabilities for a random variable are usually determined from a model that describes the random experiment. Several models will be considered in the following sections. Before that, several general probability properties are stated that can be understood from the relative frequency interpretation of probability. The properties do *not* determine probabilities; probabilities are assigned based on our knowledge of the system under study. However, the properties enable us to easily calculate some probabilities from knowledge of others.

The following terms are used. Given a set E, the complement of E is the set of elements that are not in E. The **complement** is denoted as E'. The set of real numbers is denoted as R. The sets $E_1, E_2, \ldots, E_k$ are **mutually exclusive** if the intersection of any pair is empty. That is, each element is in one and only one of the sets $E_1, E_2, \ldots, E_k$.

Probability Properties

1. $P(X \in R) = 1$, where R is the set of real numbers.
2. $0 \le P(X \in E) \le 1$ for any set E.   (3-1)
3. If $E_1, E_2, \ldots, E_k$ are mutually exclusive sets,
   $P(X \in E_1 \cup E_2 \cup \ldots \cup E_k) = P(X \in E_1) + \cdots + P(X \in E_k)$.

Property 1 can be used to show that the maximum value for a probability is 1. Property 2 implies that a probability can't be negative. Property 3 states that the proportion of measurements that fall in $E_1 \cup E_2 \cup \cdots \cup E_k$ is the sum of the proportions that fall in E_1 and $E_2, \ldots,$ and E_k, whenever the sets are mutually exclusive. For example,

$$P(X \le 10) = P(X \le 0) + P(0 < X \le 5) + P(5 < X \le 10)$$

Property 3 is also used to relate the probability of a set E and its complement E'. Because E and E' are mutually exclusive and $E \cup E' = R$, $1 = P(X \in R) = P(X \in E \cup E') = P(X \in E) + P(X \in E')$. Consequently,

$$P(X \in E') = 1 - P(X \in E)$$

For example, $P(X \le 2) = 1 - P(X > 2)$. In general, for any fixed number x,

$$P(X \le x) = 1 - P(X > x)$$

Let $\varnothing$ denote the null set. Because the complement of R is $\varnothing$, $P(X \in \varnothing) = 0$.

Applying Probability Properties

Assume that the following probabilities apply to the random variable X that denotes the life in hours of standard fluorescent tubes: $P(X \le 5000) = 0.1$, $P(5000 < X \le 6000) = 0.3$, $P(X > 8000) = 0.4$. The following results can be determined from the probability properties. It may be helpful to graphically display the different sets.

The probability that the life is less than or equal to 6000 hours is

$$P(X \le 6000) = P(X \le 5000) + P(5000 < X \le 6000) = 0.1 + 0.3$$
$$= 0.4$$

from Property 3. The probability that the life exceeds 6000 hours is

$$P(X > 6000) = 1 - P(X \le 6000) = 1 - 0.4 = 0.6$$

The probability that the life is greater than 6000 and less than or equal to 8000 hours is determined from the fact that the sum of the probabilities for this interval and the other three intervals must equal 1. That is, the union of other three intervals is the complement of the set $\{x \mid 6000 < x \le 8000\}$. Therefore,

$$P(6000 < X \le 8000) = 1 - (0.1 + 0.3 + 0.4) = 0.2$$

The probability that the life is less than or equal to 5500 hours cannot be determined exactly. The best we can state is that

$$P(X \le 5500) \le P(X \le 6000) = 0.4 \quad \text{and} \quad 0.1 = P(X \le 5000) \le P(X \le 5500)$$

If it were also known that $P(5500 < X \le 6000) = 0.15$, we could state that

$$P(X \le 5500) = P(X \le 5000) + P(5000 < X \le 6000) - P(5500 < X \le 6000)$$
$$= 0.1 + 0.3 - 0.15 = 0.25$$

Events

A measured value is not always obtained from an experiment. Sometimes, the result is only classified (into one of several possible categories). For example, the current measurement might only be recorded as *low, medium,* or *high*; a manufactured electronic component might be classified only as defective or not; and a bit transmitted through a digital communication channel is received either in error or not. The possible categories are usually referred to as outcomes, and a set of one or more outcomes is called an **event.** The concept of probability can be applied to events and the relative frequency interpretation is still appropriate.

If 1% of the bits transmitted through a digital communications channel are received in error, the probability of an error would be assigned 0.01. If we let E denote the event that a bit is received in error, we would write

$$P(E) = 0.01$$

Probabilities assigned to events satisfy properties analogous to those in equation 3-1 so that they can be interpreted as relative frequencies. Let Ω denote the set of all possible outcomes from the experiment. Then,

1. $P(\Omega) = 1$.
2. $0 \le P(E) \le 1$ for any event E.
3. If $E_1, E_2, \ldots, E_k$ are mutually exclusive events,

$$P(E_1 \cup E_2 \cdots \cup E_k) = P(E_1 \text{ or } E_2 \cdots \text{ or } E_k) = P(E_1) + P(E_2) + \cdots + P(E_k).$$

The events, $E_1, E_2, \ldots, E_k$, are mutually exclusive when the intersection of each pair is null. Suppose that the probability of *low, medium,* and *high* results are 0.1, 0.7, and 0.2, respectively, and that the events are mutually exclusive. The probability of a *medium* or *high* result is denoted as $P(medium \text{ or } high)$ and

$$P(medium \text{ or } high) = P(medium) + P(high) = 0.7 + 0.2 = 0.9$$

EXERCISES FOR SECTION 3-3

3-10. State the complement of each of the following sets:
(a) Engineers with less than 36 months of full-time employment.
(b) Samples of cement blocks with compressive strength less than 6000 kilograms per square centimeter.
(c) Measurements of the diameter of forged pistons that do not conform to engineering specifications.
(d) Cholesterol levels that measure greater than 180 and less than 220.

3-11. If $P(X \in A) = 0.4$, and $P(X \in B) = 0.6$ and the intersection of sets A and B is empty,
(a) Are sets A and B mutually exclusive?
(b) Find $P(X \in A')$.
(c) Find $P(X \in B')$.
(d) Find $P(X \in A \cup B)$.

3-12. If $P(X \in A) = 0.3$, $P(X \in B) = 0.25$, $P(X \in C) = 0.60$, $P(X \in A \cup B) = 0.55$, and $P(X \in B \cup C) = 0.70$, determine the following probabilities.
(a) $P(X \in A')$ (b) $P(X \in B')$ (c) $P(X \in C')$
(d) Are A and B mutually exclusive?
(e) Are B and C mutually exclusive?

3-13. Let $P(X \le 15) = 0.3$, $P(15 < X \le 24) = 0.6$, and $P(X > 20) = 0.5$.
(a) Find $P(X > 15)$.
(b) Find $P(X \le 24)$.
(c) Find $P(15 < X \le 20)$.
(d) If $P(18 < X \le 24) = 0.4$, find $P(X \le 18)$.

3-14. Suppose that an ink cartridge is classified as being overfilled, medium filled, or underfilled with a probability of 0.40, 0.45, and 0.15, respectively.
(a) What is the probability that a cartridge is classified as not underfilled?
(b) What is the probability that a cartridge is either overfilled or underfilled?

3-15. Let X denote the life of a semiconductor laser (in hours) with the following probabilities: $P(X \le 5000) = 0.05$ and $P(X > 7000) = 0.45$.
(a) What is the probability that the life is less than or equal to 7000 hours?
(b) What is the probability that the life is greater than 5000 hours?
(c) What is $P(5000 < X \le 7000)$?

3-16. Let E_1 denote the event that a structural component fails during a test and E_2 denote the event that the component shows some strain, but does not fail. Given $P(E_1) = 0.15$ and $P(E_2) = 0.30$,

(a) What is the probability that a structural component does not fail during a test?

(b) What is the probability that a component either fails or shows strain during a test?

(c) What is the probability that a component neither fails nor shows strain during a test?

3-17. Let X denote the number of bars of service on your cell phone whenever you are at an intersection with the following probabilities:

x	0	1	2	3	4	5
$P(X = x)$	0.1	0.15	0.25	0.25	0.15	0.1

Determine the following probabilities:

(a) Two or three bars

(b) Fewer than two bars

(c) More than three bars

(d) At least one bar

3-18. Let X denote the number of patients who suffer an infection within a floor of a hospital per month with the following probabilities:

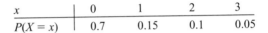

x	0	1	2	3
$P(X = x)$	0.7	0.15	0.1	0.05

Determine the following probabilities:

(a) Less than one infection

(b) More than three infections

(c) At least one infection

(d) No infections

3-4   CONTINUOUS RANDOM VARIABLES

3-4.1   Probability Density Function

The **probability distribution** or simply **distribution** of a random variable X is a description of the set of the probabilities associated with the possible values for X. The probability distribution of a random variable can be specified in more than one way.

Density functions are commonly used in engineering to describe physical systems. For example, consider the density of a loading on a long, thin beam as shown in Fig. 3-5. For any point x along the beam, the density can be described by a function (in grams/cm). Intervals with large loadings correspond to large values for the function. The total loading between points a and b is determined as the integral of the density function from a to b. This integral is the area under the density function over this interval, and it can be loosely interpreted as the sum of all the loadings over this interval.

Similarly, a **probability density function** $f(x)$ can be used to describe the probability distribution of a continuous random variable X. The probability that X is between a and b is determined as the integral of $f(x)$ from a to b. See Fig. 3-6. The notation follows.

Probability Density Function

The **probability density function** (or pdf) $f(x)$ of a continuous random variable is used to determine probabilities from areas as follows:

$$P(a < X < b) = \int_a^b f(x)\, dx \qquad (3\text{-}2)$$

The properties of the pdf are

(1) $f(x) \geq 0$

(2) $\int_{-\infty}^{\infty} f(x) = 1$

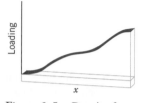

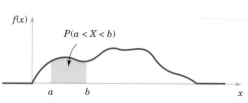

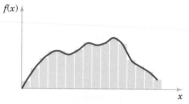

Figure 3-5   Density function of a loading on a long,
thin beam.

Figure 3-6   Probability determined from the area
under $f(x)$.

Figure 3-7   A histogram approximates a
probability density function. The area of
each bar equals the relative frequency of
the interval. The area under $f(x)$ over any interval equals the probability of the interval.

A histogram is an approximation to a pdf. See Fig. 3-7. For each interval of the histogram,
the area of the bar equals the relative frequency (proportion) of the measurements in the interval.
The relative frequency is an estimate of the probability that a measurement falls in the interval.
Similarly, the area under $f(x)$ over any interval equals the true probability that a measurement
falls in the interval.

A probability density function provides a simple description of the probabilities associated with a random variable. As long as $f(x)$ is nonnegative and $\int_{-\infty}^{\infty} f(x)\,dx = 1$,
$0 \le P(a < X < b) \le 1$, so the probabilities are properly restricted. A pdf is zero for x values
that cannot occur, and it is assumed to be zero wherever it is not specifically defined.

The important point is that $\boldsymbol{f(x)}$ **is used to calculate an area** that represents the probability that X assumes a value in $[a, b]$. For the current measurements of Section 3-1, the probability that X results in [14 mA, 15 mA] is the integral of the probability density function of X,
$f(x)$, over this interval. The probability that X results in [14.5 mA, 14.6 mA] is the integral of
the same function, $f(x)$, over the smaller interval. By appropriate choice of the shape of $f(x)$,
we can represent the probabilities associated with any random variable X. The shape of $f(x)$ determines how the probability that X assumes a value in [14.5 mA, 14.6 mA] compares to the
probability of any other interval of equal or different length.

For the density function of a loading on a long, thin beam, because every point has zero
width, the integral that determines the loading at any point is zero. Similarly, for a continuous
random variable X and *any* value x,

$$P(X = x) = 0$$

Based on this result, it might appear that our model of a continuous random variable is useless.
However, in practice, when a particular current measurement is observed, such as 14.47 milliamperes, this result can be interpreted as the rounded value of a current measurement that is
actually in a range such as $14.465 \le x \le 14.475$. Therefore, the probability that the rounded
value 14.47 is observed as the value for X is the probability that X assumes a value in the
interval [14.465, 14.475], which is not zero. Similarly, our model of a continuous random
variable implies the following.

If X is a continuous random variable, for any x_1 and x_2,

$$P(x_1 \le X \le x_2) = P(x_1 < X \le x_2) = P(x_1 \le X < x_2) = P(x_1 < X < x_2)$$

To determine a probability for a random variable, it can be helpful to apply three steps:

1. Determine the random variable and distribution of the random variable.
2. Write the probability statement in terms of the random variable.
3. Compute the probability using the probability statement and the distribution.

These steps are shown in the solutions of some examples in this chapter. In other examples and exercises you might use these steps on your own.

EXAMPLE 3-1

Current in a Wire

Let the continuous random variable X denote the current measured in a thin copper wire in milliamperes. Assume that the range of X is [0, 20 mA], and assume that the probability density function of X is $f(x)$ = 0.05 for $0 \leq x \leq 20$. What is the probability that a current measurement is less than 10 milliamperes?

Define the random variable and distribution.

Solution. The random variable is the current measurement with distribution given by $f(x)$. The pdf is shown in Fig. 3-8. It is assumed that $f(x) = 0$ wherever it is not specifically defined. The probability requested is indicated by the shaded area in Fig. 3-8.

Write the probability statement.

$$P(X < 10) = \int_0^{10} f(x)\, dx = 0.5$$

Compute the probability.

As another example,

$$P(5 < X < 10) = \int_5^{10} f(x)\, dx = 0.25$$

∎

EXAMPLE 3-2

Flaw on a Magnetic Disk

Let the continuous random variable X denote the distance in micrometers from the start of a track on a magnetic disk until the first flaw. Historical data show that the distribution of X can be modeled by a pdf $f(x) = \dfrac{1}{2000} e^{-x/2000}, x \geq 0$. For what proportion of disks is the distance to the first flaw greater than 1000 micrometers?

Solution. The density function and the requested probability are shown in Fig. 3-9. Now,

$$P(X > 1000) = \int_{1000}^{\infty} f(x)\, dx = \int_{1000}^{\infty} \frac{e^{-x/2000}}{2000}\, dx = -e^{-x/2000}\Big|_{1000}^{\infty} = e^{-1/2} = 0.607$$

What proportion of parts is between 1000 and 2000 micrometers?

Solution. Now,

$$P(1000 < X < 2000) = \int_{1000}^{2000} f(x)\, dx = -e^{-x/2000}\Big|_{1000}^{2000} = e^{-1/2} - e^{-1} = 0.239$$

Because the total area under $f(x)$ equals 1, we can also calculate $P(X < 1000) = 1 - P(X > 1000) = 1 - 0.607 = 0.393$. ∎

3-4.2 Cumulative Distribution Function

Another way to describe the probability distribution of a random variable is with a function of a real number x that provides the probability that X is less than or equal to x.

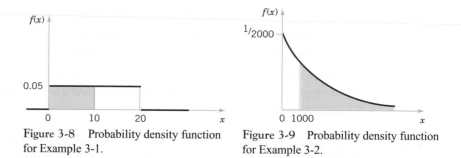

Figure 3-8   Probability density function for Example 3-1.

Figure 3-9   Probability density function for Example 3-2.

Cumulative Distribution Function of a Continuous Random Variable

The **cumulative distribution function** (or cdf) of a continuous random variable X with probability density function $f(x)$ is

$$F(x) = P(X \le x) = \int_{-\infty}^{x} f(u)\, du$$

for $-\infty < x < \infty$.

For a continuous random variable X, the definition can also be $F(x) = P(X < x)$ because $P(X = x) = 0$.

The cumulative distribution function $F(x)$ can be related to the probability density function $f(x)$ and can be used to obtain probabilities as follows.

$$P(a < X < b) = \int_{a}^{b} f(x)\, dx = \int_{-\infty}^{b} f(x)\, dx - \int_{-\infty}^{a} f(x)\, dx = F(b) - F(a)$$

Furthermore, the graph of a cdf has specific properties. Because $F(x)$ provides probabilities, it is always nonnegative. Furthermore, as x increases, $F(x)$ is nondecreasing. Finally, as x tends to ∞, $F(x) = P(X \le x)$ tends to 1.

EXAMPLE 3-3
Flaw on a Magnetic Disk Distribution Function

Define the random variable and distribution.

Write the probability statement.

Compute the probability.

The distance in micrometers from the start of a track on a magnetic disk until the first surface flaw is a random variable with the cdf

$$F(x) = 1 - \exp\left(-\frac{x}{2000}\right) \text{ for } x > 0$$

A graph of $F(x)$ is shown in Fig. 3-10. Note that $F(x) = 0$ for $x \le 0$. Also, $F(x)$ increases to 1 as mentioned. The following probabilities should be compared to the results in Example 3-2. Determine the probability that the distance until the first surface flaw is less than 1000 micrometers.

Solution. The random variable is the distance until the first surface flaw with distribution given by $F(x)$. The requested probability is

$$P(X < 1000) = F(1000) = 1 - \exp\left(-\frac{1}{2}\right) = 0.393$$

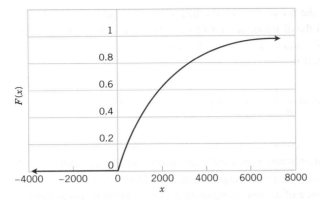

Figure 3-10 Cumulative distribution function for Example 3-3.

Determine the probability that the distance until the first surface flaw exceeds 2000 micrometers.

Solution. Now we use

$$P(2000 < X) = 1 - P(X \leq 2000) = 1 - F(2000) = 1 - [1 - \exp(-1)]$$
$$= \exp(-1) = 0.368$$

Determine the probability that the distance is between 1000 and 2000 micrometers.

Solution. The requested probability is

$$P(1000 < X < 2000) = F(2000) - F(1000) = 1 - \exp(-1) - [1 - \exp(-0.5)]$$
$$= \exp(-0.5) - \exp(-1) = 0.239$$

The cumulative distribution function is often tabulated to present probabilities. It is convenient to list $F(x)$ for selected values of x. Then additional probabilities can be determined as in the previous example.

3-4.3 Mean and Variance

Just as it is useful to summarize a sample of data by the mean and variance, we can summarize the probability distribution of X by its mean and variance. For sample data $x_1, x_2, \ldots, x_n$, the sample mean can be written as

$$\bar{x} = \frac{1}{n} x_1 + \frac{1}{n} x_2 + \cdots + \frac{1}{n} x_n$$

That is, $\bar{x}$ uses equal weights of $1/n$ as the multiplier of each measured value x_i. The mean of a random variable X uses the probability model to weight the possible values of X. The **mean** or **expected value** of X, denoted as μ or $E(X)$, is

$$\mu = E(X) = \int_{-\infty}^{\infty} xf(x) \, dx$$

The integral in $E(X)$ is analogous to the sum that is used to calculate $\bar{x}$.

Recall that $\bar{x}$ is the balance point when an equal weight is placed at the location of each measurement along a number line. Similarly, if $f(x)$ is the density function of a loading on a long, thin beam, $E(X)$ is the point at which the beam balances. See Fig. 3-5. Consequently, $E(X)$ describes the "center" of the distribution of X in a manner similar to the balance point of a loading.

For sample data $x_1, x_2, \ldots, x_n$, the sample variance is a summary of the dispersion or scatter in the data. It is

$$s^2 = \frac{1}{n-1}(x_1 - \bar{x})^2 + \frac{1}{n-1}(x_2 - \bar{x})^2 + \cdots + \frac{1}{n-1}(x_n - \bar{x})^2$$

That is, s^2 uses equal weights of $1/(n-1)$ as the multiplier of each squared deviation $(x_i - \bar{x})^2$. As mentioned previously, deviations calculated from $\bar{x}$ tend to be smaller than those calculated from μ, and the weight is adjusted from $1/n$ to $1/(n-1)$ to compensate.

The variance of a random variable X is a measure of dispersion or scatter in the possible values for X. The **variance** of X, denoted as σ^2 or $V(X)$, is

$$\sigma^2 = V(X) = \int_{-\infty}^{\infty} (x - \mu)^2 f(x)\, dx$$

$V(X)$ uses weight $f(x)$ as the multiplier of each possible squared deviation $(x - \mu)^2$. The integral in $V(X)$ is analogous to the sum that is used to calculate s^2.

Properties of integrals and the definition of μ can be used to show that

$$\begin{aligned}
V(X) &= \int_{-\infty}^{\infty} (x - \mu)^2 f(x)\, dx \\
&= \int_{-\infty}^{\infty} x^2 f(x)\, dx - 2\mu \int_{-\infty}^{\infty} x f(x)\, dx + \int_{-\infty}^{\infty} \mu^2 f(x)\, dx \\
&= \int_{-\infty}^{\infty} x^2 f(x)\, dx - 2\mu^2 + \mu^2 = \int_{-\infty}^{\infty} x^2 f(x)\, dx - \mu^2
\end{aligned}$$

so an alternative formula for $V(x)$ can be used.

Mean and Variance of a Continuous Random Variable

Suppose X is a continuous random variable with pdf $f(x)$. The **mean** or **expected value** of X, denoted as μ or $E(X)$, is

$$\mu = E(X) = \int_{-\infty}^{\infty} x f(x)\, dx \qquad (3\text{-}3)$$

The **variance** of X, denoted as $V(X)$ or σ^2, is

$$\sigma^2 = V(X) = \int_{-\infty}^{\infty} (x - \mu)^2 f(x)\, dx = \int_{-\infty}^{\infty} x^2 f(x)\, dx - \mu^2$$

The **standard deviation** of X is σ.

EXAMPLE 3-4

Current in a Wire: Mean

For the copper current measurement in Example 3-1, the mean of X is

$$E(x) = \int_{-\infty}^{\infty} xf(x)\,dx = \int_{0}^{20} x\left(\frac{1}{20}\right)dx = 0.05x^2/2 \Big|_{0}^{20} = 10$$

The variance of X is

$$V(x) = \int_{-\infty}^{\infty} (x - \mu)^2 f(x)\,dx = \int_{0}^{20} (x - 10)^2\left(\frac{1}{20}\right)dx = 0.05(x - 10)^3/3 \Big|_{0}^{20} = 33.33$$ ■

EXAMPLE 3-5

Current in a Wire: Variance

For the distance to a flaw in Example 3-2, the mean of X is

$$E(x) = \int_{0}^{\infty} xf(x)\,dx = \int_{0}^{\infty} x\frac{e^{-x/2000}}{2000}\,dx$$

A table of integrals or integration by parts can be used to show that

$$E(x) = -xe^{-x/2000}\Big|_{0}^{\infty} + \int_{0}^{\infty} e^{-x/2000}\,dx = 0 - 2000\,e^{-x/2000}\Big|_{0}^{\infty} = 2000$$

The variance of X is

$$V(X) = \int_{-\infty}^{\infty} (x - \mu)^2 f(x)\,dx = \int_{0}^{\infty} (x - 2000)^2\frac{e^{-x/2000}}{2000}\,dx$$

A table of integrals or integration by parts can be applied twice to show that

$$V(X) = 2000^2 = 4{,}000{,}000$$ ■

EXERCISES FOR SECTION 3-4

3-19. Show that the following functions are probability density functions for some value of k and determine k. Then, determine the mean and variance of X.

(a) $f(x) = kx^2$ for $0 < x < 4$
(b) $f(x) = k(1 + 2x)$ for $0 < x < 2$
(c) $f(x) = ke^{-x}$ for $0 < x$
(d) $f(x) = k$ where $k > 0$ and $100 < x < 100 + k$

3-20. For each of the density functions in Exercise 3-19, perform the following.

(a) Graph the density function and mark the location of the mean on the graph.
(b) Find the cumulative distribution function.
(c) Graph the cumulative distribution function.

3-21. Suppose that $f(x) = e^{-(x-6)}$ for $6 < x$ and $f(x) = 0$ for $x \le 6$. Determine the following probabilities.

(a) $P(X > 6)$ (b) $P(6 \le X < 8)$
(c) $P(X < 8)$ (d) $P(X > 8)$
(e) Determine x such that $P(X < x) = 0.95$.

3-22. Suppose that $f(x) = 1.5x^2$ for $-1 < x < 1$ and $f(x) = 0$ otherwise. Determine the following probabilities.

(a) $P(0 < X)$ (b) $P(0.5 < X)$
(c) $P(-0.5 \le X \le 0.5)$ (d) $P(X < -2)$
(e) $P(X < 0 \text{ or } X > -0.5)$
(f) Determine x such that $P(x < X) = 0.05$.

3-23. The pdf of the time to failure of an electronic component in a copier (in hours) is $f(x) = [\exp(-x/3000)]/3000$ for $x > 0$ and $f(x) = 0$ for $x \le 0$. Determine the probability that

(a) A component lasts more than 1000 hours before failure.
(b) A component fails in the interval from 1000 to 2000 hours.

(c) A component fails before 3000 hours.
(d) Determine the number of hours at which 10% of all components have failed.
(e) Determine the mean.

3-24. The probability density function of the net weight in ounces of a packaged compound is $f(x) = 2.0$ for $19.75 < x < 20.25$ ounces and $f(x) = 0$ for x elsewhere.

(a) Determine the probability that a package weighs less than 20 ounces.
(b) Suppose that the packaging specifications require that the weight be between 19.9 and 20.1 ounces. What is the probability that a randomly selected package will have a weight within these specifications?
(c) Determine the mean and variance.
(d) Find and graph the cumulative distribution function.

3-25. The temperature readings from a thermocouple in a furnace fluctuate according to a cumulative distribution function

$$F(x) = \begin{cases} 0 & x < 800°C \\ 0.1x - 80 & 800°C \leq x < 810°C \\ 1 & x > 810°C \end{cases}$$

Determine the following.

(a) $P(X < 805)$ (b) $P(800 < X \leq 805)$ (c) $P(X > 808)$
(d) If the specifications for the process require that the furnace temperature be between 802 and 808°C, what is the probability that the furnace will operate outside of the specifications?

3-26. The thickness measurement of a wall of plastic tubing, in millimeters, varies according to a cumulative distribution function

$$F(x) = \begin{cases} 0 & x < 2.0050 \\ 200x - 401 & 2.0050 \leq x \leq 2.0100 \\ 1 & x > 2.0100 \end{cases}$$

Determine the following.

(a) $P(X \leq 2.0080)$ (b) $P(X > 2.0055)$
(c) If the specification for the tubing requires that the thickness measurement be between 2.0090 and 2.0100 millimeters, what is the probability that a single measurement will indicate conformance to the specification?

3-27. Suppose that contamination particle size (in micrometers) can be modeled as $f(x) = 2x^{-3}$ for $1 < x$ and $f(x) = 0$ for $x \leq 1$.

(a) Confirm that $f(x)$ is a probability density function.
(b) Give the cumulative distribution function.
(c) Determine the mean.
(d) What is the probability that the size of a random particle will be less than 5 micrometers?
(e) An optical device is being marketed to detect contamination particles. It is capable of detecting particles exceeding 7 micrometers in size. What proportion of the particles will be detected?

3-28. (Integration by parts is required in this exercise.) The probability density function for the diameter of a drilled hole in millimeters is $10e^{-10(x-5)}$ for $x > 5$ mm and zero for $x \leq 5$ mm. Although the target diameter is 5 millimeters, vibrations, tool wear, and other factors can produce diameters larger than 5 millimeters.

(a) Determine the mean and variance of the diameter of the holes.
(b) Determine the probability that a diameter exceeds 5.1 millimeters.

3-29. Suppose the cumulative distribution function of the length (in millimeters) of computer cables is

$$F(x) = \begin{cases} 0 & x \leq 1200 \\ 0.1x - 120 & 1200 < x \leq 1210 \\ 1 & x > 1210 \end{cases}$$

(a) Determine $P(x < 1208)$.
(b) If the length specifications are $1195 < x < 1205$ millimeters, what is the probability that a randomly selected computer cable will meet the specification requirement?

3-30. The thickness of a conductive coating in micrometers has a density function of $600x^{-2}$ for $100\ \mu m < x < 120\ \mu m$ and zero for x elsewhere.

(a) Determine the mean and variance of the coating thickness.
(b) If the coating costs $0.50 per micrometer of thickness on each part, what is the average cost of the coating per part?

3-31. A medical linear accelerator is used to accelerate electrons to create high-energy beams that can destroy tumors with minimal impact on surrounding healthy tissue. The beam energy fluctuates between 200 and 210 MeV (million electron volts). Given the cumulative distribution function

$$F(x) = \begin{cases} 0 & x < 200 \\ 0.1x - 20 & 200 \leq x \leq 210 \\ 1 & x > 210 \end{cases}$$

Determine the following.

(a) $P(X < 209)$ (b) $P(200 < X < 208)$ (c) $P(X > 209)$
(d) What is the probability density function?
(e) Graph the probability density function and the cumulative distribution function.
(f) Determine the mean and variance of the beam energy.

3-32. The probability density function of the time a customer arrives at a terminal (in minutes after 8:00 A.M.) is $f(x) = 0.1e^{-x/10}$ for $x > 0$. Determine the probability that

(a) The customer arrives by 9:00 A.M.
(b) The customer arrives between 8:15 A.M. and 8:30 A.M.

(c) Determine the time at which the probability of an earlier arrival is 0.5.

(d) Determine the cumulative distribution function and use the cumulative distribution function to determine the probability that the customer arrives between 8:15 A.M. and 8:30 A.M.

(e) Determine the mean and standard deviation of the number of minutes until the customer arrives.

3-33. The probability density function of the weight of packages delivered by a post office is $f(x) = 70/(69x^2)$ for $1 < x < 70$ pounds.

(a) What is the probability a package weighs less than 10 pounds?

(b) Determine the mean and variance of package weight.

(c) If the shipping cost is \$3 per pound, what is the average shipping cost of a package?

3-5 IMPORTANT CONTINUOUS DISTRIBUTIONS

3-5.1 Normal Distribution

Undoubtedly, the most widely used model for the distribution of a random variable is a **normal distribution.** In Chapter 2, several histograms are shown with similar symmetric, bell shapes. A fundamental result, known as the **central limit theorem,** implies that histograms often have this characteristic shape, at least approximately. Whenever a random experiment is replicated, the random variable that equals the average (or total) result over the replicates tends to have a normal distribution as the number of replicates becomes large. De Moivre presented this result in 1733. Unfortunately, his work was lost for some time, and Gauss independently developed a normal distribution nearly 100 years later. Although De Moivre was later credited with the derivation, a normal distribution is also referred to as a **Gaussian** distribution.

When do we average (or total) results? Almost always. In Example 2-1 the average of the eight tensile strength measurements was calculated to be 1055.0 psi. If we assume that each measurement results from a replicate of a random experiment, the normal distribution can be used to make approximate conclusions about this average. These conclusions are the primary topics in the subsequent chapters of this book.

Furthermore, sometimes the central limit theorem is less obvious. For example, assume that the deviation (or error) in the length of a machined part is the sum of a large number of infinitesimal (small) effects, such as temperature and humidity drifts, vibrations, cutting angle variations, cutting tool wear, bearing wear, rotational speed variations, mounting and fixturing variations, variations in numerous raw material characteristics, and variation in levels of contamination. If the component errors are independent and equally likely to be positive or negative, the total error can be shown to have an approximate normal distribution. Furthermore, the normal distribution arises in the study of numerous basic physical phenomena. For example, the physicist Maxwell developed a normal distribution from simple assumptions regarding the velocities of molecules.

The theoretical basis of a normal distribution is mentioned to justify the somewhat complex form of the probability density function. Our objective now is to calculate probabilities for a normal random variable. The central limit theorem will be stated more carefully later in this chapter.

Random variables with different means and variances can be modeled by normal probability density functions with appropriate choices of the center and width of the curve. The value of $E(X) = \mu$ determines the center of the probability density function and the value of $V(X) = \sigma^2$ determines the width. Figure 3-11 illustrates several normal probability density functions with selected values of μ and σ^2. Each has the characteristic symmetric, bell-shaped curve, but the centers and dispersions differ. The following definition provides the formula for normal pdfs.

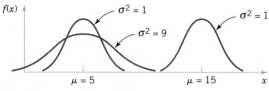
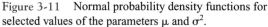

Figure 3-11   Normal probability density functions for selected values of the parameters μ and σ^2.

Normal Distribution

A random variable X with probability density function

$$f(x) = \frac{1}{\sqrt{2\pi}\sigma} e^{\frac{-(x-\mu)^2}{2\sigma^2}} \qquad \text{for} \quad -\infty < x < \infty \qquad (3\text{-}4)$$

has a **normal distribution** (and is called a **normal random variable**) with parameters μ and σ, where $-\infty < \mu < \infty$, and $\sigma > 0$. Also,

$$E(X) = \mu \quad \text{and} \quad V(X) = \sigma^2$$

The mean and variance of the normal distribution are derived at the end of this section.

The notation $N(\mu, \sigma^2)$ is often used to denote a normal distribution with mean μ and variance σ^2.

EXAMPLE 3-6

Current in a Wire: Normal Distribution

Assume that the current measurements in a strip of wire follow a normal distribution with a mean of 10 milliamperes and a variance of 4 milliamperes^2. What is the probability that a measurement exceeds 13 milliamperes?

Solution. Let X denote the current in milliamperes. The requested probability can be represented as $P(X > 13)$. This probability is shown as the shaded area under the normal probability density function in Fig. 3-12. Unfortunately, there is no closed-form expression for the integral of a normal pdf, and probabilities based on the normal distribution are typically found numerically or from a table (which we will introduce later). ■

Some useful results concerning a normal distribution are summarized in Fig. 3-13. For any normal random variable,

$$P(\mu - \sigma < X < \mu + \sigma) = 0.6827$$
$$P(\mu - 2\sigma < X < \mu + 2\sigma) = 0.9545$$
$$P(\mu - 3\sigma < X < \mu + 3\sigma) = 0.9973$$

From the symmetry of $f(x)$, $P(X > \mu) = P(X < \mu) = 0.5$. Because $f(x)$ is positive for all x, this model assigns some probability to each interval of the real line. However, the probability density function decreases as x moves farther from μ. Consequently, the probability that a measurement falls far from μ is small, and at some distance from μ the probability of an interval can be approximated as zero. The area under a normal pdf beyond 3σ from the mean is quite small. This fact is convenient for quick, rough sketches of a normal probability density function. The sketches help us determine probabilities. Because more than 0.9973 of the

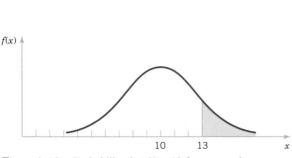

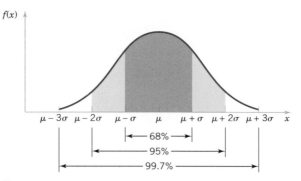

Figure 3-12   Probability that $X > 13$ for a normal random variable with $\mu = 10$ and $\sigma^2 = 4$ in Example 3-6.

Figure 3-13   Probabilities associated with a normal distribution.

probability of a normal distribution is within the interval $(\mu - 3\sigma, \mu + 3\sigma)$, 6σ *is often referred to as the width of a normal distribution.* Numerical integration can be used to show that the area under the normal pdf from $-\infty < x < \infty$ is 1.

Standard Normal Random Variable	A normal random variable with $\mu = 0$ and $\sigma^2 = 1$ is called a **standard normal** random variable. A standard normal random variable is denoted as Z.

Appendix A Table I provides cumulative probabilities for a standard normal random variable. The use of Table I is illustrated by the following example.

EXAMPLE 3-7
Standard Normal Distribution

Assume that Z is a standard normal random variable. Appendix A Table I provides probabilities of the form $P(Z \le z)$. The use of Table I to find $P(Z \le 1.5)$ is illustrated in Fig. 3-14. We read down the z column to the row that equals 1.5. The probability is read from the adjacent column, labeled 0.00 to be 0.93319.

The column headings refer to the hundredth's digit of the value of z in $P(Z \le z)$. For example, $P(Z \le 1.53)$ is found by reading down the z column to the row 1.5 and then selecting the probability from the column labeled 0.03 to be 0.93699. ∎

Standard Normal Cumulative Distribution Function	The function $$\Phi(z) = P(Z \le z)$$ is used to denote a probability from Appendix A Table I. It is the **cumulative distribution function** of a standard normal random variable. A table (or computer software) is required because the probability can't be determined by elementary methods.

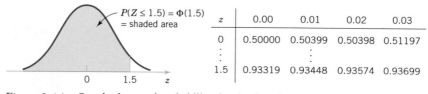

z	0.00	0.01	0.02	0.03
0	0.50000	0.50399	0.50398	0.51197
⋮		⋮		
1.5	0.93319	0.93448	0.93574	0.93699

$P(Z \le 1.5) = \Phi(1.5)$ = shaded area

Figure 3-14 Standard normal probability density function.

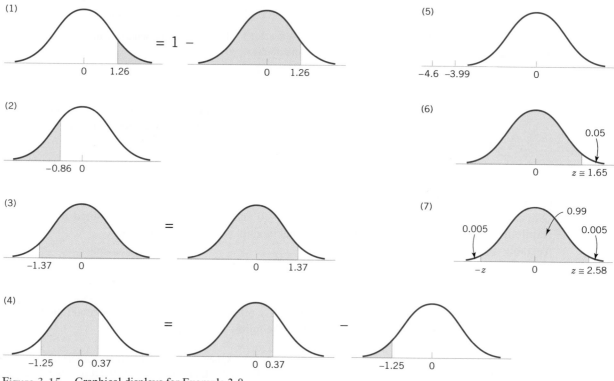

Figure 3-15 Graphical displays for Example 3-8.

Cumulative distribution functions for other random variables are widely available in computer software packages. They can be used in the same manner as $\Phi(z)$ to obtain probabilities for these random variables.

Probabilities that are not of the form $P(Z \le z)$ are found by using the basic rules of probability and the symmetry of the normal distribution along with Appendix A Table I. The following examples illustrate the method.

EXAMPLE 3-8

Normal Distribution Probabilities

The following calculations are shown pictorially in Fig. 3-15. In practice, a probability is often rounded to one or two significant digits.

(1) $P(Z > 1.26) = 1 - P(Z \le 1.26) = 1 - 0.89616 = 0.10384$

(2) $P(Z < -0.86) = 0.19490$

(3) $P(Z > -1.37) = P(Z < 1.37) = 0.91465$

(4) $P(-1.25 < Z < 0.37)$. This probability can be found from the difference of two areas, $P(Z < 0.37) - P(Z < -1.25)$. Now,

$$P(Z < 0.37) = 0.64431 \quad \text{and} \quad P(Z < -1.25) = 0.10565$$

Therefore,

$$P(-1.25 < Z < 0.37) = 0.64431 - 0.10565 = 0.53866$$

(5) $P(Z \leq -4.6)$ cannot be found exactly from Table I. However, the last entry in the table can be used to find that $P(Z \leq -3.99) = 0.00003$. Because $P(Z \leq -4.6) < P(Z \leq -3.99)$, $P(Z \leq -4.6)$ is nearly zero.

(6) Find the value z such that $P(Z > z) = 0.05$. This probability equation can be written as $P(Z \leq z) = 0.95$. Now, Table I is used in reverse. We search through the probabilities to find the value that corresponds to 0.95. The solution is illustrated in Fig. 3-15b. We do not find 0.95 exactly; the nearest value is 0.95053, corresponding to $z = 1.65$.

(7) Find the value of z such that $P(-z < Z < z) = 0.99$. Because of the symmetry of the normal distribution, if the area of the shaded region in Fig. 3-15(7) is to equal 0.99, the area in each tail of the distribution must equal 0.005. Therefore, the value for z corresponds to a probability of 0.995 in Table I. The nearest probability in Table I is 0.99506, when $z = 2.58$. ■

The preceding examples show how to calculate probabilities for standard normal random variables. Using the same approach for an arbitrary normal random variable would require a separate table for every possible pair of values for μ and σ. Fortunately, all normal probability distributions are related algebraically, and Appendix A Table I can be used to find the probabilities associated with an arbitrary normal random variable by first using a simple transformation.

Standard Normal Random Variable

If X is a normal random variable with $E(X) = \mu$ and $V(X) = \sigma^2$, the random variable

$$Z = \frac{X - \mu}{\sigma}$$

is a normal random variable with $E(Z) = 0$ and $V(Z) = 1$. That is, Z is a **standard normal** random variable.

Creating a new random variable by this transformation is referred to as **standardizing.** The random variable Z represents the distance of X from its mean in terms of standard deviations. It is the key step in calculating a probability for an arbitrary normal random variable.

EXAMPLE 3-9

Current in a Wire: Normal Distribution Probability

Define the random variable and distribution.

Suppose the current measurements in a strip of wire are assumed to follow a normal distribution with a mean of 10 milliamperes and a variance of 4 milliamperes^2. What is the probability that a measurement will exceed 13 milliamperes?

Solution. Let X denote the current in milliamperes. The requested probability can be represented as $P(X > 13)$.

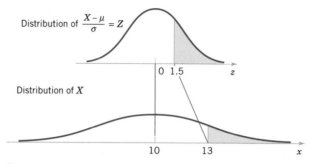

Figure 3-16   Standardizing a normal random variable.

Write the probability statement.

Let $Z = (X - 10)/2$. The relationship between several values of X and the transformed values of Z are shown in Fig. 3-16. We note that $X > 13$ corresponds to $Z > 1.5$. Therefore, from Table I,

Compute the probability.

$$P(X > 13) = P(Z > 1.5) = 1 - P(Z \le 1.5) = 1 - 0.93319 = 0.06681$$

Rather than using Fig. 3-16, the probability can be found from the inequality $X > 13$. That is,

$$P(X > 13) = P\left(\frac{X - 10}{2} > \frac{13 - 10}{2}\right) = P(Z > 1.5) = 0.06681 \qquad \blacksquare$$

In the preceding example, the value 13 is transformed to 1.5 by standardizing, and 1.5 is often referred to as the **z-value** associated with a probability. The following box summarizes the calculation of probabilities derived from normal random variables.

Standardizing

Suppose X is a normal random variable with mean μ and variance σ^2. Then,

$$P(X \le x) = P\left(\frac{X - \mu}{\sigma} \le \frac{x - \mu}{\sigma}\right) = P(Z \le z) \qquad (3\text{-}5)$$

where

Z is a **standard normal** random variable, and

$z = (x - \mu)/\sigma$ is the **z-value** obtained by **standardizing** x.

The probability is obtained by entering **Appendix A Table I** with $z = (x - \mu)/\sigma$.

EXAMPLE 3-10
Current in a Wire:
Normal
Distribution
Probability

Continuing the previous example, what is the probability that a current measurement is between 9 and 11 milliamperes?

Solution. The probability statement is $P(9 < x < 11)$. Proceeding algebraically, we have

Define the random variable and distribution.

Write the probability statement.

$$P(9 < X < 11) = P\left(\frac{9 - 10}{2} < \frac{X - 10}{2} < \frac{11 - 10}{2}\right) = P(-0.5 < Z < 0.5)$$
$$= P(Z < 0.5) - P(Z < -0.5) = 0.69146 - 0.30854 = 0.38292$$

Compute the probability.

As a second exercise, determine the value for which the probability that a current measurement is below this value is 0.98.

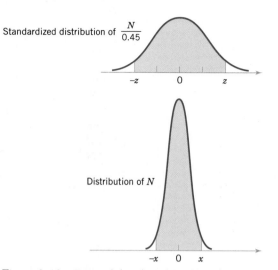

Standardized distribution of $\dfrac{N}{0.45}$

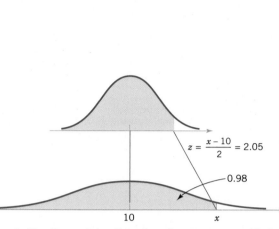

$z = \dfrac{x - 10}{2} = 2.05$

0.98

Distribution of N

Figure 3-17   Determining the value of x to meet a specified probability, Example 3-10.

Figure 3-18   Determining the value of x to meet a specified probability, Example 3-11.

Define the random variable and distribution.

Solution. The requested value is shown graphically in Fig. 3-17. We need the value of x such that $P(X < x) = 0.98$. By standardizing, this probability expression can be written as

Write the probability statement.

$$P(X < x) = P\left(\frac{X - 10}{2} < \frac{x - 10}{2}\right) = P\left(Z < \frac{x - 10}{2}\right) = 0.98$$

Table I is used to find the z value such that $P(Z < z) = 0.98$. The nearest probability from Table I results in

Compute the probability.

$$P(Z < 2.05) = 0.97982$$

Therefore, $(x - 10)/2 = 2.05$, and the standardizing transformation is used in reverse to solve for x. The result is

$$x = 2(2.05) + 10 = 14.1 \text{ milliamperes}$$ ∎

EXAMPLE 3-11
Voltage of Noise

In the transmission of a digital signal, assume that the background noise follows a normal distribution with a mean of 0 volt and standard deviation of 0.45 volt. If the system assumes that a digital 1 has been transmitted when the voltage exceeds 0.9, what is the probability of detecting a digital 1 when none was sent?

Solution. Let the random variable N denote the voltage of noise. The requested probability is

$$P(N > 0.9) = P\left(\frac{N}{0.45} > \frac{0.9}{0.45}\right) = P(Z > 2) = 1 - 0.97725 = 0.02275$$

This probability can be described as the probability of a false detection.
Determine symmetric bounds about 0 that include 99% of all noise readings.

Solution. The question requires us to find x such that $P(-x < N < x) = 0.99$. A graph is shown in Fig. 3-18. Now,

$$P(-x < N < x) = P\left(-\frac{x}{0.45} < \frac{N}{0.45} < \frac{x}{0.45}\right) = P\left(-\frac{x}{0.45} < Z < \frac{x}{0.45}\right) = 0.99$$

From Table I

$$P(-2.58 < Z < 2.58) = 0.99$$

Therefore,

$$\frac{x}{0.45} = 2.58 \qquad \text{and} \qquad x = 2.58(0.45) = 1.16$$

Suppose a digital 1 is represented as a shift in the mean of the noise distribution to 1.8 volts. What is the probability that a digital 1 is not detected? Let the random variable S denote the voltage when a digital 1 is transmitted.

Solution. Then,

$$P(S < 0.9) = P\left(\frac{S - 1.8}{0.45} < \frac{0.9 - 1.8}{0.45}\right) = P(Z < -2) = 0.02275$$

This probability can be interpreted as the probability of a missed signal. ∎

EXAMPLE 3-12
Diameter of
a Shaft

The diameter of a shaft in an optical storage drive is normally distributed with mean 0.2508 inch and standard deviation 0.0005 inch. The specifications on the shaft are 0.2500 ± 0.0015 inch. What proportion of shafts conforms to specifications?

Solution. Let X denote the shaft diameter in inches. The requested probability is shown in Fig. 3-19 and

$$P(0.2485 < X < 0.2515) = P\left(\frac{0.2485 - 0.2508}{0.0005} < Z < \frac{0.2515 - 0.2508}{0.0005}\right)$$
$$= P(-4.6 < Z < 1.4) = P(Z < 1.4) - P(Z < -4.6)$$
$$= 0.91924 - 0.0000 = 0.91924$$

Most of the nonconforming shafts are too large, because the process mean is located very near to the upper specification limit. If the process is centered so that the process mean is equal to the target value of 0.2500,

$$P(0.2485 < X < 0.2515) = P\left(\frac{0.2485 - 0.2500}{0.0005} < Z < \frac{0.2515 - 0.2500}{0.0005}\right)$$
$$= P(-3 < Z < 3) = P(Z < 3) - P(Z < -3)$$
$$= 0.99865 - 0.00135 = 0.9973$$

By recentering the process, the yield is increased to approximately 99.73%. ∎

Software such as Minitab can also be used to calculate probabilities. For example, to obtain the probability in Example 3-9 we set the mean, standard deviation, and value for the probability as follows:

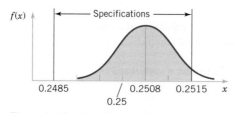

Figure 3-19   Distribution for Example 3-12.

Normal Distribution ☒

- ○ Probability density
- ◉ Cumulative probability
- ○ Inverse cumulative probability

Mean: `10.0`
Standard deviation: `2.0`

- ○ Input column:
  Optional storage:
- ◉ Input constant: `13`
  Optional storage:

Select

Help   OK   Cancel

The result is the probability $X < 13$ as shown:

Cumulative Distribution Function
```
Normal with mean = 10 and standard deviation = 2
x  P(X <= x)
13  0.933193
```

Also, a value that solves a probability equation can be determined as in Example 3-10. The Minitab input is:

Normal Distribution ☒

- ○ Probability density
- ○ Cumulative probability
- ◉ Inverse cumulative probability

Mean: `10.0`
Standard deviation: `2.0`

- ○ Input column:
  Optional storage:
- ◉ Input constant: `0.98`
  Optional storage:

Select

Help   OK   Cancel

The result from Minitab is shown:

```
Inverse Cumulative Distribution Function
Normal with mean = 10 and standard deviation = 2
P(X <= x)        x
    0.98  14.1075
```

Mean and Variance of the Normal Distribution

We show that the mean and variance of a normal random variable are μ and σ^2, respectively. The mean of x is

$$E(X) = \int_{-\infty}^{\infty} x \frac{e^{-(x-\mu)^2/2\sigma^2}}{\sqrt{2\pi}\sigma} \, dx$$

By making the change of variable $y = (x - \mu)/\sigma$, the integral becomes

$$E(X) = \mu \int_{-\infty}^{\infty} \frac{e^{-y^2/2}}{\sqrt{2\pi}} \, dy + \sigma \int_{-\infty}^{\infty} y \frac{e^{-y^2/2}}{\sqrt{2\pi}} \, dy$$

The first integral in the previous expression equals 1 because $\dfrac{e^{-y^2/2}}{\sqrt{2\pi}}$ is the standard normal pdf, and the second integral is found to be 0 by either formally making the change of variable $u = -y^2/2$ or noticing the symmetry of the integrand about $y = 0$. Therefore, $E(X) = \mu$.

The variance of X is

$$V(X) = \int_{-\infty}^{\infty} (x - \mu)^2 \frac{e^{-(x-\mu)^2/2\sigma^2}}{\sqrt{2\pi}\sigma} \, dx$$

By making the change of variable $y = (x - \mu)/\sigma$, the integral becomes

$$V(X) = \sigma^2 \int_{-\infty}^{\infty} y^2 \frac{e^{-y^2/2}}{\sqrt{2\pi}} \, dy$$

From a table of integrals or upon integrating by parts with $u = y$ and $dv = y \dfrac{e^{-y^2/2}}{\sqrt{2\pi}} \, dy$, $V(X)$ is found to be σ^2.

Animation 5: Understanding the Normal Distribution

3-5.2   Lognormal Distribution

Variables in a system sometimes follow an exponential relationship as $x = \exp(w)$. If the exponent is a random variable, say W, $X = \exp(W)$ is a random variable and the distribution of X is of interest. An important special case occurs when W has a normal distribution. In that case, the distribution of X is called a **lognormal distribution.** The name follows from the transformation $\ln(X) = W$. That is, the natural logarithm of X is normally distributed.

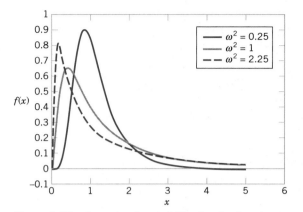

Figure 3-20   Lognormal probability density functions with $\theta = 0$ for selected values of ω^2.

Probabilities for X are obtained from the transformation to W, but we need to recognize that the range of X is $(0, \infty)$. Suppose that W is normally distributed with mean θ and variance ω^2; then the cumulative distribution function for X is

$$F(x) = P[X \leq x] = P[\exp(W) \leq x] = P[W \leq \ln(x)]$$

$$= P\left[Z \leq \frac{\ln(x) - \theta}{\omega}\right] = \Phi\left[\frac{\ln(x) - \theta}{\omega}\right]$$

for $x > 0$, where Z is a standard normal random variable. Therefore, Appendix A Table I can be used to determine the probability. Also, $F(x) = 0$, for $x \leq 0$.

The lognormal pdf can be obtained from the derivative of $F(x)$. This derivative is applied to the last term in the expression for $F(x)$, the integral of the standard normal density function. Once we know the pdf, the mean and variance of X can be derived. The details are omitted, but a summary of results follows.

Lognormal Distribution

Let W have a normal distribution with mean θ and variance ω^2; then $X = \exp(W)$ is a **lognormal random variable** with probability density function

$$f(x) = \frac{1}{x\omega\sqrt{2\pi}} \ \exp\left[-\frac{(\ln(x) - \theta)^2}{2\omega^2}\right] \qquad 0 < x < \infty \qquad (3\text{-}6)$$

The mean and variance of X are

$$E(X) = e^{\theta + \omega^2/2} \qquad \text{and} \qquad V(X) = e^{2\theta + \omega^2}(e^{\omega^2} - 1) \qquad (3\text{-}7)$$

The parameters of a lognormal distribution are θ and ω^2, but care is needed to interpret that these are the mean and variance of the normal random variable W. The mean and variance of X are the functions of these parameters shown in (3-7). Figure 3-20 illustrates lognormal distributions for selected values of the parameters.

The lifetime of a product that degrades over time is often modeled by a lognormal random variable. For example, this is a common distribution for the lifetime of a semiconductor laser.

Other continuous distributions can also be used in this type of application. However, because the lognormal distribution is derived from a simple exponential function of a normal random variable, it is easy to understand and easy to evaluate probabilities.

EXAMPLE 3-13
Lifetime of a Laser

The lifetime of a semiconductor laser has a lognormal distribution with $\theta = 10$ and $\omega = 1.5$ hours. What is the probability the lifetime exceeds 10,000 hours?

Solution. The random variable X is the lifetime of a semiconductor laser. From the cumulative distribution function for X

$$P(X > 10{,}000) = 1 - P[\exp(W) \le 10{,}000] = 1 - P[W \le \ln(10{,}000)]$$

$$= \Phi\left(\frac{\ln(10{,}000) - 10}{1.5}\right) = 1 - \Phi(-0.52) = 1 - 0.30 = 0.70$$

What lifetime is exceeded by 99% of lasers?

Solution. Now the question is to determine x such that $P(X > x) = 0.99$. Therefore,

$$P(X > x) = P[\exp(W) > x] = P[W > \ln(x)] = 1 - \Phi\left(\frac{\ln(x) - 10}{1.5}\right) = 0.99$$

From Appendix Table I, $1 - \Phi(z) = 0.99$ when $z = -2.33$. Therefore,

$$\frac{\ln(x) - 10}{1.5} = -2.33 \quad \text{and} \quad x = \exp(6.505) = 668.48 \text{ hours}$$

Determine the mean and standard deviation of lifetime.

Solution. Now,

$$E(X) = e^{\theta + \omega^2/2} = \exp(10 + 1.125) = 67{,}846.3$$

$$V(X) = e^{2\theta + \omega^2}(e^{\omega^2} - 1) = \exp(20 + 2.25)[\exp(2.25) - 1] = 39{,}070{,}059{,}886.6$$

so the standard deviation of X is 197,661.5 hours. Notice that the standard deviation of lifetime is large relative to the mean. ∎

3-5.3 Gamma Distribution

To define the gamma distribution, we require a generalization of the factorial function.

Gamma Function

The **gamma function** is

$$\Gamma(r) = \int_0^\infty x^{r-1} e^{-x} \, dx, \quad \text{for } r > 0 \tag{3-8}$$

It can be shown that the integral in the definition of $\Gamma(r)$ is finite. Furthermore, by using integration by parts it can be shown that

$$\Gamma(r) = (r - 1)\Gamma(r - 1)$$

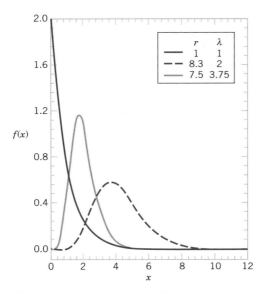

Figure 3-21 Gamma probability density functions for selected values of λ and r.

This result is left as an exercise. Therefore, if r is a positive integer $\Gamma(r) = (r - 1)!$. Also, $\Gamma(1) = 0! = 1$ and it can be shown that $\Gamma(1/2) = \pi^{1/2}$.

Now the gamma pdf can be stated.

Gamma Distribution

The random variable X with probability density function

$$f(x) = \frac{\lambda^r x^{r-1} e^{-\lambda x}}{\Gamma(r)}, \quad \text{for } x > 0 \tag{3-9}$$

is a **gamma random variable** with parameters $\lambda > 0$ and $r > 0$. The mean and variance are

$$\mu = E(X) = r/\lambda \quad \text{and} \quad \sigma^2 = V(X) = r/\lambda^2 \tag{3-10}$$

Sketches of the gamma distribution for several values of λ and r are shown in Fig. 3-21.

The gamma distribution is very useful for modeling a variety of random experiments. Furthermore, the **chi-squared distribution** is a special case of the gamma distribution in which $\lambda = 1/2$ and r equals one of the values $1/2, 1, 3/2, 2, \ldots$. This distribution is used extensively in interval estimation and tests of hypotheses that are discussed in Chapters 4 and 5. When the parameter r is an integer, the gamma distribution is called the Erlang distribution (after A. K. Erlang, who first used the distribution in the telecommunications field).

3-5.4 Weibull Distribution

The Weibull distribution is often used to model the time until failure of many different physical systems. The parameters in the distribution provide a great deal of flexibility to model systems in which the number of failures increases with time (bearing wear), decreases with time (some semiconductors), or remains constant with time (failures caused by external shocks to the system).

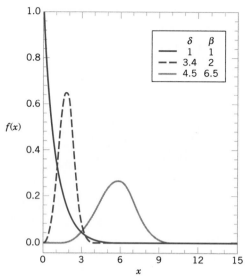

Figure 3-22   Weibull probability density functions for selected values of δ and β.

The flexibility of the Weibull distribution is illustrated by the graphs of selected probability density functions in Fig. 3-22.

The cumulative distribution function is often used to compute probabilities. The following result can be obtained.

Weibull Distribution

The random variable X with probability density function

$$f(x) = \frac{\beta}{\delta}\left(\frac{x}{\delta}\right)^{\beta-1} \exp\left[-\left(\frac{x}{\delta}\right)^{\beta}\right], \quad \text{for } x > 0 \qquad (3\text{-}11)$$

is a **Weibull random variable** with scale parameter $\delta > 0$ and shape parameter $\beta > 0$.

Weibull Cumulative Distribution Function

If X has a Weibull distribution with parameters δ and β, the cumulative distribution function of X is

$$F(x) = 1 - \exp\left[-\left(\frac{x}{\delta}\right)^{\beta}\right]$$

The mean and variance of the Weibull distribution are as follows.

If X has a Weibull distribution with parameters δ and β,

$$\mu = \delta\Gamma\left(1 + \frac{1}{\beta}\right) \quad \text{and} \quad \sigma^2 = \delta^2\Gamma\left(1 + \frac{2}{\beta}\right) - \delta^2\left[\Gamma\left(1 + \frac{1}{\beta}\right)\right]^2 \qquad (3\text{-}12)$$

EXAMPLE 3-14

Lifetime of a
Bearing

The time to failure (in hours) of a bearing in a mechanical shaft is satisfactorily modeled as a Weibull random variable with $\beta = 1/2$ and $\delta = 5000$ hours. Determine the mean time until failure.

Solution. From the expression for the mean,

$$E(X) = 5000\Gamma[1 + (1/0.5)] = 5000\Gamma[3] = 5000 \times 2! = 10,000 \text{ hours}$$

Determine the probability that a bearing lasts at least 6000 hours.

Solution. Now

$$P(X > 6000) = 1 - F(6000) = \exp\left[-\left(\frac{6000}{5000}\right)^{1/2}\right] = e^{-1.095} = 0.334$$

Consequently, only 33.4% of all bearings last at least 6000 hours.

EXERCISES FOR SECTION 3-5

3-34. Use Appendix A Table I to determine the following probabilities for the standard normal random variable Z.

(a) $P(-1 < Z < 1)$ (b) $P(-2 < Z < 2)$
(c) $P(-3 < Z < 3)$ (d) $P(Z < -3)$
(e) $P(0 < Z \le 3)$

3-35. Assume that Z has a standard normal distribution. Use Appendix A Table I to determine the value for z that solves each of the following.

(a) $P(Z < z) = 0.50000$ (b) $P(Z < z) = 0.001001$
(c) $P(Z > z) = 0.881000$ (d) $P(Z > z) = 0.866500$
(e) $P(-1.3 < Z < z) = 0.863140$

3-36. Assume that Z has a standard normal distribution. Use Appendix A Table I to determine the value for z that solves each of the following.

(a) $P(-z < Z < z) = 0.95$ (b) $P(-z < Z < z) = 0.99$
(c) $P(-z < Z < z) = 0.68$ (d) $P(-z < Z < z) = 0.9973$

3-37. Assume that X is normally distributed with a mean of 20 and a standard deviation of 2. Determine the following.

(a) $P(X < 24)$ (b) $P(X > 18)$
(c) $P(18 < X < 22)$ (d) $P(14 < X < 26)$
(e) $P(16 < X < 20)$ (f) $P(20 < X < 26)$

3-38. Assume that X is normally distributed with a mean of 20 and a standard deviation of 2. Determine the value for x that solves each of the following.

(a) $P(X > x) = 0.5$ (b) $P(X > x) = 0.95$
(c) $P(x < X < 20) = 0.2$

3-39. Assume that X is normally distributed with a mean of 37 and a standard deviation of 2. Determine the following.

(a) $P(X < 31)$ (b) $P(X > 30)$
(c) $P(33 < X < 37)$ (d) $P(32 < X < 39)$
(e) $P(30 < X < 38)$

3-40. Assume that X is normally distributed with a mean of 6 and a standard deviation of 3. Determine the value for x that solves each of the following.

(a) $P(X > x) = 0.5$ (b) $P(X > x) = 0.95$
(c) $P(x < X < 9) = 0.2$ (d) $P(3 < X < x) = 0.8$

3-41. The compressive strength of samples of cement can be modeled by a normal distribution with a mean of 6000 kilograms per square centimeter and a standard deviation of 100 kilograms per square centimeter.

(a) What is the probability that a sample's strength is less than 6250 kg/cm²?
(b) What is the probability that a sample's strength is between 5800 and 5900 kg/cm²?
(c) What strength is exceeded by 95% of the samples?

3-42. The tensile strength of paper is modeled by a normal distribution with a mean of 35 pounds per square inch and a standard deviation of 2 pounds per square inch.

(a) What is the probability that the strength of a sample is less than 39 lb/in²?
(b) If the specifications require the tensile strength to exceed 29 lb/in², what proportion of the sample is scrapped?

3-43. The line width of a tool used for semiconductor manufacturing is assumed to be normally distributed with a mean of 0.5 micrometer and a standard deviation of 0.05 micrometer.

(a) What is the probability that a line width is greater than 0.62 micrometer?
(b) What is the probability that a line width is between 0.47 and 0.63 micrometer?
(c) The line width of 90% of samples is below what value?

3-44. The fill volume of an automated filling machine used for filling cans of carbonated beverage is normally distributed with a mean of 12.4 fluid ounces and a standard deviation of 0.1 fluid ounce.

(a) What is the probability that a fill volume is less than 12 fluid ounces?

(b) If all cans less than 12.1 or greater than 12.6 ounces are scrapped, what proportion of cans is scrapped?

(c) Determine specifications that are symmetric about the mean that include 99% of all cans.

3-45.   Consider the filling machine in Exercise 3-44. Suppose that the mean of the filling operation can be adjusted easily, but the standard deviation remains at 0.1 ounce.

(a) At what value should the mean be set so that 99.9% of all cans exceed 12 ounces?

(b) At what value should the mean be set so that 99.9% of all cans exceed 12 ounces if the standard deviation can be reduced to 0.05 fluid ounce?

3-46.   The reaction time of a driver to visual stimulus is normally distributed with a mean of 0.4 second and a standard deviation of 0.05 second.

(a) What is the probability that a reaction requires more than 0.5 second?

(b) What is the probability that a reaction requires between 0.4 and 0.5 second?

(c) What is the reaction time that is exceeded 90% of the time?

3-47.   The length of an injected-molded plastic case that holds tape is normally distributed with a mean length of 90.2 millimeters and a standard deviation of 0.1 millimeter.

(a) What is the probability that a part is longer than 90.3 millimeters or shorter than 89.7 millimeters?

(b) What should the process mean be set at to obtain the greatest number of parts between 89.7 and 90.3 millimeters?

(c) If parts that are not between 89.7 and 90.3 millimeters are scrapped, what is the yield for the process mean that you selected in part (b)?

3-48.   Operators of a medical linear accelerator are interested in estimating the number of hours until the first software failure. Prior experience has shown that the time until failure is normally distributed with mean 1000 hours and standard deviation 60 hours.

(a) Find the probability that the software will not fail before 1140 hours of operation.

(b) Find the probability that the software will fail within 900 hours of operation.

3-49.   A device that monitors the levels of pollutants has sensors that detect the amount of CO in the air. Placed in a particular location, it is known that the amount of CO is normally distributed with a mean of 6.23 ppm and a variance of 4.26 ppm^2.

(a) What is the probability that the CO level exceeds 9 ppm?

(b) What is the probability that the CO level is between 5.5 ppm and 8.5 ppm?

(c) An alarm is to be activated if the CO levels exceed a certain threshold. Specify the threshold such that it is 3.75 standard deviations above the mean.

3-50.   The life of a semiconductor laser at a constant power is normally distributed with a mean of 7000 hours and a standard deviation of 600 hours.

(a) What is the probability that a laser fails before 5000 hours?

(b) What is the life in hours that 95% of the lasers exceed?

3-51.   The diameter of the dot produced by a printer is normally distributed with a mean diameter of 0.002 inch and a standard deviation of 0.0004 inch.

(a) What is the probability that the diameter of a dot exceeds 0.0026 inch?

(b) What is the probability that a diameter is between 0.0014 and 0.0026 inch?

(c) What standard deviation of diameters is needed so that the probability in part (b) is 0.995?

3-52.   The weight of a human joint replacement part is normally distributed with a mean of 2 ounces and a standard deviation of 0.05 ounce.

(a) What is the probability that a part weighs more than 2.10 ounces?

(b) What must the standard deviation of weight be for the company to state that 99.9% of its parts are less than 2.10 ounces?

(c) If the standard deviation remains at 0.05 ounce, what must the mean weight be for the company to state that 99.9% of its parts are less than 2.10 ounces?

3-53.   Suppose that X has a lognormal distribution with parameters $\theta = 5$ and $\omega^2 = 9$. Determine the following:

(a) $P(X < 13{,}300)$

(b) The value for x such that $P(X \le x) = 0.95$

(c) The mean and variance of X

3-54.   Suppose that X has a lognormal distribution with parameters $\theta = 2$ and $\omega^2 = 4$. Determine the following:

(a) $P(X < 500)$ (b) $P(500 < X < 1000)$

(c) $P(1500 < X < 2000)$

(d) What does the difference of the probabilities in parts (a), (b), and (c) imply about the probabilities of lognormal random variables?

3-55.   The length of time (in seconds) that a user views a page on a Web site before moving to another page is a lognormal random variable with parameters $\theta = 0.5$ and $\omega^2 = 1$.

(a) What is the probability that a page is viewed for more than 10 seconds?

(b) What is the length of time that 50% of users view the page?

(c) What is the mean and standard deviation of the time until a user moves from the page?

3-56.   The lifetime of a semiconductor laser has a lognormal distribution, and it is known that the mean and standard deviation of lifetime are 10,000 and 20,000, respectively.

(a) Calculate the parameters of the lognormal distribution.

(b) Determine the probability that a lifetime exceeds 10,000 hours.

(c) Determine the lifetime that is exceeded by 90% of lasers.

3-57.   Suppose that X has a Weibull distribution with $\beta = 0.2$ and $\delta = 100$ hours. Determine the mean and variance of X.

3-58. Suppose that X has a Weibull distribution with $\beta = 0.2$ and $\delta = 100$ hours. Determine the following:

(a) $P(X < 10{,}000)$ (b) $P(X > 5000)$

3-59. Assume that the life of a roller bearing follows a Weibull distribution with parameters $\beta = 2$ and $\delta = 10{,}000$ hours.

(a) Determine the probability that a bearing lasts at least 8000 hours.

(b) Determine the mean time until failure of a bearing.

(c) If 10 bearings are in use and failures occur independently, what is the probability that all 10 bearings last at least 8000 hours?

3-60. The life (in hours) of a computer processing unit (CPU) is modeled by a Weibull distribution with parameters $\beta = 3$ and $\delta = 900$ hours.

(a) Determine the mean life of the CPU.

(b) Determine the variance of the life of the CPU.

(c) What is the probability that the CPU fails before 500 hours?

3-61. An article in the *Journal of the Indian Geophysical Union*, titled "Weibull and Gamma Distributions for Wave Parameter Predictions" (Vol. 9, 2005, 55–64), used the Weibull distribution to model ocean wave heights. Assume that the mean wave height at the observation station is 2.5 m and the shape parameter equals 2. Determine the standard deviation of wave height.

3-62. Use integration by parts to show that $\Gamma(r) = (r - 1) \cdot \Gamma(r - 1)$.

3-63. Use the properties of the gamma function to evaluate the following:

(a) $\Gamma(6)$ (b) $\Gamma(5/2)$ (c) $\Gamma(9/2)$

3-64. Suppose that X has a gamma distribution with $\lambda = 3$ and $r = 6$. Determine the mean and variance of X.

3-65. Suppose that X has a gamma distribution with $\lambda = 2.5$ and $r = 3.2$. Determine the mean and variance of X.

3-66. Suppose that X represents diameter measurements from a gamma distribution with a mean of 3 mm and a variance of 1.5 mm^2. Find the parameters λ and r.

3-67. Suppose that X represents length measurements from a gamma distribution with a mean of 4.5 inches and a variance of 6.25 inches^2. Find the parameters λ and r.

3-68. Suppose that X represents time measurements from a gamma distribution with a mean of 4 minutes and a variance of 2 minutes^2. Find the parameters λ and r.

3-6 PROBABILITY PLOTS

3-6.1 Normal Probability Plots

How do we know whether a normal distribution is a reasonable model for data? **Probability plotting** is a graphical method for determining whether sample data conform to a hypothesized distribution based on a subjective visual examination of the data. The general procedure is very simple and can be performed quickly. Probability plotting typically uses special graph paper, known as **probability paper,** that has been designed for the hypothesized distribution. Probability paper is widely available for the normal, lognormal, Weibull, and various chi-square and gamma distributions. In this section we illustrate the **normal probability plot.** Section 3-6.2 discusses probability plots for some other continuous distributions.

To construct a probability plot, the observations in the sample are first ranked from smallest to largest. That is, the sample $x_1, x_2, \ldots, x_n$ is arranged as $x_{(1)}, x_{(2)}, \ldots, x_{(n)}$, where $x_{(1)}$ is the smallest observation, $x_{(2)}$ is the second smallest observation, and so forth, with $x_{(n)}$ the largest. The ordered observations $x_{(j)}$ are then plotted against their observed cumulative frequency $(j - 0.5)/n$ on the appropriate probability paper. If the hypothesized distribution adequately describes the data, the plotted points will fall approximately along a straight line; if the plotted points deviate significantly and systematically from a straight line, the hypothesized model is not appropriate. Usually, the determination of whether or not the data plot as a straight line is subjective. The procedure is illustrated in the following example.

EXAMPLE 3-15

Service Life of a Battery

Ten observations on the effective service life in minutes of batteries used in a portable personal computer are as follows: 176, 191, 214, 220, 205, 192, 201, 190, 183, 185. We hypothesize that battery life is adequately modeled by a normal distribution. To use probability plotting to investigate this hypothesis, first arrange the observations in ascending order and calculate their cumulative frequencies $(j - 0.5)/10$ as shown in the following table.

The pairs of values $x_{(j)}$ and $(j - 0.5)/10$ are now plotted on normal probability paper. This plot is shown in Fig. 3-23. Most normal probability paper plots $100(j - 0.5)/n$ on the left vertical scale and

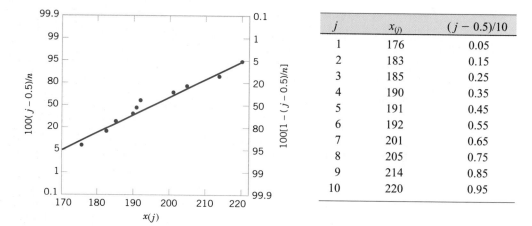

j	$x_{(j)}$	$(j - 0.5)/10$
1	176	0.05
2	183	0.15
3	185	0.25
4	190	0.35
5	191	0.45
6	192	0.55
7	201	0.65
8	205	0.75
9	214	0.85
10	220	0.95

Figure 3-23   Normal probability plot for the battery life.

$100[1 - (j - 0.5)/n]$ on the right vertical scale, with the variable value plotted on the horizontal scale. A straight line, chosen subjectively as a "best fit" line, has been drawn through the plotted points. In drawing the straight line, you should be influenced more by the points near the middle of the plot than by the extreme points. A good rule of thumb is to draw the line approximately between the 25th and 75th percentile points. This is how the line in Fig. 3-23 was determined. In assessing the systematic deviation of the points from the straight line, imagine a fat pencil lying along the line. If all the points are covered by this imaginary pencil, a normal distribution adequately describes the data. Because the points in Fig. 3-23 would pass the fat pencil test, we conclude that the normal distribution is an appropriate model.   ■

A normal probability plot can also be constructed on ordinary graph paper by plotting the standardized normal scores z_j against $x_{(j)}$, where the standardized normal scores satisfy

$$\frac{j - 0.5}{n} = P(Z \le z_j) = \Phi(z_j)$$

For example, if $(j - 0.5)/n = 0.05$, $\Phi(z_j) = 0.05$ implies that $z_j = -1.64$. To illustrate, consider the data from the previous example. In the following table we show the standardized normal scores in the last column. Figure 3-24 presents the plot of z_j versus $x_{(j)}$. This normal probability plot is equivalent to the one in Fig. 3-23.

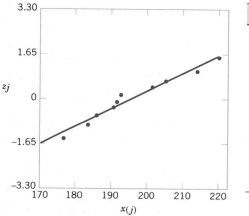

j	$x_{(j)}$	$(j - 0.5)/10$	z_j
1	176	0.05	−1.64
2	183	0.15	−1.04
3	185	0.25	−0.67
4	190	0.35	−0.39
5	191	0.45	−0.13
6	192	0.55	0.13
7	201	0.65	0.39
8	205	0.75	0.67
9	214	0.85	1.04
10	220	0.95	1.64

Figure 3-24   Normal probability plot obtained from standardized normal scores.

Table 3-1 Crack Length (mm) for an Aluminum Alloy

81	98	291	101	98	118	158	197	139	249
249	135	223	205	80	177	82	64	137	149
117	149	127	115	198	342	83	34	342	185
227	225	185	240	161	197	98	65	144	151
134	59	181	151	240	146	104	100	215	200

A very important application of normal probability plotting is in *verification of assumptions* when using statistical inference procedures that require the normality assumption.

3-6.2 Other Probability Plots

Probability plots are extremely useful and are often the first technique used when we need to determine which probability distribution is likely to provide a reasonable model for data. In using probability plots, usually the distribution is chosen by subjective assessment of the probability plot. More formal **goodness-of-fit** techniques can be used in conjunction with probability plotting. We will describe a very simple goodness-of-fit test in Section 4-10.

Interpetting a
Probability Plot

To illustrate how probability plotting can be useful in determining the appropriate distribution for data, consider the data on crack elongation in an aluminum alloy shown in Table 3-1. Figure 3-25 is a normal probability plot of the crack-length data. Notice how the data in the tails of the plot bend away from the straight line; this is an indication that the normal distribution is not a good model for the data. Figure 3-26 is a lognormal probability plot of the crack-length data, obtained from Minitab. The data fall much closer to the straight line in this plot, particularly the observations in the tails, suggesting that the lognormal distribution is more likely to provide a reasonable model for the crack-length data than is the normal distribution.

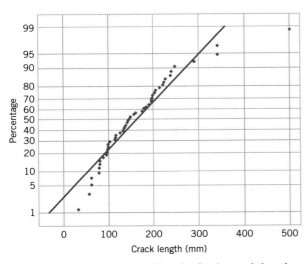

Figure 3-25 Normal probability plot for the crack-length data in Table 3-1.

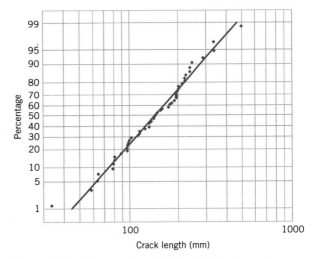

Figure 3-26 Lognormal probability plot for the crack-length data in Table 3-1.

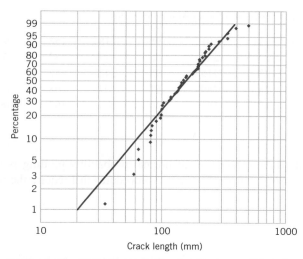

Figure 3-27   Weibull probability plot for the crack-length data in Table 3-1.

Finally, Fig. 3-27 is a Weibull probability plot for the crack-length data (also generated by Minitab). The observations in the lower tail of this plot are not very close to the straight line, suggesting that the Weibull is not a very good model for the data. Therefore, based on the three probability plots that we have constructed, the lognormal distribution appears to be the most appropriate choice as a model for the crack-length data.

EXERCISES FOR SECTION 3-6

3-69. A soft-drink bottler is studying the internal pressure strength of 1-liter glass bottles. A random sample of 16 bottles is tested, and the pressure strengths are obtained. The data are shown next. Plot these data on normal probability paper. Does it seem reasonable to conclude that pressure strength is normally distributed?

226.16 psi	208.15 psi	211.14 psi	221.31 psi
202.20 psi	195.45 psi	203.62 psi	204.55 psi
219.54 psi	193.71 psi	188.12 psi	202.21 psi
193.73 psi	200.81 psi	224.39 psi	201.63 psi

3-70. Samples of 20 parts are selected from two machines, and a critical dimension is measured on each part. The data are shown next. Plot the data on normal probability paper. Does this dimension seem to have a normal distribution? What tentative conclusions can you draw about the two machines?

Machine 1			
99.4	101.5	102.3	96.7
99.1	103.8	100.4	100.9
99.0	99.6	102.5	96.5
98.9	99.4	99.7	103.1
99.6	104.6	101.6	96.8

Machine 2			
90.9	100.7	95.0	98.8
99.6	105.5	92.3	115.5
105.9	104.0	109.5	87.1
91.2	96.5	96.2	109.8
92.8	106.7	97.6	106.5

3-71. After examining the data from the two machines in Exercise 3-70, the process engineer concludes that machine 2 has higher part-to-part variability. She makes some adjustments to the machine that should reduce the variability, and she obtains another sample of 20 parts. The measurements on those parts are shown next. Plot these data on normal probability paper and compare them with the normal probability plot of the data from machine 2 in Exercise 3-70. Is the normal distribution a reasonable model for the data? Does it appear that the variance has been reduced?

103.4	107.0	107.7	104.5
108.1	101.5	106.2	106.6
103.1	104.1	106.3	105.6
108.2	106.9	107.8	103.7
103.9	103.3	107.4	102.6

3-72. In studying the uniformity of polysilicon thickness on a wafer in semiconductor manufacturing, Lu, Davis, and Gyurcsik (*Journal of the American Statistical Association,* Vol. 93, 1998) collected data from 22 independent wafers: 494, 853, 1090, 1058, 517, 882, 732, 1143, 608, 590, 940, 920, 917, 581, 738, 732, 750, 1205, 1194, 1221, 1209, 708. Is it reasonable to model these data using a normal probability distribution?

3-73. A quality control inspector is interested in maintaining a flatness specification for the surface of metal disks. Thirty flatness measurements in (0.001 inches) were collected. Which probability density model—normal, lognormal, or Weibull—appears to provide the more suitable fit to the data?

2.49	2.14	1.63
4.46	3.69	4.58
1.28	1.28	1.59
0.82	2.23	7.55
2.20	4.78	5.24
1.54	3.81	2.13
1.45	2.21	6.65
6.40	2.06	4.06
2.66	1.66	2.38
6.04	2.85	3.70

3-74. Twenty-five measurements of the time a client waits for a server is recorded in seconds. Which probability density model—normal, lognormal, or Weibull—appears to provide the more suitable fit to the data?

1.21	4.19	1.95	6.88	3.97
9.09	6.91	1.90	10.60	0.51
2.23	13.99	8.22	8.08	4.70
4.67	0.50	0.92	4.15	7.24
4.86	1.89	6.44	0.15	17.34

3-75. The duration of an inspection task is recorded in minutes. Determine which probability density model—normal, lognormal, or Weibull—appears to provide the more suitable fit to the data.

5.15	0.30	6.66	3.76
4.29	9.54	4.38	0.60
7.06	4.34	0.80	5.12
3.69	5.94	3.18	4.47
4.65	8.93	4.70	1.04

3-76. Thirty measurements of the time-to-failure of a critical component in an electronics assembly are recorded. Determine which probability density model—normal, lognormal, or Weibull—appears to provide the more suitable fit to the data.

1.9	20.7
3.0	11.9
6.3	0.4
8.3	2.3
1.6	5.3
4.6	1.9
5.1	4.0
1.9	3.8
4.1	0.9
10.9	9.0
6.6	1.3
0.5	2.9
2.1	1.2
1.2	2.5
0.8	4.4

3-77. The following data are direct solar intensity measurements (watts/m^2) on different days at a location in southern Spain that was analyzed in Chapter 2: 562, 869, 708, 775, 775, 704, 809, 856, 655, 806, 878, 909, 918, 558, 768, 870, 918, 940, 946, 661, 820, 898, 935, 952, 957, 693, 835, 905, 939, 955, 960, 498, 653, 730, 753. Does a normal distribution provide a reasonable model for these data? Why or why not?

3-78. The following data are the temperatures of effluent at discharge from a sewage treatment facility on consecutive days:

43	47	51	48	52	50	46	49
45	52	46	51	44	49	46	51
49	45	44	50	48	50	49	50

Determine which of the probability models studied appears to provide the more suitable fit to the data.

3-7 DISCRETE RANDOM VARIABLES

For a discrete random variable, only measurements at discrete points are possible. Examples were provided previously in this chapter and others follow. This section presents properties for discrete random variables that are analogous to those presented for continuous random variables.

EXAMPLE 3-16
Voice Network

A voice communication network for a business contains 48 external lines. At a particular time, the system is observed and some of the lines are being used. Let the random variable X denote the number of lines in use. Then X can assume any of the integer values 0 through 48. ∎

EXAMPLE 3-17

Semiconductor
Wafer
Contamination

The analysis of the surface of a semiconductor wafer records the number of particles of contamination that exceed a certain size. Define the random variable X to equal the number of particles of contamination.

The possible values of X are integers from 0 up to some large value that represents the maximum number of these particles that can be found on one of the wafers. If this maximum number is very large, it might be convenient to assume that any integer from zero to ∞ is possible. ∎

3-7.1 Probability Mass Function

As mentioned previously, the probability distribution of a random variable X is a description of the probabilities associated with the possible values of X. For a discrete random variable, the distribution is often specified by just a list of the possible values along with the probability of each. In some cases, it is convenient to express the probability in terms of a formula.

EXAMPLE 3-18

Bit Transmission
Errors

There is a chance that a bit transmitted through a digital transmission channel is received in error. Let X equal the number of bits in error in the next 4 bits transmitted. The possible values for X are {0, 1, 2, 3, 4}. Based on a model for the errors that is presented in the following section, probabilities for these values will be determined. Suppose that the probabilities are

$$P(X = 0) = 0.6561 \qquad P(X = 1) = 0.2916 \qquad P(X = 2) = 0.0486$$
$$P(X = 3) = 0.0036 \qquad P(X = 4) = 0.0001$$

The probability distribution of X is specified by the possible values along with the probability of each. A graphical description of the probability distribution of X is shown in Fig. 3-28. ∎

Suppose a loading on a long, thin beam places mass only at discrete points. See Fig. 3-29. The loading can be described by a function that specifies the mass at each of the discrete points. Similarly, for a discrete random variable X, its distribution can be described by a function that specifies the probability at each of the possible discrete values for X.

**Probability
Mass
Function**

For a discrete random variable X with possible values $x_1, x_2, \ldots, x_n$, the **probability mass function** (or pmf) is

$$f(x_i) = P(X = x_i) \tag{3-13}$$

Because $f(x_i)$ is defined as a probability, $f(x_i) \geq 0$ for all x_i and

$$\sum_{i=1}^{n} f(x_i) = 1.$$

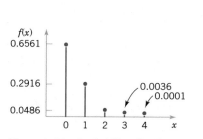

Figure 3-28   Probability distribution for X in Example 3-18.

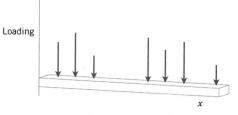

Figure 3-29   Loadings at discrete points on a long, thin beam.

The reader should check that the sum of the probabilities in the previous example equals 1.

The steps to determine a probability for a random variable described in Section 3-4.1 apply equally well to discrete random variables. The steps are repeated here:

1. Determine the random variable and distribution of the random variable.
2. Write the probability statement in terms of the random variable.
3. Compute the probability using the probability statement and the distribution.

These steps are shown in the solutions of several examples in this chapter. In other examples and exercises you might use these steps on your own.

3-7.2 Cumulative Distribution Function

A cumulative distribution function (cdf) can also be used to provide the probability distribution of a discrete random variable. The cdf at a value x is the sum of the probabilities at all points less than or equal to x.

Cumulative Distribution Function of a Discrete Random Variable

The **cumulative distribution function** of a discrete random variable X is

$$F(x) = P(X \le x) = \sum_{x_i \le x} f(x_i)$$

EXAMPLE 3-19

Bit Transmission Errors

In the previous example, the probability mass function of X is

$$P(X = 0) = 0.6561 \qquad P(X = 1) = 0.2916 \qquad P(X = 2) = 0.0486$$
$$P(X = 3) = 0.0036 \qquad P(X = 4) = 0.0001$$

Therefore,

$$F(0) = 0.6561 \qquad F(1) = 0.9477 \qquad F(2) = 0.9963 \qquad F(3) = 0.9999 \qquad F(4) = 1$$

Even if the random variable can assume only integer values, the cdf is defined at noninteger values. For example,

$$F(1.5) = P(X \le 1.5) = P(X \le 1) = 0.9477$$

The graph of $F(x)$ is shown in Figure 3-30. Note that the graph has discontinuities (jumps) at the discrete values for X. It is a piece-wise continuous function. The size of the jump at a point x equals the probability at x. For example, consider $x = 1$. Here, $F(1) = 0.9477$, but for $0 \le x < 1$, $F(x) = 0.6561$. The change is $P(X = 1) = 0.2916$. ∎

3-7.3 Mean and Variance

The mean and variance of a discrete random variable are defined similarly to a continuous random variable. Summation replaces integration in the definitions.

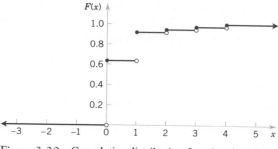

Figure 3-30   Cumulative distribution function for x in Example 3-19.

Mean and Variance of a Discrete Random Variable

Let the possible values of the random variable X be denoted as $x_1, x_2, \ldots, x_n$. The pmf of X is $f(x)$, so $f(x_i) = P(X = x_i)$.

The **mean** or **expected value** of the discrete random variable X, denoted as μ or $E(X)$, is

$$\mu = E(X) = \sum_{i=1}^{n} x_i f(x_i) \tag{3-14}$$

The **variance** of X, denoted as σ^2 or $V(X)$, is

$$\sigma^2 = V(X) = E(X - \mu)^2 = \sum_{i=1}^{n} (x_i - \mu)^2 f(x_i) = \sum_{i=1}^{n} x_i^2 f(x_i) - \mu^2$$

The **standard deviation** of X is σ.

The mean of X can be interpreted as the center of mass of the range of values of X. That is, if we place mass equal to $f(x_i)$ at each point x_i on the real line, $E(X)$ is the point at which the real line is balanced. Therefore, the term "probability mass function" can be interpreted by this analogy with mechanics.

EXAMPLE 3-20

Bit Transmission Errors: Mean and Variance

For the number of bits in error in the previous example,

$$\begin{aligned} \mu = E(X) &= 0f(0) + 1f(1) + 2f(2) + 3f(3) + 4f(4) \\ &= 0(0.6561) + 1(0.2916) + 2(0.0486) + 3(0.0036) + 4(0.0001) \\ &= 0.4 \end{aligned}$$

Although X never assumes the value 0.4, the weighted average of the possible values is 0.4. To calculate $V(X)$, a table is convenient.

x	$x - 0.4$	$(x - 0.4)^2$	$f(x)$	$f(x)(x - 0.4)^2$
0	−0.4	0.16	0.6561	0.104976
1	0.6	0.36	0.2916	0.104976
2	1.6	2.56	0.0486	0.124416
3	2.6	6.76	0.0036	0.024336
4	3.6	12.96	0.0001	0.001296

$$V(X) = \sigma^2 = \sum_{i=1}^{5} f(x_i)(x_i - 0.4)^2 = 0.36$$

■

EXAMPLE 3-21
Product Revenue

Two new product designs are to be compared on the basis of revenue potential. Marketing feels that the revenue from design A can be predicted quite accurately to be $3 million. The revenue potential of design B is more difficult to assess. Marketing concludes that there is a probability of 0.3 that the revenue from design B will be $7 million, but there is a 0.7 probability that the revenue will be only $2 million. Which design would you choose?

Solution. Let X denote the revenue from design A. Because there is no uncertainty in the revenue from design A, we can model the distribution of the random variable X as $3 million with probability one. Therefore, $E(X) = \$3$ million. ∎

As an extension of this example, let Y denote the revenue from design B. The expected value of Y in millions of dollars is

$$E(Y) = \$7(0.3) + \$2(0.7) = \$3.5$$

Because $E(Y)$ exceeds $E(X)$, we might choose design B. However, the variability of the result from design B is larger. That is,

$$\sigma^2 = (7 - 3.5)^2(0.3) + (2 - 3.5)^2(0.7) = 5.25 \text{ millions of dollars squared}$$

and

$$\sigma = \sqrt{5.25} = 2.29 \text{ millions of dollars}$$

EXERCISES FOR SECTION 3-7

Verify that the functions in Exercises 3-79 through 3-82 are probability mass functions, and determine the requested values.

3-79.

x	1	2	3	4
$f(x)$	0.326	0.088	0.019	0.251

x	5	6	7
$f(x)$	0.158	0.140	0.018

(a) $P(X \le 3)$ (b) $P(3 < X < 5.1)$
(c) $P(X > 4.5)$ (d) Mean and variance
(e) Graph $F(x)$.

3-80.

x	0	1	2	3
$f(x)$	0.025	0.041	0.049	0.074

x	4	5	6	7
$f(x)$	0.098	0.205	0.262	0.123

x	8	9
$f(x)$	0.074	0.049

(a) $P(X \le 1)$ (b) $P(2 < X < 7.2)$
(c) $P(X \ge 6)$ (d) Mean and variance
(e) Graph $F(x)$.

3-81. $f(x) = (8/7)(1/2)^x, x = 1, 2, 3$
(a) $P(X \le 1)$ (b) $P(X > 1)$
(c) Mean and variance   (d) Graph $F(x)$.

3-82. $f(x) = (1/2)(x/5), x = 1, 2, 3, 4$
(a) $P(X = 2)$ (b) $P(X \le 3)$
(c) $P(X > 2.5)$ (d) $P(X \ge 1)$
(e) Mean and variance   (f) Graph $F(x)$.

3-83. Customers purchase a particular make of automobile with a variety of options. The probability mass function of the number of options selected is

x	7	8	9	10
$f(x)$	0.040	0.130	0.190	0.240

x	11	12	13
$f(x)$	0.300	0.050	0.050

(a) What is the probability that a customer will choose fewer than 9 options?
(b) What is the probability that a customer will choose more than 11 options?
(c) What is the probability that a customer will choose between 8 and 12 options, inclusively?
(d) What is the expected number of options chosen? What is the variance?

3-84. Marketing estimates that a new instrument for the analysis of soil samples will be very successful, moderately successful, or unsuccessful, with probabilities 0.4, 0.5, and 0.1, respectively. The yearly revenue associated with a very successful, moderately successful, or unsuccessful product is $10 million, $5 million, and $1 million, respectively. Let the random variable X denote the yearly revenue of the product.

(a) Determine the probability mass function of X.
(b) Determine the expected value and the standard deviation of the yearly revenue.

(c) Plot the pmf and mark the location of the expected value.

(d) Graph $F(x)$.

3-85. Let X denote the number of bars of service on your cell phone whenever you are at an intersection with the following probabilities:

x	0	1	2	3	4	5
$P(X = x)$	0.1	0.15	0.25	0.25	0.15	0.1

Determine the following:

(a) $F(x)$

(b) Mean and variance

(c) $P(X < 2)$

(d) $P(X \le 3.5)$

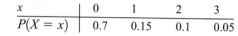

3-86. Let X denote the number of patients who suffer an infection within a floor of a hospital per month with the following probabilities:

x	0	1	2	3
$P(X = x)$	0.7	0.15	0.1	0.05

Determine the following:

(a) $F(x)$

(b) Mean and variance

(c) $P(X > 1.5)$

(d) $P(X \le 2.0)$

3-8 BINOMIAL DISTRIBUTION

A widely used discrete random variable is introduced in this section. Consider the following random experiments and random variables.

1. Flip a fair coin 10 times. Let $X =$ the number of heads obtained.

2. A worn machine tool produces 1% defective parts. Let $X =$ the number of defective parts in the next 25 parts produced.

3. Water quality samples contain high levels of organic solids in 10% of the tests. Let $X =$ number of samples high in organic solids in the next 18 tested.

4. Of all bits transmitted through a digital transmission channel, 10% are received in error. Let $X =$ the number of bits in error in the next 4 bits transmitted.

5. A multiple choice test contains 10 questions, each with four choices, and you guess at each question. Let $X =$ the number of questions answered correctly.

6. In the next 20 births at a hospital, let $X =$ the number of female births.

7. Of all patients suffering a particular illness, 35% experience improvement from a particular medication. In the next 30 patients administered the medication, let $X =$ the number of patients who experience improvement.

These examples illustrate that a general probability model that includes these experiments as particular cases would be very useful.

Each of these random experiments can be thought of as consisting of a series of repeated, random trials: 10 flips of the coin in experiment 1, the production of 25 parts in experiment 2, and so forth. The random variable in each case is a count of the number of trials that meet a specified criterion. The outcome from each trial either meets the criterion that X counts or it does not; consequently, each trial can be summarized as resulting in either a success or a failure, respectively. For example, in the multiple choice experiment, for each question, only the choice that is correct is considered a **success.** Choosing any one of the three incorrect choices results in the trial being summarized as a failure.

The terms "success" and "failure" are merely labels. We can just as well use "A" and "B" or "0" or "1." Unfortunately, the usual labels can sometimes be misleading. In experiment 2, because X counts defective parts, the production of a defective part is called a success.

A trial with only two possible outcomes is used so frequently as a building block of a random experiment that it is called a **Bernoulli trial.** It is usually assumed that the trials that constitute the random experiment are **independent.** This implies that the outcome from one trial has no effect on the outcome to be obtained from any other trial. Furthermore, it is often reasonable to assume that the **probability of a success on each trial is constant.**

In item 5, the multiple choice experiment, if the test taker has no knowledge of the material and just guesses at each question, we might assume that the probability of a correct answer is 1/4 *for each question.*

To analyze X, recall the relative frequency interpretation of probability. The proportion of times that question 1 is expected to be correct is 1/4 and the proportion of times that question 2 is expected to be correct is 1/4. For simple guesses, the proportion of times both questions are correct is expected to be

$$(1/4)(1/4) = 1/16$$

Furthermore, if one merely guesses, the proportion of times question 1 is correct and question 2 is incorrect is expected to be

$$(1/4)(3/4) = 3/16$$

Similarly, if one merely guesses, then the proportion of times question 1 is incorrect and question 2 is correct is expected to be

$$(3/4)(1/4) = 3/16$$

Finally, if one merely guesses, then the proportion of times question 1 is incorrect and question 2 is incorrect is expected to be

$$(3/4)(3/4) = 9/16$$

We have accounted for all of the possible correct and incorrect combinations for these two questions, and the four probabilities associated with these possibilities sum to 1:

$$1/16 + 3/16 + 3/16 + 9/16 = 1$$

This approach is used to derive the binomial distribution in the following example.

EXAMPLE 3-22
Bit Transmission
Errors

In Example 3-18, assume that the chance that a bit transmitted through a digital transmission channel is received in error is 0.1. Also, assume that the transmission trials are independent. Let X = the number of bits in error in the next 4 bits transmitted. Determine $P(X = 2)$.

Solution. Let the letter E denote a bit in error, and let the letter O denote that the bit is okay—that is, received without error. We can represent the outcomes of this experiment as a list of four letters that indicate the bits that are in error and okay. For example, the outcome $OEOE$ indicates that the second and fourth bits are in error and the other 2 bits are okay. The corresponding values for x are

Outcome	x	Outcome	x
OOOO	0	EOOO	1
OOOE	1	EOOE	2
OOEO	1	EOEO	2
OOEE	2	EOEE	3
OEOO	1	EEOO	2
OEOE	2	EEOE	3
OEEO	2	EEEO	3
OEEE	3	EEEE	4

The event that $X = 2$ consists of the six outcomes:

$$\{EEOO, EOEO, EOOE, OEEO, OEOE, OOEE\}$$

Using the assumption that the trials are independent, the probability of $\{EEOO\}$ is

$$P(EEOO) = P(E)P(E)P(O)P(O) = (0.1)^2(0.9)^2 = 0.0081$$

Also, any one of the six mutually exclusive outcomes for which $X = 2$ has the same probability of occurring. Therefore,

$$P(X = 2) = 6(0.0081) = 0.0486$$

In general,

$$P(X = x) = (\text{number of outcomes that result in } x \text{ errors}) \times (0.1)^x(0.9)^{4-x}$$

To complete a general probability formula, only an expression for the number of outcomes that contain x errors is needed. An outcome that contains x errors can be constructed by partitioning the four trials (letters) in the outcome into two groups. One group is of size x and contains the errors, and the other group is of size $n - x$ and consists of those trials that are okay. The number of ways of partitioning four objects into two groups, one of which is of size x, is $\binom{4}{x} = 4!/[x!(4 - x)!]$. Therefore, in this example

$$P(X = x) = \binom{4}{x}(0.1)^x(0.9)^{4-x}$$

Note that $\binom{4}{2} = 4!/[2!\,2!] = 6$, as was found previously. The probability mass function of X was shown in Fig. 3-28. ∎

The previous example motivates the following result.

Binomial Distribution

A random experiment consisting of n repeated trials such that

1. the trials are independent,
2. each trial results in only two possible outcomes, labeled as *success* and *failure,* and
3. the probability of a success on each trial, denoted as p, remains constant

is called a *binomial experiment.*

The random variable X that equals the number of trials that result in a *success* has a **binomial distribution** with parameters p and n where $0 < p < 1$ and $n = \{1, 2, 3, \ldots\}$. The pmf of X is

$$f(x) = \binom{n}{x}p^x(1 - p)^{n-x}, \quad x = 0, 1, \ldots, n \qquad (3\text{-}15)$$

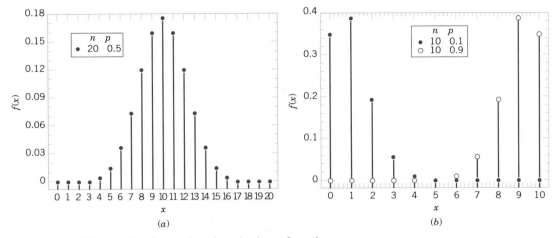

Figure 3-31   Binomial distribution for selected values of n and p.

As before, $\binom{n}{x}$ equals the number of sequences (orderings) of outcomes that contain x *successes* and $n - x$ *failures*. The number of sequences that contain x *successes* and $n - x$ *failures* times the probability of each sequence equals $P(X = x)$.

It can be shown (by using the binomial expansion formula) that the sum of the probabilities for a binomial random variable is 1. Furthermore, because each trial in the experiment is classified into two outcomes, {*success, failure*}, the distribution is called a "bi"-nomial. A more general distribution with multiple (two *or more*) outcomes is by analogy called the multinomial distribution. The multinomial distribution is covered by Montgomery and Runger (2006).

Examples of binomial distributions are shown in Fig. 3-31. For a fixed n, the distribution becomes more symmetric as p increases from 0 to 0.5 or decreases from 1 to 0.5. For a fixed p, the distribution becomes more symmetric as n increases.

EXAMPLE 3-23

Binomial Coefficient

Several examples using the binomial coefficient $\binom{n}{x}$ follow.

$$\binom{10}{3} = 10!/[3! \ 7!] = (10 \cdot 9 \cdot 8)/(3 \cdot 2) = 120$$

$$\binom{15}{10} = 15!/[10! \ 5!] = (15 \cdot 14 \cdot 13 \cdot 12 \cdot 11)/(5 \cdot 4 \cdot 3 \cdot 2) = 3003$$

$$\binom{100}{4} = 100!/[4! \ 96!] = (100 \cdot 99 \cdot 98 \cdot 97)/(4 \cdot 3 \cdot 2) = 3,921,225 \qquad \blacksquare$$

EXAMPLE 3-24

Organic Solids

Define the random variable and distribution.

Each sample of water has a 10% chance of containing high levels of organic solids. Assume that the samples are independent with regard to the presence of the solids. Determine the probability that in the next 18 samples, exactly 2 contain high solids.

Solution. Let X = the number of samples that contain high solids in the next 18 samples analyzed. Then X is a binomial random variable with $p = 0.1$ and $n = 18$. Therefore,

Write the probability statement.

$$P(X = 2) = \binom{18}{2}(0.1)^2(0.9)^{16}$$

Compute the
probability.

Now $\binom{18}{2} = (18!/[2!\ 16!]) = 18(17)/2 = 153$. Therefore,

$$P(X = 2) = 153(0.1)^2(0.9)^{16} = 0.284$$

Determine the probability that at least four samples contain high solids.

Solution. The requested probability is

$$P(X \geq 4) = \sum_{x=4}^{18} \binom{18}{x}(0.1)^x(0.9)^{18-x}$$

However, it is easier to use the complementary event,

$$P(X \geq 4) = 1 - P(X < 4) = 1 - \sum_{x=0}^{3} \binom{18}{x}(0.1)^x(0.9)^{18-x}$$

$$= 1 - [0.150 + 0.300 + 0.284 + 0.168] = 0.098$$

Determine, the probability that $3 \leq X < 7$.

Solution.

$$P(3 \leq X < 7) = \sum_{x=3}^{6} \binom{18}{x}(0.1)^x(0.9)^{18-x}$$

$$= 0.168 + 0.070 + 0.022 + 0.005 = 0.265$$ ■

Software such as Minitab is useful for binomial probability calculations. In Example 3-24, $P(X < 4)$ is determined in Minitab as $P(X \leq 3)$ as shown:

The result is:

> **Cumulative Distribution Function**
> ```
> Binomial with n = 18 and p = 0.1
> X  P(x < = x)
> 3    0.901803
> ```

This probability is subtracted from one to obtain the result for $P(X \geq 4)$ in the example.

The mean and variance of a binomial random variable depend only on the parameters p and n. The following result can be shown.

> If X is a binomial random variable with parameters p and n,
>
> $$\mu = E(X) = np \quad \text{and} \quad \sigma^2 = V(X) = np(1-p) \qquad (3\text{-}16)$$

EXAMPLE 3-25

Bit Transmission Errors: Binomial Mean and Variance

For the number of transmitted bits received in error in Example 3-18, $n = 4$ and $p = 0.1$ so

$$E(X) = 4(0.1) = 0.4$$

The variance of the number of defective bits is

$$V(X) = 4(0.1)(0.9) = 0.36$$

These results match those that were calculated directly from the probabilities in Example 3-20. ■

Animation 4: Understanding the Binomial Distribution

EXERCISES FOR SECTION 3-8

3-87. For each scenario described, state whether or not the binomial distribution is a reasonable model for the random variable and why. State any assumptions you make.

(a) A production process produces thousands of temperature transducers. Let X denote the number of nonconforming transducers in a sample of size 30 selected at random from the process.

(b) From a batch of 50 temperature transducers, a sample of size 30 is selected without replacement. Let X denote the number of nonconforming transducers in the sample.

(c) Four identical electronic components are wired to a controller. Let X denote the number of components that have failed after a specified period of operation.

(d) Let X denote the number of express mail packages received by the post office in a 24-hour period.

(e) Let X denote the number of correct answers by a student taking a multiple choice exam in which a student can eliminate some of the choices as being incorrect in some questions and all of the incorrect choices in other questions.

(f) Forty randomly selected semiconductor chips are tested. Let X denote the number of chips in which the test finds at least one contamination particle.

(g) Let X denote the number of contamination particles found on 40 randomly selected semiconductor chips.

(h) A filling operation attempts to fill detergent packages to the advertised weight. Let X denote the number of detergent packages that are underfilled.

(i) Errors in a digital communication channel occur in bursts that affect several consecutive bits. Let X denote the number of bits in error in a transmission of 100,000 bits.

(j) Let X denote the number of surface flaws in a large coil of galvanized steel.

3-88. The random variable X has a binomial distribution with $n = 10$ and $p = 0.5$.

(a) Sketch the probability mass function of X.
(b) Sketch the cumulative distribution.
(c) What value of X is most likely?
(d) What value(s) of X is (are) least likely?

3-89. The random variable X has a binomial distribution with $n = 20$ and $p = 0.5$. Determine the following probabilities.

(a) $P(X = 15)$ (b) $P(X \le 12)$
(c) $P(X \ge 19)$ (d) $P(13 \le X < 15)$
(e) Sketch the cumulative distribution fuction.

3-90. Given that X has a binomial distribution with $n = 10$ and $p = 0.01$

(a) Sketch the probability mass function.
(b) Sketch the cumulative distribution function.
(c) What value of X is most likely?
(d) What value of X is least likely?

3-91. The random variable X has a binomial distribution with $n = 10$ and $p = 0.1$. Determine the following probabilities.

(a) $P(X = 5)$ (b) $P(X \le 2)$
(c) $P(X \ge 9)$ (d) $P(3 \le X < 5)$

3-92. An electronic product contains 40 integrated circuits. The probability that any integrated circuit is defective is 0.01, and the integrated circuits are independent. The product operates only if there are no defective integrated circuits. What is the probability that the product operates?

3-93. A hip joint replacement part is being stress-tested in a laboratory. The probability of successfully completing the test is 0.80. Seven randomly and independently chosen parts are tested. What is the probability that exactly two of the seven parts successfully complete the test?

3-94. The phone lines to an airline reservation system are occupied 45% of the time. Assume that the events that the lines are occupied on successive calls are independent. Assume that eight calls are placed to the airline.

(a) What is the probability that for exactly two calls the lines are occupied?
(b) What is the probability that for at least one call the lines are occupied?
(c) What is the expected number of calls in which the lines are occupied?

3-95. Batches that consist of 50 coil springs from a production process are checked for conformance to customer requirements. The mean number of nonconforming coil springs in a batch is five. Assume that the number of nonconforming springs in a batch, denoted as X, is a binomial random variable.

(a) What are n and p?
(b) What is $P(X \le 2)$?
(c) What is $P(X \ge 49)$?

3-96. A statistical process control chart example. Samples of 20 parts from a metal punching process are selected every hour. Typically, 1% of the parts require rework. Let X denote the number of parts in the sample of 20 that require rework. A process problem is suspected if X exceeds its mean by more than three standard deviations.

(a) If the percentage of parts that require rework remains at 1%, what is the probability that X exceeds its mean by more than three standard deviations?
(b) If the rework percentage increases to 4%, what is the probability that X exceeds 1?

(c) If the rework percentage increases to 4%, what is the probability that X exceeds 1 in at least one of the next 5 hours of samples?

3-97. Because not all airline passengers show up for their reserved seat, an airline sells 125 tickets for a flight that holds only 120 passengers. The probability that a passenger does not show up is 0.10, and the passengers behave independently.

(a) What is the probability that every passenger who shows up gets a seat?
(b) What is the probability that the flight departs with empty seats?
(c) What are the mean and standard deviation of the number of passengers who show up?

3-98. This exercise illustrates that poor quality can affect schedules and costs. A manufacturing process has 100 customer orders to fill. Each order requires one component part that is purchased from a supplier. However, typically, 2% of the components are identified as defective, and the components can be assumed to be independent.

(a) If the manufacturer stocks 100 components, what is the probability that the 100 orders can be filled without reordering components?
(b) If the manufacturer stocks 102 components, what is the probability that the 100 orders can be filled without reordering components?
(c) If the manufacturer stocks 105 components, what is the probability that the 100 orders can be filled without reordering components?

3-99. The probability of successfully landing a plane using a flight simulator is given as 0.80. Nine randomly and independently chosen student pilots are asked to try to fly the plane using the simulator.

(a) What is the probability that all the student pilots successfully land the plane using the simulator?
(b) What is the probability that none of the student pilots successfully lands the plane using the simulator?
(c) What is the probability that exactly eight of the student pilots successfully land the plane using the simulator?

3-100. Traffic engineers install 10 street lights with new bulbs. The probability that a bulb fails within 50,000 hours of operation is 0.25. Assume that each of the bulbs fails independently.

(a) What is the probability that fewer than two of the original bulbs fail within 50,000 hours of operation?
(b) What is the probability that no bulbs will have to be replaced within 50,000 hours of operation?
(c) What is the probability that more than four of the original bulbs will need replacing within 50,000 hours?

3-101. An article in *Information Security Technical Report*, "Malicious Software—Past, Present and Future" (Vol. 9, 2004, pp. 6–18), provided the following data on the top 10 malicious software instances for 2002. The clear leader in the number of registered incidences for the year 2002 was the Internet worm "Klez," and it is still one of the most widespread threats. This virus was first detected on October 26,

2001, and it has held the top spot among malicious software for the longest period in the history of virology.

The 10 Most Widespread Malicious Programs for 2002.

Place	Name	% Instances
1	I-Worm.Klez	61.22
2	I-Worm.Lentin	20.52
3	I-Worm.Tanatos	2.09
4	I-Worm.BadtransII	1.31
5	Macro.Word97.Thus	1.19
6	I-Worm.Hybris	0.60
7	I-Worm.Bridex	0.32
8	I-Worm.Magistr	0.30
9	Win95.CIH	0.27
10	I-Worm.Sircam	0.24

(*Source:* Kaspersky Labs.)

Suppose that 20 malicious software instances are reported. Assume that the malicious sources can be assumed to be independent.

(a) What is the probability at least one instance is "Klez"?
(b) What is the probability that three or more instances are "Klez"?
(c) What is the mean and standard deviation of the number of "Klez" instances among the 20 reported?

3-102.   Heart failure is due to either natural occurrences (87%) or outside factors (13%). Outside factors are related to induced substances or foreign objects. Natural occurrences are caused by arterial blockage, disease, and infection. Suppose that 20 patients will visit an emergency room with heart failure. Assume that causes of heart failure between individuals are independent.

(a) What is the probability that three individuals have conditions caused by outside factors?
(b) What is the probability that three or more individuals have conditions caused by outside factors?
(c) What is the mean and standard deviation of the number of individuals with conditions caused by outside factors?

3-9   POISSON PROCESS

Consider e-mail messages that arrive at a mail server on a computer network. This is an example of events (such as message arrivals) that occur randomly in an interval (such as time). The number of events over an interval (such as the number of messages that arrive in 1 hour) is a discrete random variable that is often modeled by a Poisson distribution. The length of the interval between events (such as the time between messages) is often modeled by an exponential distribution. These distributions are related; they provide probabilities for different random variables in the same random experiment. Figure 3-32 provides a graphical description of a Poisson process.

3-9.1   Poisson Distribution

We introduce the Poisson distribution with an example.

EXAMPLE 3-26
Limit of Bit
Errors

Consider the transmission of n bits over a digital communication channel. Let the random variable X equal the number of bits in error. When the probability that a bit is in error is constant and the transmissions are independent, X has a binomial distribution. Let p denote the probability that a bit is in error. Then, $E(X) = pn$. Now, suppose that the number of bits transmitted increases and the probability of an error decreases exactly enough that pn remains equal to a constant—say, λ. That is, n increases and p decreases accordingly, such that $E(X)$ remains constant. Then,

$$P(X = x) = \binom{n}{x} p^x (1 - p)^{n-x}$$

$$= \frac{n(n-1)(n-2)\cdots(n-x+1)}{n^x\, x!}(np)^x(1-p)^n(1-p)^{-x}$$

With some work, it can be shown that

$$\lim_{n \to \infty} P(X = x) = \frac{e^{-\lambda}\lambda^x}{x!}, \qquad x = 0, 1, 2, \ldots$$

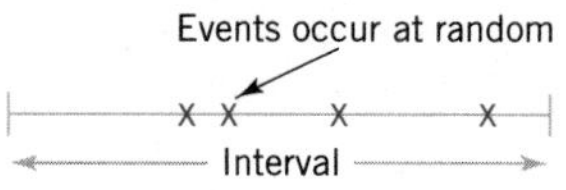

Figure 3-32   In a Poisson process, events occur at random in an interval.

Events occur at random

x x   x   x

Interval

Also, because the number of bits transmitted tends to infinity, the number of errors can equal any nonnegative integer. Therefore, the possible values for X are the integers from zero to infinity. ■

The distribution obtained as the limit in the previous example is more useful than the derivation implies. The following example illustrates the broader applicability.

EXAMPLE 3-27   Flaws occur at random along the length of a thin copper wire. Let X denote the random variable that counts the number of flaws in a length of L millimeters of wire and suppose that the average number of flaws in L millimeters is λ.

The probability distribution of X can be found by reasoning in a manner similar to Example 3-26. Partition the length of wire into n subintervals of small length—say, 1 micrometer each. If the subinterval chosen is small enough, the probability that more than one flaw occurs in the subinterval is negligible. Furthermore, we can interpret the assumption that flaws occur at random to imply that every subinterval has the same probability of containing a flaw—say, p. Finally, if we assume that the probability that a subinterval contains a flaw is independent of other subintervals, we can model the distribution of X as approximately a binomial random variable. Because

$$E(X) = \lambda = np$$

we obtain

$$p = \lambda/n$$

That is, the probability that a subinterval contains a flaw is λ/n. With small enough subintervals, n is very large and p is very small. Therefore, the distribution of X is obtained as the number of subintervals tends to infinity, as in the previous example. ■

Clearly, the previous example can be generalized to include a broad array of random experiments. The interval that was partitioned was a length of wire. However, the same reasoning can be applied to any interval, including an interval of time, an area, or a volume.

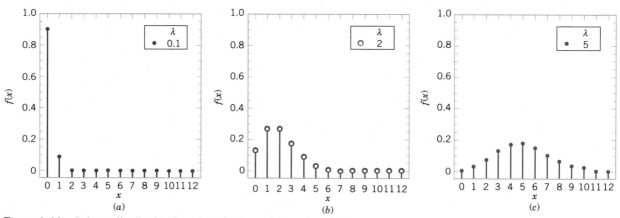

Figure 3-33   Poisson distribution for selected values of the parameter λ.

For example, counts of (1) particles of contamination in semiconductor manufacturing, (2) flaws in rolls of textiles, (3) calls to a telephone exchange, (4) power outages, and (5) atomic particles emitted from a specimen have all been successfully modeled by the pmf in the following definition.

Poisson Distribution

Assume that events occur at random throughout the interval. If the interval can be partitioned into subintervals of small enough length such that

1. The probability of more than one event in a subinterval is zero,
2. The probability of one event in a subinterval is the same for all subintervals and proportional to the length of the subinterval, and
3. The event occurrence in each subinterval is independent of other subintervals, the random experiment is called a *Poisson process.*

If the mean number of events in the interval is $\lambda > 0$, the random variable X that equals the number of events in the interval has a **Poisson distribution** with parameter λ, and the pmf of X is

$$f(x) = \frac{e^{-\lambda}\lambda^x}{x!}, \qquad x = 0, 1, 2, \ldots \qquad (3\text{-}17)$$

The mean and variance of X are

$$E(X) = \lambda \qquad \text{and} \qquad V(X) = \lambda \qquad (3\text{-}18)$$

Historically, the term "process" has been used to suggest the observation of a system over time. In our example with the copper wire, we showed that the Poisson distribution could also apply to intervals such as lengths. Figure 3-33 shows graphs of selected Poisson distributions.

It is important to **use consistent units** in the calculation of probabilities, means, and variances involving Poisson random variables. The following example illustrates unit conversions. For example, if the

Mean number of flaws per millimeter of wire is 3.4, then the

Mean number of flaws in 10 millimeters of wire is 34, and the

Mean number of flaws in 100 millimeters of wire is 340.

If a Poisson random variable represents the number of events in some interval, the mean of the random variable must be the expected number of events in the same length of interval.

EXAMPLE 3-28
Flaws in Wire

For the case of the thin copper wire, suppose that the number of flaws follows a Poisson distribution with a mean of 2.3 flaws per millimeter. Determine the probability of exactly 2 flaws in 1 millimeter of wire.

Solution. Let X denote the number of flaws in 1 millimeter of wire. Then X has a Poisson distribution and $E(X) = 2.3$ flaws and

$$P(X = 2) = \frac{e^{-2.3}2.3^2}{2!} = 0.265$$

Determine the probability of 10 flaws in 5 millimeters of wire.

Solution. Let X denote the number of flaws in 5 millimeters of wire. Then X has a Poisson distribution with

$$E(X) = 5 \text{ mm} \times 2.3 \text{ flaws/mm} = 11.5 \text{ flaws}$$

Therefore,

$$P(X = 10) = e^{-11.5} 11.5^{10}/10! = 0.113$$

Determine the probability of at least one flaw in 2 millimeters of wire.

Solution. Let X denote the number of flaws in 2 millimeters of wire. Then X has a Poisson distribution with

$$E(X) = 2 \text{ mm} \times 2.3 \text{ flaws/mm} = 4.6 \text{ flaws}$$

Therefore,

$$P(X \geq 1) = 1 - P(X = 0) = 1 - e^{-4.6} = 0.9899 \qquad \blacksquare$$

The next example uses a computer program to sum Poisson probabilities.

EXAMPLE 3-29
Contamination
on Optical Disks

Contamination is a problem in the manufacture of optical storage disks. The number of particles of contamination that occur on an optical disk has a Poisson distribution, and the average number of particles per centimeter squared of media surface is 0.1. The area of a disk under study is 100 squared centimeters. Determine the probability that 12 particles occur in the area of a disk under study.

Solution. Let X denote the number of particles in the area of a disk under study. Because the mean number of particles is 0.1 particles per cm^2,

$$E(X) = 100 \text{ cm}^2 \times 0.1 \text{ particles/cm}^2$$
$$= 10 \text{ particles}$$

Therefore,

$$P(X = 12) = \frac{e^{-10} 10^{12}}{12!} = 0.095$$

Determine the probability that zero particles occur in the area of the disk under study.

Solution. Now, $P(X = 0) = e^{-10} = 4.54 \times 10^{-5}$.

Determine the probability that 12 or fewer particles occur in the area of a disk under study.

Solution. This probability is

$$P(X \leq 12) = P(X = 0) + P(X = 1) + \cdots + P(X = 12)$$
$$= \sum_{i=0}^{12} \frac{e^{-10} 10^i}{i!}$$

Because this sum is tedious to compute, many computer programs calculate cumulative Poisson probabilities. From Minitab, we obtain $P(X \leq 12) = 0.7916$. $\qquad \blacksquare$

```
┌─────────────────────────────────────────────────────────────┐
│ Poisson Distribution                                    [X]   │
│                                                               │
│  ┌──────────────────┐   ○ Probability                        │
│  │                  │   ● Cumulative probability             │
│  │                  │   ○ Inverse cumulative probability     │
│  │                  │                                         │
│  │                  │   Mean: │10        │                    │
│  │                  │                                         │
│  │                  │                                         │
│  │                  │   ○ Input column:    │            │     │
│  │                  │     Optional storage: │            │    │
│  │                  │   ● Input constant:   │12          │    │
│  │                  │     Optional storage: │            │    │
│  │   ┌──────────┐   │                                         │
│  │   │  Select  │   │                                         │
│  └───┴──────────┴───┘                                         │
│   ┌──────────┐          ┌──────────┐    ┌──────────┐          │
│   │   Help   │          │    OK    │    │  Cancel  │          │
│   └──────────┘          └──────────┘    └──────────┘          │
└─────────────────────────────────────────────────────────────┘
```

Cumulative Distribution Function
```
Poisson with mean = 10

  x   P(X < = x)
 12   0.791556
```

The variance of a Poisson random variable was stated to equal its mean. For example, if particle counts follow a Poisson distribution with a mean of 25 particles per square centimeter, the standard deviation of the counts is 5 per square centimeter. Consequently, information on the variability is very easily obtained. Conversely, if the variance of count data is much greater than the mean of the same data, the Poisson distribution is not a good model for the distribution of the random variable.

Animation 3: Understanding the Poisson Distribution

EXERCISES FOR SECTION 3-9.1

3-103. Suppose X has a Poisson distribution with a mean of 0.3. Determine the following probabilities.

(a) $P(X = 0)$ (b) $P(X \le 3)$
(c) $P(X = 6)$ (d) $P(X = 2)$

3-104. Suppose X has a Poisson distribution with a mean of 5. Determine the following probabilities.

(a) $P(X = 0)$ (b) $P(X \le 3)$
(c) $P(X = 6)$ (d) $P(X = 9)$

3-105. Suppose that the number of customers who enter a post office in a 30-minute period is a Poisson random variable and that $P(X = 0) = 0.02$. Determine the mean and variance of X.

3-106. Suppose that the number of customers who enter a bank in an hour is a Poisson random variable and that $P(X = 0) = 0.04$. Determine the mean and variance of X.

3-107. The number of telephone calls that arrive at a phone exchange is often modeled as a Poisson random variable. Assume that on the average there are 20 calls per hour.

(a) What is the probability that there are exactly 18 calls in 1 hour?
(b) What is the probability that there are 3 or fewer calls in 30 minutes?
(c) What is the probability that there are exactly 30 calls in 2 hours?

(d) What is the probability that there are exactly 10 calls in 30 minutes?

3-108. The number of earthquake tremors in a 12-month period appears to be distributed as a Poisson random variable with a mean of 6. Assume the number of tremors from one 12-month period is independent of the number in the next 12-month period.

(a) What is the probability that there are 10 tremors in 1 year?
(b) What is the probability that there are 18 tremors in 2 years?
(c) What is the probability that there are no tremors in a 1-month period?
(d) What is the probability that there are more than 5 tremors in a 6-month period?

3-109. The number of cracks in a section of interstate highway that are significant enough to require repair is assumed to follow a Poisson distribution with a mean of two cracks per mile.

(a) What is the probability that there are no cracks that require repair in 5 miles of highway?
(b) What is the probability that at least one crack requires repair in $\frac{1}{2}$ mile of highway?
(c) If the number of cracks is related to the vehicle load on the highway and some sections of the highway have a heavy load of vehicles and other sections carry a light load, how do you feel about the assumption of a Poisson distribution for the number of cracks that require repair for all sections?

3-110. The number of surface flaws in a plastic roll used in the interior of automobiles has a Poisson distribution with a mean of 0.05 flaw per square foot of plastic roll. Assume an automobile interior contains 10 square feet of plastic roll.

(a) What is the probability that there are no surface flaws in an auto's interior?
(b) If 10 cars are sold to a rental company, what is the probability that none of the 10 cars has any surface flaws?
(c) If 10 cars are sold to a rental company, what is the probability that at most 1 car has any surface flaws?

3-111. The number of failures of a testing instrument from contamination particles on the product is a Poisson random variable with a mean of 0.04 failure per hour.

(a) What is the probability that the instrument does not fail in an 8-hour shift?
(b) What is the probability of at least three failures in a 24-hour day?

3-112. When network cards are communicating, bits can occasionally be corrupted in transmission. Engineers have determined that the number of bits in error follows a Poisson distribution with mean of 3.2 bits/kb (per kilobyte).

(a) What is the probability of 5 bits being in error during the transmission of 1 kb?
(b) What is the probability of 8 bits being in error during the transmission of 2 kb?
(c) What is the probability of no error bits in 3 kb?

3-113. A telecommunication station is designed to receive a maximum of 10 calls per $\frac{1}{2}$ second. If the number of calls to the station is modeled as a Poisson random variable with a mean of 9 calls per $\frac{1}{2}$ second, what is the probability that the number of calls will exceed the maximum design constraint of the station?

3-114. Flaws occur in mylar material according to a Poisson distribution with a mean of 0.01 flaw per square yard.

(a) If 25 square yards are inspected, what is the probability that there are no flaws?
(b) What is the probability that a randomly selected square yard has no flaws?
(c) Suppose that the mylar material is cut into 10 pieces, each being 1 yard square. What is the probability that 8 or more of the 10 pieces will have no flaws?

3-115. Messages arrive to a computer server according to a Poisson distribution with a mean rate of 10 per hour.

(a) What is the probability that three messages will arrive in 1 hour?
(b) What is the probability that six messages arrive in 30 minutes?

3-116. Data from www.centralhudsonlab.com determined that the mean number of insect fragments in 225-gram chocolate bars was 14.4, but three brands had insect contamination more than twice the average. See the U.S. Food and Drug Administration—Center for Food Safety and Applied Nutrition for Defect Action Levels for food products. Assume the number of fragments (contaminants) follows a Poisson distribution.

(a) If you consume a 225-gram bar from a brand at the mean contamination level, what is the probability of no insect contaminants?
(b) Suppose you consume a bar that is one-fifth the size tested (45 grams) from a brand at the mean contamination level. What is the probability of no insect contaminants?
(c) If you consume seven 28.35-gram (one-ounce) bars this week from a brand at the mean contamination level, what is the probability that you consume one or more insect fragments in more than one bar?
(d) Is the probability of a test result more than twice the mean of 14.4 unusual, or can it be considered typical variation? Explain.

3-117. In 1898 L. J. Bortkiewicz published a book entitled *The Law of Small Numbers.* He used data collected over 20 years to show that the number of soldiers killed by horse kicks each year in each corps in the Prussian cavalry followed a Poisson distribution with a mean of 0.61.

(a) What is the probability of more than one death in a corps in a year?
(b) What is the probability of no deaths in a corps over five years?

3-9.2   Exponential Distribution

The discussion of the Poisson distribution defined a random variable to be the number of flaws along a length of copper wire. The distance between flaws is another random variable that is often of interest. Let the random variable X denote the length from any starting point on the wire until a flaw is detected.

As you might expect, the distribution of X can be obtained from knowledge of the distribution of the number of flaws. The key to the relationship is the following concept: the distance to the first flaw exceeds 3 millimeters if and only if there are no flaws within a length of 3 millimeters—simple, but sufficient for an analysis of the distribution of X.

In general, let the random variable N denote the number of flaws in x millimeters of wire. Assume that the mean number of flaws is λ per millimeter, so that N has a Poisson distribution with mean λx. Now,

$$P(X > x) = P(N = 0) = \frac{e^{-\lambda x}(\lambda x)^0}{0!} = e^{-\lambda x}$$

and

$$P(X \le x) = 1 - e^{-\lambda x}$$

for $x \ge 0$. If $f(x)$ is the pdf of X, the cumulative distribution function is

$$F(x) = P(X \le x) = \int_{-\infty}^{x} f(u)\, du$$

From the fundamental theorem of calculus, the derivative of $F(x)$ (with respect to x) is $f(x)$. Therefore, the pdf of X is

$$f(x) = \frac{d}{dx}(1 - e^{-\lambda x}) = \lambda e^{-\lambda x} \qquad \text{for } x \ge 0$$

The distribution of X depends only on the assumption that the flaws in the wire follow a Poisson process. Also, the starting point for measuring X doesn't matter because the probability of the number of flaws in an interval of a Poisson process depends only on the length of the interval, not on the location. For any Poisson process, the following general result applies.

Exponential Distribution

> The random variable X that equals the distance between successive events of a Poisson process with mean $\lambda > 0$ has an **exponential distribution** with parameter λ. The pdf of X is
>
> $$f(x) = \lambda e^{-\lambda x}, \qquad \text{for } 0 \le x < \infty \tag{3-19}$$
>
> The mean and variance of X are
>
> $$E(X) = \frac{1}{\lambda} \quad \text{and} \quad V(X) = \frac{1}{\lambda^2} \tag{3-20}$$

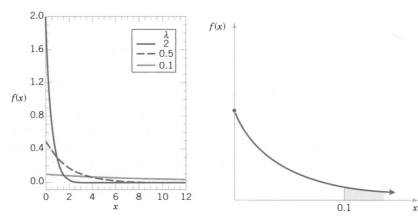

Figure 3-34   Probability density function of an exponential random variable for selected values of λ.

Figure 3-35   Probability for the exponential distribution in Example 3-30.

The exponential distribution obtains its name from the exponential function in the pdf. Plots of the exponential distribution for selected values of λ are shown in Fig. 3-34. For any value of λ, the exponential distribution is quite skewed. The formulas for the mean and variance are obtained by integration (by parts). Note also that the exponential distribution is a special case of two continuous distributions that we have studied previously. The Weibull distribution with $\beta = 1$ reduces to the exponential distribution, and the gamma distribution with $r = 1$ is an exponential distribution. The gamma distribution can be derived as the sum of r independent exponential random variables.

It is important to **use consistent units** in the calculation of probabilities, means, and variances involving exponential random variables. The following example illustrates unit conversions.

EXAMPLE 3-30

Network Log-ons

Define the random variable and distribution.

In a large corporate computer network, user log-ons to the system can be modeled as a Poisson process with a mean of 25 log-ons per hour. What is the probability that there are no log-ons in an interval of 6 minutes?

Solution. Let X denote the time in hours from the start of the interval until the first log-on. Then X has an exponential distribution with $\lambda = 25$ log-ons per hour. We are interested in the probability that X exceeds 6 minutes. Because λ is given in log-ons per hour, we express all time units in hours; that is, 6 minutes = 0.1 hour. The probability requested is shown as the shaded area under the probability density function in Fig. 3-35. Therefore,

Write the probability statement.

$$P(X > 0.1) = \int_{0.1}^{\infty} 25e^{-25x}\, dx = e^{-25(0.1)} = 0.082$$

Compute the probability.

An identical answer is obtained by expressing the mean number of log-ons as 0.417 log-ons per minute and computing the probability that the time until the next log-on exceeds 6 minutes. Try it!

What is the probability that the time until the next log-on is between 2 and 3 minutes?

Solution. On converting all units to hours,

$$P(0.033 < X < 0.05) = \int_{0.033}^{0.05} 25e^{-25x}\, dx = -e^{-25x}\, \Big|_{0.033}^{0.05} = 0.152$$

Determine the interval of time such that the probability that no log-on occurs in the interval is 0.90.

Solution. The question asks for the length of time x such that $P(X > x) = 0.90$. At the start of this section we determined that $P(X > x) = e^{-\lambda x}$. Now,

$$P(X > x) = e^{-25x} = 0.90$$

Therefore, from logarithms of both sides,

$$x = 0.00421 \text{ hour} = 0.25 \text{ minute}$$

Furthermore, the mean time until the next log-on is

$$E(X) = 1/25 = 0.04 \text{ hour} = 2.4 \text{ minutes}$$

The standard deviation of the time until the next log-on is

$$\sigma_X = 1/25 \text{ hours} = 2.4 \text{ minutes} \qquad\blacksquare$$

In the previous example, the probability that there are no log-ons in a 6-minute interval is 0.082 regardless of the starting time of the interval. A Poisson process assumes that events occur independently, with constant probability, throughout the interval of observation; that is, there is no clustering of events. If the log-ons are well modeled by a Poisson process, the probability that the first log-on after noon occurs after 12:06 P.M. is the same as the probability that the first log-on after 3:00 P.M. occurs after 3:06 P.M. If someone logs on at 2:22 P.M., the probability that the next log-on occurs after 2:28 P.M. is still 0.082.

Our starting point for observing the system does not matter. However, if there are high-use periods during the day, such as right after 8:00 A.M., followed by a period of low use, a Poisson process is not an appropriate model for log-ons, and the distribution is not appropriate for computing probabilities. It might be reasonable to model each of the high- and low-use periods by a separate Poisson process, employing a larger value for λ during the high-use periods and a smaller value otherwise. Then, an exponential distribution with the corresponding value of λ can be used to calculate log-on probabilities for the high- and low-use periods.

An even more interesting property of an exponential random variable is the **lack of memory property**. Suppose that there are no log-ons from 12:00 to 12:15; the probability that there are no log-ons from 12:15 to 12:21 is still 0.082. Because we have already been waiting for 15 minutes, we feel that we are "due." That is, the probability of a log-on in the next 6 minutes should be greater than 0.082. However, for an exponential distribution, this is not true. The lack of memory property is not that surprising when you consider the development of a Poisson process. In that development, we assumed that an interval could be partitioned into small intervals that were independent. The presence or absence of events in subintervals is similar to independent Bernoulli trials that comprise a binomial process; knowledge of previous results does not affect the probabilities of events in future subintervals.

The exponential distribution is often used in **reliability studies** as the model for the time until failure of a device. For example, the lifetime of a semiconductor device might be modeled as an exponential random variable with a mean of 40,000 hours. The lack of memory property of the exponential distribution implies that the device does not wear out. That is, regardless of how long the device has been operating, the probability of a failure in the next 1000 hours is the same as the probability of a failure in the first 1000 hours of operation. The lifetime of a device with failures caused by random shocks might be appropriately modeled as an exponential random variable. However, the lifetime of a device that suffers slow mechanical wear, such as bearing wear, is better modeled by a distribution that does not lack memory, such as the Weibull distribution (with $\beta \neq 1$).

EXERCISES FOR SECTION 3-9.2

3-118. Suppose X has an exponential distribution with $\lambda = 3$. Determine the following.

(a) $P(X \le 0)$ (b) $P(X \ge 3)$
(c) $P(X \le 2)$ (d) $P(2 < X < 3)$
(e) Find the value of x such that $P(X < x) = 0.05$.

3-119. Suppose X has an exponential distribution with mean equal to 5. Determine the following.

(a) $P(X > 5)$ (b) $P(X > 15)$ (c) $P(X > 20)$
(d) Find the value of x such that $P(X < x) = 0.95$.

3-120. Suppose the counts recorded by a Geiger counter follow a Poisson process with an average of three counts per minute.

(a) What is the probability that there are no counts in a 30-second interval?
(b) What is the probability that the first count occurs in less than 10 seconds?
(c) What is the probability that the first count occurs between 1 and 2 minutes after start-up?
(d) What is the mean time between counts?
(e) What is the standard deviation of the time between counts?
(f) Determine x, such that the probability that at least one count occurs before time x minutes is 0.95.

3-121. The time between calls to a health-care provider is exponentially distributed with a mean time between calls of 12 minutes.

(a) What is the probability that there are no calls within a 30-minute interval?
(b) What is the probability that at least one call arrives within a 10-minute interval?
(c) What is the probability that the first call arrives within 5 and 10 minutes after opening?
(d) Determine the length of an interval of time such that the probability of at least one call in the interval is 0.90.

3-122. A remote-controlled undersea vehicle detects debris from a sunken craft at a rate of 50 pieces per hour. The time to detect debris can be modeled using an exponential distribution.

(a) What is the probability that the time to detect the next piece of debris is less than 2 minutes?
(b) What is the probability that the time to detect the next piece of debris is between 2.5 and 4.5 minutes?
(c) What is the expected number of detected pieces of debris in a 20-minute interval?
(d) What is the probability that 2 pieces of debris are found in a 20-minute interval?

3-123. The distance between major cracks in a highway follows an exponential distribution with a mean of 5 miles.

(a) What is the probability that there are no major cracks in a 10-mile stretch of the highway?

(b) What is the probability that there are two major cracks in a 10-mile stretch of the highway?
(c) What is the standard deviation of the distance between major cracks?
(d) What is the probability that the first major crack occurs between 12 and 15 miles of the start of inspection?
(e) What is the probability that there are no major cracks in two separate 5-mile stretches of the highway?
(f) Given that there are no cracks in the first 5 miles inspected, what is the probability that there are no major cracks in the next 10 miles inspected?

3-124. The time to failure of a certain type of electrical component is assumed to follow an exponential distribution with a mean of 4 years. The manufacturer replaces free all components that fail while under guarantee.

(a) What percentage of the components will fail in 1 year?
(b) What is the probability that a component will fail in 2 years?
(c) What is the probability that a component will fail in 4 years?
(d) If the manufacturer wants to replace a maximum of 3% of the components, for how long should the manufacturer's stated guarantee on the component be?
(e) By redesigning the component, the manufacturer could increase the life. What does the mean time to failure have to be so that the manufacturer can offer a 1-year guarantee, yet still replace at most 3% of the components?

3-125. The time between the arrival of electronic messages at your computer is exponentially distributed with a mean of 2 hours.

(a) What is the probability that you do not receive a message during a 2-hour period?
(b) If you have not had a message in the last 4 hours, what is the probability that you do not receive a message in the next 2 hours?
(c) What is the expected time between your fifth and sixth message?

3-126. The time between calls to a corporate office is exponentially distributed with a mean of 10 minutes.

(a) What is the probability that there are more than three calls in $\frac{1}{2}$ hour?
(b) What is the probability that there are no calls within $\frac{1}{2}$ hour?
(c) Determine x such that the probability that there are no calls within x hours is 0.01.
(d) What is the probability that there are no calls within a 2-hour interval?
(e) If four nonoverlapping $\frac{1}{2}$-hour intervals are selected, what is the probability that none of these intervals contains any call?

3-10 NORMAL APPROXIMATION TO THE BINOMIAL AND POISSON DISTRIBUTIONS

Because a binomial random variable is the total count of *successes* from repeated independent trials, the central limit theorem can be applied. Consequently, it should not be surprising to use the normal distribution to approximate binomial probabilities for cases in which n is large. The following example illustrates that for many physical systems, the binomial model is appropriate with an extremely large value for n. In these cases, it is difficult to calculate probabilities by using the binomial distribution. Fortunately, the normal approximation is most effective in these cases. An illustration is provided in Fig. 3-36. Each bar in the figure has unit width, so the area of the bar over a value x equals the binomial probability of x. A normal distribution with $\mu = np = 5$ and $\sigma^2 = np(1 - p) = 2.5$ is superimposed. Note that the area of bars (binomial probability) can be approximated by the area under the normal curve (probability obtained from the normal distribution).

EXAMPLE 3-31

Bit Transmission Errors: Large Sample Size

In a digital communication channel, assume that the number of bits received in error can be modeled by a binomial random variable, and assume that the probability that a bit is received in error is 1×10^{-5}. If 16 million bits are transmitted, what is the probability that more than 150 errors occur?

Solution. Let the random variable X denote the number of errors. Then X is a binomial random variable and

$$P(X > 150) = 1 - P(X \le 150)$$
$$= 1 - \sum_{x=0}^{150} \binom{16,000,000}{x}(10^{-5})^x(1 - 10^{-5})^{16,000,000 - x} \qquad \blacksquare$$

Clearly, the probability in the previous example is difficult to compute. Fortunately, the normal distribution can be used to provide an excellent approximation in this example.

If X is a binomial random variable,

$$Z = \frac{X - np}{\sqrt{np(1 - p)}} \qquad (3\text{-}21)$$

is approximately a standard normal random variable. Consequently, probabilities computed from Z can be used to approximate probabilities for X.

Recall that for a binomial variable X, $E(X) = np$ and $V(X) = np(1 - p)$. Consequently, the normal approximation is nothing more than the formula for standardizing the random variable X. Probabilities involving X can be approximated by using a standard normal random variable. The normal approximation to the binomial distribution is good *if n is large enough relative to p*; in particular, whenever

$$np > 5 \quad \text{and} \quad n(1 - p) > 5$$

A correction factor can be used to further improve the approximation. Notice in Fig. 3-36 that the area of bars that represent a binomial probability such as $P(4 < X \le 7) = P(X = 5) + P(X = 6) + P(X = 7)$ is well approximated by the area under the normal curve

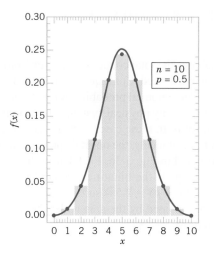

Figure 3-36   Normal approximation to
the binomial distribution.

between 4.5 and 7.5. Also notice that $P(X = 6)$ is well approximated by the area under the normal curve from 6.5 to 7.5. Consequently, $\pm\frac{1}{2}$ is added to the binomial values to improve the approximation. The rule of thumb is to apply the $\pm\frac{1}{2}$ correction factor in a manner that increases the values in the binomial probability that is to be approximated.

The digital communication problem is solved as follows:

$$P(X > 150) = P(X \geq 151) = P\left(\frac{X - 160}{\sqrt{160(1 - 10^{-5})}} > \frac{150.5 - 160}{\sqrt{160(1 - 10^{-5})}}\right)$$

$$= P(Z > -0.75) = P(Z < 0.75) = 0.773$$

EXAMPLE 3-32   Again consider the transmission of bits in the previous example. To judge how well the normal approximation works, assume that only $n = 50$ bits are to be transmitted and that the probability of an error is $p = 0.1$. The exact probability that 2 or fewer errors occur is

$$P(X \leq 2) = \binom{50}{0}0.9^{50} + \binom{50}{1}0.1(0.9^{49}) + \binom{50}{2}0.1^2(0.9^{48}) = 0.11$$

Based on the normal approximation,

$$P(X \leq 2) = P\left(\frac{X - 5}{\sqrt{50(0.1)(0.9)}} < \frac{2.5 - 5}{\sqrt{50(0.1)(0.9)}}\right) = P(Z < -1.18) = 0.12$$ ∎

For a sample as small as 50 bits, with $np = 5$, the normal approximation is reasonable. However, if np or $n(1 - p)$ is small, the binomial distribution is quite skewed and the symmetric normal distribution is not a good approximation. Two cases are illustrated in Fig. 3-37.

Recall that the Poisson distribution was developed as the limit of a binomial distribution as the number of trials increased to infinity. Consequently, the normal distribution can also be used to approximate probabilities of a Poisson random variable. The approximation is good for

$$\lambda > 5$$

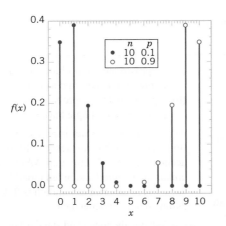

Figure 3-37   Binomial distribution is not symmetrical if p is near 0 or 1.

and a continuity correction can also be applied.

> If X is a Poisson random variable with $E(X) = \lambda$ and $V(X) = \lambda$,
>
> $$Z = \frac{X - \lambda}{\sqrt{\lambda}} \qquad (3\text{-}22)$$
>
> is approximately a standard normal random variable.

EXAMPLE 3-33
Water
Contaminants

Assume that the number of contamination particles in a liter water sample follows a Poisson distribution with a mean of 1000. If a sample is analyzed, what is the probability that fewer than 950 particles are found?

Solution. This probability can be expressed exactly as

$$P(X \le 950) = \sum_{x=0}^{950} \frac{e^{-1000} 1000^x}{x!}$$

The computational difficulty is clear. The probability can be approximated as

$$P(X \le x) \cong P\left(Z \le \frac{950.5 - 1000}{\sqrt{1000}}\right) = P(Z \le -1.57) = 0.059$$

Approximate the probability of more than 25 particles in 20 milliliters of water.

Solution. If the mean number of particles per liter is 1000, the mean per milliliter is 1, and the mean per 20 milliliters is 20. Let X denote the number of particles in 20 milliliters. Then X has a Poisson distribution with a mean of 20 and the requested probability is

$$P(X > 25) \approx P\left(Z > \frac{25.5 - 20}{\sqrt{20}}\right) = P(Z > 1.22) = 0.109$$

Animation 6: Binomial and Normal Distributions

EXERCISES FOR SECTION 3-10

3-127. Suppose that X has a binomial distribution with $n = 300$ and $p = 0.4$. Approximate the following probabilities.

(a) $P(X \le 100)$ (b) $P(80 \le X < 100)$
(c) $P(X > 130)$

3-128. Suppose that X has a Poisson distribution with mean of 50. Approximate the following probabilities.

(a) $P(X < 45)$ (b) $P(X \le 45)$
(c) $P(40 < X \le 60)$ (d) $P(X > 55)$

3-129. A particular vendor produces parts with a defect rate of 8%. Incoming inspection to a manufacturing plant samples 100 delivered parts from this vendor and rejects the delivery if 8 defective parts are discovered.

(a) Compute the exact probability that the inspector accepts delivery.
(b) Approximate the probability of acceptance and compare the result to part (a).

3-130. A large electronic office product contains 2000 electronic components. Assume that the probability that each component operates without failure during the useful life of the product is 0.995, and assume that the components fail independently. Approximate the probability that 5 or more of the original 2000 components fail during the useful life of the product.

3-131. The manufacturing of semiconductor chips produces 2% defective chips. Assume that the chips are independent and that a lot contains 1000 chips.

(a) Approximate the probability that more than 25 chips are defective.
(b) Approximate the probability that between 20 and 30 chips are defective.

3-132. There were 49.7 million people with some type of long-lasting condition or disability living in the United States in 2000. This represented 19.3 percent of the majority of civilians aged five and over (http://factfinder.census.gov). A sample of 1000 persons is selected and, it can be assumed the disability statuses of these individuals are independent.

(a) Approximate the probability that more than 200 persons in the sample have a disability.
(b) Approximate the probability that between 180 and 300 people in the sample have a disability.

3-133. Phoenix water is provided to approximately 1.4 million people, who are served through more than 362,000 accounts (http://phoenix.gov/WATER/wtrfacts.html). All accounts are metered and billed monthly. The probability that an account has an error in a month is 0.001, and accounts can be assumed to be independent.

(a) What is the mean and standard deviation of the number of account errors each month?
(b) Approximate the probability of fewer than 350 errors in a month.
(c) Approximate a value so that the probability that the number of errors exceeding this value is 0.05.

3-134. Suppose that the number of asbestos particles in a sample of 1 squared centimeter of dust is a Poisson random variable with a mean of 1000. Apporximate the probability that 10 squared centimeters of dust contain more than 10,000 particles.

3-135. The number of spam emails received each day follows a Poisson distribution with a mean of 50. Approximate the following probabilities.

(a) More than 40 and less than 60 spam emails in a day.
(b) At least 40 spam emails in a day.
(c) Less than 40 spam emails in a day.
(d) Approximate the probability that the total number of spam emails exceeds 340 in a seven-day week.

3-136. The number of calls to a health-care provider follows a Poisson distribution with a mean of 36 per hour. Approximate the following probabilities.

(a) More than 42 calls in an hour
(b) Less than 30 calls in an hour
(c) More than 300 calls in an eight-hour day

3-11 MORE THAN ONE RANDOM VARIABLE AND INDEPENDENCE

3-11.1 Joint Distributions

In many experiments, more than one variable is measured. For example, suppose both the diameter and thickness of an injection-molded disk are measured and denoted by X and Y, respectively. These two random variables are often related. If pressure in the mold increases, there might be an increase in the fill of the cavity that results in larger values for both X and Y. Similarly, a pressure decrease might result in smaller values for both X and Y. Suppose that diameter and thickness measurements from many parts are plotted in an X–Y plane (scatter diagram). As shown in Fig. 3-38, the relationship between X and Y implies that some

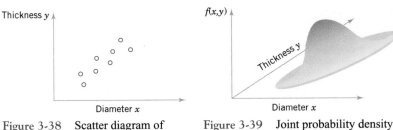

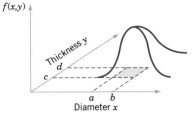

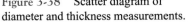

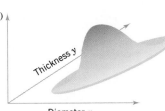

Figure 3-38   Scatter diagram of diameter and thickness measurements.

Figure 3-39   Joint probability density function of x and y.

Figure 3-40   Probability of a region is the volume enclosed by $f(x, y)$ over the region.

regions of the X–Y plane are more likely to contain measurements than others. This concept was discussed in Section 2-6 when the sample correlation coefficient was defined.

This tendency can be modeled by a probability density function [denoted as $f(x, y)$] over the X–Y plane as shown in Fig. 3-39. The analogies that related a probability density function to the loading on a long, thin beam can be applied to relate this two-dimensional probability density function to the density of a loading over a large, flat surface. The probability that the random experiment (part production) generates measurements in a region of the X–Y plane is determined from the integral of $f(x, y)$ over the region as shown in Fig. 3-40. This integral equals the volume enclosed by $f(x, y)$ over the region. Because $f(x, y)$ determines probabilities for two random variables, it is referred to as a **joint probability density function.** From Fig. 3-40, the probability that a part is produced in the region shown is

$$P(a < X < b, c < Y < d) = \int_a^b \int_c^d f(x, y)\, dy\, dx$$

Similar concepts can be applied to discrete random variables. For example, suppose the quality of each bit received through a digital communications channel is categorized into one of four classes, "excellent," "good," "fair," and "poor," denoted by E, G, F, and P, respectively. Let the random variables X, Y, W, and Z denote the numbers of bits that are E, G, F, and P, respectively, in a transmission of 20 bits. In this example, we are interested in the joint probability distribution of four random variables. To simplify, we only consider X and Y. The joint probability distribution of X and Y can be specified by a **joint probability mass function** $f(x, y) = P(X = x, Y = y)$. Because each of the 20 bits is categorized into one of the four classes, $X + Y + W + Z = 20$, only integers such that $X + Y \le 20$ have positive probability in the joint probability mass function of X and Y. The joint probability mass function is zero elsewhere. For a general discussion of joint distributions, we refer the interested reader to Montgomery and Runger (2006). Instead, we focus here on the important special case of independent random variables.

3-11.2   Independence

If we make some assumptions regarding our probability models, a probability involving more than one random variable can often be simplified. In Example 3-12, the probability that a diameter meets specifications was determined to be 0.919. What can we say about 10 such diameters? What is the probability that they all meet specifications? This is the type of question of interest to a customer of optical drives.

Such questions lead to an important concept and definition. To accommodate more than just the two random variables X and Y, we adopt the notation $X_1, X_2, \ldots, X_n$ to represent n random variables.

Independence

> The random variables $X_1, X_2, \ldots, X_n$ are **independent** if
>
> $$P(X_1 \in E_1, X_2 \in E_2, \ldots, X_n \in E_n) = P(X_1 \in E_1)P(X_2 \in E_2) \cdots P(X_n \in E_n)$$
>
> for *any* sets $E_1, E_2, \ldots, E_n$.

The importance of independence is illustrated in the following example.

EXAMPLE 3-34

Optical Drive Diameters

In Example 3-12, the probability that a diameter meets specifications was determined to be 0.919. What is the probability that 10 diameters all meet specifications, assuming that the diameters are independent?

Solution. Denote the diameter of the first shaft as X_1, the diameter of the second shaft as X_2, and so forth, so that the diameter of the tenth shaft is denoted as X_{10}. The probability that all shafts meet specifications can be written as

$$P(0.2485 < X_1 < 0.2515, 0.2485 < X_2 < 0.2515, \ldots, 0.2485 < X_{10} < 0.2515)$$

In this example, the only set of interest is

$$E_1 = (0.2485, 0.2515)$$

With respect to the notation used in the definition of independence,

$$E_1 = E_2 = \cdots = E_{10}$$

Recall the relative frequency interpretation of probability. The proportion of times that shaft 1 is expected to meet the specifications is 0.919, the proportion of times that shaft 2 is expected to meet the specifications is 0.919, and so forth. If the random variables are independent, the proportion of times in which we measure 10 shafts that we expect all to meet the specifications is

$$P(0.2485 < X_1 < 0.2515, 0.2485 < X_2 < 0.2515, \ldots, 0.2485 < X_{10} < 0.2515)$$
$$= P(0.2485 < X_1 < 0.2515) \times P(0.2485 < X_2 < 0.2515) \times \cdots \times P(0.2485 < X_{10} < 0.2515) = 0.919^{10} = 0.430$$

Independent random variables are fundamental to the analyses in the remainder of the book. We often assume that random variables that record the replicates of a random experiment are independent. Really what we assume is that the ϵ_i disturbances (for $i = 1, 2, \ldots, n$ replicates) in the model

$$X_i = \mu_i + \epsilon_i$$

are independent because it is the disturbances that generate the randomness and the probabilities associated with the measurements.

Note that independence implies that the probabilities can be multiplied for *any* sets E_1, $E_2, \ldots, E_n$. Therefore, it should not be surprising to learn that an equivalent definition of independence is that the joint probability density function of the random variables equals the product of the probability density function of each random variable. This definition also holds for the joint probability mass function if the random variables are discrete.

EXAMPLE 3-35

Coating
Thickness

Suppose X_1, X_2, and X_3 represent the thickness in micrometers of a substrate, an active layer, and a coating layer of a chemical product, respectively. Assume that X_1, X_2, and X_3 are independent and normally distributed with $\mu_1 = 10000$, $\mu_2 = 1000$, $\mu_3 = 80$, $\sigma_1 = 250$, $\sigma_2 = 20$, and $\sigma_3 = 4$. The specifications for the thickness of the substrate, active layer, and coating layer are $9200 < x_1 < 10800$, $950 < x_2 < 1050$, and $75 < x_3 < 85$, respectively. What proportion of chemical products meets all thickness specifications? Which one of the three thicknesses has the least probability of meeting specifications?

Solution. The requested probability is $P(9200 < X_1 < 10800, 950 < X_2 < 1050, 75 < X_3 < 85)$. Using the notation in the definition of independence, $E_1 = (9200, 10,800)$, $E_2 = (950, 1050)$, and $E_3 = (75, 85)$ in this example. Because the random variables are independent,

$$P(9200 < X_1 < 10800, 950 < X_2 < 1050, 75 < X_3 < 85)$$
$$= P(9200 < X_1 < 10800)P(950 < X_2 < 1050)P(75 < X_3 < 85)$$

After standardizing, the above equals

$$P(-3.2 < Z < 3.2)P(-2.5 < Z < 2.5)P(-1.25 < Z < 1.25)$$

where Z is a standard normal random variable. From the table of the standard normal distribution, the above equals

$$(0.99862)(0.98758)(0.78870) = 0.7778$$

The thickness of the coating layer has the least probability of meeting specifications. Consequently, a priority should be to reduce variability in this part of the process. ∎

The concept of independence can also be applied to experiments that classify results. We used this concept to derive the binomial distribution. Recall that a test taker who just guesses from four multiple choices has probability 1/4 that any question is answered correctly. If it is assumed that the correct or incorrect outcome from one question is independent of others, the probability that, say, five questions are answered correctly can be determined by multiplication to equal

$$(1/4)^5 = 0.00098$$

Some additional applications of independence frequently occur in the area of system analysis. Consider a system that consists of devices that are either functional or failed. It is assumed that the devices are independent.

EXAMPLE 3-36

Series System

The system shown here operates only if there is a path of functional components from left to right. The probability that each component functions is shown in the diagram. Assume that the components function or fail independently. What is the probability that the system operates?

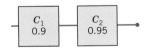

Solution. Let C_1 and C_2 denote the events that components 1 and 2 are functional, respectively. For the system to operate, both components must be functional. The probaility that the system operates is

$$P(C_1, C_2) = P(C_1)P(C_2) = (0.9)(0.95) = 0.855$$

Note that the probability that the system operates is smaller than the probability that any component operates. This system fails whenever *any* component fails. A system of this type is called a **series system.** ∎

EXAMPLE 3-37
Parallel System

The system shown here operates only if there is a path of functional components from left to right. The probability that each component functions is shown. Assume that the components function or fail independently. What is the probability that the system operates?

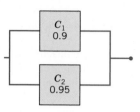

Solution. Let C_1 and C_2 denote the events that components 1 and 2 are functional, respectively. Also, C_1' and C_2' denote the events that components 1 and 2 fail, respectively, with associated probabilities $P(C_1')$ = 1 − 0.9 = 0.1 and $P(C_2')$ = 1 − 0.95 = 0.05. The system will operate if either component is functional. The probability that the system operates is 1 minus the probability that the system fails, and this occurs whenever both independent components fail. Therefore, the requested probability is

$$P(C_1 \text{ or } C_2) = 1 - P(C_1', C_2') = 1 - P(C_1')P(C_2') = 1 - (0.1)(0.05) = 0.995$$

Note that the probability that the system operates is greater than the probability that any component operates. This is a useful design strategy to decrease system failures. This system fails only if *all* components fail. A system of this type is called a **parallel system.** ∎

More general results can be obtained. The probability that a component does not fail over the time of its mission is called its **reliability.** Suppose that r_i denotes the reliability of component i in a system that consists of k components and that r denotes the probability that the system does not fail over the time of the mission. That is, r can be called the system reliability. The previous examples can be extended to obtain the following for a series system

$$r = r_1 r_2 \cdots r_k$$

and for a parallel system

$$r = 1 - (1 - r_1)(1 - r_2) \cdots (1 - r_k)$$

The analysis of a complex system can be accomplished by a partition into subsystems, which are sometimes called blocks.

EXAMPLE 3-38
Complex System

The system shown here operates only if there is a path of functional components from left to right. The probability that each component functions is shown. Assume that the components function or fail independently. What is the probability that the system operates?

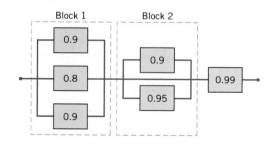

Solution. The system can be partitioned into blocks such that exclusively parallel subsystems exist in each block. The result for a parallel system can be applied to each block, and the block results can be combined by the analysis for a series system. For block 1, the reliability is obtained from the result for a parallel system to be

$$1 - (0.1)(0.2)(0.1) = 0.998$$

Similarly, for block 2, the reliability is

$$1 - (0.1)(0.05) = 0.995$$

The system reliability is determined from the result for a series system to be

$$(0.998)(0.995)(0.99) = 0.983$$

■

EXERCISES FOR SECTION 3-11

3-137. Let X be a normal random variable with $\mu = 10$ and $\sigma = 1.5$ and Y be a normal random variable with $\mu = 2$ and $\sigma = 0.25$. Assume X and Y are independent. Find the following probabilities.

(a) $P(X < 9, Y < 2.5)$
(b) $P(X > 8, Y < 2.25)$
(c) $P(8.5 \leq X \leq 11.5, Y > 1.75)$
(d) $P(X < 13, 1.5 \leq Y \leq 1.8)$

3-138. Let X be a normal random variable with $\mu = 15.0$ and $\sigma = 3$ and Y be a normal random variable with $\mu = 20$ and $\sigma = 1$. Assume X and Y are independent. Find the following probabilities.

(a) $P(X < 12, Y < 19)$
(b) $P(X > 16, Y < 18)$
(c) $P(14 \leq X < 16, Y > 22)$
(d) $P(11 \leq X \leq 20, 17.5 \leq Y \leq 21)$

3-139. Let X be a Poisson random variable with $\lambda = 2$ and Y be a Poisson random variable with $\lambda = 4$. Assume X and Y are independent. Find the following probabilities.

(a) $P(X < 4, Y < 4)$ (b) $P(X > 2, Y < 4)$
(c) $P(2 \leq X < 4, Y \geq 3)$ (d) $P(X < 5, 1 \leq Y \leq 4)$

3-140. Let X be an exponential random variable with mean equal to 5 and Y be an exponential random variable with mean equal to 8. Assume X and Y are independent. Find the following probabilities.

(a) $P(X \leq 5, Y \leq 8)$ (b) $P(X > 5, Y \leq 6)$
(c) $P(3 < X \leq 7, Y > 7)$ (d) $P(X > 7, 5 < Y \leq 7)$

3-141. Two independent vendors supply cement to a highway contractor. Through previous experience it is known that the compressive strength of samples of cement can be modeled by a normal distribution, with $\mu_1 = 6000$ kilograms per square centimeter and $\sigma_1 = 100$ kilograms per square centimeter for vendor 1, and $\mu_2 = 5825$ and $\sigma_2 = 90$ for vendor 2. What is the probability that both vendors supply a sample with compressive strength

(a) Less than 6100 kg/cm²?
(b) Between 5800 and 6050?
(c) In excess of 6200?

3-142. The time between surface finish problems in a galvanizing process is exponentially distributed with a mean of 40 hours. A single plant operates three galvanizing lines that are assumed to operate independently.

(a) What is the probability that none of the lines experiences a surface finish problem in 40 hours of operation?
(b) What is the probability that all three lines experience a surface finish problem between 20 and 40 hours of operation?

3-143. The inside thread diameter of plastic caps made using an injection molding process is an important quality characteristic. The mold has four cavities. A cap made using cavity i is considered independent from any other cap and can have one of three quality levels: first, second, or third (worst). The notation $P(F_i)$, $P(S_i)$, and $P(T_i)$ represents the probability that a cap made using cavity i has first, second, or third quality, respectively. Given $P(F_1) = 0.4$, $P(S_1) = 0.25$, $P(F_2) = 0.25$, $P(S_2) = 0.30$, $P(F_3) = 0.35$, $P(S_3) = 0.40$, $P(F_4) = 0.5$, and $P(S_4) = 0.40$,

(a) List the probability of third-level quality for each of the cavities.
(b) What is the probability that one production lot (i.e., a cap from each cavity) has four caps of first quality?
(c) What is the probability that one production lot has four caps with quality at the first or second level?

3-144. The yield in pounds from a day's production is normally distributed with a mean of 1500 pounds and a variance of 10,000 pounds squared. Assume that the yields on different days are independent random variables.

(a) What is the probability that the production yield exceeds 1400 pounds on each of 5 days?
(b) What is the probability that the production yield exceeds 1400 pounds on none of the next 5 days?

3-145.   Consider the series system described in Example 3-36. Suppose that the probability that component C_1 functions is 0.95 and that the probability that component C_2 functions is 0.92. What is the probability that the system operates?

3-146.   Suppose a series system has three components C_1, C_2, and C_3 with the probability that each component functions equal to 0.90, 0.99, and 0.95, respectively. What is the probability that the system operates?

3-147.   Consider the parallel system described in Example 3-37. Suppose the probability that component C_1 functions is 0.85 and the probability that component C_2 functions is 0.92. What is the probability that the system operates?

3-148.   Suppose a parallel system has three components C_1, C_2, and C_3, in parallel, with the probability that each component functions equal to 0.90, 0.99, and 0.95, respectively. What is the probability that the system operates?

3-149.   The following circuit operates if and only if there is a path of functional devices from left to right. The probability that each device functions is as shown. Assume that the probability that a device is functional does not depend on whether or not other devices are functional. What is the probability that the circuit operates?

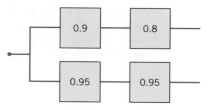

3-150.   The following circuit operates if and only if there is a path of functional devices from left to right. The probability that each device functions is as shown. Assume that the probability that a device functions does not depend on whether or not other devices are functional. What is the probability that the circuit operates?

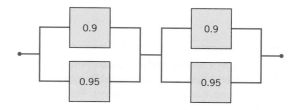

3-12   FUNCTIONS OF RANDOM VARIABLES

In many practical problems, a random variable is defined as a function of one or more other random variables. There are methods to determine the probability distribution of a function of one or more random variables and to find important properties such as the mean and variance. A more complete treatment of this general topic is in Montgomery and Runger (2006). In this section, we present a few of the most useful results.

We begin with a few simple properties. Let X be a random variable with mean μ and variance σ^2, and let c be a constant. Define a new random variable Y as

$$Y = X + c$$

From the definition of expectation and variance (see equation 3-3), it follows that

$$E(Y) = E(X) + c = \mu + c \qquad (3\text{-}23)$$
$$V(Y) = V(X) + 0 = \sigma^2 \qquad (3\text{-}24)$$

That is, adding a constant to a random variable increases the mean by the value of the constant, but the variance of the random variable is unchanged.

Now suppose that the random variable X is multiplied by a constant, resulting in

$$Y = cX$$

In this case we have

$$E(Y) = E(cX) = cE(X) = c\mu \qquad (3\text{-}25)$$
$$V(Y) = V(cX) = c^2 V(X) = c^2 \sigma^2 \qquad (3\text{-}26)$$

So the mean of a random variable multiplied by a constant is equal to the constant times the mean of the original random variable, but the variance of a random variable multiplied by a constant is the *square* of the constant times the variance of the original random variable. We now consider some additional cases involving several random variables. The results in equations 3-23 through 3-26 will be useful.

3-12.1 Linear Functions of Independent Random Variables

Many situations involve a linear function of random variables. For example, suppose that the random variables X_1 and X_2 represent the length and width, respectively, of a manufactured part. For X_1, suppose that we know that $\mu_1 = 2$ centimeters and $\sigma_1 = 0.1$ centimeters and for X_2, we know that $\mu_5 = 5$ centimeters and $\sigma_2 = 0.2$ centimeters. Also, assume that X_1 and X_2 are independent. We wish to determine the mean and standard deviation of the perimeter of the part. We assume that the width of both sides is the same and that the length of the top and bottom is the same so that the part is always rectangular.

Now the perimeter of the part is

$$Y = 2X_1 + 2X_2$$

and we need to find the mean and standard deviation of Y. This problem is a special case of finding the mean and variance (or equivalently, the standard deviation) of a linear function of **independent** random variables.

Let $c_0, c_1, c_2, \ldots, c_n$ be constants, and let $X_1, X_2, \ldots, X_n$ be independent random variables with means $E(X_i) = \mu_i, i = 1, 2, \ldots, n$ and variances $V(X_i) = \sigma_i^2, i = 1, 2, \ldots, n$.

Mean and Variance of a Linear Function: Independent Random Variables

The mean and variance of the linear function of **independent** random variables are

$$Y = c_0 + c_1 X_1 + c_2 X_2 + \cdots + c_n X_n$$
$$E(Y) = c_0 + c_1 \mu_1 + c_2 \mu_2 + \cdots + c_n \mu_n \qquad (3\text{-}27)$$

and

$$V(Y) = c_1^2 \sigma_1^2 + c_2^2 \sigma_2^2 + \cdots + c_n^2 \sigma_n^2 \qquad (3\text{-}28)$$

EXAMPLE 3-39

Perimeter of a Molded Part

Reconsider the manufactured part described above where the random variables X_1 and X_2 represent the length and width, respectively. For the length we know that $\mu_1 = 2$ centimeters and $\sigma_1 = 0.1$ centimeters and for the width X_2 we know that $\mu_5 = 5$ centimeters and $\sigma_2 = 0.2$ centimeters. The perimeter of the part $Y = 2X_1 + 2X_2$ is just a linear combination of the length and width. Using equations 3-27 and 3-28, the mean of the perimeter is

$$E(Y) = 2E(X_1) + 2E(X_2) = 2(2) + 2(5) = 14 \text{ centimeters}$$

and the variance of the perimeter is

$$V(Y) = 2^2(0.1^2) + 2^2(0.2^2) = 0.2 \text{ square centimeters}$$

Therefore, the standard deviation of the perimeter of the part is

$$\sigma_Y = \sqrt{V(Y)} = \sqrt{0.2} = 0.447 \text{ centimeters}$$ ■

A very important case occurs when all of the random variables $X_1, X_2, \ldots, X_n$ in the linear function are **independent** and **normally distributed.**

Linear Function of Independent Normal Random Variables

Let $X_1, X_2, \ldots, X_n$ be independent, normally distributed random variables with means $E(X_i) = \mu_i$, $i = 1, 2, \ldots, n$ and variances $V(X_i) = \sigma_i^2$, $i = 1, 2, \ldots, n$. Then the linear function

$$Y = c_0 + c_1 X_1 + c_2 X_2 + \cdots + c_n X_n$$

is normally distributed with mean

$$E(Y) = c_0 + c_1\mu_1 + c_2\mu_2 + \cdots + c_n\mu_n$$

and variance

$$V(Y) = c_1^2\sigma_1^2 + c_2^2\sigma_2^2 + \cdots + c_n^2\sigma_n^2$$

EXAMPLE 3-40

Perimeter of a Molded Part: Normal Distribution

Once again, consider the manufactured part described previously. Now suppose that the length X_1 and the width X_2 are normally and independently distributed with $\mu_1 = 2$ centimeters, $\sigma_1 = 0.1$ centimeters, $\mu_5 = 5$ centimeters, and $\sigma_2 = 0.2$ centimeters. In the previous example we determined that the mean and variance of the perimeter of the part $Y = 2X_1 + 2X_2$ were $E(Y) = 14$ centimeters and $V(Y) = 0.2$ square centimeters, respectively. Determine the probability that the perimeter of the part exceeds 14.5 centimeters.

Solution. From the above result, Y is also a normally distributed random variable, so we may calculate the desired probability as follows:

$$P(Y > 14.5) = P\left(\frac{Y - \mu_Y}{\sigma_Y} > \frac{14.5 - 14}{0.447}\right) = P(Z > 1.12) = 0.13$$

Therefore, the probability is 0.13 that the perimeter of the part exceeds 14.5 centimeters.   ■

3-12.2   What If the Random Variables Are Not Independent?

After reading the previous section, a very logical question arises: What if the random variables in the linear function are not independent? The independence assumption is quite important. Let's consider a very simple case

$$Y = X_1 + X_2$$

where the two random variables X_1 and X_2 have means μ_1 and μ_2 and variances σ_1^2 and σ_2^2 but where X_1 and X_2 are not independent. The mean of Y is still

$$E(Y) = E(X_1 + X_2) = E(X_1) + E(X_2) = \mu_1 + \mu_2$$

That is, the mean of Y is just the sum of the means of the two random variables X_1 and X_2. The variance of Y, using equation 3-3, is

$$\begin{aligned} V(Y) &= E(Y^2) - E(Y)^2 \\ &= E[(X_1 + X_2)^2] - [E(X_1 + X_2)]^2 \end{aligned}$$

Now the $E(X_1 + X_2) = \mu_1 + \mu_2$, so this last equation becomes

$$\begin{aligned} V(Y) &= E(X_1^2 + X_2^2 + 2X_1X_2) - \mu_1^2 - \mu_2^2 - 2\mu_1\mu_2 \\ &= E(X_1^2) + E(X_2^2) + 2E(X_1X_2) - \mu_1^2 - \mu_2^2 - 2\mu_1\mu_2 \\ &= [E(X_1^2) - \mu_1^2] + [E(X_2^2) - \mu_2^2] + 2E(X_1X_2) - 2\mu_1\mu_2 \\ &= \sigma_1^2 + \sigma_2^2 + 2[E(X_1X_2) - \mu_1\mu_2] \end{aligned}$$

The quantity $E(X_1X_2) - \mu_1\mu_2$ is called the **covariance** of the random variables X_1 and X_2. When the two random variables X_1 and X_2 are independent, the covariance $E(X_1X_2) - \mu_1\mu_2 = 0$, and we get the familiar result for the special case in equation 3-28, namely, $V(Y) = \sigma_1^2 + \sigma_2^2$. The covariance is a measure of the linear relationship between the two random variables X_1 and X_2. When the covariance is not zero, the random variables X_1 and X_2 are not independent. The covariance is closely related to the **correlation** between the random variables X_1 and X_2; in fact, the correlation between X_1 and X_2 is defined as follows.

Correlation

> The correlation between two random variables X_1 and X_2 is
>
> $$\rho_{X_1X_2} = \frac{E(X_1X_2) - \mu_1\mu_2}{\sqrt{\sigma_1^2\sigma_2^2}} = \frac{\text{Cov}(X_1, X_2)}{\sqrt{\sigma_1^2\sigma_2^2}} \qquad (3\text{-}29)$$
>
> with $-1 \le \rho_{X_1X_2} \le +1$, and $\rho_{X_1X_2}$ is usually called the **correlation coefficient.**

Because the variances are always positive, if the covariance between X_1 and X_2 is negative, zero, or positive, the correlation between X_1 and X_2 is also negative, zero, or positive, respectively. However, because the correlation coefficient lies in the interval from -1 to $+1$, it is easier to interpret than the covariance. Furthermore, the **sample correlation coefficient** introduced in Section 2-6 (see equation 2-6) is usually employed to estimate the correlation coefficient from sample data. You may find it helpful to reread the discussion on the sample correlation coefficient in Section 2-6.

A general result for a linear function of random variables uses the covariances between pairs of variables.

**Mean and
Variance of a
Linear
Function:
General Case**

Let $X_1, X_2, \ldots, X_n$ be random variables with means $E(X_i) = \mu_i$ and variances $V(X_i) = \sigma_i^2$, $i = 1, 2, \ldots, n$, and covariances $\text{Cov}(X_1, X_2)$, $i, j = 1, 2, \ldots, n$ with $i < j$. Then the mean of the linear combination

$$Y = c_0 + c_1 X_1 + c_2 X_2 + \cdots + c_n X_n$$

is

$$E(Y) = c_0 + c_1\mu_1 + c_2\mu_2 + \cdots + c_n\mu_n \qquad (3\text{-}30)$$

and the variance is

$$V(Y) = c_1^2\sigma_1^2 + c_2^2\sigma_2^2 + \cdots + c_n^2\sigma_n^2 + 2\sum_{i<j}\sum c_i c_j \,\text{Cov}(X_i, X_j) \qquad (3\text{-}31)$$

3-12.3 What If the Function Is Nonlinear?

Many problems in engineering involve nonlinear functions of random variables. For example, the power P dissipated by the resistance R in an electrical circuit is given by the relationship

$$P = I^2 R$$

where I is the current. If the resistance is a known constant and the current is a random variable, the power is a random variable that is a nonlinear function of the current. As another example, the period T of a pendulum is given by

$$T = 2\pi \sqrt{L/g}$$

where L is the length of the pendulum and g is the acceleration due to gravity. If g is a constant and the length L is a random variable, the period of the pendulum is a nonlinear function of a random variable. Finally, we can experimentally measure the acceleration due to gravity by dropping a baseball and measuring the time T is takes for the ball to travel a known distance d. The relationship is

$$G = 2d/T^2$$

Because the time T in this experiment is measured with error, it is a random variable. Therefore, the acceleration due to gravity is a nonlinear function of the random variable T.

In general, suppose that the random variable Y is a function of the random variable X, say,

$$Y = h(X)$$

then a general solution for the mean and variance of Y can be difficult. It depends on the complexity of the function $h(X)$. However, if a linear approximation to $h(X)$ can be used, an approximate solution is available.

Propagation
of Error
Formula:
Single
Variable

If X has mean μ_X and variance σ_X^2, the approximate mean and variance of Y can be computed using the following result:

$$E(Y) = \mu_Y \simeq h(\mu_X) \qquad (3\text{-}32)$$

$$V(Y) = \sigma_Y^2 \simeq \left(\frac{dh}{dX}\right)^2 \sigma_X^2 \qquad (3\text{-}33)$$

where the derivative dh/dX is evaluated at μ_X.

Engineers usually call equation 3-33 the **transmission of error** or **propagation of error formula.**

EXAMPLE 3-41

Power in a
Circuit

The power P dissipated by the resistance R in an electrical circuit is given by $P = I^2R$ where I, the current, is a random variable with mean $\mu_I = 20$ amperes and standard deviation $\sigma_I = 0.1$ amperes. The resistance $R = 80$ ohms is a constant. We want to find the approximate mean and standard deviation of the power. In this problem the function $h = I^2R$, so taking the derivative $dh/dI = 2IR = 2I(80)$ and applying the equations 3-32 and 3-33, we find that the approximate mean power is

$$E(P) = \mu_P \simeq h(\mu_I) = \mu_I R = 20(80) = 1600 \text{ watts}$$

and the approximate variance of power is

$$V(P) = \sigma_P^2 \simeq \left(\frac{dh}{dI}\right)^2 \sigma_I^2 = [2(20)(80)]^2 0.1^2 = 102{,}400 \text{ square watts}$$

So the standard deviation of the power is $\sigma_P \simeq 320$ watts. Remember that the derivative dh/dI is evaluated at $\mu_I = 20$ amperes. ∎

Equations 3-32 and 3-33 are developed by approximating the nonlinear function h with a linear function. The linear approximation is found by using a first-order Taylor series. Assuming that $h(X)$ is differentiable, the first-order Taylor series approximation for $Y = h(X)$ around the point μ_X is

$$Y \simeq h(\mu_X) + \frac{dh}{dX}(X - \mu_X) \qquad (3\text{-}34)$$

Now dh/dX is a constant when it is evaluated at μ_X, $h(\mu_X)$ is a constant, and $E(X) = \mu_X$, so when we take the expected value of Y, the second term in equation 3-34 is zero and consequently

$$E(Y) \simeq h(\mu_X)$$

The approximate variance of Y is

$$V(Y) \simeq V[h(\mu_x)] + V\left[\frac{dh}{dX}(X - \mu_X)\right] = \left(\frac{dh}{dX}\right)^2 \sigma_X^2$$

which is the transmission of error formula in equation 3-33. The Taylor series method that we used to find the approximate mean and variance of Y is usually called the **delta method.**

Sometimes the variable Y is a nonlinear function of several random variables, say,

$$Y = h(X_1, X_2, \ldots, X_n) \tag{3-35}$$

where $X_1, X_2, \ldots, X_n$ are assumed to be independent random variables with means $E(X_i) = \mu_i$ and variances $V(X_i) = \sigma_i^2, i = 1, 2, \ldots, n$. The delta method can be used to find approximate expressions for the mean and variance of Y. The first-order Taylor series expansion of equation 3-35 is

$$Y \simeq h(\mu_1, \mu_2, \ldots, \mu_n) + \frac{\partial h}{\partial X_1}(X_1 - \mu_1) + \frac{\partial h}{\partial X_2}(X_2 - \mu_2) + \cdots + \frac{\partial h}{\partial X_n}(X_n - \mu_n)$$

$$= h(\mu_1, \mu_2, \ldots, \mu_n) + \sum_{i=1}^{n} \frac{\partial h}{\partial X_i}(X_i - \mu_i) \tag{3-36}$$

Taking the expectation and variance of Y in equation 3-36 (with use of the linear combination formulas in equations 3-27 and 3-28) produces the following results.

Propagation of Error Formula: Multiple Variables

Let

$$Y = h(X_1, X_2, \ldots, X_n)$$

for independent random variables $X_i, i = 1,2,\ldots, n$, each with mean μ_i and variance σ_i^2, the approximate mean and variance of Y are

$$E(Y) = \mu_Y \simeq h(\mu_1, \mu_2, \ldots, \mu_n) \tag{3-37}$$

$$V(Y) = \sigma_Y^2 \simeq \sum_{i=1}^{n} \left(\frac{\partial h}{\partial X_i}\right)^2 \sigma_i^2 \tag{3-38}$$

where the partial derivatives $\partial h / \partial X_i$ are evaluated at $\mu_1, \mu_2, \ldots, \mu_n$.

EXAMPLE 3-42

Resistances in Parallel

Two resistors are connected in parallel. The resistances R_1 and R_2 are random variables with $E(R_1) = \mu_{R_1} = 20$ ohms, $V(R_1) = \sigma_{R_1}^2 = 0.5$ square ohms, and $E(R_2) = \mu_{R_2} = 50$ ohms, $V(R_2) = \sigma_{R_2}^2 = 1$ square ohm. Determine the mean and standard deviation of the combined resistance, which is given by

$$R = \frac{R_1 R_2}{R_1 + R_2}$$

The approximate mean of R is

$$E(R) = \mu_R \simeq \frac{20(50)}{20 + 50} = 14.29 \text{ ohms}$$

The partial derivatives, evaluated at μ_{R_1} and μ_{R_2} are

$$\frac{\partial R}{\partial R_1} = \left(\frac{R_2}{R_1 + R_2}\right)^2 = \left(\frac{50}{20 + 50}\right)^2 = 0.5102$$

$$\frac{\partial R}{\partial R_2} = \left(\frac{R_1}{R_1 + R_2}\right)^2 = \left(\frac{20}{20 + 50}\right)^2 = 0.0816$$

From equation 3-38, the approximate variance of R is

$$V(R) = \sigma_R^2 \approx \left(\frac{\partial R}{\partial R_1}\right)^2 \sigma_{R_1}^2 + \left(\frac{\partial R}{\partial R_2}\right)^2 \sigma_{R_2}^2$$

$$\approx (0.5102)^2(0.5) + (0.0816)^2(1)$$

$$\approx 0.1367\Omega^2$$

The standard deviation of R is $\sigma_R \approx 0.3698$ ohms. ■

EXERCISES FOR SECTION 3-12

3-151. If X_1 and X_2 are independent random variables with $E(X_1) = 2$, $E(X_2) = 5$, $V(X_1) = 2$, $V(X_2) = 10$, and $Y = 3X_1 + 5X_2$, determine the following.

(a) $E(Y)$ (b) $V(Y)$

3-152. If X_1, X_2, and X_3 are independent random variables with $E(X_1) = 4$, $E(X_2) = 3$, $E(X_3) = 2$, $V(X_1) = 1$, $V(X_2) = 5$, $V(X_3) = 2$, and $Y = 2X_1 + X_2 - 3X_3$, determine the following.

(a) $E(Y)$ (b) $V(Y)$

3-153. If X_1 and X_2 are independent random variables with $\mu_1 = 6$, $\mu_2 = 1$, $\sigma_1 = 2$, $\sigma_2 = 4$, and $Y = 4X_1 - 2X_2$, determine the following.

(a) $E(Y)$ (b) $V(Y)$ (c) $E(2Y)$ (d) $V(2Y)$

3-154. If X_1, X_2, and X_3 are independent random variables with $\mu_1 = 1.2$, $\mu_2 = 0.8$, $\mu_3 = 0.5$, $\sigma_1 = 1$, $\sigma_2 = 0.25$, $\sigma_3 = 2.2$, and $Y = 2.5X_1 - 0.5X_2 + 1.5X_3$, determine the following.

(a) $E(Y)$ (b) $V(Y)$ (c) $E(-3Y)$ (d) $V(-3Y)$

3-155. Consider the variables defined in Exercise 3-151. Assume that X_1 and X_2 are normal random variables. Compute the following probabilities.

(a) $P(Y \leq 50)$
(b) $P(25 \leq Y \leq 37)$
(c) $P(14.63 \leq Y \leq 47.37)$

3-156. Consider the variables defined in Exercise 3-152. Assume that X_1, X_2, and X_3 are normal random variables. Compute the following probabilities.

(a) $P(Y > 2.0)$ (b) $P(1.3 \leq Y \leq 8.3)$

3-157. A plastic casing for a magnetic disk is composed of two halves. The thickness of each half is normally distributed with a mean of 1.5 millimeters and a standard deviation of 0.1 millimeter, and the halves are independent.

(a) Determine the mean and standard deviation of the total thickness of the two halves.
(b) What is the probability that the total thickness exceeds 3.3 millimeters?

3-158. The width of a casing for a door is normally distributed with a mean of 24 inches and a standard deviation of $\frac{1}{8}$ inch. The width of a door is normally distributed with a mean of 23 and $\frac{7}{8}$ inches and a standard deviation of $\frac{1}{16}$ inch. Assume independence.

(a) Determine the mean and standard deviation of the difference between the width of the casing and the width of the door.
(b) What is the probability that the width of the casing minus the width of the door exceeds $\frac{1}{4}$ inch?
(c) What is the probability that the door does not fit in the casing?

3-159. A U-shaped assembly is to be formed from the three parts A, B, and C. The picture is shown in Fig. 3-41. The length of A is normally distributed with a mean of 10 millimeters and a standard deviation of 0.1 millimeter. The thickness of part B is normally distributed with a mean of 2 millimeters and a standard deviation of 0.05 millimeter. The thickness of C is normally distributed with mean of 2 millimeters and a standard deviation of 0.10 millimeter. Assume that all dimensions are independent.

(a) Determine the mean and standard deviation of the length of the gap D.
(b) What is the probability that the gap D is less than 5.9 millimeters?

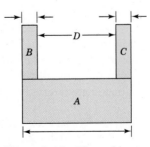

Figure 3-41   Figure for Exercise 3-159.

3-160. Consider the random variables in Exercise 3-151, but assume that the variables are dependent with covariance 3 and that they are normally distributed. Determine the following.

(a) $V(Y)$ (b) $P(Y > 10)$

3-161. Consider the random variables in Exercise 3-152, but assume that the variables are dependent with $\text{Cov}(X_1, X_2) = 1$, $\text{Cov}(X_1, X_3) = 1$, $\text{Cov}(X_2, X_3) = 2$ and that they are normally distributed. Determine the following.

(a) $V(Y)$ (b) $P(Y > 12)$

3-162. Consider the random variables defined in Exercise 3-153. Assume that the random variables are not independent and have $\text{Cov}(X_1, X_2) = 5$. Compute the mean and variance of Y.

3-163. Let X have mean 20 and variance 9. Define $Y = 2X^2$. Compute the mean and variance of Y.

3-164. Let X have mean 100 and variance 25. Define $Y = X^2 + 2X + 1$. Compute the mean and variance of Y.

3-165. Consider Example 3-41. Let the current have mean of 40 amperes and a standard deviation of 0.5 amperes. If the electrical circuit has a resistance of 100 ohms, compute the mean and variance of P.

3-166. Consider the equation for the period T of a pendulum given in Section 3-12.3. Suppose that the length L is a random variable with mean 30 feet and standard deviation 0.02 feet. Compute the mean and variance of T.

3-167. Consider the equation for the acceleration due to gravity, G, given in Section 3-12.3. Suppose that $E(T) = 5.2$ seconds and $V(T) = 0.0004$ square seconds. Compute the mean and variance of G.

3-168. Consider X_1 and X_2 given in Exercise 3-149. Define $Y = X_1 X_2$. Compute the mean and variance of Y.

3-169. Consider X_1, X_2, and X_3 given in Exercise 3-150. Define $Y = X_1 X_2 X_3$. Compute the mean and variance of Y.

3-170. The volume V of a cube is defined as the product of the length, L, the width, W, and the height, H. Assume that each of these dimensions is a random variable with mean 2 inches and standard deviation 0.1 inch. Assume independence and compute the mean and variance of V.

3-13   RANDOM SAMPLES, STATISTICS, AND THE CENTRAL LIMIT THEOREM

Previously in this chapter it was mentioned that data are the observed values of random variables obtained from replicates of a random experiment. Let the random variables that represent the observations from the n replicates be denoted by $X_1, X_2, \ldots, X_n$. Because the replicates are identical, each random variable has the same distribution. Furthermore, the random variables are often assumed to be independent. That is, the results from some replicates do not affect the results from others. Throughout the remainder of the book, a common model is that data are observations from independent random variables with the same distribution. That is, data are observations from independent replicates of a random experiment. This model is used so frequently that we provide a definition.

Random Sample

> Independent random variables $X_1, X_2, \ldots, X_n$ with the same distribution are called a **random sample.**

The term "random sample" stems from the historical use of statistical methods. Suppose that from a large population of objects, a sample of n objects is selected randomly. Here, "randomly" means that each subset of size n is equally likely to be selected. If the number of objects in the population is much larger than n, the random variables $X_1, X_2, \ldots, X_n$ that represent the observations from the sample can be shown to be approximately independent random variables with the same distribution. Consequently, independent random variables with the same distribution are referred to as a random sample.

EXAMPLE 3-43
Strength of O-Rings

In Example 2-1 in Chapter 2, the average tensile strength of eight rubber O-rings was 1055 psi. Two obvious questions are the following: What can we conclude about the average tensile strength of future O-rings? How wrong might we be if we concluded that the average tensile strength of this future population of O-rings is 1055?

There are two important issues to be considered in the answer to these questions.

1. First, because a conclusion is needed for a **future population,** this is an example of an **analytic study.** Certainly, we need to assume that the current specimens are representative of the O-rings that will be produced. This is related to the issue of stability in analytic studies that we discussed in Chapter 1. The usual approach is to assume that these O-rings are a random sample from the future population. Suppose that the mean of this future population is denoted as μ. The objective is to estimate μ.

2. Second, even if we assume that these O-rings are a random sample from future production, the average of these eight items might not equal the average of future production. However, the error that can occur can be quantified.

The key concept is the following: The average is a function of the individual tensile strengths of the eight O-rings. That is, the average is a function of a random sample. Consequently, the average is a random variable with its own distribution. Recall that the distribution of an individual random variable can be used to determine the probability that a measurement is more than one, two, or three standard deviations from the mean of the distribution. In the same manner, the distribution of an average provides the probability that the average is more than a specified distance from μ. Consequently, if we estimate that μ is 1055 psi, the error is determined by the distribution of the average. We discuss this distribution in the remainder of the section. ■

Example 3-43 illustrates that a typical summary of data, such as an average, can be thought of as a function of a random sample. Many other summaries are often used, and this leads to an important definition.

Statistic

> A **statistic** is a function of the random variables in a random sample.

Given data, we calculate statistics all the time. All of the numerical summaries in Chapter 2 such as the sample mean $\overline{X}$, the sample variance S^2, and the sample standard deviation S are statistics. Although the definition of a statistic might seem overly complex, this is because we do not usually consider the distribution of a statistic. However, once we ask how wrong we might be, we are forced to think of a statistic as a function of random variables. Consequently, each statistic has a distribution. It is the distribution of a statistic that determines how well it estimates a quantity such as μ. Often, the probability distribution of a statistic can be determined from the probability distribution of a member of the random sample and the sample size. Another definition is in order.

Sampling Distribution

> The probability distribution of a statistic is called its **sampling distribution.**

Consider the sampling distribution of the sample mean $\overline{X}$. Suppose that a random sample of size n is taken from a normal population with mean μ and variance σ^2. Now each random variable in this sample—say, $X_1, X_2, \ldots, X_n$—is a normally and independently distributed random variable with mean μ and variance σ^2. Then from the results in Section 3-12.1 on linear functions of normally and independently distributed random variables, we conclude that the sample mean

$$\overline{X} = \frac{X_1 + X_2 + \cdots + X_n}{n}$$

has a normal distribution with mean

$$E(\overline{X}) = \frac{\mu + \mu + \cdots + \mu}{n} = \mu$$

and variance

$$V(\overline{X}) = \frac{\sigma^2 + \sigma^2 + \cdots + \sigma^2}{n^2} = \frac{\sigma^2}{n}$$

The mean and variance of $\overline{X}$ are denoted as $\mu_{\overline{X}}$ and $\sigma_{\overline{X}}^2$, respectively.

EXAMPLE 3-44

Fill Volumes

Soft-drink cans are filled by an automated filling machine. The mean fill volume is 12.1 fluid ounces, and the standard deviation is 0.05 fluid ounce. Assume that the fill volumes of the cans are independent, normal random variables. What is the probability that the average volume of 10 cans selected from this process is less than 12 fluid ounces?

Let $X_1, X_2, \ldots, X_{10}$ denote the fill volumes of the 10 cans. The average fill volume (denoted as $\overline{X}$) is a normal random variable with

$$E(\overline{X}) = 12.1 \quad \text{and} \quad V(\overline{X}) = \frac{0.05^2}{10} = 0.00025$$

Consequently, $\sigma_{\overline{X}} = \sqrt{0.00025} = 0.0158$ and

$$P(\overline{X} < 12) = P\left(\frac{\overline{X} - \mu_{\overline{X}}}{\sigma_{\overline{X}}} < \frac{12 - 12.1}{0.0158}\right)$$
$$= P(Z < -6.32) \approx 0 \qquad \blacksquare$$

If we are sampling from a population that has an unknown probability distribution, the sampling distribution of the sample mean is still approximately normal with mean μ and variance σ^2/n, if the sample size n is large. This is one of the most useful theorems in statistics. It is called the **central limit theorem.** The statement is as follows.

Central Limit Theorem

If $X_1, X_2, \ldots, X_n$ is a random sample of size n taken from a population with mean μ and variance σ^2, and if $\overline{X}$ is the sample mean, the limiting form of the distribution of

$$Z = \frac{\overline{X} - \mu}{\sigma/\sqrt{n}} \qquad (3\text{-}39)$$

as $n \to \infty$, is the standard normal distribution.

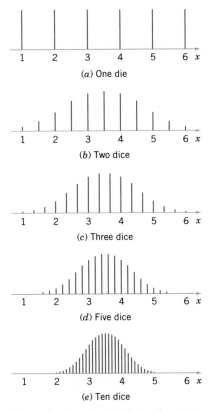

Figure 3-42   Distributions of average scores from throwing dice. [Adapted with permission from Box, Hunter, and Hunter (1978).]

The normal approximation for $\overline{X}$ depends on the sample size n. Figure 3-42a shows the distribution obtained for throws of a single, six-sided true die. The probabilities are equal (1/6) for all the values obtained, 1, 2, 3, 4, 5, or 6. Figure 3-42b shows the distribution of the average score obtained when tossing 2 dice, and Figs. 3-42c, 3-42d, and 3-42e show the distributions of average scores obtained when tossing 3, 5, and 10 dice, respectively. Notice that, while the distribution of a single die is relatively far from normal, the distribution of averages is approximated reasonably well by the normal distribution for sample sizes as small as 5. (The dice throw distributions are discrete, however, whereas the normal is continuous.) Although the central limit theorem will work well for small samples ($n = 4, 5$) in most cases—particularly where the population is continuous, unimodal, and symmetric—larger samples will be required in other situations, depending on the shape of the population. In many cases of practical interest, if $n \geq 30$, the normal approximation will be satisfactory regardless of the shape of the population. If $4 \leq n$, the central limit theorem will work if the distribution of the population is not severely nonnormal.

EXAMPLE 3-45

Average
Resistance

An electronics company manufactures resistors that have a mean resistance of 100 Ω and a standard deviation of 10 Ω. Find the probability that a random sample of $n = 25$ resistors will have an average resistance less than 95 Ω.

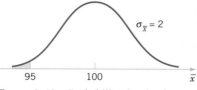

Figure 3-43 Probability density function of average resistance.

Note that the sampling distribution of $\overline{X}$ is approximately normal, with mean $\mu_{\overline{X}} = 100 \ \Omega$ and a standard deviation of

$$\sigma_{\overline{X}} = \frac{\sigma}{\sqrt{n}} = \frac{10}{\sqrt{25}} = 2$$

Therefore, the desired probability corresponds to the shaded area in Fig. 3-43. Standardizing the point $\overline{X} = 95$ in Fig. 3-43, we find that

$$z = \frac{95 - 100}{2} = -2.5$$

and therefore,

$$P(\overline{X} < 95) = P(Z < -2.5) = 0.0062 \qquad \blacksquare$$

Animation 7: Understanding Sampling Distributions and the Central Limit Theorem

EXERCISES FOR SECTION 3-13

3-171. Given that X is normally distributed with mean 100 and standard deviation 9, compute the following for $n = 16$.
(a) Mean and variance of $\overline{X}$
(b) $P(\overline{X} \le 98)$
(c) $P(\overline{X} > 103)$
(d) $P(96 \le \overline{X} \le 102)$

3-172. Given that X is normally distributed with mean 50 and standard deviation 4, compute the following for $n = 25$.
(a) Mean and variance of $\overline{X}$
(b) $P(\overline{X} \le 49)$
(c) $P(\overline{X} > 52)$
(d) $P(49 \le \overline{X} \le 51.5)$

3-173. Assume a sample of 40 observations are drawn from a population with mean 20 and variance 2. Compute the following.
(a) Mean and variance of $\overline{X}$
(c) $P(\overline{X} \le 19)$
(d) $P(\overline{X} > 22)$
(e) $P(19 \le \overline{X} \le 21.5)$

3-174. Intravenous fluid bags are filled by an automated filling machine. Assume that the fill volumes of the bags are independent, normal random variables with a standard deviation of 0.08 fluid ounces.

(a) What is the standard deviation of the average fill volume of 20 bags?
(b) If the mean fill volume of the machine is 6.16 ounces, what is the probability that the average fill volume of 20 bags is below 5.95 ounces?
(c) What should the mean fill volume equal in order that the probability that the average of 20 bags is below 6 ounces is 0.001?

3-175. The photoresist thickness in semiconductor manufacturing has a mean of 10 micrometers and a standard deviation of 1 micrometer. Assume that the thickness is normally distributed and that the thicknesses of different wafers are independent.

(a) Determine the probability that the average thickness of 10 wafers is either greater than 11 or less than 9 micrometers.
(b) Determine the number of wafers that need to be measured such that the probability that the average thickness exceeds 11 micrometers is 0.01.

3-176. The time to complete a manual task in a manufacturing operation is considered a normally distributed random variable with mean of 0.50 minute and a standard deviation of 0.05 minute. Find the probability that the average time to complete the manual task, after 49 repetitions, is less than 0.465 minute.

3-177. A synthetic fiber used in manufacturing carpet has tensile strength that is normally distributed with mean 75.5 psi and standard deviation 3.5 psi. Find the probability that a random sample of $n = 6$ fiber specimens will have sample mean tensile strength that exceeds 75.75 psi.

3-178.   The compressive strength of concrete has a mean of 2500 psi and a standard deviation of 50 psi. Find the probability that a random sample of $n = 5$ specimens will have a sample mean strength that falls in the interval from 2490 psi to 2510 psi.

3-179.   The amount of time that a customer spends waiting at an airport check-in counter is a random variable with mean 8.2 minutes and standard deviation 1.5 minutes. Suppose that a random sample of $n = 49$ customers is observed. Find the probability that the average time waiting in line for these customers is

(a) Less than 8 minutes
(b) Between 8 and 9 minutes
(c) Less than 7.5 minutes

3-180.   Suppose that X has the following discrete distribution

$$f(x) = \begin{cases} \frac{1}{3}, & x = 1, 2, 3 \\ 0, & \text{otherwise} \end{cases}$$

A random sample of $n = 36$ is selected from this population. Approximate the probability that the sample mean is greater than 2.1 but less than 2.5.

3-181.   The viscosity of a fluid can be measured in an experiment by dropping a small ball into a calibrated tube containing the fluid and observing the random variable X, the time it takes for the ball to drop the measured distance. Assume that X is normally distributed with a mean of 20 seconds and a standard deviation of 0.5 second for a particular type of liquid.

(a) What is the standard deviation of the average time of 40 experiments?
(b) What is the probability that the average time of 40 experiments will exceed 20.1 seconds?
(c) Suppose the experiment is repeated only 20 times. What is the probability that the average value of X will exceed 20.1 seconds?
(d) Is the probability computed in part (b) greater than or less than the probability computed in part (c)? Explain why this inequality occurs.

3-182.   A random sample of $n = 9$ structural elements is tested for compressive strength. We know that the true mean compressive strength $\mu = 5500$ psi and the standard deviation is $\sigma = 100$ psi. Find the probability that the sample mean compressive strength exceeds 4985 psi.

SUPPLEMENTAL EXERCISES

3-183.   Suppose that $f(x) = e^{-x}$ for $0 < x$ and $f(x) = 0$ for $x < 0$. Determine the following probabilities.

(a) $P(X \le 1.5)$
(b) $P(X < 1.5)$
(c) $P(1.5 < X < 3)$
(d) $P(X = 3)$
(e) $P(X > 3)$

3-184.   Suppose that $f(x) = e^{-x/2}$ for $0 < x$ and $f(x) = 0$ for $x < 0$.

(a) Determine x such that $P(x < X) = 0.20$.
(b) Determine x such that $P(X \le x) = 0.75$.

3-185.   The random variable X has the following probability distribution.

x	2	3	5	8
Probability	0.2	0.4	0.3	0.1

Determine the following.

(a) $P(X \le 3)$
(b) $P(X > 2.5)$
(c) $P(2.7 < X < 5.1)$
(d) $E(X)$
(e) $V(X)$

3-186.   A driveshaft will suffer fatigue failure with a mean time-to-failure of 40,000 hours of use. If it is known that the probability of failure before 36,000 hours is 0.04 and that the distribution governing time-to-failure is a normal distribution, what is the standard deviation of the time-to-failure distribution?

3-187.   A standard fluorescent tube has a life length that is normally distributed with a mean of 7000 hours and a standard deviation of 1000 hours. A competitor has developed a compact fluorescent lighting system that will fit into incandescent sockets. It claims that a new compact tube has a normally distributed life length with a mean of 7500 hours and a standard deviation of 1200 hours. Which fluorescent tube is more likely to have a life length greater than 9000 hours? Justify your answer.

3-188.   The average life of a certain type of compressor is 10 years with a standard deviation of 1 year. The manufacturer replaces free all compressors that fail while under guarantee. The manufacturer is willing to replace 3% of all compressors sold. For how many years should the guarantee be in effect? Assume a normal distribution.

3-189.   The probability that a call to an emergency help line is answered in less than 15 seconds is 0.85. Assume that all calls are independent.

(a) What is the probability that exactly 7 of 10 calls are answered within 15 seconds?
(b) What is the probability that at least 16 of 20 calls are answered in less than 15 seconds?
(c) For 50 calls, what is the mean number of calls that are answered in less than 15 seconds?
(d) Repeat parts (a)–(c) using the normal approximation.

3-190.   The number of messages sent to a computer bulletin board is a Poisson random variable with a mean of five messages per hour.

(a) What is the probability that 5 messages are received in 1 hour?
(b) What is the probability that 10 messages are received in 1.5 hours?

(c) What is the probability that fewer than 2 messages are received in $\frac{1}{2}$ hour?

3-191.   Continuation of Exercise 3-190. Let Y be the random variable defined as the time between messages arriving to the computer bulletin board.

(a) What is the distribution of Y? What is the mean of Y?

(b) What is the probability that the time between messages exceeds 15 minutes?

(c) What is the probability that the time between messages is less than 5 minutes?

(d) Given that 10 minutes have passed without a message arriving, what is the probability that there will not be a message in the next 10 minutes?

3-192.   The number of errors in a textbook follows a Poisson distribution with mean of 0.01 error per page.

(a) What is the probability that there are three or fewer errors in 100 pages?

(b) What is the probability that there are four or more errors in 100 pages?

(c) What is the probability that there are three or fewer errors in 200 pages?

3-193.   Continuation of Exercise 3-192. Let Y be the random variable defined as the number of pages between errors.

(a) What is the distribution of Y? What is the mean of Y?

(b) What is the probability that there are fewer than 100 pages between errors?

(c) What is the probability that there are no errors in 200 consecutive pages?

(d) Given that there are 100 consecutive pages without errors, what is the probability that there will not be an error in the next 50 pages?

3-194.   Polyelectrolytes are typically used to separate oil and water in industrial applications. The separation process is dependent on controlling the pH. Fifteen pH readings of wastewater following these processes were recorded. Is it reasonable to model these data using a normal distribution?

| 6.2 | 6.5 | 7.6 | 7.7 | 7.1 | 7.1 | 7.9 | 8.4 |
| 7.0 | 7.3 | 6.8 | 7.6 | 8.0 | 7.1 | 7.0 | |

3-195.   The lifetimes of six major components in a copier are independent exponential random variables with means of 8000, 10,000, 10,000, 20,000, 20,000, and 25,000 hours, respectively.

(a) What is the probability that the lifetimes of all the components exceed 5000 hours?

(b) What is the probability that none of the components has a lifetime that exceeds 5000 hours?

(c) What is the probability that the lifetimes of all the components are less than 3000 hours?

3-196.   A random sample of 36 observations has been drawn. Find the probability that the sample mean is in the interval $47 < X < 53$ for each of the following population distributions and population parameter values.

(a) Normal with mean 50 and standard deviation 12.

(b) Exponential with mean 50.

(c) Poisson with mean 50.

(d) Compare the probabilities obtained in parts (a)–(c) and explain why the probabilities differ.

3-197.   From contractual commitments and extensive past laboratory testing, we know that compressive strength measurements are normally distributed with the true mean compressive strength $\mu = 5500$ psi and standard deviation $\sigma = 100$ psi. A random sample of structural elements is tested for compressive strength at the customer's receiving location.

(a) What is the standard deviation of the sampling distribution of the sample mean for this problem if $n = 9$?

(b) What is the standard deviation of the sampling distribution of the sample mean for this problem if $n = 20$?

(c) Compare your results of parts (a) and (b), and comment on why they are the same or different.

3-198.   The weight of adobe bricks for construction is normally distributed with a mean of 3 pounds and a standard deviation of 0.25 pound. Assume that the weights of the bricks are independent and that a random sample of 25 bricks is chosen. What is the probability that the mean weight of the sample is less than 2.95 pounds?

3-199.   A disk drive assembly consists of one hard disk and spacers on each side, as shown in Fig. 3-44. The height of the top spacer, W, is normally distributed with mean 120 millimeters and standard deviation 0.5 millimeters; the height of the disk, X, is normally distributed with mean 20 millimeters and standard deviation 0.1 millimeters; and the height of the bottom spacer, Y, is normally distributed with mean 100 millimeters and standard deviation 0.4 millimeters.

(a) What are the distribution, the mean, and the variance of the height of the stack?

(b) Assume that the stack must fit into a space with a height of 242 millimeters. What is the probability that the stack height will exceed the space height?

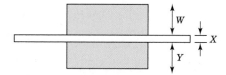

Figure 3-44   Figure for Exercise 3-199.

3-200.   The time for an automated system in a warehouse to locate a part is normally distributed with mean 45 seconds and standard deviation 30 seconds. Suppose that independent requests are made for 10 parts.

(a) What is the probability that the average time to locate 10 parts exceeds 60 seconds?

(b) What is the probability that the total time to locate 10 parts exceeds 600 seconds?

3-201.   A mechanical assembly used in an automobile engine contains four major components. The weights of the components are independent and normally distributed with the following means and standard deviations (in ounces).

Component	Mean	Standard Deviation
Left case	4	0.4
Right case	5.5	0.5
Bearing assembly	10	0.2
Bolt assembly	8	0.5

(a) What is the probability that the weight of an assembly exceeds 29.5 ounces?
(b) What is the probability that the mean weight of eight independent assemblies exceeds 29 ounces?

3-202. A bearing assembly contains 10 bearings. The bearing diameters are assumed to be independent and normally distributed with a mean of 1.5 millimeters and a standard deviation of 0.025 millimeter. What is the probability that the maximum diameter bearing in the assembly exceeds 1.6 millimeters?

3-203. A process is said to be of **six-sigma quality** if the process mean is at least six standard deviations from the nearest specification. Assume a normally distributed measurement.

(a) If a process mean is centered between the upper and lower specifications at a distance of six standard deviations from each, what is the probability that a product does not meet specifications? Using the result that 0.000001 equals one part per million, express the answer in parts per million.
(b) Because it is difficult to maintain a process mean centered between the specifications, the probability of a product not meeting specifications is often calculated after assuming the process shifts. If the process mean positioned as in part (a) shifts upward by 1.5 standard deviations, what is the probability that a product does not meet specifications? Express the answer in parts per million.

3-204. Continuation of Exercise 3-69. Recall that it was determined that a normal distribution adequately fit the internal pressure strength data. Use this distribution and suppose that the sample mean of 206.04 and standard deviation of 11.57 are used to estimate the population parameters. Estimate the following probabilities.

(a) What is the probability that the internal pressure strength measurement will be between 210 and 220 psi?
(b) What is the probability that the internal pressure strength measurement will exceed 228 psi?
(c) Find x such that $P(X \geq x) = 0.02$, where X is the internal pressure strength random variable.

3-205. Continuation of Exercise 3-70. Recall that it was determined that a normal distribution adequately fit the dimensional measurements for parts from two different machines. Using this distribution, suppose that $\bar{x}_1 = 100.27$, $s_1 = 2.28$, $\bar{x}_2 = 100.11$, and $s_2 = 7.58$ are used to estimate the

population parameters. Estimate the following probabilities. Assume that the engineering specifications indicate that acceptable parts measure between 96 and 104.

(a) What is the probability that machine 1 produces acceptable parts?
(b) What is the probability that machine 2 produces acceptable parts?
(c) Use your answers from parts (a) and (b) to determine which machine is preferable.
(d) Recall that the data reported in Exercise 3-71 were a result of a process engineer making adjustments to machine 2. Use the new sample mean 105.39 and sample standard deviation 2.08 to estimate the population parameters. What is the probability that the newly adjusted machine 2 will produce acceptable parts? Did adjusting machine 2 improve its overall performance?

3-206. Continuation of Exercise 2-1.
(a) Plot the data on normal probability paper. Does concentration appear to have a normal distribution?
(b) Suppose it has been determined that the largest observation, 68.7, was suspected to be an outlier. Consequently, it can be removed from the data set. Does this improve the fit of the normal distribution to the data?

3-207. Continuation of Exercise 2-2.
(a) Plot the data on normal probability paper. Do these data appear to have a normal distribution?
(b) Remove the largest observation from the data set. Does this improve the fit of the normal distribution to the data?

3-208. The weight of a certain type of brick has an expectation of 1.12 kilograms with a variance of 0.0009 kilogram. How many bricks would need to be selected so that their average weight has a standard deviation of no more than 0.005 kilogram?

3-209. The thickness of glass sheets produced by a certain process are normally distributed with a mean of $\mu = 3.00$ millimeters and a standard deviation of $\sigma = 0.12$ millimeters. What is the value of c for which there is a 99% probability that a glass sheet has a thickness within the interval $[3.00 - c, 3.00 + c]$?

3-210. The weights of bags filled by a machine are normally distributed with standard deviation 0.05 kilograms and mean that can be set by the operator. At what level should the mean be set if it is required that only 1% of the bags weigh less than 10 kilograms?

3-211. The research and development team of a medical device manufacturer is designing a new diagnostic test strip to detect the breath alcohol level. The materials used to make the device are listed here together with their mean and standard deviation of their thickness.

The materials are stacked as shown in the following figure. Assuming that the thickness of each material is independent and normally distributed, answer the following questions.

Material	Random Variable	Mean Thickness m, mm	Standard Deviation of Thickness s, mm
Protective layer 1	W	10	2
Absorbant pad	X	50	10
Reaction layer	Y	5	1
Protective layer 2	Z	8	1

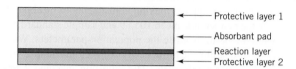

Protective layer 1
Absorbant pad
Reaction layer
Protective layer 2

(a) Using the random variables W, X, Y, and Z, give the equation representing the thickness of the layered strip.
(b) What is the mean thickness of the strip?
(c) What is the variance of the thickness of the strip?
(d) What is the probability that the thickness of the strip will be greater than 75 millimeters?

3-212.   Overheating is a major problem in microprocessor operation. After much testing, it has been determined that the operating temperature is normally distributed with a mean of 150 degrees and a standard deviation of 7 degrees. The processor will malfunction at 165 degrees.

(a) What is the probability of a malfunction?
(b) A newer fan useful for cooling the processor is being considered. With the new fan, the operating temperature has a mean of 144 degrees and a standard deviation of 9 degrees. What is the probability of a malfunction with the new fan?
(c) Suppose that all processors are sold for $1200. The cost of the original system is $1000, whereas the cost with the new fan is $1050. Assume that 1000 units are planned to be produced and sold. Also assume that there is a money-back guarantee for all systems that malfunction. Under these assumptions, which system will generate the most revenue?

3-213.   Manufacturers need to determine that each medical linear accelerator works within proper parameters before shipping to hospitals. An individual machine is known to have a probability of failure during initial testing of 0.10. Eight accelerators are tested.

(a) What is the probability that at most two fail?
(b) What is the probability that none fails?

3-214.   A keyboard for a personal computer is known to have a mean life of 5 years. The life of the keyboard can be modeled using an exponential distribution.

(a) What is the probability that a keyboard will have a life between 2 and 4 years?
(b) What is the probability that the keyboard will still function after 1 year?
(c) If a warranty is set at 6 months, what is the probability that a keyboard will need to replaced under warranty?

3-215.   A cartridge company develops ink cartridges for a printer company and supplies both the ink and the cartridge. The following is the probability mass function of the number of cartridges over the life of the printer.

x	5	6	7	8	9
$f(x)$	0.04	0.19	0.61	0.13	0.03

(a) What is the expected number of cartridges used?
(b) What is the probability that more than six cartridges are used?
(c) What is the probability that 9 out of 10 randomly selected printers use more than 6 cartridges?

3-216.   Consider the following system made up of functional components in parallel and series. The probability that each component functions is shown in Fig. 3-45.

(a) What is the probability that the system operates?
(b) What is the probability that the system fails due to the components in series? Assume parallel components do not fail.
(c) What is the probability that the system fails due to the components in parallel? Assume series components do not fail.
(d) Compute the probability that the system fails using the following formula:

$$[1 - P(C_1) \cdot P(C_4)] \cdot [1 - P(C_2')P(C_4')]$$
$$+ P(C_1) \cdot P(C_4) \cdot P(C_2') \cdot P(C_3')$$
$$+ [1 - P(C_1)P(C_4)] \cdot P(C_2') \cdot P(C_3')$$

(e) Describe in words the meaning of each of the terms in the formula in part (d).
(f) Use part (a) to compute the probability that the system fails.
(g) Improve the probability that component C_1 functions to a value of 0.95 and recompute parts (a), (b), (c), and (f).
(h) Alternatively, do not change the original probability associated with C_1; rather, increase the probability that

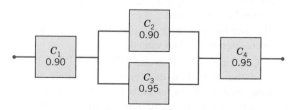

C_2
0.90

C_1
0.90

C_3
0.95

C_4
0.95

Figure 3-45   Figure for Exercise 3-216.

component C_2 functions to a value of 0.95 and recompute parts (a), (b), (c), and (f).

(i) Based on your answers in parts (a) and (b) of this exercise, comment on whether you would recommend increasing the reliability of a series component or a parallel component to increase overall system reliability.

3-217. Show that the gamma density function integrates to 1.

3-218. To illustrate the effect of a log transformation, consider the following data, which represent cycles to failure for a yarn product:

$$675, 3650, 175, 1150, 290, 2000, 100, 375$$

(a) Using a normal probability plot, comment on the adequacy of the fit.

(b) Transform the data using logarithms, that is, let y^* (new value) $= \log y$ (old value). Perform a normal probability plot on the transformed data and comment on the adequacy of the fit.

(c) Engineering has specified that acceptable yarn strength should exceed 200 cycles prior to failure. Use your results from part (b) to estimate the proportion of acceptable yarn. (*Hint:* Be sure to transform the lower specification limit, 200, prior to computing the proportion. Suppose that the sample mean and sample standard deviation are used to estimate the population parameters in your calculations.)

3-219. Consider the following data, which represent the number of hours of operation of a surveillance camera until failure:

246,785	183,424	1060	22,310	921	35,659
127,015	10,649	13,859	53,731	10,763	1456
189,880	2414	21,414	411,884	29,644	1473

(a) Perform a normal probability plot and comment on the adequacy of the fit.

(b) Transform the data using logarithms, that is, let y^* (new value) $= \log y$ (old value). Perform a normal probability plot and comment on the adequacy of the fit.

(c) The manufacturer of the cameras is interested in defining a warranty limit such that not more than 2% of the cameras will need to be replaced. Use your fitted model from part (b) to propose a warranty limit on the time to failure of the number of hours of a random surveillance camera. (*Hint:* Be sure to give the warranty limit in the original units of hours. Suppose that the sample mean and sample standard deviation are used to estimate the population parameters in your calculations.)

3-220. Consider the following data, which represent the life of roller bearings (in hours).

7203	3917	7476	5410	7891	10,033
4484	12,539	2933	16,710	10,702	16,122
13,295	12,653	5610	6466	5263	2,504
9098	7759				

(a) Perform a Weibull probability plot and determine the adequacy of the fit.

(b) Using the estimated shape parameter $= 2.2$ and the estimated scale parameter 9525, estimate the probability that a bearing lasts at least 7500 hours.

(c) If five bearings are in use and failures occur independently, what is the probability that all five bearings last at least 7500 hours?

3-221. Consider the following data that represent the life of packaged magnetic disks exposed to corrosive gases (in hours):

$$4, 86, 335, 746, 80, 1510, 195, 562, 137, 1574, 7600, 4394, 4, 98, 1196, 15, 934, 11$$

(a) Perform a Weibull probability plot and determine the adequacy of the fit.

(b) Using the estimated shape parameter $= 0.53$ and the estimated scale parameter 604, estimate the probability that a disk fails before 150 hours.

(c) If a warranty is planned to cover no more than 10% of the manufactured disks, at what value should the warranty level be set?

3-222. Nonuniqueness of Probability Models. It is possible to fit more than one model to a set of data. Consider the life data given in Exercise 3-221.

(a) Transform the data using logarithms; that is, let y^* (new value) $= \log y$ (old value). Perform a normal probability plot on the transformed data and comment on the adequacy of the fit.

(b) Use the fitted normal distribution from part (a) to estimate the probability that the disk fails before 150 hours. Compare your results with your answer in part (b) of Exercise 3-221.

3-223. Cholesterol is a fatty substance that is an important part of the outer lining (membrane) of cells in the body of animals. Suppose that the mean and standard deviation for a population of individuals are 180 mg/dl and 20 mg/dl, respectively. Samples are obtained from 25 individuals, and these are considered to be independent.

(a) What is the probability that the average of the 25 measurements exceeds 185 mg/dl?

(b) Determine symmetric limits around 180 such that the probability that the sample average is within the limits equals 0.95.

3-224. Arthroscopic meniscal repair was successful 70% of the time for tears greater than 25 millimeters (in 50 surgeries) and 76% of the time for shorter tears (in 100 surgeries).

(a) Describe the random variable used in these probability statements.

(b) Is the random variable continuous or discrete?

(c) Explain why these probabilities do not add to 1.

3-225. The lifetime of a mechanical assembly in a vibration test is exponentially distributed with a mean of 400 hours.

(a) What is the probability that an assembly on test fails in less than 100 hours?

(b) What is the probability that an assembly operates for more than 500 hours before failure?

(c) If an assembly has been on test for 400 hours without a failure, what is the probability of a failure in the next 100 hours?

3-226. An article in *Knee Surgery, Sports Traumatology, Arthroscopy,* "Effect of Provider Volume on Resource Utilization for Surgical Procedures" (Vol. 13, 2005, pp. 273–279), showed a mean time of 129 minutes and a standard deviation of 14 minutes for ACL reconstruction surgery at high-volume hospitals (with more than 300 such surgeries per year).

(a) What is the probability that your ACL surgery at a high-volume hospital is completed in less than 100 minutes?

(b) What is the probability that your surgery time is greater than two standard deviations above the mean?

(c) The probability of a completed ACL surgery at a high-volume hospital is equal to 95% at what time?

TEAM EXERCISES

3-227. Using the data set that you found or collected in the first team exercise of Chapter 2, or another data set of interest, answer the following questions:

(a) Is a continuous or discrete distribution model more appropriate to model your data? Explain.

(b) You have studied the normal, exponential, Poisson, and binomial distributions in this chapter. Based on your recommendation in part (a), attempt to fit at least one model to your data set. Report on your results.

3-228. Computer software can be used to simulate data from a normal distribution. Use a package such as Minitab to simulate dimensions for parts *A*, *B*, and *C* in Fig. 3-43 of Exercise 3-159.

(a) Simulate 500 assemblies from simulated data for parts *A*, *B*, and *C* and calculate the length of gap *D* for each.

(b) Summarize the data for gap *D* with a histogram and relevant summary statistics.

(c) Compare your simulated results with those obtained in Exercise 3-159.

(d) Describe a problem for which simulation is a good method of analysis.

3-229. Consider the data on weekly waste (percent) as reported for five suppliers of the Levi-Strauss clothing plant in Albuquerque and reported on the Web site http://lib.stat.cmu.edu/DASL/Stories/wasterunup.html.

Test each of the data sets for conformance to a normal probability model using a normal probability plot. For those that do not pass the test for normality, delete any outliers (these can be identified using a box plot) and replot the data. Summarize your findings.

IMPORTANT TERMS AND CONCEPTS

Binomial distribution	Gamma distribution	Poisson distribution	Random variable
Central limit theorem	Independence	Poisson process	Sampling distribution
Continuity correction	Joint probability	Probability	Standard deviation of a
Continuous random	distribution	Probability density	random variable
variable	Lognormal distribution	function	Standard normal
Cumulative distribution	Mean of a random	Probability distribution	distribution
function	variable	Probability mass	Statistic
Delta method	Normal approximations	function	Variance of a random
Discrete random	to binomial and	Probability plots	variable
variable	Poisson distributions	Propagation of error	Weibull distribution
Events	Normal distribution	Random experiment	
Exponential distribution	Normal probability plot	Random sample	

4 Decision Making for a Single Sample

LEARNING OBJECTIVES

After careful study of this chapter, you should be able to do the following:

1. Perform hypothesis tests and construct confidence intervals on the mean of a normal distribution.

2. Perform hypothesis tests and construct confidence intervals on the variance of a normal distribution.

3. Perform hypothesis tests and construct confidence intervals on a population proportion.

4. Compute power and type II error, and make sample-size selection decisions for hypothesis tests and confidence intervals.

5. Explain and use the relationship between confidence intervals and hypothesis tests.

6. Construct a prediction interval for a future observation.

7. Construct a tolerance interval for a normal population.

8. Explain the difference between confidence intervals, prediction intervals, and tolerance intervals.

9. Use the chi-square goodness-of-fit test to check distributional assumptions.

4-1 STATISTICAL INFERENCE

The field of statistical inference consists of those methods used to make decisions or to draw conclusions about a **population.** These methods utilize the information contained in a **random sample** from the population in drawing conclusions. Figure 4-1 illustrates the relationship between the population and the sample. This chapter begins our study of the statistical methods used for inference and decision making.

Statistical inference may be divided into two major areas: **parameter estimation** and **hypothesis testing.** As an example of a parameter estimation problem, suppose that a structural engineer is analyzing the tensile strength of a component used in an automobile chassis. Because variability in tensile strength is naturally present between the individual components because of differences in raw material batches, manufacturing processes, and measurement procedures (for example), the engineer is interested in estimating the mean tensile strength of the components. Knowledge of the statistical sampling properties of the estimator used would enable the engineer to establish the precision of the estimate.

Now consider a situation in which two different reaction temperatures can be used in a chemical process—say, t_1 and t_2. The engineer conjectures that t_1 results in higher yields than does t_2. Statistical hypothesis testing is a framework for solving problems of this type. In this

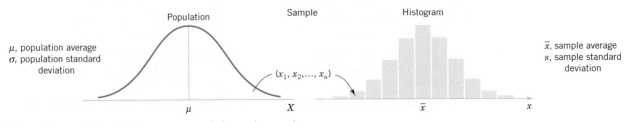

Figure 4-1   Relationship between a population and a sample.

case, the hypothesis would be that the mean yield using temperature t_1 is greater than the mean yield using temperature t_2. Note that there is no emphasis on estimating yields; instead, the focus is on drawing conclusions about a stated hypothesis.

This chapter begins by discussing methods for estimating parameters. Then we introduce the basic principles of hypothesis testing. Once these statistical fundamentals have been presented, we will apply them to several situations that arise frequently in engineering practice. These include inference on the mean of a population, the variance of a population, and a population proportion.

4-2 POINT ESTIMATION

One very important application of statisitics is in obtaining **point estimates** of parameters such as the population mean and the population variance. When discussing inference problems, it is convenient to have a general symbol to represent the parameter of interest. We will use the Greek symbol θ (theta) to represent the parameter. The objective of point estimation is to select a single number, based on the data in a random sample, that is the most plausible value for θ. A numerical value of a sample statistic will be used as the point estimate.

Sample Mean

For example, suppose that the random variable X is normally distributed with unknown mean μ. The sample mean is a point estimator of the unknown population mean μ; that is, $\hat{\mu} = \overline{X}$. After the sample has been selected, the numerical value $\overline{x}$ is the point estimate of μ. Thus, if $x_1 = 25$, $x_2 = 30$, $x_3 = 29$, and $x_4 = 31$, the point estimate of μ is

$$\overline{x} = \frac{25 + 30 + 29 + 31}{4} = 28.75$$

Similarly, if the population variance σ^2 is also unknown, a point estimator for σ^2 is the sample variance S^2, and the numerical value $s^2 = 6.9$ calculated from the sample data is called the point estimate of σ^2.

In general, if X is a random variable with probability distribution $f(x)$, characterized by the unknown parameter θ, and if $X_1, X_2, \ldots, X_n$ is a random sample of size n from $f(x)$, the statistic $\hat{\Theta} = h(X_1, X_2, \ldots, X_n)$ is called a **point estimator** of θ. Here h is just a function of observations in the random sample. Note that $\hat{\Theta}$ is a random variable because it is a function of random variables. After the sample has been selected, $\hat{\Theta}$ takes on a particular numerical value $\hat{\theta}$ called the **point estimate** of θ.

Point Estimate

> A **point estimate** of some population parameter θ is a single numerical value $\hat{\theta}$ of a statistic $\hat{\Theta}$.

Estimation problems occur frequently in engineering. We often need to estimate

- The mean μ of a single population
- The variance σ^2 (or standard deviation σ) of a single population
- The proportion p of items in a population that belong to a class of interest
- The difference in means of two populations, $\mu_1 - \mu_2$
- The difference in two population proportions, $p_1 - p_2$

Reasonable point estimates of these parameters are as follows:

- For μ, the estimate is $\hat{\mu} = \bar{x}$, the sample mean.
- For σ^2, the estimate is $\hat{\sigma}^2 = s^2$, the sample variance.
- For p, the estimate is $\hat{p} = x/n$, the sample proportion, where x is the number of items in a random sample of size n that belong to the class of interest.
- For $\mu_1 - \mu_2$, the estimate is $\hat{\mu}_1 - \hat{\mu}_2 = \bar{x}_1 - \bar{x}_2$, the difference between the sample means of two independent random samples.
- For $p_1 - p_2$, the estimate is $\hat{p}_1 - \hat{p}_2$, the difference between two sample proportions computed from two independent random samples.

The following display summarizes the relationship between the unknown parameters and their typical associated statistics and point estimates.

Unknown Parameter θ	Statistic $\hat{\Theta}$	Point Estimate $\hat{\theta}$
μ	$\bar{X} = \dfrac{\Sigma X_i}{n}$	$\bar{x}$
σ^2	$S^2 = \dfrac{\Sigma(X_i - \bar{X})^2}{n-1}$	s^2
p	$\hat{P} = \dfrac{X}{n}$	$\hat{p}$
$\mu_1 - \mu_2$	$\bar{X}_1 - \bar{X}_2 = \dfrac{\Sigma X_{1i}}{n_1} - \dfrac{\Sigma X_{2i}}{n_2}$	$\bar{x}_1 - \bar{x}_2$
$p_1 - p_2$	$\hat{P}_1 - \hat{P}_2 = \dfrac{X_1}{n_1} - \dfrac{X_2}{n_2}$	$\hat{p}_1 - \hat{p}_2$

We may have several different choices for the point estimator of a parameter. For example, if we wish to estimate the mean of a population, we might consider the sample mean, the sample median, or perhaps the average of the smallest and largest observations in the sample as point estimators. To decide which point estimator of a particular parameter is the best one to use, we need to examine their statistical properties and develop some criteria for comparing estimators.

An estimator should be "close" in some sense to the true value of the unknown parameter. Formally, we say that $\hat{\Theta}$ is an unbiased estimator of θ if the expected value of $\hat{\Theta}$ is equal to θ. This is equivalent to saying that the mean of the probability distribution of $\hat{\Theta}$ (or the mean of the sampling distribution of $\hat{\Theta}$) is equal to θ.

Unbiased Estimator

The point estimator $\hat{\Theta}$ is an **unbiased estimator** for the parameter θ if

$$E(\hat{\Theta}) = \theta \tag{4-1}$$

If the estimator is not unbiased, then the difference

$$E(\hat{\Theta}) - \theta \tag{4-2}$$

is called the **bias** of the estimator $\hat{\Theta}$.

When an estimator is unbiased, then $E(\hat{\Theta}) - \theta = 0$; that is, the bias is zero.

EXAMPLE 4-1
Unbiased
Estimators

Suppose that X is a random variable with mean μ and variance σ^2. Let $X_1, X_2, \ldots, X_n$ be a random sample of size n from the population represented by X. Show that the sample mean $\bar{X}$ and sample variance S^2 are unbiased estimators of μ and σ^2, respectively.

Solution. First consider the sample mean. In Chapter 3, we indicated that $E(\bar{X}) = \mu$. Therefore, the sample mean $\bar{X}$ is an unbiased estimator of the population mean μ.

Now consider the sample variance. We have

$$E(S^2) = E\left[\frac{\sum_{i=1}^{n}(X_i - \bar{X})^2}{n-1}\right] = \frac{1}{n-1}E\sum_{i=1}^{n}(X_i - \bar{X})^2$$

$$= \frac{1}{n-1}E\sum_{i=1}^{n}(X_i^2 + \bar{X}^2 - 2\bar{X}X_i) = \frac{1}{n-1}E\left(\sum_{i=1}^{n}X_i^2 - n\bar{X}^2\right)$$

$$= \frac{1}{n-1}\left[\sum_{i=1}^{n}E(X_i^2) - nE(\bar{X}^2)\right]$$

The last equality follows from equation 3-28. However, because $E(X_i^2) = \mu^2 + \sigma^2$ and $E(\bar{X}^2) = \mu^2 + \sigma^2/n$, we have

$$E(S^2) = \frac{1}{n-1}\left[\sum_{i=1}^{n}(\mu^2 + \sigma^2) - n\left(\mu^2 + \frac{\sigma^2}{n}\right)\right]$$

$$= \frac{1}{n-1}(n\mu^2 + n\sigma^2 - n\mu^2 - \sigma^2)$$

$$= \sigma^2$$

Therefore, the sample variance S^2 is an unbiased estimator of the population variance σ^2. However, we can show that the sample standard deviation S is a biased estimator of the population standard deviation. For large samples this bias is negligible. ■

Sometimes there are several unbiased estimators of the sample population parameter. For example, suppose we take a random sample of size $n = 10$ from a normal population and obtain the data $x_1 = 12.8$, $x_2 = 9.4$, $x_3 = 8.7$, $x_4 = 11.6$, $x_5 = 13.1$, $x_6 = 9.8$, $x_7 = 14.1$, $x_8 = 8.5$, $x_9 = 12.1$, $x_{10} = 10.3$. Now the sample mean is

Sample Mean versus
Sample Median

$$\bar{x} = \frac{12.8 + 9.4 + 8.7 + 11.6 + 13.1 + 9.8 + 14.1 + 8.5 + 12.1 + 10.3}{10} = 11.04$$

the sample median is

$$\tilde{x} = \frac{10.3 + 11.6}{2} = 10.95$$

and a single observation from this normal population, say, $x_1 = 12.8$.

We can show that all of these values result from unbiased estimators of μ. Because there is not a unique unbiased estimator, we cannot rely on the property of unbiasedness alone to select our estimator. We need a method to select among unbiased estimators.

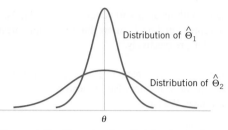

Figure 4-2   The sampling distributions of two unbiased estimators $\hat{\Theta}_1$ and $\hat{\Theta}_2$.

Suppose that $\hat{\Theta}_1$ and $\hat{\Theta}_2$ are unbiased estimators of θ. This indicates that the distribution of each estimator is centered at the true value of θ. However, the variances of these distributions may be different. Figure 4-2 illustrates the situation. Because $\hat{\Theta}_1$ has a smaller variance than $\hat{\Theta}_2$, the estimator $\hat{\Theta}_1$ is more likely to produce an estimate close to the true value θ. A logical principle of estimation, when selecting among several estimators, is to choose the estimator that has minimum variance.

Minimum Variance Unbiased Estimator

If we consider all unbiased estimators of θ, the one with the smallest variance is called the **minimum variance unbiased estimator** (MVUE).

The concepts of an unbiased estimator and an estimator with minimum variance are extremely important. There are methods for formally deriving estimates of the parameters of a probability distribution. One of these methods, the **method of maximum likelihood,** produces point estimators that are approximately unbiased and very close to the minimum variance estimator. For further information on the method of maximum likelihood, see Montgomery and Runger (2006).

In practice, one must occasionally use a biased estimator (such as S for σ). In such cases, the mean square error of the estimator can be important. The **mean square error** of an estimator $\hat{\Theta}$ is the expected squared difference between $\hat{\Theta}$ and θ.

Mean Square Error of an Estimator

The **mean square error** of an estimator $\hat{\Theta}$ of the parameter θ is defined as

$$\text{MSE}(\hat{\Theta}) = E(\hat{\Theta} - \theta)^2 \tag{4-3}$$

The mean square error can be rewritten as follows:

$$\text{MSE}(\hat{\Theta}) = E[\hat{\Theta} - E(\hat{\Theta})]^2 + [\theta - E(\hat{\Theta})]^2$$
$$= V(\hat{\Theta}) + (\text{bias})^2$$

That is, the mean square error of $\hat{\Theta}$ is equal to the variance of the estimator plus the squared bias. If $\hat{\Theta}$ is an unbiased estimator of θ, the mean square error of $\hat{\Theta}$ is equal to the variance of $\hat{\Theta}$.

The mean square error is an important criterion for comparing two estimators. Let $\hat{\Theta}_1$ and $\hat{\Theta}_2$ be two estimators of the parameter θ, and let $\text{MSE}(\hat{\Theta}_1)$ and $\text{MSE}(\hat{\Theta}_2)$ be the mean square errors of $\hat{\Theta}_1$ and $\hat{\Theta}_2$. Then the **relative efficiency** of $\hat{\Theta}_2$ to $\hat{\Theta}_1$ is defined as

$$\frac{\text{MSE}(\hat{\Theta}_1)}{\text{MSE}(\hat{\Theta}_2)} \tag{4-4}$$

If this relative efficiency is less than 1, we would conclude that $\hat{\Theta}_1$ is a more efficient estimator of θ than $\hat{\Theta}_2$, in the sense that it has smaller mean square error.

Previously, we suggested several estimators of μ: the sample average, the sample median, and a single observation. Because the variance of the sample median is somewhat awkward to work with, we consider only the sample mean $\hat{\Theta}_1 = \overline{X}$ and $\hat{\Theta}_2 = X_i$. Note that both $\overline{X}$ and X_i are unbiased estimators of μ; consequently, the mean square error of both estimators is simply the variance. For the sample mean, we have $\text{MSE}(\overline{X}) = V(\overline{X}) = \sigma^2/n$ from equation 3-28. Therefore, the **relative efficiency** of X_i to $\overline{X}$ is

$$\frac{\text{MSE}(\hat{\Theta}_1)}{\text{MSE}(\hat{\Theta}_2)} = \frac{\sigma^2/n}{\sigma^2} = \frac{1}{n}$$

Because $(1/n) < 1$ for sample sizes $n \geq 2$, we would conclude that the sample mean is a better estimator of μ than a single observation X_i. This is an important point because it illustrates why, in general, large samples are preferable to small ones for many kinds of statistics problems.

The variance of an estimator, $V(\hat{\Theta})$, can be thought of as the variance of the sampling distribution of $\hat{\Theta}$. The square root of this quantity, $\sqrt{V(\hat{\Theta})}$, is usually called the standard error of the estimator.

Standard Error

> The **standard error** of a statistic is the standard deviation of its sampling distribution. If the standard error involves unknown parameters whose values can be estimated, substitution of these estimates into the standard error results in an **estimated standard error.**

The standard error gives some idea about the **precision of estimation.** For example, if the sample mean $\overline{X}$ is used as a point estimator of the population mean μ, the standard error of $\overline{X}$ measures how precisely $\overline{X}$ estimates μ.

Suppose we are sampling from a normal distribution with mean μ and variance σ^2. Now the distribution of $\overline{X}$ is normal with mean μ and variance σ^2/n, and so the standard error of $\overline{X}$ is

$$\sigma_{\overline{X}} = \frac{\sigma}{\sqrt{n}}$$

If we did not know σ but substituted the sample standard deviation S into the above equation, the estimated standard error of $\overline{X}$ would be

$$\hat{\sigma}_{\overline{X}} = \frac{S}{\sqrt{n}}$$

Calculating the Standard Error

To illustrate this definition, an article in the *Journal of Heat Transfer (Trans. ASME,* Ses. C, 96, 1974, p. 59) described a new method of measuring the thermal conductivity of Armco iron. Using a temperature of 100°F and a power input of 550 watts, the following 10 measurements of thermal conductivity (in Btu/hr-ft-°F) were obtained: 41.60, 41.48, 42.34, 41.95, 41.86, 42.18, 41.72, 42.26, 41.81, 42.04. A point estimate of the mean thermal conductivity at 100°F and 550 watts is the sample mean or

$$\bar{x} = 41.924 \text{ Btu/hr-ft-°F}$$

The standard error of the sample mean is $\sigma_{\bar{X}} = \sigma/\sqrt{n}$, and because σ is unknown, we may replace it by the sample standard deviation $s = 0.284$ to obtain the estimated standard error of $\bar{X}$ as

$$\hat{\sigma}_{\bar{X}} = \frac{s}{\sqrt{n}} = \frac{0.284}{\sqrt{10}} = 0.0898$$

Note that the standard error is about 0.2% of the sample mean, implying that we have obtained a relatively precise point estimate of thermal conductivity.

EXERCISES FOR SECTION 4-2

4-1. The Minitab output for a random sample of data is shown below. Some of the quantities are missing. Compute the values of the missing quantities.

Variable	N	Mean	SE Mean	StDev	Variance	Min.	Max.
X	9	19.96	?	3.12	?	15.94	27.16

4-2. The Minitab output for a random sample of data is shown below. Some of the quantities are missing. Compute the values of the missing quantities.

Variable	N	Mean	SE Mean	StDev	Sum
X	16	?	0.159	?	399.851

4-3. Suppose we have a random sample of size $2n$ from a population denoted by X, and $E(X) = \mu$ and $V(X) = \sigma^2$. Let

$$\bar{X}_1 = \frac{1}{2n}\sum_{i=1}^{2n} X_i \quad \text{and} \quad \bar{X}_2 = \frac{1}{n}\sum_{i=1}^{n} X_i$$

be two estimators of μ. Which is the better estimator of μ? Explain your choice.

4-4. Let $X_1, X_2, \ldots, X_9$ denote a random sample from a population having mean μ and variance σ^2. Consider the following estimators of μ:

$$\hat{\Theta}_1 = \frac{X_1 + X_2 + \cdots + X_9}{9}$$

$$\hat{\Theta}_2 = \frac{3X_1 - X_6 + 2X_4}{2}$$

(a) Is either estimator unbiased?
(b) Which estimator is "best"? In what sense is it best?

4-5. Suppose that $\hat{\Theta}_1$ and $\hat{\Theta}_2$ are unbiased estimators of the parameter θ. We know that $V(\hat{\Theta}_1) = 2$ and $V(\hat{\Theta}_2) = 4$. Which estimator is better, and in what sense is it better?

4-6. Calculate the relative efficiency of the two estimators in Exercise 4-4.

4-7. Calculate the relative efficiency of the two estimators in Exercise 4-5.

4-8. Suppose that $\hat{\Theta}_1$ and $\hat{\Theta}_2$ are estimators of the parameter θ. We know that $E(\hat{\Theta}_1) = \theta$, $E(\hat{\Theta}_2) = \theta/2$, $V(\hat{\Theta}_1) = 10$, $V(\hat{\Theta}_2) = 4$. Which estimator is "best"? In what sense is it best?

4-9. Suppose that $\hat{\Theta}_1$, $\hat{\Theta}_2$, and $\hat{\Theta}_3$ are estimators of θ. We know that $E(\hat{\Theta}_1) = E(\hat{\Theta}_2) = \theta$, $E(\hat{\Theta}_3) \ne \theta$, $V(\hat{\Theta}_1) = 16$, $V(\hat{\Theta}_2) = 11$, and $E(\hat{\Theta}_3 - \theta)^2 = 6$. Compare these three estimators. Which do you prefer? Why?

4-10. Let three random samples of sizes $n_1 = 20$, $n_2 = 10$, and $n_3 = 8$ be taken from a population with mean μ and variance σ^2. Let S_1^2, S_2^2, and S_3^2 be the sample variances. Show that $S^2 = (20S_1^2 + 10S_2^2 + 8S_3^2)/38$ is an unbiased estimator of σ^2.

4-11. (a) Show that $\sum_{i=1}^{n}(X_i - \bar{X})^2/n$ is a biased estimator of σ^2.
(b) Find the amount of bias in the estimator.
(c) What happens to the bias as the sample size n increases?

4-12. Let $X_1, X_2, \ldots, X_n$ be a random sample of size n.
(a) Show that $\bar{X}^2$ is a biased estimator for μ^2.
(b) Find the amount of bias in this estimator.
(c) What happens to the bias as the sample size n increases?

4-3 HYPOTHESIS TESTING

4-3.1 Statistical Hypotheses

In the previous section we illustrated how a parameter can be estimated from sample data. However, many problems in engineering require that we decide whether to accept or reject a statement about some parameter. The statement is called a **hypothesis,** and the decision-making procedure about the hypothesis is called **hypothesis testing.** This is one of the most useful aspects of statistical inference because many types of decision-making problems, tests, or experiments in the engineering world can be formulated as hypothesis testing problems. We like to think of statistical hypothesis testing as the data analysis stage of a **comparative experiment** in which the engineer is interested, for example, in comparing the mean of a population to a specified value. These simple comparative experiments are frequently encountered in practice and provide a good foundation for the more complex experimental design problems that we will discuss in Chapter 7. In this chapter we discuss comparative experiments involving one population, and one area that we focus on is testing hypotheses concerning the parameters of the population.

A statistical hypothesis can arise from physical laws, theoretical knowledge, past experience, or external considerations, such as engineering requirements. We now give a formal definition of a statistical hypothesis.

Statistical Hypothesis

> A **statistical hypothesis** is a statement about the parameters of one or more populations.

Because we use probability distributions to represent populations, a statistical hypothesis may also be thought of as a statement about the probability distribution of a random variable. The hypothesis will usually involve one or more parameters of this distribution.

For example, suppose that we are interested in the burning rate of a solid propellant used to power aircrew escape systems; burning rate is a random variable that can be described by a probability distribution. Suppose that our interest focuses on the mean burning rate (a parameter of this distribution). Specifically, we are interested in deciding whether or not the mean burning rate is 50 cm/s. We may express this formally as

$$H_0: \mu = 50 \text{ cm/s}$$
$$H_1: \mu \neq 50 \text{ cm/s} \tag{4-5}$$

The statement $H_0: \mu = 50$ cm/s in equation 4-5 is called the **null hypothesis,** and the statement $H_1: \mu \neq 50$ is called the **alternative hypothesis.** Because the alternative hypothesis specifies values of μ that could be either greater or less than 50 cm/s, it is called a **two-sided alternative hypothesis.** In some situations, we may wish to formulate a **one-sided alternative hypothesis,*** as in

$$H_0: \mu = 50 \text{ cm/s} \quad H_1: \mu < 50 \text{ cm/s} \quad \text{or} \quad H_0: \mu = 50 \text{ cm/s} \quad H_1: \mu > 50 \text{ cm/s} \tag{4-6}$$

---

*There are two models that can be used for the one-sided alternative hypothesis. If $H_1: \mu > 50$ cm/s (for example), then we can write the null hypothesis as $H_0: \mu = 50$ or as $H_0: \mu \leq 50$. In the first case, we are restricting μ to be equal to 50 (the null value), and in the second we are allowing the null value to be less than 50. Either way, expression of H_0 leads to the same testing and decision-making procedures (i.e., both expressions lead to a procedure based on the equality $\mu = 50$). As the reader becomes more familiar with hypothesis testing procedures, it will become evident that a decision leading to rejection of the null hypothesis when $H_0: \mu = 50$ will necessarily also lead to rejection of the null hypothesis when $H_0: \mu < 50$. Consequently, we usually write the null hypothesis solely with the equality sign, but assume that it also represents "$\leq$" or "$\geq$" as appropriate.

It is important to remember that hypotheses are always statements about the population or distribution under study, not statements about the sample. The value of the population parameter specified in the null hypothesis (50 cm/s in the above example) is usually determined in one of three ways. First, it may result from past experience or knowledge of the process or even from previous tests or experiments. The objective of hypothesis testing then is usually to determine whether the parameter value has changed. Second, this value may be determined from some theory or model regarding the process under study. Here the objective of hypothesis testing is to verify the theory or model. A third situation arises when the value of the population parameter results from external considerations, such as design or engineering specifications, or from contractual obligations. In this situation, the usual objective of hypothesis testing is conformance testing.

A procedure leading to a decision about a particular hypothesis is called a **test of a hypothesis.** Hypothesis testing procedures rely on using the information in a random sample from the population of interest. If this information is consistent with the hypothesis, we will conclude that the hypothesis is true; however, if this information is inconsistent with the hypothesis, we will conclude that the hypothesis is false. We emphasize that the truth or falsity of a particular hypothesis can never be known with certainty unless we can examine the entire population. This is usually impossible in most practical situations. Therefore, a hypothesis testing procedure should be developed with the probability of reaching a wrong conclusion in mind.

The structure of hypothesis testing problems is identical in all the applications that we will consider. The null hypothesis is the hypothesis we wish to test. Rejection of the null hypothesis always leads to accepting the alternative hypothesis. In our treatment of hypothesis testing, the null hypothesis will always be stated so that it specifies an exact value of the parameter (as in the statement H_0: $\mu = 50$ cm/s in equation 4-5). The alternative hypothesis will allow the parameter to take on several values (as in the statement H_1: $\mu \neq 50$ cm/s in equation 4-5). Testing the hypothesis involves taking a random sample, computing a **test statistic** from the sample data, and then using the test statistic to make a decision about the null hypothesis.

4-3.2 Testing Statistical Hypotheses

To illustrate the general concepts, consider the propellant burning rate problem introduced earlier. The null hypothesis is that the mean burning rate is 50 cm/s, and the alternative is that it is not equal to 50 cm/s. That is, we wish to test

$$H_0: \mu = 50 \text{ cm/s}$$
$$H_1: \mu \neq 50 \text{ cm/s}$$

Suppose that a sample of $n = 10$ specimens is tested and that the sample mean burning rate $\bar{x}$ is observed. The sample mean is an estimate of the true population mean μ. A value of the sample mean $\bar{x}$ that falls close to the hypothesized value of $\mu = 50$ cm/s is evidence that the true mean μ is really 50 cm/s; that is, such evidence supports the null hypothesis H_0. On the other hand, a sample mean that is considerably different from 50 cm/s is evidence in support of the alternative hypothesis H_1. Thus, the sample mean is the test statistic in this case.

The sample mean can take on many different values. Suppose that if $48.5 \leq \bar{x} \leq 51.5$, we will not reject the null hypothesis H_0: $\mu = 50$, and if either $\bar{x} < 48.5$ or $\bar{x} > 51.5$, we will reject the null hypothesis in favor of the alternative hypothesis H_1: $\mu \neq 50$. This situation is illustrated in Fig. 4-3. The values of $\bar{x}$ that are less than 48.5 and greater than 51.5 constitute the **critical region** for the test, whereas all values that are in the interval $48.5 \leq \bar{x} \leq 51.5$ form a region for which we will fail to reject the null hypothesis. The boundaries that define the critical regions are called the **critical values.** In our example the critical values are 48.5

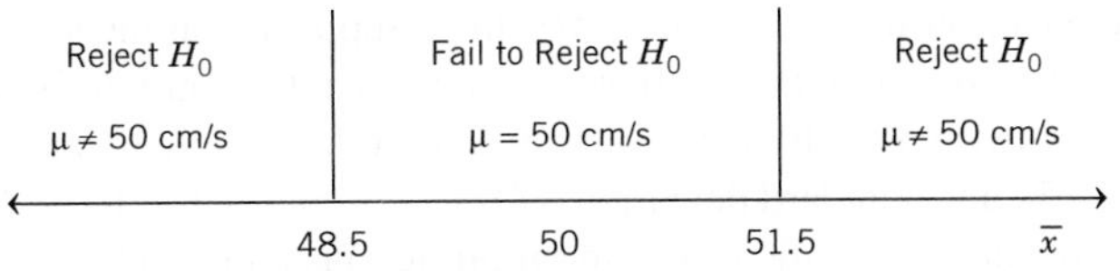

Reject H_0	Fail to Reject H_0	Reject H_0
$\mu \neq 50$ cm/s	$\mu = 50$ cm/s	$\mu \neq 50$ cm/s

Figure 4-3 Decision criteria for testing H_0: $\mu = 50$ cm/s versus H_1: $\mu \neq 50$ cm/s.

and 51.5. It is customary to state conclusions relative to the null hypothesis H_0. Therefore, we reject H_0 in favor of H_1 if the test statistic falls in the critical region and fail to reject H_0 otherwise.

This decision procedure can lead to either of two wrong conclusions. For example, the true mean burning rate of the propellant could be equal to 50 cm/s. However, for the randomly selected propellant specimens that are tested, we could observe a value of the test statistic $\bar{x}$ that falls into the critical region. We would then reject the null hypothesis H_0 in favor of the alternative H_1 when, in fact, H_0 is really true. This type of wrong conclusion is called a **type I error.**

Type I Error

> Rejecting the null hypothesis H_0 when it is true is defined as a **type I error.**

Now suppose that the true mean burning rate is different from 50 cm/s, yet the sample mean $\bar{x}$ does not fall in the critical region. In this case we would fail to reject H_0 when it is false. This type of wrong conclusion is called a **type II error.**

Type II Error

> Failing to reject the null hypothesis when it is false is defined as a **type II error.**

Thus, in testing any statistical hypothesis, four different situations determine whether the final decision is correct or in error. These situations are presented in Table 4-1.

Because our decision is based on random variables, probabilities can be associated with the type I and type II errors in Table 4-1. The probability of making a type I error is denoted by the Greek letter α (alpha). That is,

$$\alpha = P(\text{type I error}) = P(\text{reject } H_0 \text{ when } H_0 \text{ is true}) \tag{4-7}$$

Sometimes the type I error probability is called the **significance level** or **size** of the test. In the propellant burning rate example, a type I error will occur when either $\bar{x} > 51.5$ or $\bar{x} < 48.5$

Table 4-1 Decisions in Hypothesis Testing

Decision	H_0 Is True	H_0 Is False
Fail to reject H_0	No error	Type II error
Reject H_0	Type I error	No error

when the true mean burning rate is $\mu = 50$ cm/s. Suppose that the standard deviation of burning rate is $\sigma = 2.5$ cm/s and that the burning rate has a distribution for which the conditions of the central limit theorem apply, so the distribution of the sample mean is approximately normal with mean $\mu = 50$ and standard deviation $\sigma/\sqrt{n} = 2.5/\sqrt{10} = 0.79$. The probability of making a type I error (or the significance level of our test) is equal to the sum of the areas that have been shaded in the tails of the normal distribution in Fig. 4-4. We may find this probability as

$$\alpha = P(\overline{X} < 48.5 \text{ when } \mu = 50) + P(\overline{X} > 51.5 \text{ when } \mu = 50)$$

Computing the Significance Level α

The z-values that correspond to the critical values 48.5 and 51.5 are

$$z_1 = \frac{48.5 - 50}{0.79} = -1.90 \quad \text{and} \quad z_2 = \frac{51.5 - 50}{0.79} = 1.90$$

Therefore,

$$\alpha = P(Z < -1.90) + P(Z > 1.90)$$
$$= 0.0287 + 0.0287 = 0.0574$$

This implies that 5.74% of all random samples would lead to rejection of the hypothesis H_0: $\mu = 50$ cm/s when the true mean burning rate is really 50 cm/s.

From inspection of Fig. 4-4, note that we can reduce α by pushing the critical regions further into the tails of the distribution. For example, if we make the critical values 48 and 52, the value of α is

$$\alpha = P\left(Z < \frac{48 - 50}{0.79}\right) + P\left(Z > \frac{52 - 50}{0.79}\right) = P(Z < -2.53) + P(Z > 2.53)$$
$$= 0.0057 + 0.0057 = 0.0114$$

Measuring the Effect of Sample Size

We could also reduce α by increasing the sample size, assuming that the critical values of 48.5 and 51.5 do not change. If $n = 16$, $\sigma/\sqrt{n} = 2.5/\sqrt{16} = 0.625$, and using the original critical region in Fig. 4-3, we find

$$z_1 = \frac{48.5 - 50}{0.625} = -2.40 \quad \text{and} \quad z_2 = \frac{51.5 - 50}{0.625} = 2.40$$

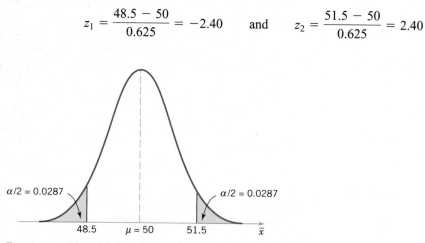

Figure 4-4   The critical region for H_0: $\mu = 50$ versus H_1: $\mu \neq 50$ and $n = 10$.

Therefore,

$$\alpha = P(Z < -2.40) + P(Z > 2.40) = 0.0082 + 0.0082 = 0.0164$$

In evaluating a hypothesis testing procedure, it is also important to examine the probability of a type II error, which we will denote by β (beta). That is,

$$\beta = P(\text{type II error}) = P(\text{fail to reject } H_0 \text{ when } H_0 \text{ is false}) \qquad (4\text{-}8)$$

To calculate β, we must have a specific alternative hypothesis; that is, we must have a particular value of μ. For example, suppose that it is important to reject the null hypothesis H_0: μ = 50 whenever the mean burning rate μ is greater than 52 cm/s or less than 48 cm/s. We could calculate the probability of a type II error β for the values μ = 52 and μ = 48 and use this result to tell us something about how the test procedure would perform. Specifically, how will the test procedure work if we wish to detect—that is, reject H_0—for a mean value of μ = 52 or μ = 48? Because of symmetry, it is only necessary to evaluate one of the two cases—say, find the probability of not rejecting the null hypothesis H_0: μ = 50 cm/s when the true mean is μ = 52 cm/s.

Compute the Probability of a Type II Error β Figure 4-5 will help us calculate the probability of type II error β. The normal distribution on the left in Fig. 4-5 is the distribution of the test statistic $\overline{X}$ when the null hypothesis H_0: μ = 50 is true (this is what is meant by the expression "under H_0: μ = 50"), and the normal distribution on the right is the distribution of $\overline{X}$ when the alternative hypothesis is true and the value of the mean is 52 (or "under H_1: μ = 52"). Now a type II error will be committed if the sample mean $\overline{x}$ falls between 48.5 and 51.5 (the critical region boundaries) when μ = 52. As seen in Fig. 4-5, this is just the probability that $48.5 \leq \overline{X} \leq 51.5$ when the true mean is μ = 52, or the shaded area under the normal distribution on the right. Therefore, referring to Fig. 4-5, we find that

$$\beta = P(48.5 \leq \overline{X} \leq 51.5 \text{ when } \mu = 52)$$

The z-values corresponding to 48.5 and 51.5 when μ = 52 are

$$z_1 = \frac{48.5 - 52}{0.79} = -4.43 \qquad \text{and} \qquad z_2 = \frac{51.5 - 52}{0.79} = -0.63$$

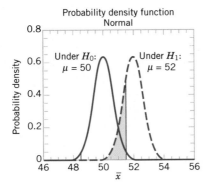

Figure 4-5   The probability of type II error when μ = 52 and $n = 10$.

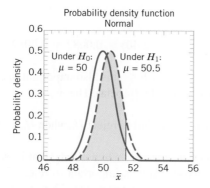

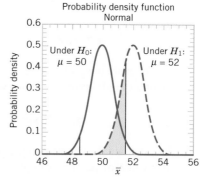

Figure 4-6   The probability of type II error when $\mu = 50.5$ and $n = 10$.

Figure 4-7   The probability of type II error when $\mu = 52$ and $n = 16$.

Therefore,

$$\beta = P(-4.43 \le Z \le -0.63) = P(Z \le -0.63) - P(Z \le -4.43)$$
$$= 0.2643 - 0.000 = 0.2643$$

Thus, if we are testing H_0: $\mu = 50$ against H_1: $\mu \neq 50$ with $n = 10$, and the true value of the mean is $\mu = 52$, the probability that we will fail to reject the false null hypothesis is 0.2643. By symmetry, if the true value of the mean is $\mu = 48$, the value of β will also be 0.2643.

The probability of making a type II error β increases rapidly as the true value of μ approaches the hypothesized value. For example, see Fig. 4-6, where the true value of the mean is $\mu = 50.5$ and the hypothesized value is H_0: $\mu = 50$. The true value of μ is very close to 50, and the value for β is

$$\beta = P(48.5 \le \overline{X} \le 51.5 \text{ when } \mu = 50.5)$$

The z-values corresponding to 48.5 and 51.5 when $\mu = 50.5$ are

$$z_1 = \frac{48.5 - 50.5}{0.79} = -2.53 \quad \text{and} \quad z_2 = \frac{51.5 - 50.5}{0.79} = 1.27$$

Therefore,

$$\beta = P(-2.53 \le Z \le 1.27) = P(Z \le 1.27) - P(Z \le -2.53)$$
$$= 0.8980 - 0.0057 = 0.8923$$

Thus, the type II error probability is much higher for the case in which the true mean is 50.5 cm/s than for the case in which the mean is 52 cm/s. Of course, in many practical situations we would not be as concerned with making a type II error if the mean were "close" to the hypothesized value. We would be much more interested in detecting large differences between the true mean and the value specified in the null hypothesis.

The type II error probability also depends on the sample size n. Suppose that the null hypothesis is H_0: $\mu = 50$ cm/s and that the true value of the mean is $\mu = 52$. If the sample size is increased from $n = 10$ to $n = 16$, the situation of Fig. 4-7 results. The normal distribution on the left is the distribution of $\overline{X}$ when the mean $\mu = 50$, and the normal distribution on the right is the distribution of $\overline{X}$ when $\mu = 52$. As shown in Fig. 4-7, the type II error probability is

$$\beta = P(48.5 \le \overline{X} \le 51.5 \text{ when } \mu = 52)$$

When $n = 16$, the standard deviation of $\overline{X}$ is $\sigma/\sqrt{n} = 2.5/\sqrt{16} = 0.625$, and the z-values corresponding to 48.5 and 51.5 when $\mu = 52$ are

$$z_1 = \frac{48.5 - 52}{0.625} = -5.60 \quad \text{and} \quad z_2 = \frac{51.5 - 52}{0.625} = -0.80$$

Therefore,

$$\beta = P(-5.60 \leq Z \leq -0.80) = P(Z \leq -0.80) - P(Z \leq -5.60)$$
$$= 0.2119 - 0.000 = 0.2119$$

Recall that when $n = 10$ and $\mu = 52$, we found that $\beta = 0.2643$; therefore, increasing the sample size results in a decrease in the probability of type II error.

The results from this section and a few other similar calculations are summarized next:

Fail to Reject H_0 When	Sample Size	α	β at $\mu = 52$	β at $\mu = 50.5$
$48.5 < \bar{x} < 51.5$	10	0.0574	0.2643	0.8923
$48 < \bar{x} < 52$	10	0.0114	0.5000	0.9705
$48.5 < \bar{x} < 51.5$	16	0.0164	0.2119	0.9445
$48 < \bar{x} < 52$	16	0.0014	0.5000	0.9918

Understanding the Relationship Between α and β

The results in boxes were not calculated in the text but can be easily verified by the reader. This display and the preceding discussion reveal four important points:

1. The size of the critical region, and consequently the probability of a type I error α, can always be reduced by appropriate selection of the critical values.

2. Type I and type II errors are related. A decrease in the probability of one type of error always results in an increase in the probability of the other, provided that the sample size n does not change.

3. An increase in sample size will generally reduce both α and β, provided that the critical values are held constant.

4. When the null hypothesis is false, β increases as the true value of the parameter approaches the value hypothesized in the null hypothesis. The value of β decreases as the difference between the true mean and the hypothesized value increases.

Generally, the analyst controls the type I error probability α when he or she selects the critical values. Thus, it is usually easy for the analyst to set the type I error probability at (or near) any desired value. Because the analyst can directly control the probability of wrongly rejecting H_0, we always think of rejection of the null hypothesis H_0 as a **strong conclusion.**

Because we can control the probability of making a type I error (or significance level) α, a logical question is what value should be used. The type I error probability is a measure of risk, specifically, the risk of concluding that the null hypothesis is false when it really isn't. So, the value of α should be chosen to reflect the consequences (economic, social, etc.) of incorrectly rejecting H_0. Smaller values of α would reflect more serious consequences and larger values of α would be consistent with less severe consequences. This is often hard to do, and what has evolved in much of scientific and engineering practice is to use the value $\alpha = 0.05$ in most situations, unless there is information available that indicates that this is an inappropriate choice. In the rocket propellant problem with $n = 10$, this would correspond to critical values of 48.45 and 51.55.

> A widely used procedure in hypothesis testing is to use a type I error or significance level of $\alpha = 0.05$. This value has evolved through experience, and may not be appropriate for all situations.

In contrast, the probability of type II error β is not a constant, but depends on both the true value of the parameter and the sample size that we have selected. Because the type II error probability β is a function of both the sample size and the extent to which the null hypothesis H_0 is false, it is customary to think of the decision not to reject H_0 as a **weak conclusion,** unless we know that β is acceptably small. Therefore, rather than saying we "accept H_0," we prefer the terminology "fail to reject H_0." Failing to reject H_0 implies that we have not found sufficient evidence to reject H_0—that is, to make a strong statement. Failing to reject H_0 does not necessarily mean there is a high probability that H_0 is true. It may simply mean that more data are required to reach a strong conclusion. This can have important implications for the formulation of hypotheses.

An important concept that we will make use of is the **power** of a statistical test.

Definition

> The **power** of a statistical test is the probability of rejecting the null hypothesis H_0 when the alternative hypothesis is true.

The power is computed as $1 - \beta$, and power can be interpreted as *the probability of correctly rejecting a false null hypothesis.* We often compare statistical tests by comparing their power properties. For example, consider the propellant burning rate problem when we are testing H_0: $\mu = 50$ cm/s against H_1: $\mu \neq 50$ cm/s. Suppose that the true value of the mean is $\mu = 52$. When $n = 10$, we found that $\beta = 0.2643$, so the power of this test is $1 - \beta = 1 - 0.2643 = 0.7357$ when $\mu = 52$.

Power is a very descriptive and concise measure of the **sensitivity** of a statistical test, where by sensitivity we mean the ability of the test to detect differences. In this case, the sensitivity of the test for detecting the difference between a mean burning rate of 50 and 52 cm/s is 0.7357. That is, if the true mean is really 52 cm/s, this test will correctly reject H_0: $\mu = 50$ and "detect" this difference 73.57% of the time. If this value of power is judged to be too low, the analyst can increase either α or the sample size n.

4-3.3 P-Values in Hypothesis Testing

The approach to hypothesis testing that we have outlined in the previous sections has emphasized using a **fixed significance level** α, and this value will often be $\alpha = 0.05$. The fixed significance level or fixed type I error rate approach to hypothesis testing is very nice because it leads directly to the definitions of type II error and power, which are very useful concepts and of considerable value in determining appropriate sample sizes for hypothesis testing.

Fixed significance level testing does have some disadvantages. For example, suppose that you learn that the null hypothesis regarding the mean burning rate of the rocket propellant has been rejected at the $\alpha = 0.05$ level of significance. This statement may be inadequate, because it gives you no idea about whether the sample average burning rate was just barely in the critical region or whether it was very far into this region. This relates to the strength of the evidence against H_0. Furthermore, stating the results this way imposes the predefined fixed level of significance on other users of the information. This may be unsatisfactory because some decision makers might be uncomfortable with the risks imposed by choosing $\alpha = 0.05$.

To avoid these potential difficulties, the **P-value approach** to hypothesis testing has been widely adopted in practice. the P-value is the probability that the sample average will take on a value that is at least as extreme as the observed value when the null hypothesis H_0 is true. In other words, the P-value conveys information about the weight of evidence against H_0. The smaller the P-value, the greater the evidence against H_0. When the P-value is small enough, we reject the null hypothesis in favor of the alternative. The P-value approach allows a decision maker to draw conclusions at any level of significance that is appropriate. We now give a formal definition of a P-value.

Definition

> The **P-value** is the smallest level of significance that would lead to rejection of the null hypothesis H_0.

To illustrate the P-value concept, let's consider the propellant burning rate situation, for which the hypotheses are

$$H_0: \mu = 50 \text{ cm/s}$$
$$H_1: \mu \neq 50 \text{ cm/s}$$

where we know that $\sigma = 2.5$ cm/s. Suppose that a random sample of $n = 10$ propellant specimens results in a sample average of $\bar{x} = 51.8$ cm/s. Figure 4-8 illustrates how the P-value is computed. The normal curve in this figure is the distribution of the sample average under the null hypothesis; normal with mean $\mu = 50$ and standard deviation $\sigma/\sqrt{n} = 2.5/\sqrt{10} = 0.79$. The value 51.8 is the observed value of the sample average. The probability of observing a value of the sample average that is at least as large as 51.8 is found by computing the z-value

$$z = \frac{51.8 - 50}{0.79} = 2.28$$

and the probability that the standard normal random variable equals or exceeds 2.28 is 0.0113. Because the null hypothesis is two-sided, this is half of the P-value. We must also consider the case where the z-value could have been negative; that is, $z = -2.28$ (this would correspond to the point 48.2 shown in Fig. 4-8). Because the normal curve is symmetric, the probability that the standard normal random variable is less than or equal to -2.28 is 0.0113. Therefore, the P-value for this hypothesis testing problem is

$$P = 0.0113 + 0.0113 = 0.0226$$

Interpretting the P-value

The P-value tells us that if the null hypothesis H_0 is true, the probability of obtaining a random sample whose mean is at least as far from 50 as 51.8 (or 48.2) is 0.0226. Therefore, an observed sample mean of 51.8 is a pretty rare event if the null hypothesis is really true. Compared to the "standard" level of significance 0.05, our observed P-value is smaller, so if we were using a fixed significance level of 0.05, the null hypothesis would be rejected. In fact, H_0 would be rejected at *any* level of significance greater than or equal to 0.0226. This illustrates the boxed definition above; the P-value is the smallest level of significance that would lead to rejection of H_0.

Operationally, once a P-value is computed, we typically compare it to a predefined significance level to make a decision. Often this predefined significance level is 0.05. However, in presenting results and conclusions, it is standard practice to report the observed P-value along with the decision that is made regarding the null hypothesis.

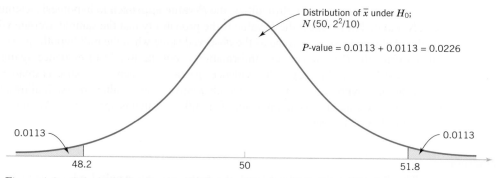

Distribution of $\bar{x}$ under H_0;
$N(50, 2^2/10)$

P-value $= 0.0113 + 0.0113 = 0.0226$

0.0113

0.0113

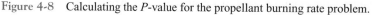

48.2   50   51.8

Figure 4-8   Calculating the P-value for the propellant burning rate problem.

Clearly, the P-value provides a measure of the credibility of the null hypothesis. Specifically, it is the risk that you have made an incorrect decision if you reject H_0. The P-value is *not* the probability that the null hypothesis is false, nor is $1 - P$ the probability that the null hypothesis is true. The null hypothesis is either true or false (there is no probability associated with this), and so the proper interpretation of the P-value is in terms of the risk of wrongly rejecting H_0.

We will use the P-value approach extensively. Modern statistics software packages report the results of hypothesis testing problems almost exclusively in terms of P-values.

4-3.4 One-Sided and Two-Sided Hypotheses

A test of any hypothesis such as

$$H_0\text{: } \mu = \mu_0$$
$$H_1\text{: } \mu \neq \mu_0$$

is called a **two-sided** test because it is important to detect differences from the hypothesized value of the mean μ_0 that lie on either side of μ_0. In such a test, the critical region is split into two parts, with (usually) equal probability placed in each tail of the distribution of the test statistic.

Many hypothesis testing problems naturally involve a **one-sided** alternative hypothesis, such as

$$H_0\text{: } \mu = \mu_0 \qquad \qquad H_0\text{: } \mu = \mu_0$$
$$H_1\text{: } \mu > \mu_0 \qquad \text{or} \qquad H_1\text{: } \mu < \mu_0$$

If the alternative hypothesis is H_1: $\mu > \mu_0$, the P-value should be calculated from the upper tail of the distribution of the test statistic, whereas if the alternative hypothesis is H_1: $\mu < \mu_0$, the P-value should be calculated from the lower tail of the distribution. Consequently, these tests are sometimes called **one-tailed** tests. Determining the P-value for one-sided tests is usually easy. Simply visualize the behavior of the test statistic if the null hypothesis is true and calculate the P-value from the appropriate end or tail of the distribution. Generally, the inequality in the alternative hypothesis "points" to the tail of the curve where the P-value is computed. If a fixed significance level test is used, the inequality in the alternative hypothesis points in the direction of the critical region.

Formulating the Null Hypothesis   In constructing hypotheses, we will always state the null hypothesis as an equality, so that the probability of type I error α can be controlled at a specific value (refer to the footnote regarding one-sided alternative hypothesis on page 145). The alternative hypothesis might be either one- or two-sided, depending on the conclusion to be drawn if H_0 is rejected. If the objective is to make a claim involving statements such as "greater than," "less than," "superior to," "exceeds," "at least," and so forth, a one-sided alternative is appropriate. If no direction is implied by the claim or if the claim "not equal to" is to be made, a two-sided alternative should be used.

EXAMPLE 4-2

The Rocket Propellant

Consider the propellant burning rate problem. Suppose that if the burning rate is less than 50 cm/s, we wish to show this with a strong conclusion. The hypotheses should be stated as

$$H_0: \mu = 50 \text{ cm/s}$$
$$H_1: \mu < 50 \text{ cm/s}$$

Here the P-value would be calculated by finding the probability that the normal random variable is less than the observed value of $\bar{X}$. That is, the P-value is calculated from the lower tail of the null distribution of $\bar{X}$. Because the rejection of H_0 is always a strong conclusion, this statement of the hypotheses will produce the desired outcome if H_0 is rejected. Note that although the null hypothesis is stated with an equal sign, it is understood to include any value of μ not specified by the alternative hypothesis. Therefore, failing to reject H_0 does not mean that $\mu = 50$ cm/s exactly, but only that we do not have strong evidence in support of H_1. ■

In some real-world problems where one-sided test procedures are indicated, it is occasionally difficult to choose an appropriate formulation of the alternative hypothesis. For example, suppose that a soft-drink beverage bottler purchases 2-liter bottles from a glass company. The bottler wants to be sure that the bottles meet the specification on mean internal pressure or bursting strength, which for 2-liter bottles is a minimum strength of 200 psi. The bottler has decided to formulate the decision procedure for a specific lot of bottles as a hypothesis problem. There are two possible formulations for this problem, either

$$H_0: \mu = 200 \text{ psi}$$
$$H_1: \mu > 200 \text{ psi} \tag{4-9}$$

or

$$H_0: \mu = 200 \text{ psi}$$
$$H_1: \mu < 200 \text{ psi} \tag{4-10}$$

Consider the upper-tailed formulation in equation 4-9. If the null hypothesis is rejected, the bottles will be judged satisfactory, whereas if H_0 is not rejected, the implication is that the bottles do not conform to specifications and should not be used. Because rejecting H_0 is a strong conclusion, this formulation forces the bottle manufacturer to "demonstrate" that the mean bursting strength of the bottles exceeds the specifications. Now consider the lower-tailed formulation in equation 4-10. In this situation, the bottles will be judged satisfactory unless H_0 is rejected. That is, we conclude that the bottles are satisfactory unless there is strong evidence to the contrary.

Which formulation is correct, the upper-tailed test in equation 4-9 or the lower-tailed test in equation 4-10? The answer is, "it depends." For the upper-tailed test, there is some probability that H_0 will not be rejected (i.e., we would decide that the bottles are not satisfactory) even though the true mean is slightly greater than 200 psi. This formulation implies that we want the bottle manufacturer to demonstrate that the product meets or exceeds our specifications. Such a formulation could be appropriate if the manufacturer has experienced difficulty in meeting specifications in the past or if product safety considerations force us to hold tightly to the 200 psi specification. On the other hand, for the lower-tailed test of equation 4-10 there is some probability that H_0 will be accepted and the bottles judged satisfactory even though the true mean is slightly less than 200 psi. We would conclude that the bottles are unsatisfactory only when there is strong evidence that the mean does not exceed 200 psi—that is, when H_0: $\mu = 200$ psi is rejected. This formulation assumes that we are relatively happy with the bottle manufacturer's past performance and that small deviations from the specification of $\mu \geq 200$ psi are not harmful.

In formulating one-sided alternative hypotheses, we should remember that rejecting H_0 is always a strong conclusion. Consequently, **we should put the statement about which it is important to make a strong conclusion in the alternative hypothesis.** In real-world problems, this will often depend on our point of view and experience with the situation.

4-3.5 General Procedure for Hypothesis Testing

This chapter develops hypothesis testing procedures for many practical problems. Use of the following sequence of steps in applying hypothesis testing methodology is recommended:

1. **Parameter of interest:** From the problem context, identify the parameter of interest.
2. **Null hypothesis, H_0:** State the null hypothesis, H_0.
3. **Alternative hypothesis, H_1:** Specify an appropriate alternative hypothesis, H_1.
4. **Test statistic:** State an appropriate test statistic.
5. **Reject H_0 if:** Define the criteria that will lead to rejection of H_0.
6. **Computations:** Compute any necessary sample quantities, substitute these into the equation for the test statistic, and compute that value.
7. **Conclusions:** Decide whether or not H_0 should be rejected and report that in the problem context. This could involve computing a P-value or comparing the test statistic to a set of critical values.

Steps 1–4 should be completed prior to examination of the sample data. This sequence of steps will be illustrated in subsequent sections.

EXERCISES FOR SECTION 4-3

4-13.   A textile fiber manufacturer is investigating a new drapery yarn, which has a standard deviation of 0.3 kg. The company wishes to test the hypothesis H_0: $\mu = 14$ against H_1: $\mu < 14$, using a random sample of five specimens.
(a) What is the P-value if the sample average is $\bar{x} = 13.7$ kg?
(b) Find β for the case where the true mean elongation force is 13.5 kg and we assume that $\alpha = 0.05$.
(c) What is the power of the test from part (b)?

4-14.   Repeat Exercise 4-13 using a sample size of $n = 16$ and the same critical region.

4-15.   In Exercise 4-13 with $n = 5$:
(a) Find the boundary of the critical region if the type I error probability is specified to be $\alpha = 0.01$.
(b) Find β for the case when the true mean elongation force is 13.5 kg.
(c) What is the power of the test?

4-16.   In Exercise 4-14 with $n = 16$:
(a) Find the boundary of the critical region if the type I error probability is specified to be 0.05.
(b) Find β for the case when the true mean elongation force is 13.0 kg.
(c) What is the power of the test from part (b)?

4-17.   The heat evolved in calories per gram of a cement mixture is approximately normally distributed. The mean is thought to be 100 and the standard deviation is 2. We wish to test H_0: $\mu = 100$ versus H_1: $\mu \neq 100$ with a sample of $n = 9$ specimens.
(a) If the rejection region is defined as $\bar{x} < 98.5$ or $\bar{x} > 101.5$, find the type I error probability α.
(b) Find β for the case where the true mean heat evolved is 103.
(c) Find β for the case where the true mean heat evolved is 105. This value of β is smaller than the one found in part (b). Why?

4-18. Repeat Exercise 4-17 using a sample size of $n = 5$ and the same critical region.

4-19. A consumer products company is formulating a new shampoo and is interested in foam height (in mm). Foam height is approximately normally distributed and has a standard deviation of 20 mm. The company wishes to test $H_0: \mu = 175$ mm versus $H_1: \mu > 175$ mm, using the results of $n = 10$ samples.

(a) Find the P-value if the sample average is $\bar{x} = 185$.

(b) What is the probability of type II error if the true mean foam height is 200 mm and we assume that $\alpha = 0.05$?

(c) What is the power of the test from part (b)?

4-20. In Exercise 4-19, suppose that the sample data result in $\bar{x} = 190$ mm.

(a) What conclusion would you reach in a fixed-level test with $\alpha = 0.05$?

(b) How "unusual" is the sample value $\bar{x} = 190$ mm if the true mean is 175 mm? That is, what is the probability that you would observe a sample average as large as 190 mm (or larger), if the true mean foam height was 175 mm?

4-21. Repeat Exercise 4-19 assuming that the sample size is $n = 16$.

4-22. Consider Exercise 4-19, and suppose that the sample size is increased to $n = 16$.

(a) Where would the boundary of the critical region be placed if the type I error probability is 0.05?

(b) Using $n = 16$ and the critical region found in part (a), find the type II error probability β if the true mean foam height is 195 mm.

(c) Compare the value of β obtained in part (b) with the value from Exercise 4-19 (b). What conclusions can you draw? Which has higher power?

4-23. A manufacturer is interested in the output voltage of a power supply used in a PC. Output voltage is assumed to be normally distributed, with standard deviation 0.25 V, and the manufacturer wishes to test $H_0: \mu = 9$ V against $H_1: \mu \neq 9$ V, using $n = 10$ units.

(a) The critical region is $\bar{x} < 8.85$ or $\bar{x} > 9.15$. Find the value of α.

(b) Find the power of the test for detecting a true mean output voltage of 9.1 V.

4-24. Rework Exercise 4-23 when $n = 16$ batches and the boundaries of the critical region do not change.

4-25. Consider Exercise 4-23, and suppose that the process engineer wants the type I error probability for the test to be $\alpha = 0.05$. Where should the critical region be located?

4-4   INFERENCE ON THE MEAN OF A POPULATION, VARIANCE KNOWN

In this section, we consider making inferences about the mean μ of a single population where the variance of the population σ^2 is known.

Assumptions

1. $X_1, X_2, \ldots, X_n$ is a random sample of size n from a population.

2. The population is normally distributed, or if it is not, the conditions of the central limit theorem apply.

Based on our previous discussion in Section 4.2, the sample mean $\overline{X}$ is an **unbiased point estimator** of μ. With these assumptions, the distribution of $\overline{X}$ is approximately normal with mean μ and variance σ^2/n.

Under the previous assumptions, the quantity

$$Z = \frac{\overline{X} - \mu}{\sigma/\sqrt{n}} \tag{4-11}$$

has a standard normal distribution, $N(0, 1)$.

4-4.1   Hypothesis Testing on the Mean

Suppose that we wish to test the hypotheses

$$H_0: \mu = \mu_0$$
$$H_1: \mu \neq \mu_0 \tag{4-12}$$

where μ_0 is a specified constant. We have a random sample $X_1, X_2, \ldots, X_n$ from the population. Because $\overline{X}$ has an approximate normal distribution (i.e., the **sampling distribution** of $\overline{X}$ is approximately normal) with mean μ_0 and standard deviation $\sigma/\sqrt{n}$, if the null hypothesis is true, we could either calculate a P-value, or if we wanted to use fixed significance level testing, we could construct a critical region for the computed value of the sample mean $\overline{x}$, as we discussed in Sections 4-3.2 and 4-3.3.

It is usually more convenient to *standardize* the sample mean and use a test statistic based on the standard normal distribution. That is, the test procedure for $H_0: \mu = \mu_0$ uses the **test statistic**

$$Z_0 = \frac{\overline{X} - \mu_0}{\sigma/\sqrt{n}} \tag{4-13}$$

If the null hypothesis $H_0: \mu = \mu_0$ is true, $E(\overline{X}) = \mu_0$, and the distribution of the test statistic Z_0 is the standard normal distribution [denoted $N(0, 1)$].

Suppose that we take a random sample of size n and the observed value of the sample mean is $\overline{x}$. To test the null hypothesis using the P-value approach, we would find the probability of observing a value of the sample mean that is at least as extreme as $\overline{x}$, given that the null hypothesis is true. The standard normal z-value that corresponds to $\overline{x}$ is found from the test statistic in equation 4-13:

$$z_0 = \frac{\overline{x} - \mu_0}{\sigma/\sqrt{n}}$$

In terms of the standard normal cumulative distribution function (cdf), the probability we are seeking is $1 - \Phi(|z_0|)$. The reason that the argument of the standard normal cdf is $|z_0|$ is that the value of z_0 could be either positive or negative, depending on the observed sample mean. Because this is a two-tailed test, this is only one-half of the P-value. Therefore, for the two-sided alternative hypothesis, the P-value is

$$P = 2\big[1 - \Phi(|z_0|)\big] \tag{4-14}$$

This is illustrated in Fig. 4-9a.

Now let's consider the one-sided alternatives. Suppose that we are testing

$$H_0: \mu = \mu_0$$
$$H_1: \mu > \mu_0 \tag{4-15}$$

Once again, suppose that we have a random sample of size n and that the sample mean is $\overline{x}$. We compute the test statistic from equation 4-13 and obtain z_0. Because the test is an upper-tailed test, only values of $\overline{x}$ that are greater than μ_0 are consistent with the alternative hypothesis. Therefore, the P-value would be the probability that the standard normal random variable is greater than the value of the test statistic z_0. This P-value is computed as

$$P = 1 - \Phi(z_0) \tag{4-16}$$

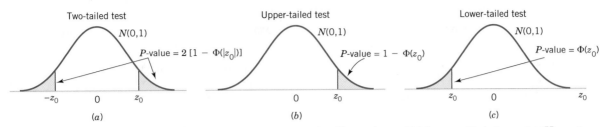

Figure 4-9   The P-value for a z-test. (a) The two-sided alternative H_1: $\mu \neq \mu_0$. (b) The one-sided alternative H_1: $\mu > \mu_0$. (c) The one-sided alternative H_1: $\mu < \mu_0$.

This P-value is shown in Fig. 4-9b.

The lower-tailed test involves the hypotheses

$$H_0\text{: } \mu = \mu_0$$
$$H_1\text{: } \mu < \mu_0 \tag{4-17}$$

Suppose that we have a random sample of size n and that the sample mean is $\bar{x}$. We compute the test statistic from equation 4-13 and obtain z_0. Because the test is a lower-tailed test, only values of $\bar{x}$ that are less than μ_0 are consistent with the alternative hypothesis. Therefore, the P-value would be the probability that the standard normal random variable is less than the value of the test statistic z_0. This P-value is computed as

$$P = \Phi(z_0) \tag{4-18}$$

and shown in Fig. 4-9c.

It is not always easy to compute the exact P-value for a statistical test. However, most modern computer programs for statistical analysis report P-values, and they can be obtained on some handheld calculators. We will also show how to approximate P-values.

We can also use fixed significance level testing. All we have to do is determine where to place the critical regions for the two-sided and one-sided alternative hypotheses. First consider the two-sided alternative in equation 4-12. Now if H_0: $\mu = \mu_0$ is true, the probability is $1 - \alpha$ that the test statistic Z_0 falls between $-z_{\alpha/2}$ and $z_{\alpha/2}$, where $z_{\alpha/2}$ is the $100\alpha/2$ percentage point of the standard normal distribution. The regions associated with $z_{\alpha/2}$ and $-z_{\alpha/2}$ are illustrated in Fig. 4-10a. Note that the probability is α that the test statistic Z_0 will fall in the region $Z_0 > z_{\alpha/2}$ or $Z_0 < -z_{\alpha/2}$ when H_0: $\mu = \mu_0$ is true. Clearly, a sample producing a value of the test statistic that falls in the tails of the distribution of Z_0 would be unusual if H_0: $\mu = \mu_0$ is true; therefore, it is an indication that H_0 is false. Thus, we should reject H_0 if either

$$z_0 > z_{\alpha/2} \tag{4-19}$$

or

$$z_0 < -z_{\alpha/2} \tag{4-20}$$

and we should fail to reject H_0 if

$$-z_{\alpha/2} \leq z_0 \leq z_{\alpha/2} \tag{4-21}$$

Equations 4-19 and 4-20 define the **critical region** or **rejection region** for the test. The type I error probability for this test procedure is α.

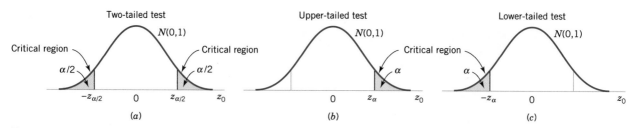

Figure 4-10   The distribution of Z_0 when H_0: $\mu = \mu_0$ is true, with critical region for (a) The two-sided alternative H_1: $\mu \neq \mu_0$. (b) The one-sided alternative H_1: $\mu > \mu_0$. (c) The one-sided alternative H_1: $\mu < \mu_0$.

We may also develop fixed significance level testing procedures for the one-sided alternatives. Consider the upper-tailed case in equation 4-15.

In defining the critical region for this test, we observe that a negative value of the test statistic Z_0 would never lead us to conclude that H_0: $\mu = \mu_0$ is false. Therefore, we would place the critical region in the upper tail of the standard normal distribution and reject H_0 if the computed value z_0 is too large. Refer to Figure 4-10b. That is, we would reject H_0 if

$$z_0 > z_\alpha \qquad (4\text{-}22)$$

Similarly, to test the lower-tailed case in equation 4-17, we would calculate the test statistic Z_0 and reject H_0 if the value of Z_0 is too small. That is, the critical region is in the lower tail of the standard normal distribution as in Fig. 4-10c, and we reject H_0 if

$$z_0 < -z_\alpha \qquad (4\text{-}23)$$

Summary

Testing Hypotheses on the Mean, Variance Known

Null hypothesis: H_0: $\mu = \mu_0$

Test statistic: $Z_0 = \dfrac{\overline{X} - \mu_0}{\sigma/\sqrt{n}}$

Alternative Hypotheses	P-Value	Rejection Criterion for Fixed-Level Tests
H_1: $\mu \neq \mu_0$	Probability above z_0 and probability below $- z_0$, $P = 2[1 - \Phi(|z_0|)]$	$z_0 > z_{\alpha/2}$ or $z_0 < -z_{\alpha/2}$
H_1: $\mu > \mu_0$	Probability above z_0, $P = 1 - \Phi(z_0)$	$z_0 > z_\alpha$
H_1: $\mu < \mu_0$	Probability below z_0, $P = \Phi(z_0)$	$z_0 < -z_\alpha$

The P-values and critical regions for these situations are shown in Figs. 4-9 and 4-10.

EXAMPLE 4-3

Propellant
Burning Rate

Aircrew escape systems are powered by a solid propellant. The burning rate of this propellant is an important product characteristic. Specifications require that the mean burning rate must be 50 cm/s. We know that the standard deviation of burning rate is $\sigma = 2$ cm/s. The experimenter decides to specify a type I error probability or significance level of $\alpha = 0.05$. He selects a random sample of $n = 25$ and obtains a sample average burning rate of $\bar{x} = 51.3$ cm/s. What conclusions should he draw?

Solution. We may solve this problem by following the seven-step procedure outlined in Section 4-3.5. This results in the following:

1. **Parameter of interest:** The parameter of interest is μ, the mean burning rate.

2. **Null hypothesis, H_0:** $\mu = 50$ cm/s

3. **Alternative hypothesis, H_1:** $\mu \neq 50$ cm/s

4. **Test statistic:** The test statistic is

$$z_0 = \frac{\bar{x} - \mu_0}{\sigma/\sqrt{n}}$$

5. **Reject H_0 if:** Reject H_0 if the P-value is smaller than 0.05. (Note that the corresponding critical region boundaries for fixed significance level testing would be $-z_{0.025} = -1.96$ and $z_{0.025} = 1.96$.)

6. **Computations:** Since $\bar{x} = 51.3$ and $\sigma = 2$,

$$z_0 = \frac{51.3 - 50}{2/\sqrt{25}} = \frac{1.3}{0.4} = 3.25$$

7. **Conclusions:** The P-value is $P = 2[1 - \Phi(3.25)] = 0.0012$. Since $P = 0.0012 < 0.05$, we reject H_0: $\mu = 50$. Stated more completely, we conclude that the mean burning rate differs from 50 cm/s, based on a sample of 25 measurements. In fact, there is strong evidence that the mean burning rate exceeds 50 cm/s. ∎

The procedure we have just described is sometimes called the **z-test.** Minitab will perform the z-test. The output below results from using Minitab for the propellant burning rate problem in Example 4-3. Notice that Minitab reports the standard error of the mean $(\sigma/\sqrt{n} = 0.4)$. This is the denominator of the z-test statistic. The P-value for the test statistic is also provided.

One-Sample Z

Test of mu = 50 vs not = 50
The assumed standard deviation = 2

N	Mean	SE Mean	95% CI	Z	P
25	51.3000	0.4000	(50.5160, 52.0840)	3.25	0.001

Minitab also reports a **confidence interval** (CI) estimate of the mean burning rate. In Section 4-4.5 we will describe how this interval is computed and how it is interpreted.

4-4.2   Type II Error and Choice of Sample Size

In testing hypotheses, the analyst directly selects the type I error probability. However, the probability of type II error β depends on the choice of sample size. In this section, we will show how to calculate the probability of type II error β. We will also show how to select the sample size to obtain a specified value of β.

Finding the Probability of Type II Error β
Consider the two-sided hypothesis

$$H_0: \mu = \mu_0$$
$$H_1: \mu \neq \mu_0$$

Suppose that the null hypothesis is false and that the true value of the mean is $\mu = \mu_0 + \delta$, say, where $\delta > 0$. The expected value of the test statistic Z_0 is

$$E(Z_0) = E\left(\frac{\overline{X} - \mu_0}{\sigma/\sqrt{n}}\right) = \frac{\mu_0 + \delta - \mu_0}{\sigma/\sqrt{n}} = \frac{\delta\sqrt{n}}{\sigma}$$

and the variance of Z_0 is unity. Therefore, the distribution of Z_0 when H_1 is true is

$$Z_0 \sim N\left(\frac{\delta\sqrt{n}}{\sigma}, 1\right)$$

Here the notation "$\sim$" means "is distributed as." The distribution of the test statistic Z_0 under both the null hypothesis H_0 and the alternative hypothesis H_1 is shown in Fig. 4-11. From examining this figure, we note that if H_1 is true, a type II error will be made only if $-z_{\alpha/2} \leq Z_0 \leq z_{\alpha/2}$ where $Z_0 \sim N(\delta\sqrt{n}/\sigma, 1)$. That is, the probability of the type II error β is the probability that Z_0 falls between $-z_{\alpha/2}$ and $z_{\alpha/2}$ *given that H_1 is true*. This probability is shown as the shaded portion of Fig. 4-11 and is expressed mathematically in the following equation.

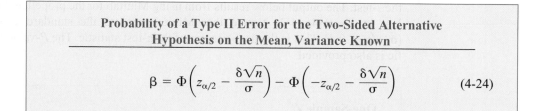

Probability of a Type II Error for the Two-Sided Alternative Hypothesis on the Mean, Variance Known

$$\beta = \Phi\left(z_{\alpha/2} - \frac{\delta\sqrt{n}}{\sigma}\right) - \Phi\left(-z_{\alpha/2} - \frac{\delta\sqrt{n}}{\sigma}\right) \qquad (4\text{-}24)$$

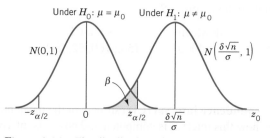

Figure 4-11   The distribution of Z_0 under H_0 and H_1.

where $\Phi(z)$ denotes the probability to the left of z in the standard normal distribution. Note that equation 4-24 was obtained by evaluating the probability that Z_0 falls in the interval $[-z_{\alpha/2}, z_{\alpha/2}]$ when H_1 is true. Furthermore, note that equation 4-24 also holds if $\delta < 0$, due to the symmetry of the normal distribution. It is also possible to derive an equation similar to equation 4-24 for a one-sided alternative hypothesis.

Sample Size Formulas

One may easily obtain formulas that determine the appropriate sample size to obtain a particular value of β for a given δ and α. For the two-sided alternative hypothesis, we know from equation 4-24 that

$$\beta = \Phi\left(z_{\alpha/2} - \frac{\delta\sqrt{n}}{\sigma}\right) - \Phi\left(-z_{\alpha/2} - \frac{\delta\sqrt{n}}{\sigma}\right)$$

or if $\delta > 0$,

$$\beta \simeq \Phi\left(z_{\alpha/2} - \frac{\delta\sqrt{n}}{\sigma}\right) \tag{4-25}$$

because $\Phi(-z_{\alpha/2} - \delta\sqrt{n}/\sigma) \simeq 0$ when δ is positive. Let z_β be the 100β upper percentile of the standard normal distribution. Then $\beta = \Phi(-z_\beta)$. From equation 4-25

$$-z_\beta \simeq z_{\alpha/2} - \frac{\delta\sqrt{n}}{\sigma}$$

which leads to the following equation.

Sample Size for Two-Sided Alternative Hypothesis on the Mean, Variance Known

---

For the two-sided alternative hypothesis on the mean with variance known and significance level α, the sample size required to detect a difference between the true and hypothesized mean of δ with power at least $1 - \beta$ is

$$n \simeq \frac{(z_{\alpha/2} + z_\beta)^2 \sigma^2}{\delta^2} \tag{4-26}$$

where

$$\delta = \mu - \mu_0$$

If n is not an integer, the convention is to always round the sample size up to the next integer.

This approximation is good when $\Phi(-z_{\alpha/2} - \delta\sqrt{n}/\sigma)$ is small compared to β. For either of the one-sided alternative hypotheses, the sample size required to produce a specified type II error with probability β given δ and α is as follows.

Sample Size for One-Sided Alternative Hypothesis on the Mean, Variance Known

For the one-sided alternative hypothesis on the mean with variance known and significance level α, the sample size required to detect a difference between the true and hypothesized mean of δ with power at least $1 - \beta$ is

$$n = \frac{(z_\alpha + z_\beta)^2 \sigma^2}{\delta^2} \tag{4-27}$$

where

$$\delta = \mu - \mu_0$$

If n is not an integer, the convention is to round the sample size up to the next integer.

EXAMPLE 4-4

Sample Size for the Propellant Burning Rate Problem

Consider the propellant burning rate problem of Example 4-3. Suppose that the analyst wishes to design the test so that if the true mean burning rate differs from 50 cm/s by as much as 1 cm/s, the test will detect this (i.e., reject H_0: $\mu = 50$) with a high probability—say, 0.90.

Solution. Note that $\sigma = 2$, $\delta = 51 - 50 = 1$, $\alpha = 0.05$, and $\beta = 0.10$. Because $z_{\alpha/2} = z_{0.025} = 1.96$ and $z_\beta = z_{0.10} = 1.28$, the sample size required to detect this departure from H_0: $\mu = 50$ is found by equation 4-26 as

$$n \simeq \frac{(z_{\alpha/2} + z_\beta)^2 \sigma^2}{\delta^2} = \frac{(1.96 + 1.28)^2 2^2}{(1)^2} \simeq 42$$

The approximation is good here because $\Phi(-z_{\alpha/2} - \delta\sqrt{n}/\sigma) = \Phi(-1.96 - (1)\sqrt{42}/2) = \Phi(-5.20) \simeq 0$, which is small relative to β. ∎

Minitab Sample Size and Power Calculations

Many statistics software packages will calculate sample sizes and type II error probabilities. To illustrate, Table 4-2 shows some computations from Minitab for the propellant burning rate problem.

In the first part of Table 4-2, we asked Minitab to work Example 4-4, that is, to find the sample size n that would allow detection of a difference from $\mu_0 = 50$ cm/s of 1 cm/s with power of 0.9 and $\alpha = 0.05$. The answer, $n = 43$, agrees closely with the calculated value from equation 4-26 in Example 4-4, which was $n = 42$. The difference is due to Minitab using a value of z_β that has more than two decimal places. The second part of the computer output relaxes the power requirement to 0.75. Note that the effect is to reduce the required sample size to $n = 28$. The third part of the output is the situation of Example 4-4, but now we wish to determine the type II error probability (β) or the power $= 1 - \beta$ for the sample size $n = 25$.

4-4.3 Large-Sample Test

Although we have developed the test procedure for the null hypothesis H_0: $\mu = \mu_0$ assuming that σ^2 is known, in many if not most practical situations σ^2 will be unknown. In general, if $n \geq 30$, the sample variance s^2 will be close to σ^2 for most samples, and so s can be substituted for σ in the test procedures with little harmful effect. Thus, although we have given a test for

Table 4-2 Minitab Computations

1-Sample Z Test

Testing mean = null (versus not = null)
Calculating power for mean = null + difference
Alpha = 0.05 Sigma = 2

Difference	Sample Size	Target Power	Actual Power
1	43	0.9000	0.9064

1-Sample Z Test

Testing mean = null (versus not = null)
Calculating power for mean = null + difference
Alpha = 0.05 Sigma = 2

Difference	Sample Size	Target Power	Actual Power
1	28	0.7500	0.7536

1-Sample Z Test

Testing mean = null (versus not = null)
Calculating power for mean = null + difference
Alpha = 0.05 Sigma = 2

Difference	Sample Size	Power
1	25	0.7054

known σ^2, it can be easily converted into a *large-sample test procedure for unknown* σ^2. Exact treatment of the case where σ^2 is unknown and n is small involves use of the t distribution and will be deferred until Section 4-5.

4-4.4 Some Practical Comments on Hypothesis Testing

The Seven-Step Procedure

In Section 4-3.5 we described a seven-step procedure for statistical hypothesis testing. This procedure was illustrated in Example 4-3 and will be encountered many times in this chapter. In practice, such a formal and (seemingly) rigid procedure is not always necessary. Generally, once the experimenter (or decision maker) has decided on the question of interest and has determined the **design of the experiment** (that is, how the data are to be collected, how the measurements are to be made, and how many observations are required), only three steps are really required:

1. Specify the hypothesis (two-, upper-, or lower-tailed).
2. Specify the test statistic to be used (such as z_0).
3. Specify the criteria for rejection (typically, the value of α, or the P-value at which rejection should occur).

These steps are often completed almost simultaneously in solving real-world problems, although we emphasize that it is important to think carefully about each step. That is why we present and use the seven-step process: it seems to reinforce the essentials of the correct approach. Although you may not use it every time in solving real problems, it is a helpful framework when you are first learning about hypothesis testing.

Statistical versus Practical Significance

We noted previously that reporting the results of a hypothesis test in terms of a P-value is very useful because it conveys more information than just the simple statement "reject H_0" or "fail to reject H_0." That is, rejection of H_0 at the 0.05 level of significance is much more meaningful if the value of the test statistic is well into the critical region, greatly exceeding the 5% critical value, than if it barely exceeds that value.

Even a very small P-value can be difficult to interpret from a practical viewpoint when we are making decisions; although a small P-value indicates **statistical significance** in the sense that H_0 should be rejected in favor of H_1, the actual departure from H_0 that has been detected may have little (if any) **practical significance** (engineers like to say "engineering significance"). This is particularly true when the sample size n is large.

Relationship between Power and Sample Size for the z-Test

For example, consider the propellant burning rate problem of Example 4-3 where we are testing H_0: $\mu = 50$ cm/s versus H_1: $\mu \neq 50$ cm/s with $\sigma = 2$. If we suppose that the mean rate is really 50.5 cm/s, this is not a serious departure from H_0: $\mu = 50$ cm/s in the sense that if the mean really is 50.5 cm/s, there is no practical observable effect on the performance of the aircrew escape system. In other words, concluding that $\mu = 50$ cm/s when it is really 50.5 cm/s is an inexpensive error and has no practical significance. For a reasonably large sample size, a true value of $\mu = 50.5$ will lead to a sample $\bar{x}$ that is close to 50.5 cm/s, and we would not want this value of $\bar{x}$ from the sample to result in rejection of H_0. The accompanying display shows the P-value for testing H_0: $\mu = 50$ when we observe $\bar{x} = 50.5$ cm/s and the power of the test at $\alpha = 0.05$ when the true mean is 50.5 for various sample sizes n.

Sample Size n	P-Value When $\bar{x} = 50.5$	Power (at $\alpha = 0.05$) When $\mu = 50.5$
10	0.4295	0.1241
25	0.2113	0.2396
50	0.0767	0.4239
100	0.0124	0.7054
400	5.73×10^{-7}	0.9988
1000	2.57×10^{-15}	1.0000

The P-value column in this display indicates that for large sample sizes, the observed sample value of $\bar{x} = 50.5$ would strongly suggest that H_0: $\mu = 50$ should be rejected, even though the observed sample results imply that from a practical viewpoint the true mean does not differ much at all from the hypothesized value $\mu_0 = 50$. The power column indicates that if we test a hypothesis at a fixed significance level α and even if there is little practical difference between the true mean and the hypothesized value, a large sample size will almost always lead to rejection of H_0. The moral of this demonstration is clear:

> Be careful when interpreting the results from hypothesis testing when the sample size is large because any small departure from the hypothesized value μ_0 will probably be detected, even when the difference is of little or no practical significance.

4-4.5 Confidence Interval on the Mean

In many situations, a point estimate does not provide enough information about a parameter. For example, in the rocket propellant problem we have rejected the null hypothesis H_0: $\mu = 50$, and our point estimate of the mean burning rate is $\bar{x} = 51.3$ cm/s. However, the engineer would prefer to have an **interval** in which we would expect to find the true mean burning rate because it is unlikely that $\mu = 51.3$. One way to accomplish this is with an interval estimate called a **confidence interval** (CI).

An interval estimate of the unknown parameter μ is an interval of the form $l \le \mu \le u$, where the endpoints l and u depend on the numerical value of the sample mean $\bar{X}$ for a particular sample. Because different samples will produce different values of $\bar{x}$ and, consequently, different values of the endpoints l and u, these endpoints are values of random variables—say, L and U, respectively. From the sampling distribution of the sample mean $\bar{X}$ we will be able to determine values of L and U such that the following probability statement is true:

$$P(L \le \mu \le U) = 1 - \alpha \tag{4-28}$$

where $0 < \alpha < 1$. Thus, we have a probability of $1 - \alpha$ of selecting a sample that will produce an interval containing the true value of μ.

The resulting interval

$$l \le \mu \le u \tag{4-29}$$

is called a $100(1 - \alpha)\%$ CI for the parameter μ. The quantities l and u are called the **lower- and upper-confidence limits,** respectively, and $1 - \alpha$ is called the **confidence coefficient.** The interpretation of a CI is that, if an infinite number of random samples are collected and a $100(1 - \alpha)\%$ CI for μ is computed from each sample, $100(1 - \alpha)\%$ of these intervals will contain the true value of μ.

The situation is illustrated in Fig. 4-12, which shows several $100(1 - \alpha)\%$ CIs for the mean μ of a distribution. The dots at the center of each interval indicate the point estimate of μ (that is, $\bar{x}$). Note that 1 of the 20 intervals fails to contain the true value of μ. If this were a 95% CI, in the long run only 5% of the intervals would fail to contain μ.

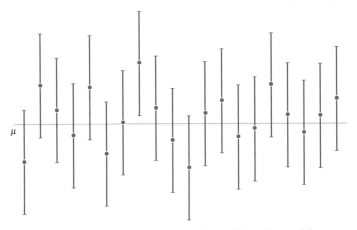

Figure 4-12   Repeated construction of a confidence interval for μ.

Now in practice, we obtain only one random sample and calculate one confidence interval. Because this interval either will or will not contain the true value of μ, it is not reasonable to attach a probability level to this specific event. The appropriate statement is that the observed interval $[l, u]$ brackets the true value of μ with confidence $100(1 - \alpha)$. This statement has a frequency interpretation; that is, we don't know whether the statement is true for this specific sample, but the method used to obtain the interval $[l, u]$ yields correct statements $100(1 - \alpha)\%$ of the time.

The CI in equation 4-29 is more properly called a **two-sided confidence interval,** because it specifies both a lower and an upper limit on μ. Occasionally, a **one-sided confidence bound** might be more appropriate. A one-sided $100(1 - \alpha)\%$ lower-confidence bound on μ is given by

$$l \leq \mu \tag{4-30}$$

where the lower-confidence bound l is chosen so that

$$P(L \leq \mu) = 1 - \alpha \tag{4-31}$$

Similarly, a one-sided $100(1 - \alpha)\%$ upper-confidence bound on μ is given by

$$\mu \leq u \tag{4-32}$$

where the upper-confidence bound u is chosen so that

$$P(\mu \leq U) = 1 - \alpha \tag{4-33}$$

The length $u - l$ of the observed two-sided CI is an important measure of the quality of the information obtained from the sample. The half-interval length $\mu - l$ or $u - \mu$ is called the **precision** of the estimator. The longer the CI, the more confident we are that the interval actually contains the true value of μ. On the other hand, the longer the interval is, the less information we have about the true value of μ. In an ideal situation, we obtain a relatively short interval with high confidence.

It is very easy to find the quantities L and U that define the two-sided CI for μ. We know that the sampling distribution of $\overline{X}$ is normal with mean μ and variance σ^2/n. Therefore, the distribution of the statistic

$$Z = \frac{\overline{X} - \mu}{\sigma/\sqrt{n}}$$

is a standard normal distribution.

The distribution of $Z = (\overline{X} - \mu)/(\sigma/\sqrt{n})$ is shown in Fig. 4-13. From an examination of this figure we see that

$$P\{-z_{\alpha/2} \leq Z \leq z_{\alpha/2}\} = 1 - \alpha$$

so that

$$P\left\{-z_{\alpha/2} \leq \frac{\overline{X} - \mu}{\sigma/\sqrt{n}} \leq z_{\alpha/2}\right\} = 1 - \alpha$$

This can be rearranged as

$$P\left\{\overline{X} - \frac{z_{\alpha/2}\sigma}{\sqrt{n}} \leq \mu \leq \overline{X} + \frac{z_{\alpha/2}\sigma}{\sqrt{n}}\right\} = 1 - \alpha \tag{4-34}$$

Figure 4-13   The distribution of Z.

From consideration of equation 4-28, the lower and upper limits of the inequalities in equation 4-34 are the lower- and upper-confidence limits L and U, respectively. This leads to the following definition.

Confidence Interval on the Mean, Variance Known

If $\bar{x}$ is the sample mean of a random sample of size n from a population with known variance σ^2, a $100(1 - \alpha)\%$ **confidence interval on μ** is given by

$$\bar{x} - \frac{z_{\alpha/2}\sigma}{\sqrt{n}} \leq \mu \leq \bar{x} + \frac{z_{\alpha/2}\sigma}{\sqrt{n}} \qquad (4\text{-}35)$$

where $z_{\alpha/2}$ is the upper $100\alpha/2$ percentage point and $-z_{\alpha/2}$ is the lower $100\alpha/2$ percentage point of the standard normal distribution in Appendix A Table I.

For samples from a normal population or for samples of size $n \geq 30$ regardless of the shape of the population, the CI in equation 4-35 will provide good results. However, for small samples from nonnormal populations we cannot expect the confidence level $1 - \alpha$ to be exact.

EXAMPLE 4-5
Propellant
Burning Rate

Consider the rocket propellant problem in Example 4-3. Find a 95% CI on the mean burning rate.

Solution. We can use equation 4-35 to construct the CI. A 95% interval implies that $1 - \alpha = 0.95$, so $\alpha = 0.05$ and from Table I in the Appendix $z_{\alpha/2} = z_{0.05/2} = z_{0.025} = 1.96$. The lower confidence limit is

$$
\begin{aligned}
l &= \bar{x} - z_{\alpha/2}\sigma/\sqrt{n} \\
&= 51.3 - 1.96(2)/\sqrt{25} \\
&= 51.3 - 0.78 \\
&= 50.52
\end{aligned}
$$

and the upper confidence limit is

$$
\begin{aligned}
u &= \bar{x} + z_{\alpha/2}\sigma/\sqrt{n} \\
&= 51.3 + 1.96(2)/\sqrt{25} \\
&= 51.3 + 0.78 \\
&= 52.08
\end{aligned}
$$

Thus, the 95% two-sided CI is

$$50.52 \leq \mu \leq 52.08$$

Remember how to interpret the CI; this specific interval either contains μ or it doesn't (and we don't know which), but because of the procedure that we use to construct the CI, in repeated sampling 95% of the intervals that we would compute will contain the true value of μ. This CI was also reported by Minitab in the output of the z-test in Section 4-4.1. ■

Relationship between Tests of Hypotheses and Confidence Intervals

There is a close relationship between the test of a hypothesis about any parameter—say, θ— and the confidence interval for θ. If $[l, u]$ is a $100(1 - \alpha)\%$ CI for the parameter θ, the test of significance level α of the hypothesis

$$H_0\colon \theta = \theta_0$$
$$H_1\colon \theta \neq \theta_0$$

will lead to rejection of H_0 if and only if θ_0 is not in the $100(1 - \alpha)\%$ CI $[l, u]$. As an illustration, consider the escape system propellant problem discussed above. The null hypothesis $H_0\colon \mu = 50$ was rejected, using $\alpha = 0.05$. The 95% two-sided confidence interval on μ is $50.52 \leq \mu \leq 52.08$. That is, the interval $[l, u]$ is $[50.52, 52.08]$, and because $\mu_0 = 50$ is not included in this interval, the null hypothesis $H_0\colon \mu = 50$ is rejected.

Confidence Level and Precision of Estimation

Note in the previous example that our choice of the 95% level of confidence was essentially arbitrary. What would have happened if we had chosen a higher level of confidence—say, 99%? In fact, doesn't it seem reasonable that we would want the higher level of confidence? At $\alpha = 0.01$, we find $z_{\alpha/2} = z_{0.01/2} = z_{0.005} = 2.58$, whereas for $\alpha = 0.05$, $z_{0.025} = 1.96$. Thus, the length of the 95% CI is

$$2(1.96 \, \sigma/\sqrt{n}) = 3.92 \, \sigma/\sqrt{n}$$

whereas the length of the 99% CI is

$$2(2.58 \, \sigma/\sqrt{n}) = 5.16 \, \sigma/\sqrt{n}$$

The 99% CI is longer than the 95% CI, which is why we have a higher level of confidence in the 99% CI. Generally, for a fixed sample size n and standard deviation σ, the higher the confidence level is, the longer is the resulting CI.

Because the half-length of the confidence interval measures the precision of estimation, we see that precision is inversely related to the confidence level. As noted earlier, it is desirable to obtain a CI that is short enough for decision-making purposes and also has adequate confidence. One way to achieve this is by choosing the sample size n to be large enough to give a CI of specified length with prescribed confidence.

Choice of Sample Size

The precision of the confidence interval in equation 4-35 is $z_{\alpha/2}\sigma/\sqrt{n}$. This means that in using $\bar{x}$ to estimate μ, the error $E = |\bar{x} - \mu|$ is less than or equal to $z_{\alpha/2}\sigma/\sqrt{n}$ with confidence

Figure 4-14 Error in estimating μ with $\bar{x}$.

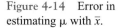

Figure 4-14 Error in estimating μ with $\bar{x}$.

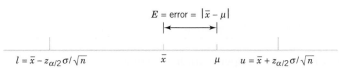

$100(1 - \alpha)$. This is shown graphically in Fig. 4-14. In situations where the sample size can be controlled, we can choose n so that we are $100(1 - \alpha)\%$ confident that the error in estimating μ is less than a specified error E. The appropriate sample size is found by choosing n such that $z_{\alpha/2}\sigma/\sqrt{n} = E$. Solving this equation gives the following formula for n.

Sample Size for a Specified E on the Mean, Variance Known

If $\bar{x}$ is used as an estimate of μ, we can be $100(1 - \alpha)\%$ confident that the error $|\bar{x} - \mu|$ will not exceed a specified amount E when the sample size is

$$n = \left(\frac{z_{\alpha/2}\sigma}{E}\right)^2 \qquad (4\text{-}36)$$

If the right-hand side of equation 4-36 is not an integer, it must be rounded up, which will ensure that the level of confidence does not fall below $100(1 - \alpha)\%$. Note that $2E$ is the length of the resulting CI.

EXAMPLE 4-6

Propellant Burning Rate

To illustrate the use of this procedure, suppose that we wanted the error in estimating the mean burning rate of the rocket propellant to be less than 1.5 cm/s, with 95% confidence. Find the required sample size.

Solution. Because $\sigma = 2$ and $z_{0.025} = 1.96$, we may find the required sample size from equation 4-36 as

$$n = \left(\frac{z_{\alpha/2}\sigma}{E}\right)^2 = \left[\frac{(1.96)2}{1.5}\right]^2 = 6.83 \cong 7$$

Note the general relationship between sample size, desired length of the confidence interval $2E$, confidence level $100(1 - \alpha)\%$, and standard deviation σ:

- As the desired length of the interval $2E$ decreases, the required sample size n increases for a fixed value of σ and specified confidence.

- As σ increases, the required sample size n increases for a fixed desired length $2E$ and specified confidence.

- As the level of confidence increases, the required sample size n increases for fixed desired length $2E$ and standard deviation σ. ∎

One-Sided Confidence Bounds

It is also possible to obtain one-sided confidence bounds for μ by setting either $l = -\infty$ or $u = \infty$ and replacing $z_{\alpha/2}$ by z_{α}.

As discussed above for the two-sided case, we can also use the one-sided confidence bound to perform hypothesis testing with a one-sided alternative hypothesis. Specifically, if u is the

---

One-Sided Confidence Bounds on the Mean, Variance Known

The $100(1 - \alpha)\%$ **upper-confidence bound** for μ is

$$\mu \leq u = \bar{x} + z_\alpha \sigma / \sqrt{n} \qquad (4\text{-}37)$$

and the $100(1 - \alpha)\%$ **lower-confidence bound** for μ is

$$\bar{x} - z_\alpha \sigma / \sqrt{n} = l \leq \mu \qquad (4\text{-}38)$$

---

upper bound of a $100(1 - \alpha)\%$ one-sided confidence bound for the parameter θ, the test of significance level α of the hypothesis

$$H_0: \theta = \theta_0$$
$$H_1: \theta < \theta_0$$

will lead to rejection if and only if $\theta_0 > u$. Similarly, if l is the lower bound of a $100(1 - \alpha)\%$ one-sided confidence bound, the test of significance level α of the hypothesis

$$H_0: \theta = \theta_0$$
$$H_1: \theta > \theta_0$$

will lead to rejection if and only if $\theta_0 < l$.

4-4.6   General Method for Deriving a Confidence Interval

It is easy to give a general method for finding a CI for an unknown parameter θ. Let $X_1, X_2, \ldots, X_n$ be a random sample of n observations. Suppose we can find a statistic $g(X_1, X_2, \ldots, X_n; \theta)$ with the following properties:

1. $g(X_1, X_2, \ldots, X_n; \theta)$ depends on both the sample and θ, and
2. the probability distribution of $g(X_1, X_2, \ldots, X_n; \theta)$ does not depend on θ or any other unknown parameter.

In the case considered in this section, the parameter $\theta = \mu$. The random variable $g(X_1, X_2, \ldots, X_n; \mu) = (\bar{X} - \mu)/(\sigma/\sqrt{n})$ and satisfies both these conditions; it depends on the sample and on μ, and it has a standard normal distribution because σ is known. Now you must find constants C_L and C_U so that

$$P[C_L \leq g(X_1, X_2, \ldots, X_n; \theta) \leq C_U] = 1 - \alpha$$

Because of property 2, C_L and C_U do not depend on θ. In our example, $C_L = -z_{\alpha/2}$ and $C_U = z_{\alpha/z}$. Finally, you must manipulate the inequalities in the probability statement so that

$$P[L(X_1, X_2, \ldots, X_n) \leq \theta \leq U(X_1, X_2, \ldots, X_n)] = 1 - \alpha$$

This gives $L(X_1, X_2, \ldots, X_n)$ and $U(X_1, X_2, \ldots, X_n)$ as the lower and upper confidence limits defining the $100(1 - \alpha)\%$ CI for θ. In our example, we found $L(X_1, X_2, \ldots, X_n) = \bar{X} - z_{\alpha/2}\sigma/\sqrt{n}$ and $U(X_1, X_2, \ldots, X_n) = \bar{X} + z_{\alpha/2}\sigma/\sqrt{n}$.

Animation 8: Confidence Intervals

EXERCISES FOR SECTION 4-4

4-26. Suppose that we are testing $H_0: \mu = \mu_0$ versus $H_1: \mu > \mu_0$. Calculate the P-value for the following observed values of the test statistic:

(a) $z_0 = 2.35$ (b) $z_0 = 1.53$
(c) $z_0 = 2.00$ (d) $z_0 = 1.85$
(e) $z_0 = -0.15$

4-27. Suppose that we are testing $H_0: \mu = \mu_0$ versus $H_1: \mu \neq \mu_0$. Calculate the P-value for the following observed values of the test statistic:

(a) $z_0 = 2.45$ (b) $z_0 = -1.53$
(c) $z_0 = 2.15$ (d) $z_0 = 1.95$
(e) $z_0 = -0.25$

4-28. Suppose that we are testing $H_0: \mu = \mu_0$ versus $H_1: \mu < \mu_0$. Calculate the P-value for the following observed values of the test statistic:

(a) $z_0 = -2.15$ (b) $z_0 = -1.80$
(c) $z_0 = -2.50$ (d) $z_0 = -1.60$
(e) $z_0 = 0.35$

4-29. Consider the Minitab output shown below.

One-Sample Z

Test of mu = 30 vs not = 30
The assumed standard deviation = 1.2

N	Mean	SE Mean	95% CI	Z	P
16	31.2000	0.3000	(30.6120, 31.7880)	?	?

(a) Fill in the missing values in the output. What conclusion would you draw?
(b) Is this a one-sided or a two-sided test?
(c) Use the output and the normal table to find a 99% CI on the mean.
(d) What is the P-value if the alternative hypothesis is $H_1: \mu > 30$?

4-30. Consider the Minitab output shown below.

One-Sample Z

Test of mu = 100 vs > 100
The assumed standard deviation = 2.4

N	Mean	SE Mean	95% Lower Bound	Z	P
25	101.560	?	100.770	3.25	?

(a) Fill in the missing values in the output. Can the null hypothesis be rejected at the 0.05 level? Why?
(b) Is this a one-sided or a two-sided test?
(c) If the hypotheses had been $H_0: \mu = 99$ versus $H_1: \mu > 99$, would you reject the null hypothesis at the 0.05 level? Can you answer this question without doing any additional calculations? Why?
(d) Use the output and the normal table to find a 95% two-sided CI on the mean.
(e) What is the P-value if the alternative hypothesis is $H_1: \mu \neq 100$?

4-31. Medical researchers have developed a new artificial heart constructed primarily of titanium and plastic. The heart will last and operate almost indefinitely once it is implanted in the patient's body, but the battery pack needs to be recharged about every four hours. A random sample of 50 battery packs is selected and subjected to a life test. The average life of these batteries is 4.05 hours. Assume that battery life is normally distributed with standard deviation $\sigma = 0.2$ hour.

(a) Is there evidence to support the claim that mean battery life exceeds 4 hours? Use $\alpha = 0.05$.
(b) What is the P-value for the test in part (a)?
(c) Compute the power of the test if the true mean battery life is 4.5 hours.
(d) What sample size would be require to detect a true mean battery life of 4.5 hours if we wanted the power of the test to be at least 0.9?
(e) Explain how the question in part (a) could be answered by constructing a one-sided confidence bound on the mean life.

4-32.* The mean breaking strength of yarn used in manufacturing drapery material is required to be at least 100 psi. Past experience has indicated that the standard deviation of breaking strength is 2 psi. A random sample of nine specimens is tested, and the average breaking strength is found to be 100.6 psi.

(a) Should the fiber be judged acceptable? Use the P-value approach.
(b) What is the probability of not rejecting the null hypothesis at $\alpha = 0.05$ if the fiber has a true mean breaking strength of 102 psi?
(c) Find a 95% one-sided lower CI on the true mean breaking strength.
(d) Use the CI found in part (d) to test the hypothesis.
(e) What sample size is required to detect a true mean breaking strength of 101 with probability 0.95?

4-33. The yield of a chemical process is being studied. From previous experience with this process the standard deviation of yield is known to be 3. The past 5 days of plant operation have

*Please remember that the web symbol indicates that the individual observations for the sample are available on the book website.

resulted in the following yields: 91.6, 88.75, 90.8, 89.95, and 91.3%. Use $\alpha = 0.05$.

(a) Is there evidence that the mean yield is not 90%? Use the *P*-value approach.

(b) What sample size would be required to detect a true mean yield of 85% with probability 0.95?

(c) What is the type II error probability if the true mean yield is 92%?

(d) Find a 95% two-sided CI on the true mean yield.

(e) Use the CI found in part (d) to test the hypothesis.

4-34. Benzene is a toxic chemical used in the manufacturing of medicinal chemicals, dyes, artificial leather, and linoleum. A manufacturer claims that its exit water meets the federal regulation with a mean of less than 7980 ppm of benzene. To assess the benzene content of the exit water, 10 independent water samples were collected and found to have an average of 7906 ppm of benzene. Assume a known standard deviation of 80 ppm and use a significance level of 0.01.

(a) Test the manufacturer's claim. Use the *P*-value approach.

(b) What is the β-value if the true mean is 7920?

(c) What sample size would be necessary to detect a true mean of 7920 with a probability of at least 0.90?

(d) Find a 99% one-sided upper confidence bound on the true mean.

(e) Use the CI found in part (d) to test the hypothesis.

4-35. In the production of airbag inflators for automotive safety systems, a company is interested in ensuring that the mean distance of the foil to the edge of the inflator is at least 2.00 cm. Measurements on 20 inflators yielded an average value of 2.02 cm. Assume a standard deviation of 0.05 on the distance measurements and a significance level of 0.01.

(a) Test for conformance to the company's requirement. Use the *P*-value approach.

(b) What is the β-value if the true mean is 2.03?

(c) What sample size would be necessary to detect a true mean of 2.03 with a probability of at least 0.90?

(d) Find a 99% one-sided lower confidence bound on the true mean.

(e) Use the CI found in part (d) to test the hypothesis.

4-36. The life in hours of a thermocouple used in a furnace is known to be approximately normally distributed, with standard deviation $\sigma = 20$ hours. A random sample of 15 thermocouples resulted in the following data: 553, 552, 567, 579, 550, 541, 537, 553, 552, 546, 538, 553, 581, 539, 529.

(a) Is there evidence to support the claim that mean life exceeds 540 hours? Use a fixed-level test with $\alpha = 0.05$.

(b) What is the *P*-value for this test?

(c) What is the β-value for this test if the true mean life is 560 hours?

(d) What sample size would be required to ensure that β does not exceed 0.10 if the true mean life is 560 hours?

(e) Construct a 95% one-sided lower CI on the mean life.

(f) Use the CI found in part (e) to test the hypothesis.

4-37. A civil engineer is analyzing the compressive strength of concrete. Compressive strength is approximately normally distributed with variance $\sigma^2 = 1000$ psi^2. A random sample of 12 specimens has a mean compressive strength of $\bar{x} = 3255.42$ psi.

(a) Test the hypothesis that mean compressive strength is 3500 psi. Use a fixed-level test with $\alpha = 0.01$.

(b) What is the smallest level of significance at which you would be willing to reject the null hypothesis?

(c) Construct a 95% two-sided CI on mean compressive strength.

(d) Construct a 99% two-sided CI on mean compressive strength. Compare the width of this confidence interval with the width of the one found in part (c). Comment.

4-38. Suppose that in Exercise 4-36 we wanted to be 95% confident that the error in estimating the mean life is less than 5 hours. What sample size should we use?

4-39. Suppose that in Exercise 4-35 we wanted to be 95% confident that the error in estimating the mean distance is less than 0.01 cm. What sample size should we use?

4-40. Suppose that in Exercise 4-37 it is desired to estimate the compressive strength with an error that is less than 15 psi at 99% confidence. What sample size is required?

4-5   INFERENCE ON THE MEAN OF A POPULATION, VARIANCE UNKNOWN

When we are testing hypotheses or constructing CIs on the mean μ of a population when σ^2 is unknown, we can use the test procedures in Section 4-4, provided that the sample size is large ($n \geq 30$, say). These procedures are approximately valid (because of the central limit theorem) regardless of whether or not the underlying population is normal. However, when the sample is small and σ^2 is unknown, we must make an assumption about the form of the underlying distribution to obtain a test procedure. A reasonable assumption in many cases is that the underlying distribution is normal.

Many populations encountered in practice are well approximated by the normal distribution, so this assumption will lead to inference procedures of wide applicability. In fact, moderate departure from normality will have little effect on validity. When the assumption is unreasonable, an alternative is to use nonparametric procedures that are valid for any underlying distribution. See Montgomery and Runger (2006) for an introduction to these techniques.

4-5.1 Hypothesis Testing on the Mean

Suppose that the population of interest has a normal distribution with unknown mean μ and variance σ^2. We wish to test the hypothesis that μ equals a constant μ_0. Note that this situation is similar to that in Section 4-4, except that now both μ and σ^2 are unknown. Assume that a random sample of size n—say, $X_1, X_2, \ldots, X_n$—is available, and let $\overline{X}$ and S^2 be the sample mean and variance, respectively.

We wish to test the two-sided alternative hypothesis

$$H_0: \mu = \mu_0$$
$$H_1: \mu \neq \mu_0$$

If the variance σ^2 is known, the test statistic is equation 4-13:

$$Z_0 = \frac{\overline{X} - \mu_0}{\sigma/\sqrt{n}}$$

When σ^2 is unknown, a reasonable procedure is to replace σ in the above expression with the sample standard deviation S. The **test statistic** is now

$$T_0 = \frac{\overline{X} - \mu_0}{S/\sqrt{n}} \qquad (4\text{-}39)$$

A logical question is: What effect does replacing σ by S have on the distribution of the statistic T_0? If n is large, the answer to this question is: Very little, and we can proceed to use the test procedure based on the normal distribution from Section 4-4. However, n is usually small in most engineering problems, and in this situation a different distribution must be employed.

Let $X_1, X_2, \ldots, X_n$ be a random sample for a normal distribution with unknown mean μ and unknown variance σ^2. The quantity

$$T = \frac{\overline{X} - \mu}{S/\sqrt{n}}$$

has a t distribution with $n - 1$ degrees of freedom.

The t probability density function is

$$f(x) = \frac{\Gamma[(k+1)/2]}{\sqrt{\pi k}\,\Gamma(k/2)} \cdot \frac{1}{[(x^2/k)+1]^{(k+1)/2}} \qquad -\infty < x < \infty \qquad (4\text{-}40)$$

where k is the number of degrees of freedom. The mean and variance of the t distribution are zero and $k/(k-2)$ (for $k > 2$), respectively. The function $\Gamma(m) = \int_0^\infty e^{-x} x^{m-1}\, dx$ is the gamma function. Recall that it was introduced previously in Section 3-5.3. Although it is defined for $m \geq 0$, in the special case that m is an integer, $\Gamma(m) = (m-1)!$. Also, $\Gamma(1) = \Gamma(0) = 1$.

Several t distributions are shown in Fig. 4-15. The general appearance of the t distribution is similar to the standard normal distribution, in that both distributions are symmetric and unimodal, and the maximum ordinate value is reached when the mean $\mu = 0$. However, the t distribution has heavier tails than the normal; that is, it has more probability in the tails than the normal distribution. As the number of degrees of freedom $k \to \infty$, the limiting form of the t distribution is the standard normal distribution. In visualizing the t distribution, it is sometimes useful to know that the ordinate of the density at the mean $\mu = 0$ is approximately 4 to 5 times larger than the ordinate at the 5th and 95th percentiles. For example, with 10 degrees of freedom for t this ratio is 4.8, with 20 degrees of freedom it is 4.3, and with 30 degrees of freedom, 4.1. By comparison, this factor is 3.9 for the normal distribution.

Appendix A Table II provides **percentage points** of the t distribution. We will let $t_{\alpha,k}$ be the value of the random variable T with k degrees of freedom above which we find an area (or probability) α. Thus, $t_{\alpha,k}$ is an upper-tail 100α percentage point of the t distribution with k degrees of freedom. This percentage point is shown in Fig. 4-16. In Appendix A Table II the α values are the column headings, and the degrees of freedom are listed in the left column. To illustrate the use of the table, note that the t-value with 10 degrees of freedom having an area of 0.05 to the right is $t_{0.05,10} = 1.812$. That is,

$$P(T_{10} > t_{0.05,10}) = P(T_{10} > 1.812) = 0.05$$

Because the t distribution is symmetric about zero, we have $t_{1-\alpha} = -t_\alpha$; that is, the t-value having an area of $1 - \alpha$ to the right (and therefore an area of α to the left) is equal to the negative of the t-value that has area α in the right tail of the distribution. See Fig. 4-16. Therefore, $t_{0.95,10} = -t_{0.05,10} = -1.812$.

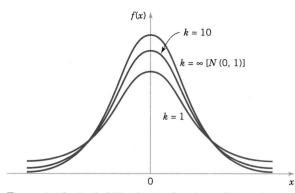

Figure 4-15   Probability density functions of several t distributions.

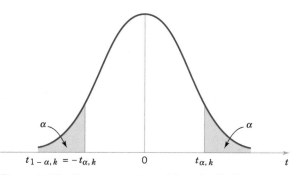

Figure 4-16   Percentage points of the t distribution.

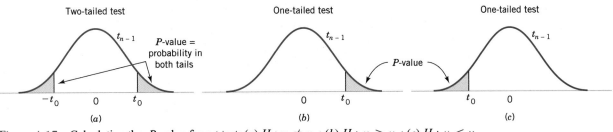

Figure 4-17   Calculating the P-value for a t-test: (a) H_1: $\mu \neq \mu_0$; (b) H_1: $\mu > \mu_0$; (c) H_1: $\mu < \mu_0$.

It is straightforward to see that the distribution of the **test statistic** in equation 4-39 is t with $n - 1$ degrees of freedom if the null hypothesis H_0: $\mu = \mu_0$ is true. To test H_0: $\mu = \mu_0$ against the two-sided alternative H_1: $\mu \neq \mu_0$, the value of the test statistic t_0 in equation 4-39 is calculated, and the P-value is found from the t distribution with $n - 1$ degrees of freedom. Because the test is two-tailed, the P-value is the sum of the probabilities in the two tails of the t distribution. Refer to Fig. 4-17a. If the test statistic is positive, then the P-value is the probability above the test statistic value t_0 plus the probability below the negative value of the test statistic value $-t_0$. Alternatively, if the test statistic is negative, then the P-value is the probability below the value of the test statistic $-t_0$ plus the probability above the absolute value of the test statistic $|-t_0| = t_0$. Because the t distribution is symmetric around zero, a simple way to write this is

$$P = 2P(T_{n-1} > |t_0|) \tag{4-41}$$

A small P-value is evidence against H_0, so if P is sufficiently small value (typically < 0.05), reject the null hypothesis.

For the one-sided alternative hypothesis

$$H_0\text{: } \mu = \mu_0$$
$$H_1\text{: } \mu > \mu_0 \tag{4-42}$$

we calculate the test statistic t_0 from equation 4-39 and calculate the P-value as

$$P = P(T_{n-1} > t_0) \tag{4-43}$$

For the other one-sided alternative

$$H_0\text{: } \mu = \mu_0$$
$$H_1\text{: } \mu < \mu_0 \tag{4-44}$$

we calculate the P-value as

$$P = P(T_{n-1} < t_0) \tag{4-45}$$

Figure 4-17b and c shows how these P-values are calculated.

Statistics software packages calculate and display P-values. However, in working problems by hand, it is useful to be able to find the P-value for a t-test. Because the t-table in Appendix A Table II contains only 10 critical values for each t distribution, determining the exact P-value from this table is usually impossible. Fortunately, it's easy to find lower and upper bounds on the P-value by using this table.

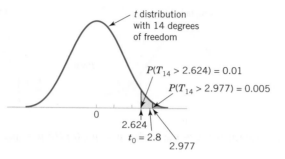

Figure 4-18 *P*-value for $t_0 = 2.8$ and an upper-tailed test is shown to be between 0.005 and 0.01.

To illustrate, suppose that we are conducting an upper-tailed *t*-test (so H_1: $\mu > \mu_0$) with 14 degrees of freedom. The relevant critical values from Appendix A Table II are as follows:

Critical Value: 0.258 0.692 1.345 1.761 2.145 2.624 2.977 3.326 3.787 4.140
Tail Area: 0.40   0.25   0.10   0.05   0.025 0.01   0.005 0.0025 0.001 0.0005

After calculating the test statistic, we find that $t_0 = 2.8$. Now, $t_0 = 2.8$ is between two tabulated values, 2.624 and 2.977. Therefore, the *P*-value must be between 0.01 and 0.005. Refer to Fig. 4-18. These are effectively the upper and lower bounds on the *P*-value.

This illustrates the procedure for an upper-tailed test. If the test is lower tailed, just change the sign on the lower and upper bounds for t_0 and proceed as above. Remember that for a two-tailed test, the level of significance associated with a particular critical value is twice the corresponding tail area in the column heading. This consideration must be taken into account when we compute the bound on the *P*-value. For example, suppose that $t_0 = 2.8$ for a two-tailed alternative based on 14 degrees of freedom. The value of the test statistic $t_0 > 2.624$ (corresponding to $\alpha = 2 \times 0.01 = 0.02$) and $t_0 < 2.977$ (corresponding to $\alpha = 2 \times 0.005 = 0.01$), so the lower and upper bounds on the *P*-value would be $0.01 < P < 0.02$ for this case.

Some statistics software packages can help you calculate *P*-values. For example, Minitab has the capability to find cumulative probabilities from many standard probability distributions, including the *t*-distribution. Simply enter the value of the test statistic t_0 along with the appropriate number of degrees of freedom. Minitab will display the probability $P(T_v \le t_0)$ where v is the degrees of freedom for the test statistic t_0. From the cumulative probability the *P*-value can be determined.

The single-sample *t*-test we have just described can also be conducted using the **fixed significance level** approach. Consider the two-sided alternative hypothesis. The null hypothesis would be rejected if the value of the test statistic t_0 falls in the critical region defined by the lower and upper $\alpha/2$ percentage points of the *t* distribution with $n - 1$ degrees of freedom. That is, reject H_0 if

$$t_0 > t_{\alpha/2, n-1} \quad \text{or} \quad t_0 < -t_{\alpha/2, n-1}$$

For the one-tailed tests, the location of the critical region is determined by the direction that the inequality in the alternative hypothesis "points." So if the alternative is H_1: $\mu > \mu_0$, reject H_0 if

$$t_0 > t_{\alpha, n-1}$$

and if the alternative is H_1: $\mu < \mu_0$, reject H_0 if

$$t_0 < -t_{\alpha, n-1}$$

Figure 4-18 shows the locations of these critical regions.

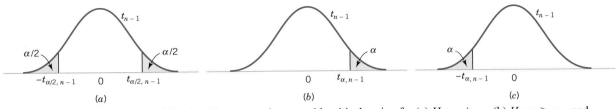

Figure 4-19   The distribution of T_0 when H_0: $\mu = \mu_0$ is true, with critical region for (a) H_1: $\mu \neq \mu_0$, (b) H_1: $\mu > \mu_0$, and (c) H_1: $\mu < \mu_0$.

Summary

Testing Hypotheses on the Mean of a Normal Distribution, Variance Unknown

Null hypothesis: H_0: $\mu = \mu_0$

Test statistic: $T_0 = \dfrac{\bar{X} - \mu_0}{S/\sqrt{n}}$

Alternative Hypotheses	P-Value	Rejection Criterion for Fixed-Level Tests
H_1: $\mu \neq \mu_0$	Sum of the probability above t_0 and the probability below $-t_0$	$t_0 > t_{\alpha/2,n-1}$ or $t_0 < -t_{\alpha/2,n-1}$
H_1: $\mu > \mu_0$	Probability above t_0	$t_0 > t_{\alpha,n-1}$
H_1: $\mu < \mu_0$	Probability below t_0	$t_0 < -t_{\alpha,n-1}$

The locations of the critical regions for these situations are shown in Fig. 4-19a, b, and c, respectively.

EXAMPLE 4-7

Golf Clubs

The increased availability of light materials with high strength has revolutionized the design and manufacture of golf clubs, particularly drivers. Clubs with hollow heads and very thin faces can result in much longer tee shots, especially for players of modest skills. This is due partly to the "spring-like effect" that the thin face imparts to the ball. Firing a golf ball at the head of the club and measuring the ratio of the outgoing velocity of the ball to the incoming velocity can quantify this spring-like effect. The ratio of velocities is called the coefficient of restitution of the club. An experiment was performed in which 15 drivers produced by a particular club maker were selected at random and their coefficients of restitution measured. In the experiment the golf balls were fired from an air cannon so that the incoming velocity and spin rate of the ball could be precisely controlled. It is of interest to determine if there is evidence (with $\alpha = 0.05$) to support a claim that the mean coefficient of restitution exceeds 0.82. The observations follow:

0.8411	0.8191	0.8182	0.8125	0.8750
0.8580	0.8532	0.8483	0.8276	0.7983
0.8042	0.8730	0.8282	0.8359	0.8660

The sample mean and sample standard deviation are $\bar{x} = 0.83725$ and $s = 0.02456$. The normal probability plot of the data in Fig. 4-20 supports the assumption that the coefficient of restitution is normally distributed. Because the objective of the experimenter is to demonstrate that the mean coefficient of restitution exceeds 0.82, a one-sided alternative hypothesis is appropriate.

Solution. The solution using the seven-step procedure for hypothesis testing is as follows:

1. **Parameter of interest:** The parameter of interest is the mean coefficient of restitution, μ.
2. **Null hypothesis, H_0:** $\mu = 0.82$

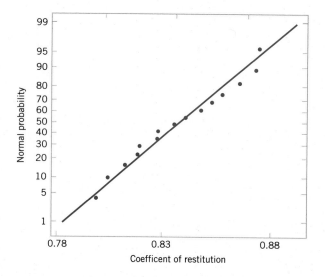

Figure 4-20   Normal probability plot of the coefficient of restitution data from Example 4-7.

3. **Alternative hypothesis, H_1:** $\mu > 0.82$. We want to reject H_0 if the mean coefficient of restitution exceeds 0.82.

4. **Test statistic:** The test statistic is

$$t_0 = \frac{\bar{x} - \mu_0}{s/\sqrt{n}}$$

5. **Reject H_0 if:** Reject H_0 if the P-value is smaller than 0.05.

6. **Computations:** Because $\bar{x} = 0.83725$, $s = 0.02456$, $\mu_0 = 0.82$, and $n = 15$, we have

$$t_0 = \frac{0.83725 - 0.82}{0.02456/\sqrt{15}} = 2.72$$

7. **Conclusions:** From Appendix A Table II we find, for a t-distribution with 14 degrees of freedom, that $t_0 = 2.72$ falls between two values: 2.624 for which $\alpha = 0.01$ and 2.977 for which $\alpha = 0.005$. Because this is a one-tailed test we know that the P-value is between those two values. That is, $0.005 < P < 0.01$. Therefore, since $P < 0.05$, we reject H_0 and conclude that the mean coefficient of restitution exceeds 0.82. To use Minitab to compute the P-value, use the Calc menu and select the probability distribution option. Then for the t-distribution, enter 14 degrees of freedom and the value of the test statistic $t_0 = 2.72$ as the input constant. Minitab returns the probability $P(T_{14} \le 2.72) = 0.991703$. The P-value is $P(T_{14} \ge 2.72)$ or $P = 1 - P(T_{14} \le 2.72) = 1 - 0.991703 = 0.008297$. ∎

Minitab will conduct the one-sample t-test. The output from this software package is in the following display:

One-Sample T: COR

Test of mu = 0.82 vs mu > 0.82

Variable	N	Mean	StDev	SE Mean
COR	15	0.83725	0.02456	0.00634

Variable	95.0%	Lower Bound	T	P
COR		0.82608	2.72	0.008

Notice that Minitab computes both the test statistic T_0 and a 95% lower confidence bound for the coefficient of restitution. We will give the confidence bound formulas in Section 4-5.3. However, recalling the discussion in Section 4-4.5 about the relationship between hypothesis tests and CIs, we observe that because the 95% lower confidence bound exceeds 0.82, we would reject the hypothesis that H_0: $\mu = 0.82$ and conclude that the alternative hypothesis H_1: $\mu > 0.82$ is more appropriate. Minitab also calculates a P-value for the test statistic T_0. The reported value is $P = 0.008$, which lies between the lower and upper bounds that we obtained from the t-table and closely agrees with the value we found directly from the cumulative t distribution function in Minitab.

Animation 9: Hypothesis Testing for Averages

4-5.2 Type II Error and Choice of Sample Size

The type II error probability for tests on the mean of a normal distribution with unknown variance depends on the distribution of the test statistic in equation 4-39 when the null hypothesis H_0: $\mu = \mu_0$ is false. When the true value of the mean is $\mu = \mu_0 + \delta$, the distribution for T_0 is called the **noncentral t distribution** with $n - 1$ degrees of freedom and noncentrality parameter $\delta \sqrt{n}/\sigma$. Note that if $\delta = 0$, the noncentral t distribution reduces to the usual **central t distribution.** Therefore, the type II error of the two-sided alternative (for example) would be

$$\begin{aligned} \beta &= P\{-t_{\alpha/2,n-1} \le T_0 \le t_{\alpha/2,n-1} \text{ when } \delta \ne 0\} \\ &= P\{-t_{\alpha/2,n-1} \le T_0' \le t_{\alpha/2,\,n-1}\} \end{aligned}$$

where T_0' denotes the noncentral t random variable. Finding the type II error probability β for the t-test involves finding the probability contained between two points on the noncentral t distribution. Because the noncentral t-random variable has a messy density function, this integration must be done numerically.

Fortunately, this unpleasant task has already been done, and the results are summarized in a series of graphs in Appendix A Charts Va, Vb, Vc, and Vd that plot β for the t-test against a parameter d for various sample sizes n. These graphics are called **operating characteristic** (or **OC) curves.** Curves are provided for two-sided alternatives in Charts Va and Vb. The abscissa scale factor d on these charts is defined as

$$d = \frac{|\mu - \mu_0|}{\sigma} = \frac{|\delta|}{\sigma} \tag{4-46}$$

For the one-sided alternative $\mu > \mu_0$ as in equation 4-42, we use Charts Vc and Vd with

$$d = \frac{\mu - \mu_0}{\sigma} = \frac{\delta}{\sigma} \tag{4-47}$$

whereas if $\mu < \mu_0$, as in equation 4-44,

$$d = \frac{\mu_0 - \mu}{\sigma} = \frac{\delta}{\sigma} \tag{4-48}$$

We note that d depends on the unknown parameter σ^2. We can avoid this difficulty in several ways. In some cases, we may use the results of a previous experiment or prior information to make a rough initial estimate of σ^2. If we are interested in evaluating test performance after the data have been collected, we could use the sample variance s^2 to estimate σ^2. If there is no previous experience on which to draw in estimating σ^2, we then define the difference in the mean d that we wish to detect relative to σ. For example, if we wish to detect a small difference in the mean, we might use a value of $d = |\delta|/\sigma \le 1$ (for example), whereas if we are interested in detecting only moderately large differences in the mean, we might select $d = |\delta|/\sigma = 2$ (for example). That is, it is the value of the ratio $|\delta|/\sigma$ that is important in determining sample size, and if it is possible to specify the relative size of the difference in means that we are interested in detecting, then a proper value of d usually can be selected.

EXAMPLE 4-8
Golf Clubs

Consider the golf club testing problem from Example 4-7. If the mean coefficient of restitution differs from 0.82 by as much as 0.02, is the sample size $n = 15$ adequate to ensure that H_0: $\mu = 0.82$ will be rejected with probability at least 0.8?

Solution. To solve this problem, we will use the sample standard deviation $s = 0.02456$ to estimate σ. Then $d = |\delta|/\sigma = 0.02/0.02456 = 0.81$. By referring to the operating characteristic curves in Appendix A Chart Vc (for $\alpha = 0.05$) with $d = 0.81$ and $n = 15$, we find that $\beta = 0.10$, approximately. Thus, the probability of rejecting H_0: $\mu = 0.82$ if the true mean exceeds this by 0.02 is approximately $1 - \beta = 1 - 0.10 = 0.90$, and we conclude that a sample size of $n = 15$ is adequate to provide the desired sensitivity. ■

Minitab will also perform power and sample size computations for the one-sample t-test. The following display shows several calculations based on the golf club testing problem.

1-Sample t Test
Testing mean = null (versus > null)
Calculating power for mean = null + difference
Alpha = 0.05 Sigma = 0.02456

Difference	Sample Size	Power
0.02	15	0.9117

1-Sample t Test
Testing mean = null (versus > null)
Calculating power for mean = null + difference
Alpha = 0.05 Sigma = 0.02456

Difference	Sample Size	Power
0.01	15	0.4425

1-Sample t Test
Testing mean = null (versus > null)
Calculating power for mean = null + difference
Alpha = 0.05 Sigma = 0.02456

Difference	Sample Size	Target Power	Actual Power
0.01	39	0.8000	0.8029

Interpretting Minitab Output

In the first portion of the display, Minitab reproduces the solution to Example 4-8, verifying that a sample size of $n = 15$ is adequate to give power of at least 0.8 if the mean coefficient of restitution exceeds 0.82 by at least 0.02. In the middle section of the output, we asked Minitab to compute the power if the difference in means we wanted to detect was 0.01. Notice that with $n = 15$, the power drops considerably to 0.4425. The third section of the output is the sample size required to give a power of at least 0.8 if the difference in means of interest is actually 0.01. A much larger sample size ($n = 39$) is required to detect this smaller difference.

4-5.3   Confidence Interval on the Mean

It is easy to find a $100(1 - \alpha)\%$ CI on the mean of a normal distribution with unknown variance by proceeding as we did in Section 4-4.5. In general, the distribution of $T = (\overline{X} - \mu)/(S/\sqrt{n})$ is t with $n - 1$ degrees of freedom. Letting $t_{\alpha/2,n-1}$ be the upper $100\alpha/2$ percentage point of the t distribution with $n - 1$ degrees of freedom, we may write:

$$P(-t_{\alpha/2,n-1} \leq T \leq t_{\alpha/2,n-1}) = 1 - \alpha$$

or

$$P\left(-t_{\alpha/2,n-1} \leq \frac{\overline{X} - \mu}{S/\sqrt{n}} \leq t_{\alpha/2,n-1}\right) = 1 - \alpha$$

Rearranging this last equation yields

$$P(\overline{X} - t_{\alpha/2,\,n-1}S/\sqrt{n} \leq \mu \leq \overline{X} + t_{\alpha/2,n-1}S/\sqrt{n}) = 1 - \alpha \tag{4-49}$$

This leads to the following definition of the $100(1 - \alpha)\%$ two-sided CI on μ.

**Confidence Interval on the Mean of
a Normal Distribution, Variance Unknown**

If $\overline{x}$ and s are the mean and standard deviation of a random sample from a normal distribution with unknown variance σ^2, a $100(1 - \alpha)\%$ CI on μ is given by

$$\overline{x} - t_{\alpha/2,n-1}s/\sqrt{n} \leq \mu \leq \overline{x} + t_{\alpha/2,n-1}s/\sqrt{n} \tag{4-50}$$

where $t_{\alpha/2,n-1}$ is the upper $100\alpha/2$ percentage point of the t distribution with $n - 1$ degrees of freedom.

One-Sided Confidence Bound

To find a $100(1 - \alpha)\%$ lower confidence bound on μ, with unknown σ^2, simply replace $-t_{\alpha/2,n-1}$ with $-t_{\alpha,n-1}$ in the lower bound of equation 4-50 and set the upper bound to ∞. Similarly, to find a $100(1 - \alpha)\%$ upper confidence bound on μ, with unknown σ^2, replace $t_{\alpha/2,n-1}$ with $t_{\alpha,n-1}$ in the upper bound and set the lower bound to $-\infty$. These formulas are given in the table on the inside front cover.

EXAMPLE 4-9
Golf Clubs

Reconsider the golf club coefficient of restitution problem in Example 4-7. We know that $n = 15, \bar{x} = 0.83725$, and $s = 0.02456$. Find a 95% CI on μ.

Solution. From equation 4-50 we find ($t_{\alpha/2,n-1} = t_{0.025,14} = 2.145$):

$$\bar{x} - t_{\alpha/2,n-1}s/\sqrt{n} \le \mu \le \bar{x} + t_{\alpha/2,n-1}s/\sqrt{n}$$
$$0.83725 - 2.145(0.02456)/\sqrt{15} \le \mu \le 0.83725 + 2.145(0.02456)/\sqrt{15}$$
$$0.83725 - 0.01360 \le \mu \le 0.83725 + 0.01360$$
$$0.82365 \le \mu \le 0.85085$$

In Example 4-7, we tested a one-sided alternative hypothesis on μ. Some engineers might be interested in a one-sided confidence bound. Recall that the Minitab output actually computed a lower confidence bound. The 95% lower confidence bound on mean coefficient of restitution is

$$\bar{x} - t_{0.05,n-1}s/\sqrt{n} \le \mu$$
$$0.83725 - 1.761(0.02456)/\sqrt{15} \le \mu$$
$$0.82608 \le \mu$$

Thus we can state with 95% confidence that the mean coefficient of restitution exceeds 0.82608. This is also the result reported by Minitab. ∎

Animation 8: Confidence Intervals

EXERCISES FOR SECTION 4-5

4-41. Suppose that we are testing $H_0: \mu = \mu_0$ versus $H_1: \mu > \mu_0$ with a sample size of $n = 15$. Calculate bounds on the P-value for the following observed values of the test statistic:

(a) $t_0 = 2.35$ (b) $t_0 = 3.55$
(c) $t_0 = 2.00$ (d) $t_0 = 1.55$

4-42. Suppose that we are testing $H_0: \mu = \mu_0$ versus $H_1: \mu \ne \mu_0$ with a sample size of $n = 10$. Calculate bounds on the P-value for the following observed values of the test statistic:

(a) $t_0 = 2.48$ (b) $t_0 = -3.95$
(c) $t_0 = 2.69$ (d) $t_0 = 1.88$
(e) $t_0 = -1.25$

4-43. Suppose that we are testing $H_0: \mu = \mu_0$ versus $H_1: \mu < \mu_0$ with a sample size of $n = 25$. Calculate bounds on the P-value for the following observed values of the test statistic:

(a) $t_0 = -2.59$ (b) $t_0 = -1.76$
(c) $t_0 = -3.05$ (d) $t_0 = -1.30$

4-44. Consider the Minitab output shown below.

One-Sample T: X

Test of mu = 91 vs not = 91

Variable	N	Mean	StDev	SE Mean	95% CI	T	P
X	25	92.5805	?	0.4673	(91.6160, ?)	3.38	0.002

(a) Fill in the missing values in the output. Can the null hypothesis be rejected at the 0.05 level? Why?
(b) Is this a one-sided or a two-sided test?
(c) If the hypotheses had been $H_0: \mu = 90$ versus $H_1: \mu \ne 90$, would you reject the null hypothesis at the 0.05 level?
(d) Use the output and the t-table to find a 99% two-sided CI on the mean.
(e) What is the P-value if the alternative hypothesis is $H_1: \mu > 91$?

4-45. Consider the Minitab output shown below.

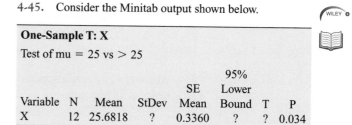

One-Sample T: X

Test of mu = 25 vs > 25

Variable	N	Mean	StDev	SE Mean	95% Lower Bound	T	P
X	12	25.6818	?	0.3360	?	?	0.034

(a) How many degrees of freedom are there on the t-test statistic?
(b) Fill in the missing information.

4-46. An article in the *ASCE Journal of Energy Engineering* (Vol. 125, 1999, pp. 59–75) describes a study of the thermal inertia properties of autoclaved aerated concrete used as a building material. Five samples of the material were tested in a structure, and the average interior temperature (°C) reported was as follows: 23.01, 22.22, 22.04, 22.62, and 22.59.

(a) Test the hypotheses H_0: $\mu = 22.5$ versus H_1: $\mu \neq 22.5$, using $\alpha = 0.05$. Use the P-value approach.

(b) Check the assumption that interior temperature is normally distributed.

(c) Find a 95% CI on the mean interior temperature.

(d) What sample size would be required to detect a true mean interior temperature as high as 22.75 if we wanted the power of the test to be a least 0.9? Use the sample standard deviation s as an estimate of σ.

4-47. A research engineer for a tire manufacturer is investigating tire life for a new rubber compound. She has built 10 tires and tested them to end-of-life in a road test. The sample mean and standard deviation are 61,492 and 3035 kilometers, respectively.

(a) The engineer would like to demonstrate that the mean life of this new tire is in excess of 60,000 km. Formulate and test appropriate hypotheses, being sure to state (test, if possible) assumptions, and draw conclusions using the P-value approach.

(b) Suppose that if the mean life is as long as 61,000 km, the engineer would like to detect this difference with probability at least 0.90. Was the sample size $n = 10$ used in part (a) adequate? Use the sample standard deviation s as an estimate of σ in reaching your decision.

(c) Find a 95% one-sided lower confidence bound on mean tire life.

(d) Use the bound found in part (c) to test the hypothesis.

4-48. An Izod impact test was performed on 20 specimens of PVC pipe. The ASTM standard for this material requires that Izod impact strength must be greater than 1.0 ft-lb/in. The sample average and standard deviation obtained were $\bar{x} = 1.121$ and $s = 0.328$, respectively. Test H_0: $\mu = 1.0$ versus H_1: $\mu > 1.0$ using $\alpha = 0.01$ and draw conclusions. State any necessary assumptions about the underlying distribution of the data.

4-49. The life in hours of a biomedical device under development in the laboratory is known to be approximately normally distributed. A random sample of 15 devices is selected and found to have an average life of 5625.1 hours and a sample standard deviation of 226.1 hours.

(a) Test the hypothesis that the true mean life of a biomedical device is greater than 5500 using the P-value approach.

(b) Construct a 95% lower confidence bound on the mean.

(c) Use the confidence bound found in part (b) to test the hypothesis.

4-50. A particular brand of diet margarine was analyzed to determine the level of polyunsaturated fatty acid (in percent). A sample of six packages resulted in the following data: 16.8, 17.2, 17.4, 16.9, 16.5, 17.1.

(a) Test the hypothesis that the mean is not 17.0, using the P-value approach. What are your conclusions? Use a normal probability plot to test the normality assumption.

(b) Suppose that if the mean polyunsaturated fatty acid content is $\mu = 17.5$, it is important to detect this with probability at least 0.90. Is the sample size $n = 6$ adequate? Use

the sample standard deviation to estimate the population standard deviation σ. Use $\alpha = 0.01$.

(c) Find a 99% two-sided CI on the mean μ. Provide a practical interpretation of this interval.

4-51. In building electronic circuitry, the breakdown voltage of diodes is an important quality characteristic. The breakdown voltage of 12 diodes was recorded as follows: 9.099, 9.174, 9.327, 9.377, 8.471, 9.575, 9.514, 8.928, 8.800, 8.920, 9.913, and 8.306.

(a) Check the normality assumption for the data.

(b) Test the claim that the mean breakdown voltage is less than 9 volts with a significance level of 0.05.

(c) Construct a 95% one-sided upper confidence bound on the mean breakdown voltage.

(d) Use the bound found in part (c) to test the hypothesis.

(e) Suppose that the true breakdown voltage is 8.8 volts; it is important to detect this with a probability of at least 0.95. Using the sample standard deviation to estimate the population standard deviation and a significance level of 0.05, determine the necessary sample size.

4-52. A machine produces metal rods used in an automobile suspension system. A random sample of 12 rods is selected, and the diameter is measured. The resulting data in millimeters are shown here.

8.23	8.31	8.42
8.29	8.19	8.24
8.19	8.29	8.30
8.14	8.32	8.40

(a) Check the assumption of normality for rod diameter.

(b) Is there strong evidence to indicate that mean rod diameter is not 8.20 mm using a fixed-level test with $\alpha = 0.05$?

(c) Find the P-value for this test.

(d) Find a 95% two-sided CI on mean rod diameter and provide a practical interpretation of this interval.

4-53. The wall thickness of 25 glass 2-liter bottles was measured by a quality-control engineer. The sample mean was $\bar{x} = 4.058$ mm, and the sample standard deviation was $s = 0.081$ mm.

(a) Suppose that it is important to demonstrate that the wall thickness exceeds 4.0 mm. Formulate and test appropriate hypotheses using these data. Draw conclusions at $\alpha = 0.05$. Calculate the P-value for this test.

(b) Find a 95% lower confidence bound for mean wall thickness. Interpret the interval you obtain.

4-54. Measurements on the percentage of enrichment of 12 fuel rods used in a nuclear reactor were reported as follows:

3.11	2.88	3.08	3.01
2.84	2.86	3.04	3.09
3.08	2.89	3.12	2.98

(a) Use a normal probability plot to check the normality assumption.

(b) Test the hypothesis H_0: $\mu = 2.95$ versus H_1: $\mu \neq 2.95$, and draw appropriate conclusions. Use the P-value approach.

(c) Find a 99% two-sided CI on the mean percentage of enrichment. Are you comfortable with the statement that the mean percentage of enrichment is 2.95%? Why?

 4-55.   A post-mix beverage machine is adjusted to release a certain amount of syrup into a chamber where it is mixed with carbonated water. A random sample of 25 beverages was found to have a mean syrup content of $\bar{x} = 1.098$ fluid ounces and a standard deviation of $s = 0.016$ fluid ounces.

(a) Do the data presented in this exercise support the claim that the mean amount of syrup dispensed is not 1.0 fluid ounce? Test this claim using $\alpha = 0.05$.

(b) Do the data support the claim that the mean amount of syrup dispensed exceeds 1.0 fluid ounce? Test this claim using $\alpha = 0.05$.

(c) Consider the hypothesis test in part (a). If the mean amount of syrup dispensed differs from $\mu = 1.0$ by as much as 0.05, it is important to detect this with a high probability (at least 0.90, say). Using s as an estimate of σ, what can you say about the adequacy of the sample size $n = 25$ used by the experimenters?

(d) Find a 95% two-sided CI on the mean amount of syrup dispensed.

WILEY 4-56.   An article in the *Journal of Composite Materials* (Vol. 23, 1989, p. 1200) describes the effect of delamination on the natural frequency of beams made from composite laminates. Five such delaminated beams were subjected to loads, and the resulting frequencies were as follows (in Hz):

$$230.66, 233.05, 232.58, 229.48, 232.58$$

Find a 90% two-sided CI on mean natural frequency. Do the results of your calculations support the claim that mean natural frequency is 235 Hz? Discuss your findings and state any necessary assumptions.

4-57.   Cloud seeding has been studied for many decades as a weather modification procedure (for an interesting study of this subject, see the article in *Technometrics*, "A Bayesian Analysis of a Multiplicative Treatment Effect in Weather Modification," Vol. 17, 1975, pp. 161–166). The rainfall in acre-feet from 20 clouds that were selected at random and seeded with silver nitrate follows: 18.0, 30.7, 19.8, 27.1, 22.3, 18.8, 31.8, 23.4, 21.2, 27.9, 31.9, 27.1, 25.0, 24.7, 26.9, 21.8, 29.2, 34.8, 26.7, and 31.6.

(a) Can you support a claim that mean rainfall from seeded clouds exceeds 25 acre-feet? Use $\alpha = 0.01$. Find the P-value.

(b) Check that rainfall is normally distributed.

(c) Compute the power of the test if the true mean rainfall is 27 acre-feet.

(d) What sample size would be required to detect a true mean rainfall of 27.5 acre-feet if we wanted the power of the test to be at least 0.9?

(e) Explain how the question in part (a) could be answered by constructing a one-sided confidence bound on the mean diameter.

4-6   INFERENCE ON THE VARIANCE OF A NORMAL POPULATION

Sometimes hypothesis tests and CIs on the population variance or standard deviation are needed. If we have a random sample $X_1, X_2, \ldots, X_n$, the sample variance S^2 is an unbiased point estimator of σ^2. When the population is modeled by a normal distribution, the tests and intervals described in this section are applicable.

4-6.1   Hypothesis Testing on the Variance of a Normal Population

Suppose that we wish to test the hypothesis that the variance of a normal population σ^2 equals a specified value—say, σ_0^2. Let $X_1, X_2, \ldots, X_n$ be a random sample of n observations from this population. To test

$$H_0: \sigma^2 = \sigma_0^2$$
$$H_1: \sigma^2 \neq \sigma_0^2 \tag{4-51}$$

we will use the following **test statistic**:

$$X_0^2 = \frac{(n-1)S^2}{\sigma_0^2} \qquad (4\text{-}52)$$

To define the test procedure, we will need to know the distribution of the test statistic X_0^2 in equation 4-52 when the null hypothesis is true.

Let $X_1, X_2, \ldots, X_n$ be a random sample from a normal distribution with unknown mean μ and unknown variance σ^2. The quantity

$$X^2 = \frac{(n-1)S^2}{\sigma^2} \qquad (4\text{-}53)$$

has a chi-square distribution with $n-1$ degrees of freedom, abbreviated as χ_{n-1}^2. In general, the probability density function of a chi-square random variable is

$$f(x) = \frac{1}{2^{k/2}\Gamma(k/2)} x^{(k/2)-1} e^{-x/2} \qquad x > 0 \qquad (4\text{-}54)$$

where k is the number of degrees of freedom and $\Gamma(k/2)$ was defined in Section 4-5.1.

The mean and variance of the χ^2 distribution are

$$\mu = k \qquad \text{and} \qquad \sigma^2 = 2k \qquad (4\text{-}55)$$

respectively. Several chi-square distributions are shown in Fig. 4-21. Note that the chi-square random variable is nonnegative and that the probability distribution is skewed to the right. However, as k increases, the distribution becomes more symmetric. As $k \to \infty$, the limiting form of the chi-square distribution is the normal distribution.

The **percentage points** of the χ^2 distribution are given in Table III of Appendix A. Define $\chi_{\alpha,k}^2$ as the percentage point or value of the chi-square random variable with k degrees of freedom such that the probability that X^2 exceeds this value is α. That is,

$$P(X^2 > \chi_{\alpha,k}^2) = \int_{\chi_{\alpha,k}^2}^{\infty} f(u)\, du = \alpha$$

Using Table III of Appendix A for the X^2 Distribution

This probability is shown as the shaded area in Fig. 4-22. To illustrate the use of Table III, note that the areas α are the column headings and the degrees of freedom k are given in the left column, labeled ν. Therefore, the value with 10 degrees of freedom having an area (probability) of 0.05 to the right is $\chi_{0.05,10}^2 = 18.31$. This value is often called an upper 5% point of chi square with 10 degrees of freedom. We may write this as a probability statement as follows:

$$P(X^2 > \chi_{0.05,10}^2) = P(X^2 > 18.31) = 0.05$$

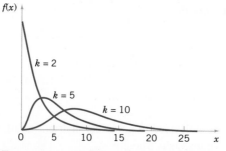

Figure 4-21   Probability density functions of several χ^2 distributions.

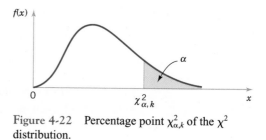

Figure 4-22   Percentage point $\chi^2_{\alpha,k}$ of the χ^2 distribution.

It is relatively easy to construct a test for the hypothesis in equation 4-51. We can use X_0^2 defined in equation 4-52 as the **test statistic.** If the null hypothesis H_0: $\sigma^2 = \sigma_0^2$ is true, the test statistic X_0^2 follows the chi-square distribution with $n - 1$ degrees of freedom. To conduct a fixed-significance-level test, calculate the value of the test statistic χ_0^2, and reject the hypothesis H_0: $\sigma^2 = \sigma_0^2$ if

$$\chi_0^2 > \chi^2_{\alpha/2,n-1} \qquad \text{or} \qquad \chi_0^2 < \chi^2_{1-\alpha/2,n-1} \tag{4-56}$$

where $\chi^2_{\alpha/2,n-1}$ and $\chi^2_{1-\alpha/2,n-1}$ are the upper and lower $100\alpha/2$ percentage points of the chi-square distribution with $n - 1$ degrees of freedom, respectively. The critical region is shown in Fig. 4-23a.

The same test statistic is used for one-sided alternative hypotheses. For the one-sided hypothesis

$$H_0\text{: } \sigma^2 = \sigma_0^2$$
$$H_1\text{: } \sigma^2 > \sigma_0^2 \tag{4-57}$$

we would reject H_0 if

$$\chi_0^2 > \chi^2_{\alpha,n-1} \tag{4-58}$$

For the other one-sided hypothesis

$$H_0\text{: } \sigma^2 = \sigma_0^2$$
$$H_1\text{: } \sigma^2 < \sigma_0^2 \tag{4-59}$$

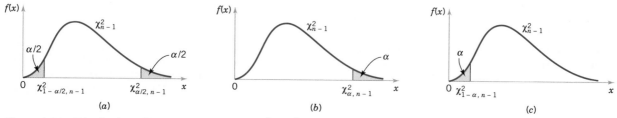

Figure 4-23   Distribution of the test statistic for H_0: $\sigma^2 = \sigma_0^2$ with critical region values for (a) H_1: $\sigma^2 \neq \sigma_0^2$, (b) H_0: $\sigma^2 > \sigma_0^2$, and (c) H_0: $\sigma^2 < \sigma_0^2$.

we would reject H_0 if

$$\chi_0^2 < \chi_{1-\alpha,n-1}^2 \tag{4-60}$$

The one-sided critical regions are shown in Fig. 4-23b and c.

Summary

Testing Hypotheses on the Variance of a Normal Distribution

Null hypothesis: $H_0: \sigma^2 = \sigma_0^2$

Test statistic: $\chi_0^2 = \dfrac{(n-1)S^2}{\sigma_0^2}$

Alternative Hypotheses	Rejection Criterion
$H_1: \sigma^2 \neq \sigma_0^2$	$\chi_0^2 > \chi_{\alpha/2,n-1}^2$ or $\chi_0^2 < \chi_{1-\alpha/2,n-1}^2$
$H_1: \sigma^2 > \sigma_0^2$	$\chi_0^2 > \chi_{\alpha,n-1}^2$
$H_1: \sigma^2 < \sigma_0^2$	$\chi_0^2 < \chi_{1-\alpha,n-1}^2$

The locations of the critical region are shown in Fig. 4-23.

EXAMPLE 4-10
Bottle Filling

An automatic filling machine is used to fill bottles with liquid detergent. A random sample of 20 bottles results in a sample variance of fill volume of $s^2 = 0.0153$ (fluid ounces)^2. If the variance of fill volume exceeds 0.01 (fluid ounces)^2 an unacceptable proportion of bottles will be under and overfilled. Is there evidence in the sample data to suggest that the manufacturer has a problem with under- and overfilled bottles? Use $\alpha = 0.05$, and assume that fill volume has a normal distribution.

Solution. Using the seven-step procedure results in the following:

1. **Parameter of interest:** The parameter of interest is the population variance σ^2.
2. **Null hypothesis, H_0:** $\sigma^2 = 0.01$
3. **Alternative hypothesis, H_1:** $\sigma^2 > 0.01$
4. **Test statistic:** The test statistic is

$$\chi_0^2 = \frac{(n-1)s^2}{\sigma_0^2}$$

5. **Reject H_0:** To use a fixed-significance-level test, reject H_0 if $\chi_0^2 > \chi_{0.05,19}^2 = 30.14$.
6. **Computations:**

$$\chi_0^2 = \frac{19(0.0153)}{0.01} = 29.07$$

7. **Conclusions:** Because $\chi_0^2 = 29.07 < \chi_{0.05,19}^2 = 30.14$, we conclude that there is no strong evidence that the variance of fill volume exceeds 0.01 (fluid ounces)^2.

Computing P-Value

P-values can also be used with chi-square tests. For example, consider Example 4–10, which involved an upper-tail one-tail test. The *P*-value is the probability to the right of the computed value of the test statistic in the χ^2_{n-1} distribution. Because Appendix A Table III contains only 11 tail areas, we usually will have to find lower and upper bounds on *P*. This is easy to do. The computed value of the test statistic in Example 4-10 is $\chi^2_0 = 29.07$. From inspection of the table, we find that $\chi^2_{0.10,19} = 27.20$ and $\chi^2_{0.05,19} = 30.14$. Because $27.20 < 29.07 < 30.14$, we conclude that the *P*-value for the test in Example 4-10 is in the interval $0.05 < P < 0.10$. The actual *P*-value can be computed from Minitab. For 19 degrees of freedom, Minitab calculates the cumulative chi-square probability that is less than or equal to the value of the test statistic $\chi^2_0 = 29.07$ as 0.935108 (use the cumulative distribution function in the Calc menu), so the *P*-value is the probability (area) beyond 29.07, or $P = 1 - 0.935108 = 0.064892$. ∎

The *P*-value for the lower-tail test would be found as the area (probability) in the lower tail of the χ^2_{n-1} distribution to the left of the computed value of the test statistic χ^2_0. For the two-sided alternative, find the tail area associated with the computed value of the test statistic and double it to obtain the *P*-value.

4-6.2 Confidence Interval on the Variance of a Normal Population

It was noted in the previous section that if the population is normal, the sampling distribution of

$$X^2 = \frac{(n-1)S^2}{\sigma^2}$$

is chi-square with $n - 1$ degrees of freedom. To develop the confidence interval, we first write

$$P(\chi^2_{1-\alpha/2,n-1} \leq X^2 \leq \chi^2_{\alpha/2,n-1}) = 1 - \alpha$$

so that

$$P\left[\chi^2_{1-\alpha/2,n-1} \leq \frac{(n-1)S^2}{\sigma^2} \leq \chi^2_{\alpha/2,n-1}\right] = 1 - \alpha$$

This last equation can be rearranged as

$$P\left[\frac{(n-1)S^2}{\chi^2_{\alpha/2,n-1}} \leq \sigma^2 \leq \frac{(n-1)S^2}{\chi^2_{1-\alpha/2,n-1}}\right] = 1 - \alpha \tag{4-61}$$

This leads to the following definition of the CI for σ^2.

Confidence Interval on the Variance of a Normal Distribution

If s^2 is the sample variance from a random sample of n observations from a normal distribution with unknown variance σ^2, a $100(1 - \alpha)\%$ CI on σ^2 is

$$\frac{(n-1)s^2}{\chi^2_{\alpha/2,n-1}} \leq \sigma^2 \leq \frac{(n-1)s^2}{\chi^2_{1-\alpha/2,n-1}} \tag{4-62}$$

where $\chi^2_{\alpha/2,n-1}$ and $\chi^2_{1-\alpha/2,n-1}$ are the upper and lower $100\alpha/2$ percentage points of the chi-square distribution with $n - 1$ degrees of freedom, respectively.

One-Sided Confidence Bounds

To find a $100(1 - \alpha)\%$ lower confidence bound on σ^2, set the upper confidence bound in equation 4-62 equal to ∞ and replace $\chi^2_{\alpha/2,n-1}$ with $\chi^2_{\alpha,n-1}$. The $100(1 - \alpha)\%$ upper confidence bound is found by setting the lower confidence limit in equation 4-62 equal to zero and replacing $\chi^2_{1-\alpha/2,n-1}$ with $\chi^2_{1-\alpha,n-1}$. For your convenience, these equations for constructing the one-sided upper and lower confidence intervals are given in the table on the inside front cover of this text.

EXAMPLE 4-11
Bottle Filling

Reconsider the bottle filling machine from Example 4-10. We will continue to assume that fill volume is approximately normally distributed. A random sample of 20 bottles results in a sample variance of $s^2 = 0.0153$ (fluid ounces)^2. We want to find the 95% upper confidence bound on σ^2.

Solution. The CI is found from equation 4-62 as follows:

$$\sigma^2 \leq \frac{(n-1)s^2}{\chi^2_{0.95,19}}$$

or

$$\sigma^2 \leq \frac{(19)0.0153}{10.12} = 0.0287 \text{ (fluid ounces)}^2$$

This last statement may be converted into a confidence bound on the standard deviation σ by taking the square root of both sides, resulting in

$$\sigma \leq 0.17 \text{ fluid ounces}$$

Therefore, at the 95% level of confidence, the data indicate that the process standard deviation could be as large as 0.17 fluid ounces. ∎

EXERCISES FOR SECTION 4-6

4-58. Suppose that we are testing $H_0: \sigma^2 = \sigma_0^2$ versus $H_1: \sigma^2 > \sigma_0^2$ with a sample size of $n = 15$. Calculate bounds on the P-value for the following observed values of the test statistic:

(a) $\chi_0^2 = 22.35$ (b) $\chi_0^2 = 23.50$
(c) $\chi_0^2 = 25.00$ (d) $\chi_0^2 = 28.55$

4-59. A rivet is to be inserted into a hole. If the standard deviation of hole diameter exceeds 0.02 mm, there is an unacceptably high probability that the rivet will not fit. A random sample of $n = 15$ parts is selected, and the hole diameter is measured. The sample standard deviation of the hole diameter measurements is $s = 0.016$ mm.

(a) Is there strong evidence to indicate that the standard deviation of hole diameter exceeds 0.02 mm? Calculate a P-value to draw conclusions. State any necessary assumptions about the underlying distribution of the data.
(b) Construct a 95% lower confidence bound for σ.
(c) Use the confidence bound in part (b) to test the hypothesis.

4-60. The sugar content of the syrup in canned peaches is normally distributed, and the variance is thought to be $\sigma^2 = 18$ (mg)^2.

(a) Test the hypothesis that the variance is not 18 (mg)^2 if a random sample of $n = 10$ cans yields a sample standard deviation of $s = 4$ mg, using a fixed-level test with $\alpha = 0.05$. State any necessary assumptions about the underlying distribution of the data.
(b) What is the P-value for this test?
(c) Find a 95% two-sided CI for σ.
(d) Use the CI in part (c) to test the hypothesis.

4-61. Consider the tire life data in Exercise 4-47.

(a) Can you conclude, using $\alpha = 0.05$, that the standard deviation of tire life exceeds 3000 km? State any necessary assumptions about the underlying distribution of the data.
(b) Find the P-value for this test.
(c) Find a 95% lower confidence bound for σ.
(d) Use the confidence bound in part (c) to test the hypothesis.

4-62. Consider the Izod impact test data in Exercise 4-48.

(a) Test the hypothesis that $\sigma^2 = 0.10$ against an alternative specifying that $\sigma^2 \neq 0.10$, using $\alpha = 0.01$, and draw a conclusion. State any necessary assumptions about the underlying distribution of the data.

(b) What is the *P*-value for this test?

(c) Find a 99% two-sided CI for σ^2.

(d) Use the CI in part (c) to test the hypothesis.

4-63.   The percentage of titanium in an alloy used in aerospace castings is measured in 51 randomly selected parts. The sample standard deviation is $s = 0.37$.

(a) Test the hypothesis $H_0: \sigma = 0.35$ versus $H_1: \sigma \neq 0.35$ using $\alpha = 0.05$. State any necessary assumptions about the underlying distribution of the data.

(b) Find the *P*-value for this test.

(c) Construct a 95% two-sided CI for σ.

(d) Use the CI in part (c) to test the hypothesis.

4-7   INFERENCE ON A POPULATION PROPORTION

It is often necessary to test hypotheses and construct CIs on a population proportion. For example, suppose that a random sample of size *n* has been taken from a large (possibly infinite) population and that $X(\leq n)$ observations in this sample belong to a class of interest. Then $\hat{P} = X/n$ is a point estimator of the proportion of the population *p* that belongs to this class. Note that *n* and *p* are the parameters of a binomial distribution. Furthermore, from Chapter 3 we know that the sampling distribution of $\hat{P}$ is approximately normal with mean *p* and variance $p(1 - p)/n$, if *p* is not too close to either 0 or 1 and if *n* is relatively large. Typically, to apply this approximation we require that np and $n(1 - p)$ be greater than or equal to 5. We will make use of the normal approximation in this section.

4-7.1   Hypothesis Testing on a Binomial Proportion

In many engineering problems, we are concerned with a random variable that follows the binomial distribution. For example, consider a production process that manufactures items classified as either acceptable or defective. It is usually reasonable to model the occurrence of defectives with the binomial distribution, where the binomial parameter *p* represents the proportion of defective items produced. Consequently, many engineering decision problems include hypothesis testing about *p*.

We will consider testing

$$H_0: p = p_0$$
$$H_1: p \neq p_0 \tag{4-63}$$

An approximate test based on the normal approximation to the binomial will be given. As noted above, this approximate procedure will be valid as long as *p* is not extremely close to 0 or 1, and if the sample size is relatively large. The following result will be used to perform hypothesis testing and to construct confidence intervals on *p*.

Let *X* be the number of observations in a random sample of size *n* that belongs to the class associated with *p*. Then the quantity

$$Z = \frac{X - np}{\sqrt{np(1 - p)}} \tag{4-64}$$

has approximately a standard normal distribution, $N(0, 1)$.

Then, if the null hypothesis H_0: $p = p_0$ is true, we have $X \sim N[np_0, np_0(1 - p_0)]$, approximately. To test H_0: $p = p_0$, calculate the **test statistic**

$$Z_0 = \frac{X - np_0}{\sqrt{np_0(1 - p_0)}}$$

and determine the P-value. Because the test statistic follows a standard normal distribution if H_0 is true, the P-value is calculated exactly like the P-value for the z-tests in Section 4-4. So for the two-sided alternative hypothesis, the P-value is the sum of the probability in the standard normal distribution above the positive computed value of the test statistic z_0 and the probability below the negative value $-z_0$, or

$$P = 2[1 - \Phi(|z_0|)]$$

For the one-sided alternative hypothesis H_0: $p > p_0$, the P-value is the probability above z_0, or

$$P = 1 - \Phi(z_0)$$

and for the one-sided alternative hypothesis H_0: $p < p_0$, the P-value is the probability below z_0, or

$$P = \Phi(z_0)$$

We can also perform a **fixed-significance-level** test. For the two-sided alternative hypothesis, we would reject H_0: $p \neq p_0$ if

$$z_0 > z_{\alpha/2} \quad \text{or} \quad z_0 < -z_{\alpha/2}$$

Critical regions for the one-sided alternative hypotheses would be constructed in the usual manner.

Summary

Testing Hypotheses on a Binomial Proportion		
Null hypotheses: $\quad H_0$: $p = p_0$		
Test statistic: $\quad Z_0 = \dfrac{X - np_0}{\sqrt{np_0(1 - p_0)}}$		

Alternative Hypotheses	**P-Value**	**Rejection Criterion for Fixed-Level Tests**
H_1: $p \neq p_0$	Probability above z_0 and probability below $-z_0$, $P = 2[1 - \Phi(|z_0|)]$	$z_0 > z_{\alpha/2}$ or $z_0 < -z_{\alpha/2}$
H_1: $p > p_0$	Probability above z_0, $P = 1 - \Phi(z_0)$	$z_0 > z_\alpha$
H_1: $p < p_0$	Probability below z_0, $P = \Phi(z_0)$	$z_0 < -z_\alpha$

EXAMPLE 4-12

Engine Controllers

A semiconductor manufacturer produces controllers used in automobile engine applications. The customer requires that the process fallout or fraction defective at a critical manufacturing step not exceed 0.05 and that the manufacturer demonstrate process capability at this level of quality using $\alpha = 0.05$. The semiconductor manufacturer takes a random sample of 200 devices and finds that 4 of them are defective. Can the manufacturer demonstrate process capability for the customer?

Solution. We may solve this problem using the seven-step hypothesis testing procedure as follows:

1. **Parameter of interest:** The parameter of interest is the process fraction defective p.
2. **Null hypothesis, H_0:** $p = 0.05$
3. **Alternative hypothesis, H_1:** $p < 0.05$
    This formulation of the problem will allow the manufacturer to make a strong claim about process capability if the null hypothesis H_0: $p = 0.05$ is rejected.
4. **Test statistic:** The test statistic is (from equation 4-64)

$$z_0 = \frac{x - np_0}{\sqrt{np_0(1 - p_0)}}$$

    where $x = 4$, $n = 200$, and $p_0 = 0.05$.
5. **Reject H_0 if:** Reject H_0: $p = 0.05$ if the P-value is less than 0.05.
6. **Computations:** The test statistic is

$$z_0 = \frac{4 - 200(0.05)}{\sqrt{200(0.05)(0.95)}} = -1.95$$

7. **Conclusions:** Because $z_0 = -1.95$, the P-value is $\Phi(-1.95) = 0.0256$; since this is less than 0.05, we reject H_0 and conclude that the process fraction defective p is less than 0.05. We conclude that the process is capable. ∎

We occasionally encounter another form of the test statistic Z_0 in equation 4-64. Note that if X is the number of observations in a random sample of size n that belongs to a class of interest, $\hat{P} = X/n$ is the sample proportion that belongs to that class. Now divide both numerator and denominator of Z_0 in equation 4-64 by n, giving

$$Z_0 = \frac{X/n - p_0}{\sqrt{p_0(1 - p_0)/n}}$$

or

$$Z_0 = \frac{\hat{P} - p_0}{\sqrt{p_0(1 - p_0)/n}} \tag{4-65}$$

This presents the test statistic in terms of the sample proportion instead of the number of items X in the sample that belongs to the class of interest.

Minitab can be used to perform the test on a binomial proportion. The following Minitab output shows the results for Example 4-12.

Test and CI for One Proportion

Test of p = 0.05 vs p < 0.05

Sample	X	N	Sample p	95% Upper Bound	Z-Value	P-Value
1	4	200	0.020000	0.036283	−1.95	0.026.

Note The normal approximation may be inaccurate for small samples.

This output also shows a 95% one-sided upper confidence bound on P. In Section 4-7.3 we will show how CIs on a binomial proportion are computed. This Minitab display shows the result of using the normal approximation for tests and CIs. When the sample size is small, this may be inappropriate.

Small Sample Tests on a Binomial Proportion
Tests on a proportion when the sample size n is small are based on the binomial distribution, not the normal approximation to the binomial. To illustrate, suppose we wish to test $H_0: p < p_0$. Let X be the number of successes in the sample. The P-value for this test would be found from the lower tail of a binomial distribution with parameters n and p_0. Specifically, the P-value would be the probability that a binomial random variable with parameters n and p_0 is less than or equal to X. P-values for the upper-tail one-sided test and the two-sided alternative are computed similarly.

Minitab will calculate the exact P-value for a binomial test. The output below contains the exact P-value results for Example 4-12.

Test of p = 0.05 vs p < 0.05

Sample	X	N	Sample p	95% Upper Bound	Exact P-Value
1	4	200	0.020000	0.045180	0.026

The P-value is the same as that reported for the normal approximation, because the sample size is pretty large. Notice that the CI is different from the one found using the normal approximation.

4-7.2 Type II Error and Choice of Sample Size

It is possible to obtain closed-form equations for the approximate β-error for the tests in Section 4-7.1. Suppose that p is the true value of the population proportion.

The approximate β-error for the two-sided alternative $H_1: p \neq p_0$ is

$$\beta = \Phi\left[\frac{p_0 - p + z_{\alpha/2}\sqrt{p_0(1 - p_0)/n}}{\sqrt{p(1 - p)/n}}\right]$$
$$- \Phi\left[\frac{p_0 - p - z_{\alpha/2}\sqrt{p_0(1 - p_0)/n}}{\sqrt{p(1 - p)/n}}\right] \tag{4-66}$$

If the alternative is $H_1: p < p_0$,

$$\beta = 1 - \Phi\left[\frac{p_0 - p - z_{\alpha}\sqrt{p_0(1 - p_0)/n}}{\sqrt{p(1 - p)/n}}\right] \tag{4-67}$$

whereas if the alternative is $H_1: p > p_0$,

$$\beta = \Phi\left[\frac{p_0 - p + z_{\alpha}\sqrt{p_0(1 - p_0)/n}}{\sqrt{p(1 - p)/n}}\right] \tag{4-68}$$

These equations can be solved to find the approximate sample size n that gives a test of level α that has a specified β risk. The sample size equation follows.

Sample Size for a Two-Sided Hypothesis Test on a Binomial Proportion

$$n = \left[\frac{z_{\alpha/2}\sqrt{p_0(1 - p_0)} + z_{\beta}\sqrt{p(1 - p)}}{p - p_0}\right]^2 \tag{4-69}$$

If n is not an integer, round the sample size up to the next larger integer.

For a one-sided alternative, replace $z_{\alpha/2}$ in equation 4-69 by z_{α}.

EXAMPLE 4-13
Engine Controllers

Consider the semiconductor manufacturer from Example 4-12. Suppose that the process fallout is really $p = 0.03$. What is the β-error for this test of process capability, which uses $n = 200$ and $\alpha = 0.05$?

Solution. The β-error can be computed using equation 4-67 as follows:

$$\beta = 1 - \Phi\left[\frac{0.05 - 0.03 - (1.645)\sqrt{0.05(0.95)/200}}{\sqrt{0.03(1 - 0.03)/200}}\right] = 1 - \Phi(-0.44) = 0.67$$

Thus, the probability is about 0.7 that the semiconductor manufacturer will fail to conclude that the process is capable if the true process fraction defective is $p = 0.03$ (3%). This appears to be a large β-error, but the difference between $p = 0.05$ and $p = 0.03$ is fairly small, and the sample size $n = 200$ is not particularly large.

Suppose that the semiconductor manufacturer was willing to accept a β-error as large as 0.10 if the true value of the process fraction defective was $p = 0.03$. If the manufacturer continues to use $\alpha = 0.05$, what sample size would be required?

The required sample size can be computed from equation 4-69 as follows:

$$n = \left[\frac{1.645\sqrt{0.05(0.95)} + 1.28\sqrt{0.03(0.97)}}{0.03 - 0.05} \right]^2 \simeq 832$$

where we have used $p = 0.03$ in equation 4-69 and $z_{\alpha/2}$ is replaced by z_α for the one-sided alternative. Note that $n = 832$ is a very large sample size. However, we are trying to detect a fairly small deviation from the null value $p_0 = 0.05$. ■

4-7.3 Confidence Interval on a Binomial Proportion

It is straightforward to find an approximate $100(1 - \alpha)\%$ CI on a binomial proportion using the normal approximation. Recall that the sampling distribution of $\hat{P}$ is approximately normal with mean p and variance $p(1 - p)/n$, if p is not too close to either 0 or 1 and if n is relatively large. Then the distribution of

$$Z = \frac{X - np}{\sqrt{np(1 - p)}} = \frac{\hat{P} - p}{\sqrt{\dfrac{p(1 - p)}{n}}} \tag{4-70}$$

is approximately standard normal.

To construct the CI on p, note that

$$P(-z_{\alpha/2} \le Z \le z_{\alpha/2}) \simeq 1 - \alpha$$

so that

$$P\left[-z_{\alpha/2} \le \frac{\hat{P} - p}{\sqrt{\dfrac{p(1 - p)}{n}}} \le z_{\alpha/2} \right] \simeq 1 - \alpha$$

This may be rearranged as

$$P\left[\hat{P} - z_{\alpha/2}\sqrt{\frac{p(1 - p)}{n}} \le p \le \hat{P} + z_{\alpha/2}\sqrt{\frac{p(1 - p)}{n}} \right] \simeq 1 - \alpha \tag{4-71}$$

The quantity $\sqrt{p(1 - p)/n}$ in equation 4-71 is called the **standard error of the point estimator** $\hat{P}$. Unfortunately, the upper and lower limits of the CI obtained from equation 4-71 contain the unknown parameter p. However, a satisfactory solution is to replace p with $\hat{P}$ in the standard error, which results in

$$P\left[\hat{P} - z_{\alpha/2}\sqrt{\frac{\hat{P}(1 - \hat{P})}{n}} \le p \le \hat{P} + z_{\alpha/2}\sqrt{\frac{\hat{P}(1 - \hat{P})}{n}} \right] \simeq 1 - \alpha \tag{4-72}$$

Equation 4-72 leads to the approximate $100(1 - \alpha)\%$ CI on p.

Confidence Interval on a Binomial Proportion

If $\hat{p}$ is the proportion of observations in a random sample of size n that belong to a class of interest, an approximate $100(1 - \alpha)\%$ CI on the proportion p of the population that belongs to this class is

$$\hat{p} - z_{\alpha/2}\sqrt{\frac{\hat{p}(1 - \hat{p})}{n}} \leq p \leq \hat{p} + z_{\alpha/2}\sqrt{\frac{\hat{p}(1 - \hat{p})}{n}} \qquad (4\text{-}73)$$

where $z_{\alpha/2}$ is the upper $100\,\alpha/2$ percentage point of the standard normal distribution.

This procedure depends on the adequacy of the normal approximation to the binomial. To be reasonably conservative, this requires that np and $n(1 - p)$ be greater than or equal to 5. In situations where this approximation is inappropriate, particularly in cases where n is small, other methods must be used. Tables of the binomial distribution could be used to obtain a confidence interval for p. However, we prefer to use numerical methods based on the binomial probability mass function that are implemented in computer programs. This method is used in Minitab and is illustrated for the situation of Example 4-12 in the boxed display on page 195.

EXAMPLE 4-14

Crankshaft Bearings

In a random sample of 85 automobile engine crankshaft bearings, 10 have a surface finish that is rougher than the specifications allow. Find a 95% confidence interval on the proportion of defective bearings.

Solution. A point estimate of the proportion of bearings in the population that exceeds the roughness specification is $\hat{p} = x/n = 10/85 = 0.12$. A 95% two-sided CI for p is computed from equation 4-73 as

$$\hat{p} - z_{0.025}\sqrt{\frac{\hat{p}(1 - \hat{p})}{n}} \leq p \leq \hat{p} + z_{0.025}\sqrt{\frac{\hat{p}(1 - \hat{p})}{n}}$$

or

$$0.12 - 1.96\sqrt{\frac{0.12(0.88)}{85}} \leq p \leq 0.12 + 1.96\sqrt{\frac{0.12(0.88)}{85}}$$

which simplifies to

$$0.05 \leq p \leq 0.19 \qquad \blacksquare$$

Choice of Sample Size

Because $\hat{P}$ is the point estimator of p, we can define the error in estimating p by $\hat{P}$ as $E = |\hat{P} - p|$. Note that we are approximately $100(1 - \alpha)\%$ confident that this error is less than $z_{\alpha/2}\sqrt{p(1 - p)/n}$. For instance, in Example 4-14, we are 95% confident that the sample proportion $\hat{p} = 0.12$ differs from the true proportion p by an amount not exceeding 0.07.

In situations where the sample size can be selected, we may choose n so that we are 100 $(1 - \alpha)\%$ confident that the error is less than some specified value E. If we set $E = z_{\alpha/2}\sqrt{p(1 - p)/n}$ and solve for n, we obtain the following formula.

Sample Size for a Specified E on a Binomial Proportion

If $\hat{P}$ is used as an estimate of p, we can be $100(1 - \alpha)\%$ confident that the error $|\hat{P} - p|$ will not exceed a specified amount E when the sample size is

$$n = \left(\frac{z_{\alpha/2}}{E}\right)^2 p(1 - p) \tag{4-74}$$

An estimate of p is required to use equation 4-74. If an estimate $\hat{p}$ from a previous sample is available, it can be substituted for p in equation 4-74, or perhaps a subjective estimate can be made. If these alternatives are unsatisfactory, a preliminary sample can be taken, $\hat{p}$ computed, and then equation 4-74 used to determine how many additional observations are required to estimate p with the desired accuracy. Another approach to choosing n uses the fact that the sample size from equation 4-74 will always be a maximum for $p = 0.5$ [that is, $p(1 - p) \leq 0.25$ with equality for $p = 0.5$], and this can be used to obtain an upper bound on n. In other words, we are at least $100(1 - \alpha)\%$ confident that the error in estimating p by $\hat{p}$ is less than E if the sample size is selected as follows.

For a specified error E, an upper bound on the sample size for estimating p is

$$n = \left(\frac{z_{\alpha/2}}{E}\right)^2 \frac{1}{4} \tag{4-75}$$

EXAMPLE 4-15
Crankshaft
Bearings

Consider the situation in Example 4-14. How large a sample is required if we want to be 95% confident that the error in using $\hat{p}$ to estimate p is less than 0.05?

Solution. Using $\hat{p} = 0.12$ as an initial estimate of p, we find from equation 4-74 that the required sample size is

$$n = \left(\frac{z_{0.025}}{E}\right)^2 \hat{p}(1 - \hat{p}) = \left(\frac{1.96}{0.05}\right)^2 0.12(0.88) \cong 163$$

If we wanted to be *at least* 95% confident that our estimate $\hat{p}$ of the true proportion p was within 0.05 regardless of the value of p, we would use equation 4-75 to find the sample size

$$n = \left(\frac{z_{0.025}}{E}\right)^2 (0.25) = \left(\frac{1.96}{0.05}\right)^2 (0.25) \cong 385$$

Note that if we have information concerning the value of p, either from a preliminary sample or from past experience, we could use a smaller sample while maintaining both the desired precision of estimation and the level of confidence. ■

One-Sided Confidence Bounds

To find an approximate $100(1 - \alpha)\%$ lower confidence bound on p, simply replace $-z_{\alpha/2}$ with $-z_{\alpha}$ in the lower bound of equation 4-73 and set the upper bound to 1. Similarly, to find an approximate $100(1 - \alpha)\%$ upper confidence bound on p, replace $z_{\alpha/2}$ with z_{α} in the upper bound of equation 4-73 and set the lower bound to 0. These formulas are given in the table on the inside front cover. Similarly, when determining sample size in the case of the one-sided confidence bounds, simply replace $z_{\alpha/2}$ with z_{α} in equations 4-74 and 4-75.

Animation 8: Confidence Intervals

EXERCISES FOR SECTION 4-7

4-64. Consider the Minitab output shown below.

Test and CI for One Proportion

Test of p = 0.3 vs p not = 0.3

Sample	X	N	Sample p	95% CI	Z-Value	P-Value
1	95	250	0.380000	(0.319832, 0.440168)	2.76	0.006

(a) Is this a one-sided or a two-sided test?
(b) Was this test conducted using the normal approximation to the binomial? Was that appropriate?
(c) Can the null hypothesis be rejected at the 0.05 level?
(d) Can the null hypothesis $H_0: p = 0.4$ versus $H_0: p \neq 0.4$ be rejected at the 0.05 level? How can you do this without performing any additional calculations?
(e) Construct an approximate 90% CI for p.

4-65. Consider the following Minitab output.

Test and CI for One Proportion

Test of p = 0.65 vs p > 0.65

Sample	X	N	Sample p	95% Lower Bound	Z-Value	P-Value
1	553	800	?	?	2.45	?

(a) Is this a one-sided or a two-sided test?
(b) Was this test conducted using the normal approximation to the binomial? Was that appropriate?
(c) Fill in the missing values.

4-66. Of 1000 randomly selected cases of lung cancer, 823 resulted in death.
(a) Test the hypotheses $H_0: p = 0.85$ versus $H_1: p \neq 0.85$ with $\alpha = 0.05$.
(b) Construct a 95% two-sided CI on the death rate from lung cancer.
(c) How large a sample would be required to be at least 95% confident that the error in estimating the death rate from lung cancer is less than 0.03?

4-67. Large passenger vans are thought to have a high propensity of rollover accidents when fully loaded. Thirty accidents of these vans were examined, and 11 vans had rolled over.
(a) Test the claim that the proportion of rollovers exceeds 0.25 with $\alpha = 0.10$.

(b) Suppose that the true $p = 0.35$ and $\alpha = 0.10$. What is the β-error for this test?
(c) Suppose that the true $p = 0.35$ and $\alpha = 0.10$. How large a sample would be required if we want $\beta = 0.10$?
(d) Find a 90% lower confidence bound on the rollover rate of these vans.
(e) Use the confidence bound found in part (d) to test the hypothesis.
(f) How large a sample would be required to be at least 95% confident that the error on p is less than 0.02? Use an initial estimate of p from this problem.

4-68. A random sample of 50 suspension helmets used by motorcycle riders and automobile race-car drivers was subjected to an impact test, and on 18 of these helmets some damage was observed.
(a) Test the hypotheses $H_0: p = 0.3$ versus $H_1: p \neq 0.3$ with $\alpha = 0.05$.
(b) Find the P-value for this test.
(c) Find a 95% two-sided CI on the true proportion of helmets of this type that would show damage from this test. Explain how this confidence interval can be used to test the hypothesis in part (a).
(d) Using the point estimate of p obtained from the preliminary sample of 50 helmets, how many helmets must be tested to be 95% confident that the error in estimating the true value of p is less than 0.02?
(e) How large must the sample be if we wish to be at least 95% confident that the error in estimating p is less than 0.02, regardless of the true value of p?

4-69. The Arizona Department of Transportation wishes to survey state residents to determine what proportion of the population would be in favor of building a citywide light-rail system. How many residents do they need to survey if they want to be at least 99% confident that the sample proportion is within 0.05 of the true proportion?

4-70. A manufacturer of electronic calculators is interested in estimating the fraction of defective units produced. A random sample of 800 calculators contains 10 defectives.
(a) Formulate and test an appropriate hypothesis to determine if the fraction defective exceeds 0.01 at the 0.05 level of significance.

(b) Suppose that the true $p = 0.02$ and $\alpha = 0.05$. What is the β-error for this test?

(c) Suppose that the true $p = 0.02$ and $\alpha = 0.05$. How large a sample would be required if we want $\beta = 0.10$?

4-71.   A study is to be conducted of the percentage of home-owners who have a high-speed Internet connection. How large a sample is required if we wish to be 95% confident that the error in estimating this quantity is less than 0.02?

4-72.   The fraction of defective integrated circuits produced in a photolithography process is being studied. A random sample of 300 circuits is tested, revealing 18 defectives.

(a) Use the data to test the hypothesis that the proportion is not 0.04. Use $\alpha = 0.05$.

(b) Find the P-value for the test.

(c) Find a 95% two-sided CI on the proportion defective.

(d) Use the CI found in part (c) to test the hypothesis.

4-73.   Consider the defective circuit data and hypotheses in Exercise 4-72.

(a) Suppose that the fraction defective is actually $p = 0.05$. What is the β-error for this test?

(b) Suppose that the manufacturer is willing to accept a β-error of 0.10 if the true value of p is 0.05. With $\alpha = 0.05$, what sample size would be required?

4-74.   An article in *Fortune* (September 21, 1992) claimed that one-half of all engineers continue academic studies beyond the B.S. degree, ultimately receiving either an M.S. or a Ph.D. degree. Data from an article in *Engineering Horizons* (Spring 1990) indicated that 117 of 484 new engineering graduates were planning graduate study.

(a) Are the data from *Engineering Horizons* consistent with the claim reported by *Fortune*? Use $\alpha = 0.05$ in reaching your conclusions.

(b) Find the P-value for this test.

4-75.   A manufacturer of interocular lenses is qualifying a new grinding machine. She will qualify the machine if the percentage of polished lenses that contain surface defects does not exceed 4%. A random sample of 300 lenses contains 11 defective lenses.

(a) Formulate and test an appropriate set of hypotheses to determine whether the machine can be qualified. Use a fixed-level test with $\alpha = 0.05$.

(b) Find the P-value for this test.

(c) Suppose that the percentage of defective lenses is actually 2%. What is the β-error for this test?

(d) Suppose that a β-error of 0.05 is acceptable if the true percentage is 2%. With $\alpha = 0.05$, what is the required sample size?

4-76.   A researcher claims that at least 10% of all football helmets have manufacturing flaws that could potentially cause injury to the wearer. A sample of 200 helmets revealed that 24 helmets contained such defects.

(a) Does this finding support the researcher's claim? Use a fixed-level test with $\alpha = 0.01$.

(b) Find the P-value for this test.

4-77.   A random sample of 500 registered voters in Phoenix is asked whether they favor the use of oxygenated fuels year round to reduce air pollution. If more than 315 voters respond positively, we will conclude that at least 60% of the voters favor the use of these fuels.

(a) Find the probability of type I error if exactly 60% of the voters favor the use of these fuels.

(b) What is the type II error probability β if 75% of the voters favor this action?

4-78.   The warranty for batteries for mobile phones is set at 400 operating hours, with proper charging procedures. A study of 2000 batteries is carried out and three stop operating prior to 400 hours. Do these experimental results support the claim that less than 0.2% of the company's batteries will fail during the warranty period, with proper charging procedures? Use a hypothesis testing procedure with $\alpha = 0.01$.

4-8   OTHER INTERVAL ESTIMATES FOR A SINGLE SAMPLE

4-8.1   Prediction Interval

In some situations, we are interested in **predicting** a future observation of a random variable. We may also want to find a range of likely values for the variable associated with making the prediction. This is a different problem than estimating the mean of that random variable, so a CI on the mean isn't really appropriate. To illustrate, let's consider the golf clubs from Example 4-7. Suppose that you plan to purchase a new driver of the type that was tested in that example. What is a reasonable prediction of the coefficient of restitution for the driver that you purchase (which is *not* one of the clubs that was tested in the study), and what is a range of likely values for the coefficient of restitution? The sample average $\overline{X}$ of the clubs that were tested is a reasonable point prediction of the coefficient of restitution of the new golf club, and we will show how to obtain a $100(1 - \alpha)\%$ **prediction interval** (PI) on the new observation.

Suppose that $X_1, X_2, \ldots, X_n$ is a random sample from a normal population with unknown mean and variance. We wish to predict the value of a single future observation, say X_{n+1}. As noted above, the average of the original sample, $\overline{X}$, is a reasonable point prediction of X_{n+1}. The expected value of the prediction error is $E(X_{n+1} - \overline{X}) = \mu - \mu = 0$ and the variance of the prediction error is

$$V(X_{n+1} - \overline{X}) = \sigma^2 + \frac{\sigma^2}{n} = \sigma^2 \left(1 + \frac{1}{n}\right)$$

because the future observation X_{n+1} is independent of the current sample mean $\overline{X}$. The prediction error is normally distributed because the original observations are normally distributed. Therefore,

$$Z = \frac{X_{n+1} - \overline{X}}{\sigma\sqrt{1 + \frac{1}{n}}}$$

has a standard normal distribution. Replacing σ with the sample standard deviation S results in

$$T = \frac{X_{n+1} - \overline{X}}{S\sqrt{1 + \frac{1}{n}}}$$

which has the t-distribution with $n - 1$ degrees of freedom. Manipulating this T-ratio as we have done previously in developing CIs leads to a prediction interval on the future observation X_{n+1}.

Prediction Interval

> **A $100(1 - \alpha)\%$ PI on a single future observation from a normal distribution** is given by
>
> $$\overline{x} - t_{\alpha/2, n-1} s \sqrt{1 + \frac{1}{n}} \leq X_{n+1} \leq \overline{x} + t_{\alpha/2, n-1} s \sqrt{1 + \frac{1}{n}} \qquad (4\text{-}76)$$

The PI for X_{n+1} will always be longer than the CI for μ because there is more variability associated with the prediction error for X_{n+1} than with the error of estimation for μ. This is easy to see intuitively because the prediction error is the difference between two random variables $(X_{n+1} - \overline{X})$, and the estimation error used in constructing a CI is the difference between one random variable and a constant $(\overline{X} - \mu)$. As n gets larger, $(n \rightarrow \infty)$ the length of the CI reduces to zero, becoming the true value of the mean, μ, but the length of the PI approaches $2z_{\alpha/2}\sigma$. So as n increases, the uncertainty in estimating μ goes to zero, but there will always be uncertainty about the future observation X_{n+1} even when there is no need to estimate any of the distribution parameters.

Finally, recall that CIs and hypothesis tests on the mean are relatively insensitive to the normality assumption. PIs, on the other hand, do not share this nice feature and are rather sensitive to the normality assumption because they are associated with a single future value drawn at random from the normal distribution.

EXAMPLE 4-16
Golf Clubs

Reconsider the golf clubs that were tested in Example 4-7. The coefficient of restitution was measured for $n = 15$ randomly selected metal drivers, and we found that $\bar{x} = 0.83725$ and $s = 0.02456$. We plan to buy a new golf club of the type tested. What is a likely range of values for the coefficient of restitution for the new club?

Solution. The normal probability plot in Fig. 4-17 does not indicate any problems with the normality assumption. A reasonable point prediction of its coefficient of restitution is the sample mean, 0.83725. A 95% PI on the coefficient of restitution for the new driver is computed from equation 4-76 as follows:

$$\bar{x} - t_{\alpha/2,\,n-1}s\sqrt{1 + \frac{1}{n}} \le X_{n+1} \le \bar{x} + t_{\alpha/2,\,n-1}s\sqrt{1 + \frac{1}{n}}$$

$$0.83725 - 2.145(0.02456)\sqrt{1 + \frac{1}{15}} \le X_{16} \le 0.83725 + 2.145(0.02456)\sqrt{1 + \frac{1}{15}}$$

$$0.78284 \le X_{16} \le 0.89166$$

So we could logically expect that the new golf club will have a coefficient of restitution between 0.78284 and 0.89166. By way of comparison, the 95% two-sided CI on the mean coefficient of restitution is $0.82365 \le \mu \le 0.85085$. Notice that the prediction interval is considerable longer than the CI on the mean. ∎

4-8.2 Tolerance Intervals for a Normal Distribution

Although confidence and prediction intervals are very useful, there is a third type of interval that finds many applications. Consider the population of golf clubs from which the sample of size $n = 15$ used in Examples 4-7 and 4-16 was selected. Suppose that we knew with certainty that the mean coefficient of restitution for the drivers in this population was $\mu = 0.83$ and that the standard deviation was $\sigma = 0.025$. Then the interval from $0.83 - 1.96(0.025) = 0.781$ to $0.83 + 1.96(0.025) = 0.879$ captures the coefficient of restitution of 95% of the drivers in this population because the interval from -1.96 to $+1.96$ captures 95% of the area (probability) under the standard normal curve. Generally, the interval from $\mu - z_{\alpha/2}\sigma$ to $\mu + z_{\alpha/2}\sigma$ is called a **100(1 − α)% tolerance interval.**

If the normal distribution parameters μ and σ are unknown, we can use the data from a random sample of size n to compute $\bar{x}$ and s and then form the interval $(\bar{x} - 1.96s, \bar{x} + 1.96s)$. However, because of sampling variability in $\bar{x}$ and s, it is likely that this interval will contain less than 95% of the values in the population. The solution is to replace 1.96 with some value that will make the proportion of the population contained in the interval 95% with some level of confidence. Fortunately, it is easy to do this.

Tolerance Interval

> A **tolerance interval** to contain at least γ% of the values in a normal population with confidence level 100(1 − α)% is
>
> $$\bar{x} - ks, \bar{x} + ks$$
>
> where k is a tolerance interval factor for the normal distribution found in Appendix A Table VI. Values of k are given for $1 - \alpha = 0.90, 0.95, 0.99$ confidence level and for $\gamma = 90, 95,$ and 99%.

One-sided tolerance bounds can also be computed. The tolerance factors for these bounds are also given in Appendix A Table VI.

EXAMPLE 4-17
Golf Clubs

Reconsider the golf clubs from Example 4-7. Recall that the sample mean and standard deviation of the coefficient of restitution for the $n = 15$ clubs tested are $\bar{x} = 0.83725$ and $s = 0.02456$. We want to find a tolerance interval for the coefficient of restitution that includes 95% of the clubs in the population with 90% confidence.

Solution. From Appendix A Table VI the tolerance factor is $k = 2.713$. The desired tolerance interval is

$$(\bar{x} - ks, \bar{x} + ks) \quad \text{or} \quad [0.83725 - (2.713)0.02456, 0.83725 + (2.713)0.02456]$$

which reduces to $(0.77062, 0.90388)$. Therefore, we can be 90% confident that at least 95% of the golf clubs in this population have a coefficient of restitution between 0.77062 and 0.90388. ∎

From Appendix A Table VI we note that as the sample size $n \to \infty$, the value of the normal distribution tolerance interval factor k goes to the z-value associated with the desired level of containment for the normal distribution. For example, if we want 95% of the population to fall inside the two-sided tolerance interval, k approaches $z_{0.05} = 1.96$ as $n \to \infty$. Note that as $n \to \infty$, a $100(1 - \alpha)\%$ prediction interval on a future observation approaches a tolerance interval that contains $100(1 - \alpha)\%$ of the distribution.

EXERCISES FOR SECTION 4-8

4-79. Consider the tire life problem described in Exercise 4-47.
(a) Construct a 95% PI on the life of a single tire.
(b) Find a tolerance interval for the tire life that includes 90% of the tires in the population with 95% confidence.

4-80. Consider the Izod impact strength problem described in Exercise 4-48.
(a) Construct a 90% PI for the impact strength of a single specimen of PVC pipe.
(b) Find a tolerance interval for the impact strength that includes 95% of the specimens in the population with 95% confidence.

4-81. Consider the life of biomedical devices described in Exercise 4-49.
(a) Construct a 99% PI for the life of a single device.
(b) Find a tolerance interval for the device life that includes 99% of the devices in the population with 90% confidence.

4-82. Consider the fatty acid content of margarine described in Exercise 4-50.
(a) Construct a 95% PI for the fatty acid content of a single package of margarine.
(b) Find a tolerance interval for the fatty acid content that includes 95% of the margarine packages with 99% confidence.

4-83. Consider the breakdown voltage of diodes described in Exercise 4-51.
(a) Construct a 99% PI for the breakdown voltage of a single diode.
(b) Find a tolerance interval for the breakdown voltage that includes 99% of the diodes with 99% confidence.

4-84. Consider the metal rods described in Exercise 4-52.
(a) Construct a 90% PI for the diameter of a single rod.
(b) Find a tolerance interval for the diameter that includes 90% of the rods with 90% confidence.

4-9 SUMMARY TABLES OF INFERENCE PROCEDURES FOR A SINGLE SAMPLE

The tables on the inside front cover present a summary of all the single-sample hypothesis testing and CI procedures from this chapter. The tables contain the null hypothesis statement, the test statistic, the various alternative hypotheses and the criteria for rejecting H_0, and the formulas for constructing the $100(1 - \alpha)\%$ confidence intervals.

4-10   TESTING FOR GOODNESS OF FIT

The hypothesis testing procedures that we have discussed in previous sections are designed for problems in which the population or probability distribution is known and the hypotheses involve the parameters of the distribution. Another kind of hypothesis is often encountered: We do not know the underlying distribution of the population, and we wish to test the hypothesis that a particular distribution will be satisfactory as a population model. For example, we might wish to test the hypothesis that the population is normal.

In Chapter 3, we discussed a very useful graphical technique for this problem called **probability plotting** and illustrated how it was applied in the case of normal, lognormal, and Weibull distributions. In this section, we describe a formal goodness-of-fit test procedure based on the chi-square distribution.

The test procedure requires a random sample of size n from the population whose probability distribution is unknown. These n observations are arranged in a histogram, having k bins or class intervals. Let O_i be the observed frequency in the ith class interval. From the hypothesized probability distribution, we compute the expected frequency in the ith class interval, denoted E_i. The test statistic is

$$X_0^2 = \sum_{i=1}^{k} \frac{(O_i - E_i)^2}{E_i} \qquad (4\text{-}77)$$

It can be shown that if the population follows the hypothesized distribution, X_0^2 has approximately a chi-square distribution with $k - p - 1$ degrees of freedom, where p represents the number of parameters of the hypothesized distribution estimated by sample statistics. This approximation improves as n increases. We would reject the hypothesis that the distribution of the population is the hypothesized distribution if the calculated value of the test statistic X_0^2 is too large. Therefore, the P-value would be the area (probability) under the chi-square distribution with $k - p - 1$ degrees of freedom above the calculated value of the test statistic X_0^2. That is, $P = P(\chi_{k-p-1}^2 > \chi_0^2)$. For a fixed-level test, we would reject the null hypothesis at the α level of significance if $\chi_0^2 > \chi_{\alpha, k-p-1}^2$.

One point to be noted in the application of this test procedure concerns the magnitude of the expected frequencies. If these expected frequencies are too small, the test statistic X_0^2 will not reflect the departure of observed from expected, but only the small magnitude of the expected frequencies. There is no general agreement regarding the minimum value of expected frequencies, but values of 3, 4, and 5 are widely used as minimal. Some writers suggest that an expected frequency could be as small as 1 or 2, so long as most of them exceed 5. Should an expected frequency be too small, it can be combined with the expected frequency in an adjacent class interval. The corresponding observed frequencies would then also be combined, and k would be reduced by 1. Class intervals are not required to be of equal width.

We now give an example of the test procedure.

EXAMPLE 4-18
Printed Circuit Boards

A Poisson Distribution

The number of defects in printed circuit boards is hypothesized to follow a Poisson distribution. A random sample of $n = 60$ printed boards has been collected and the number of defects per printed circuit board observed. The following data result:

Number of Defects	Observed Frequency
0	32
1	15
2	9
3	4

Is it reasonable to conclude that the number of defects is Poisson distributed?

Solution. The mean of the assumed Poisson distribution in this example is unknown and must be estimated from the sample data. The estimate of the mean number of defects per board is the sample average—that is, $(32 \cdot 0 + 15 \cdot 1 + 9 \cdot 2 + 4 \cdot 3)/60 = 0.75$. From the Poisson distribution with parameter 0.75, we may compute p_i, the theoretical, hypothesized probability associated with the ith class interval. Because each class interval corresponds to a particular number of defects, we may find the p_i as follows:

$$p_1 = P(X = 0) = \frac{e^{-0.75}(0.75)^0}{0!} = 0.472$$

$$p_2 = P(X = 1) = \frac{e^{-0.75}(0.75)^1}{1!} = 0.354$$

$$p_3 = P(X = 2) = \frac{e^{-0.75}(0.75)^2}{2!} = 0.133$$

$$p_4 = P(X \geq 3) = 1 - (p_1 + p_2 + p_3) = 0.041$$

The expected frequencies are computed by multiplying the sample size $n = 60$ by the probabilities p_i that is, $E_i = np_i$. The expected frequencies are shown next.

Number of Defects	Probability	Expected Frequency
0	0.472	28.32
1	0.354	21.24
2	0.133	7.98
3 (or more)	0.041	2.46

Because the expected frequency in the last cell is less than 3, we combine the last two cells:

Number of Defects	Observed Frequency	Expected Frequency
0	32	28.32
1	15	21.24
2 (or more)	13	10.44

The chi-square test statistic in equation 4-77 will have $k - p - 1 = 3 - 1 - 1 = 1$ degree of freedom because the mean of the Poisson distribution was estimated from the data.

The seven-step hypothesis testing procedure may now be applied, using $\alpha = 0.05$, as follows:

1. **Parameter of interest:** The parameter of interest is the form of the distribution of defects in printed circuit boards.

2. **Null hypothesis, H_0:** The form of the distribution of defects is Poisson.

3. **Alternative hypothesis, H_1:** The form of the distribution of defects is not Poisson.

4. **Test statistic:** The test statistic is

$$\chi_0^2 = \sum_{i=1}^{k} \frac{(O_i - E_i)^2}{E_i}$$

5. **Reject H_0 if:** Reject H_0 if the P-value is less than 0.05.

6. **Computations:**

$$\chi_0^2 = \frac{(32 - 28.32)^2}{28.32} + \frac{(15 - 21.24)^2}{21.24} + \frac{(13 - 10.44)^2}{10.44} = 2.94$$

7. **Conclusions:** From Appendix A Table III we find that $\chi_{0.10, 1}^2 = 2.71$ and $\chi_{0.05, 1}^2 = 3.84$. Because $\chi_0^2 = 2.94$ lies between these two values, we conclude that the P-value is $0.05 < P < 0.10$. Therefore, since the P-value is greater than 0.05, we are unable to reject the null hypothesis that the distribution of defects in printed circuit boards is Poisson. The exact P-value for the test is $P = 0.0864$. (This value was computed using Minitab.) ∎

EXERCISES FOR SECTION 4-10

4-85. Consider the following frequency table of observations on the random variable X.

Values	0	1	2	3	4	5
Observed Frequency	8	25	23	21	16	7

(a) Based on these 100 observations, is a Poisson distribution with a mean of 2.4 an appropriate model? Perform a goodness-of-fit procedure with $\alpha = 0.05$.
(b) Calculate the P-value for this test.

4-86. Let X denote the number of flaws observed on a large coil of galvanized steel. Seventy-five coils are inspected, and the following data were observed for the values of X.

Values	1	2	3	4	5	6	7	8
Observed Frequency	1	11	8	13	11	12	10	9

(a) Does the assumption of a Poisson distribution with a mean of 6.0 seem appropriate as a probability model for these data? Use $\alpha = 0.01$.
(b) Calculate the P-value for this test.

4-87. The number of calls arriving to a switchboard from noon to 1 P.M. during the business days Monday through Friday is monitored for 4 weeks (i.e., 30 days). Let X be defined as the number of calls during that 1-hour period. The observed frequency of calls was recorded and reported as follows:

Value	5	7	8	9	10
Observed Frequency	4	4	4	5	1
Value	11	12	13	14	15
Observed Frequency	3	3	1	4	1

(a) Does the assumption of a Poisson distribution seem appropriate as a probability model for these data? Use $\alpha = 0.05$.
(b) Calculate the P-value for this test.

4-88. The number of cars passing eastbound through the intersection of Mill Avenue and University Avenue has been tabulated by a group of civil engineering students. They have obtained the following data:

Vehicles per Minute	Observed Frequency	Vehicles per Minute	Observed Frequency
40	14	53	102
41	24	54	96
42	57	55	90
43	111	56	81
44	194	57	73
45	256	58	64
46	296	59	61
47	378	60	59
48	250	61	50
49	185	62	42
50	171	63	29
51	150	64	18
52	110	65	15

(a) Does the assumption of a Poisson distribution seem appropriate as a probability model for this process? Use $\alpha = 0.05$.
(b) Calculate the P-value for this test.

4-89.   Consider the following frequency table of observations on the random variable X.

Values	0	1	2	3	4
Observed Frequency	4	21	10	13	2

(a) Based on these 50 observations, is a binomial distribution with $n = 6$ and $p = 0.25$ an appropriate model? Perform a goodness-of-fit procedure with $\alpha = 0.05$.

(b) Calculate the P-value for this test.

4-90.   Define X as the number of underfilled bottles in a filling operation in a carton of 12 bottles. Eighty cartons are inspected, and the following observations on X are recorded.

Values	0	1	2	3	4
Observed Frequency	21	30	22	6	1

(a) Based on these 80 observations, is a binomial distribution an appropriate model? Perform a goodness-of-fit procedure with $\alpha = 0.10$.

(b) Calculate the P-value for this test.

SUPPLEMENTAL EXERCISES

4-91.   If we plot the probability of accepting H_0: $\mu = \mu_0$ versus various values of μ and connect the points with a smooth curve, we obtain the **operating characteristic curve** (or the **OC curve**) of the test procedure. These curves are used extensively in industrial applications of hypothesis testing to display the sensitivity and relative performance of the test. When the true mean is really equal to μ_0, the probability of accepting H_0 is $1 - \alpha$. Construct an OC curve for Exercise 4-19, using values of the true mean μ of 178, 181, 184, 187, 190, 193, 196, and 199.

4-92.   Convert the OC curve in the previous problem into a plot of the **power function** of the test.

4-93.   Consider the confidence interval for μ with known standard deviation σ:

$$\bar{x} - z_{\alpha_1}\sigma/\sqrt{n} \leq \mu \leq \bar{x} + z_{\alpha_2}\sigma/\sqrt{n}$$

where $\alpha_1 + \alpha_2 = \alpha$. Let $\alpha = 0.05$ and find the interval for $\alpha_1 = \alpha_2 = \alpha/2 = 0.025$. Now find the interval for the case $\alpha_1 = 0.01$ and $\alpha_2 = 0.04$. Which interval is shorter? Is there any advantage to a "symmetric" CI?

4-94.   Formulate the appropriate null and alternative hypotheses to test the following claims.

(a) A plastics production engineer claims that 99.95% of the plastic tube manufactured by her company meets the engineering specifications requiring the length to exceed 6.5 inches.

(b) A chemical and process engineering team claims that the mean temperature of a resin bath is greater than 45°C.

(c) The proportion of start-up software companies that successfully market their product within 3 years of company formation is less than 0.05.

(d) A chocolate bar manufacturer claims that, at the time of purchase by a consumer, the mean life of its product is less than 90 days.

(e) The designer of a computer laboratory at a major university claims that the standard deviation of time of a student on the network is less than 10 minutes.

(f) A manufacturer of traffic signals advertises that its signals will have a mean operating life in excess of 2160 hours.

4-95.   A normal population has known mean $\mu = 50$ and variance $\sigma^2 = 5$. What is the approximate probability that the sample variance is greater than or equal to 7.44? less than or equal to 2.56?

(a) For a random sample of $n = 16$.

(b) For a random sample of $n = 30$.

(c) For a random sample of $n = 71$.

(d) Compare your answers to parts (a)–(c) for the approximate probability that the sample variance is greater than or equal to 7.44. Explain why this tail probability is increasing or decreasing with increased sample size.

(e) Compare your answers to parts (a)–(c) for the approximate probability that the sample variance is less than or equal to 2.56. Explain why this tail probability is increasing or decreasing with increased sample size.

4-96.   An article in the *Journal of Sports Science* (Vol. 5, 1987, pp. 261–271) presents the results of an investigation of the hemoglobin level of Canadian Olympic ice hockey players. The data reported are as follows (in g/dl):

15.3	16.0	14.4	16.2	16.2
14.9	15.7	15.3	14.6	15.7
16.0	15.0	15.7	16.2	14.7
14.8	14.6	15.6	14.5	15.2

(a) Given the probability plot of the data in Fig. 4-24, what is a logical assumption about the underlying distribution of the data?

(b) Explain why this check of the distribution underlying the sample data is important if we want to construct a CI on the mean.

(c) Based on these sample data, a 95% CI for the mean is [15.04, 15.62]. Is it reasonable to infer that the true mean could be 14.5? Explain your answer.

(d) Explain why this check of the distribution underlying the sample data is important if we want to construct a CI on the variance.

(e) Based on these sample data, a 95% CI for the variance is [0.22, 0.82]. Is it reasonable to infer that the true variance could be 0.35? Explain your answer.

(f) Is it reasonable to use these CIs to draw an inference about the mean and variance of hemoglobin levels
   (i) Of Canadian doctors? Explain your answer.
   (ii) Of Canadian children ages 6–12? Explain your answer.
(g) Construct a 95% PI on the hemoglobin level of a single Canadian hockey player.
(h) Find a tolerance interval for the hemoglobin level that includes 90% of the players in the population with 95% confidence.

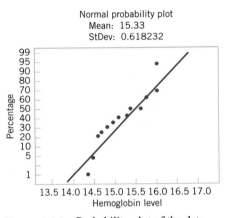

Normal probability plot
Mean: 15.33
StDev: 0.618232

Figure 4-24   Probability plot of the data for Exercise 4-96.

4-97.   The article "Mix Design for Optimal Strength Development of Fly Ash Concrete" (*Cement and Concrete Research*, Vol. 19, No. 4, 1989, pp. 634–640) investigates the compressive strength of concrete when mixed with fly ash (a mixture of silica, alumina, iron, magnesium oxide, and other ingredients). The compressive strength for nine samples in dry conditions on the twenty-eighth day are as follows (in Mpa):

40.2	30.4	28.9	30.5	22.4
25.8	18.4	14.2	15.3	

(a) Given the probability plot of the data in Fig. 4-25, what is a logical assumption about the underlying distribution of the data?
(b) Find a 99% one-sided lower confidence bound on mean compressive strength. Provide a practical interpretation of this bound.
(c) Find a 98% two-sided CI on mean compressive strength. Provide a practical interpretation of this interval and explain why the lower end point of the interval is or is not the same as in part (b).
(d) Find a 99% one-sided upper confidence bound on the variance of compressive strength. Provide a practical interpretation of this bound.
(e) Find a 98% two-sided CI on the variance of compression strength. Provide a practical interpretation of this interval and explain why the upper end point of the interval is or is not the same as in part (d).

(f) Suppose it was discovered that the largest observation 40.2 was misrecorded and should actually be 20.4. Now the sample mean $\bar{x} = 22.9$ and the sample variance $s^2 = 39.83$. Use these new values and repeat parts (c) and (e). Compare the original computed intervals and the newly computed intervals with the corrected observation value. How does this mistake affect the values of the sample mean, the sample variance, and the width of the two-sided CIs?
(g) Suppose, instead, it was discovered that the largest observation 40.2 is correct, but that the observation 25.8 is incorrect and should actually be 24.8. Now the sample mean $\bar{x} = 25.0$ and the sample variance $s^2 = 70.84$. Use these new values and repeat parts (c) and (e). Compare the original computed intervals and the newly computed intervals with the corrected observation value. How does this mistake affect the values of the sample mean, the sample variance, and the width of the two-sided CIs?
(h) Use the results from parts (f) and (g) to explain the effect of mistakenly recorded values on sample estimates. Comment on the effect when the mistaken values are near the sample mean and when they are not.
(i) Using the original data, construct a 99% PI on the compressive strength of a single sample in dry conditions.
(j) Find a tolerance interval for the compressive strength that includes 95% of the concrete in the population with 99% confidence.

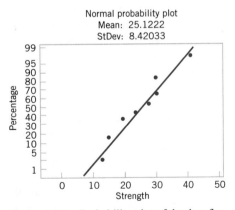

Normal probability plot
Mean: 25.1222
StDev: 8.42033

Figure 4-25   Probability plot of the data for Exercise 4-97.

4-98.   An operating system for a personal computer has been studied extensively, and it is known that the standard deviation of the response time following a particular command is $\sigma = 8$ milliseconds. A new version of the operating system is installed, and we wish to estimate the mean response time for the new system to ensure that a 95% CI for μ has length at most 5 milliseconds.

(a) If we can assume that response time is normally distributed and that $\sigma = 8$ for the new system, what sample size would you recommend?

(b) Suppose we were told by the vendor that the standard deviation of the response time of the new system is smaller—say, $\sigma = 6$; give the sample size that you recommend and comment on the effect the smaller standard deviation has on this calculation.

(c) Suppose you cannot assume that the response time of the new system is normally distributed but think that it may follow a Weibull distribution. What is the minimum sample size you would recommend to construct any CI on the true mean response time?

4-99. A manufacturer of semiconductor devices takes a random sample of size n of chips and tests them, classifying each chip as defective or nondefective. Let $X_i = 0$ if the chip is nondefective and $X_i = 1$ if the chip is defective. The sample fraction defective is

$$\hat{p}_i = \frac{X_1 + X_2 + \cdots + X_n}{n}$$

What are the sampling distribution, the sample mean, and sample variance estimates of $\hat{p}$ when
(a) The sample size is $n = 60$?
(b) The sample size is $n = 70$?
(c) The sample size is $n = 100$?
(d) Compare your answers to parts (a)–(c) and comment on the effect of sample size on the variance of the sampling distribution.

4-100. Consider the description of Exercise 4-99. After collecting a sample, we are interested in computing the error in estimating the true value p. For each of the sample sizes and estimates of p, compute the error at the 95% confidence level.

(a) $n = 60$ and $\hat{p} = 0.10$
(b) $n = 70$ and $\hat{p} = 0.10$
(c) $n = 100$ and $\hat{p} = 0.10$
(d) Compare your results from parts (a)–(c) and comment on the effect of sample size on the error in estimating the true value of p and the 95% confidence level.
(e) Repeat parts (a)–(d), this time using a 99% confidence level.
(f) Examine your results when the 95% confidence level and then the 99% confidence level are used to compute the error and explain what happens to the magnitude of the error as the percentage confidence increases.

4-101. A quality control inspector of flow metering devices used to administer fluid intravenously will perform a hypothesis test to determine whether the mean flow rate is different from the flow rate setting of 200 ml/h. Based on prior information the standard deviation of the flow rate is assumed to be known and equal to 12 ml/h. For each of the following sample sizes and a fixed $\alpha = 0.05$, find the probability of a type II error if the true mean is 205 ml/h.

(a) $n = 25$ (b) $n = 60$ (c) $n = 100$
(d) Does the probability of a type II error increase or decrease as the sample size increases? Explain your answer.

4-102. Suppose that in Exercise 4-101 the experimenter had believed that $\sigma = 14$. For each of the following sample sizes and a fixed $\alpha = 0.05$, find the probability of a type II error if the true mean is 205 ml/h.

(a) $n = 20$ (b) $n = 50$ (c) $n = 100$
(d) Comparing your answers to those in Exercise 4-101, does the probability of a type II error increase or decrease with the increase in standard deviation? Explain your answer.

4-103. The life in hours of a heating element used in a furnace is known to be approximately normally distributed. A random sample of 15 heating elements is selected and found to have an average life of 598.14 hours and a sample standard deviation of 16.93 hours.

(a) At the $\alpha = 0.05$ level of significance, test the hypotheses H_0: $\mu = 550$ versus H_1: $\mu > 550$. On completing the hypothesis test, do you believe that the true mean life of a heating element is greater than 550 hours? Clearly state your answer.
(b) Find the P-value of the test statistic.
(c) Construct a 95% lower confidence bound on the mean and describe how this interval can be used to test the alternative hypothesis of part (a).
(d) Construct a two-sided 95% CI for the underlying variance.

4-104. Suppose we wish to test the hypothesis H_0: $\mu = 85$ versus the alternative H_1: $\mu > 85$ where $\sigma = 16$. Suppose that the true mean is $\mu = 86$ and that in the practical context of the problem this is not a departure from $\mu_0 = 85$ that has practical significance.

(a) For a test with $\alpha = 0.01$, compute β for the sample sizes $n = 25, 100, 400,$ and 2500 assuming that $\mu = 86$.
(b) Suppose the sample average is $\bar{x} = 86$. Find the P-value for the test statistic for the different sample sizes specified in part (a). Would the data be statistically significant at $\alpha = 0.01$?
(c) Comment on the use of a large sample size in this exercise.

4-105. The cooling system in a nuclear submarine consists of an assembly of welded pipes through which a coolant is circulated. Specifications require that weld strength must meet or exceed 150 psi.

(a) Suppose that the design engineers decide to test the hypothesis H_0: $\mu = 150$ versus H_1: $\mu > 150$. Explain why this choice of alternative hypothesis is better than H_1: $\mu < 150$.
(b) A random sample of 20 welds results in $\bar{x} = 157.65$ psi and $s = 12.39$ psi. What conclusions can you draw about the hypothesis in part (a)? State any necessary assumptions about the underlying distribution of the data. Use the P-value approach.

4-106. Suppose we are testing H_0: $p = 0.5$ versus H_1: $p \neq 0.5$.

(a) Using $\alpha = 0.05$, find the power of the test for $n = 100$, 150, and 300, assuming the true value $p = 0.6$. Comment on the effect of sample size on the power of the test.

(b) Using $\alpha = 0.01$, find the power of the test for $n = 100$, 150, and 300, assuming the true value $p = 0.6$. Compare your answers to those from part (a) and comment on the effect of α on the power of the test for different sample sizes.

(c) Using $\alpha = 0.05$, find the power of the test for $n = 100$, assuming $p = 0.08$. Compare your answer to part (a) and comment on the effect of the true value of p on the power of the test for the same sample size and α level.

(d) Using $\alpha = 0.01$, what sample size is required if $p = 0.6$ and we want $\beta = 0.05$? What sample is required if $p = 0.8$ and we want $\beta = 0.05$? Compare the two sample sizes and comment on the effect of the true value of p on sample size required when β is held approximately constant.

4-107. Consider the biomedical device experiment described in Exercise 4-49.

(a) For this sample size $n = 15$, do the data support the claim that the standard deviation of life is less than 280 hours?

(b) Suppose that instead of $n = 15$, the sample size was 51. Repeat the analysis performed in part (a) using $n = 51$.

(c) Compare your answers and comment on how sample size affects your conclusions drawn in parts (a) and (b).

4-108. An article in *Food Testing and Analysis* ("Improving Reproducibility of Refractometry Measurements of Fruit Juices," Vol. 4, No. 4, 1999, pp. 13–17) reported the results of a study that measured the sugar concentration (Brix) in clear apple juice. All readings were taken at 20°C:

11.48	11.45	11.48	11.47	11.48
11.50	11.42	11.49	11.45	11.44
11.45	11.47	11.46	11.47	11.43
11.50	11.49	11.45	11.46	11.47

(a) Test the hypothesis H_0: $\mu = 11.5$ versus H_1: $\mu \neq 11.5$ using $\alpha = 0.05$. Find the P-value.

(b) Compute the power of the test if the true mean is 11.4.

(c) What sample size would be required to detect a true mean sugar concentration of 11.45 if we wanted the power of the test to be at least 0.9?

(d) Explain how the question in part (a) could be answered by constructing a two-sided confidence interval on the mean sugar concentration.

(e) Is there evidence to support the assumption that the sugar concentration is normally distributed?

4-109. An article in *Growth: A Journal Devoted to Problems of Normal and Abnormal Growth* ("Comparison of Measured and Estimated Fat-Free Weight, Fat, Potassium and Nitrogen of Growing Guinea Pigs," Vol. 46, No. 4, 1982, pp. 306–321) reported the results of a study that measured the body weight (grams) for guinea pigs at birth.

421.0	452.6	456.1	494.6	373.8
90.5	110.7	96.4	81.7	102.4
241.0	296.0	317.0	290.9	256.5
447.8	687.6	705.7	879.0	88.8
296.0	273.0	268.0	227.5	279.3
258.5	296.0			

(a) Test the hypothesis that mean body weight is 300 grams. Use $\alpha = 0.05$.

(b) What is the smallest level of significance at which you would be willing to reject the null hypothesis?

(c) Explain how you could answer the question in part (a) with a two-sided confidence interval on mean body weight.

4-110. An article in *Biological Trace Element Research* ("Interaction of Dietary Calcium, Manganese, and Manganese Source (Mn Oxide or Mn Methionine Complex) on Chick Performance and Manganese Utilization," Vol. 29, No. 3, 1991, pp. 217–228) showed the following results of tissue assay for liver manganese (ppm) in chicks fed high-Ca diets.

6.02	6.08	7.11	5.73	5.32	7.10
5.29	5.84	6.03	5.99	4.53	6.81

(a) Test the hypothesis H_0: $\sigma^2 = 0.6$ versus H_1: $\sigma^2 \neq 0.6$ using $\alpha = 0.01$.

(b) What is the P-value for this test?

(c) Discuss how part (a) could be answered by constructing a 99% two-sided confidence interval for σ.

4-111. An article in *Medicine and Science in Sports and Exercise* ("Maximal Leg-Strength Training Improves Cycling Economy in Previously Untrained Men," Vol. 37, 2005, pp. 131–136) reported the results of a study of cycling performance before and after eight weeks of leg-strength training. The sample size was seven and the sample mean and sample standard deviation were 315 watts and 16 watts, respectively.

(a) Is there evidence that leg strength exceeds 300 watts at significance level 0.05? Find the P-value.

(b) Compute the power of the test if the true strength is 305 watts.

(c) What sample size would be required to detect a true mean of 305 watts if the power of the test should be at least 0.90?

(d) Explain how the question in part (a) could be answered with a confidence interval.

4-112. An article in the *British Medical Journal* ("Comparison of Treatment of Renal Calculi by Operative Surgery, Percutaneous Nephrolithotomy, and Extra-corporeal Shock Wave Lithotripsy," Vol. 292, 1986, pp. 879–882) found that percutaneous nephrolithotomy (PN) had a success rate in removing kidney stones of 289 out of 350 patients. The traditional method was 78% effective.

(a) Is there evidence that the success rate for PN is greater than the historical success rate? Find the P-value.

(b) Explain how the question in part (a) could be answered with a confidence interval.

4-113. The data below are the number of earthquakes per year of magnitude 7.0 and greater since 1900. (*Source:* U.S. Geological Survey, National Earthquake Information Center, Golden, CO).

1900	13	1927	20	1954	17	1981	14
1901	14	1928	22	1955	19	1982	10
1902	8	1929	19	1956	15	1983	15
1903	10	1930	13	1957	34	1984	8
1904	16	1931	26	1958	10	1985	15
1905	26	1932	13	1959	15	1986	6
1906	32	1933	14	1960	22	1987	11
1907	27	1934	22	1961	18	1988	8
1908	18	1935	24	1962	15	1989	7
1909	32	1936	21	1963	20	1990	18
1910	36	1937	22	1964	15	1991	16
1911	24	1938	26	1965	22	1992	13
1912	22	1939	21	1966	19	1993	12
1913	23	1940	23	1967	16	1994	13
1914	22	1941	24	1968	30	1995	20
1915	18	1942	27	1969	27	1996	15
1916	25	1943	41	1970	29	1997	16
1917	21	1944	31	1971	23	1998	12
1918	21	1945	27	1972	20	1999	18
1919	14	1946	35	1973	16	2000	15
1920	8	1947	26	1974	21	2001	16
1921	11	1948	28	1975	21	2002	13
1922	14	1949	36	1976	25	2003	15
1923	23	1950	39	1977	16	2004	15
1924	18	1951	21	1978	18		
1925	17	1952	17	1979	15		
1926	19	1953	22	1980	18		

(a) Use computer software to summarize this data into a frequency distribution. Test the hypothesis that the number of earthquakes of magnitude 7.0 or greater each year follows a Poisson distribution at $\alpha = 0.05$.

(b) Calculate the P-value for the test.

WILEY **4-114.** Consider the fatty acid measurements for the diet margarine described in Exercise 4-50.

(a) For this sample size $n = 6$, using a two-sided alternative hypothesis and $\alpha = 0.01$, test $H_0: \sigma^2 = 1.0$.

(b) Suppose that instead of $n = 6$, the sample size were $n = 51$. Using the estimate s^2 from the original sample, repeat the analysis performed in part (a) using $n = 51$.

(c) Compare your answers and comment on how sample size affects your conclusions drawn in parts (a) and (b).

4-115. A manufacturer of precision measuring instruments claims that the standard deviation in the use of the instruments is at most 0.00002 mm. An analyst, who is unaware of the claim, uses the instrument eight times and obtains a sample standard deviation of 0.00001 mm.

(a) Confirm using a test procedure and an α level of 0.01 that there is insufficient evidence to support the claim that the standard deviation of the instruments is at most 0.00002.

State any necessary assumptions about the underlying distribution of the data.

(b) Explain why the sample standard deviation, $s = 0.00001$, is less than 0.00002, yet the statistical test procedure results do not support the claim.

4-116. A biotechnology company produces a therapeutic drug whose concentration has a standard deviation of 4 g/l. A new method of producing this drug has been proposed, although some additional cost is involved. Management will authorize a change in production technique only if the standard deviation of the concentration in the new process is less than 4 g/l. The researchers chose $n = 10$ and obtained the following data. Perform the necessary analysis to determine whether a change in production technique should be implemented.

16.628 g/l	16.630 g/l
16.622	16.631
16.627	16.624
16.623	16.622
16.618	16.626

4-117. A manufacturer of electronic calculators claims that less than 1% of its production output is defective. A random sample of 1200 calculators contains 8 defective units.

(a) Confirm using a test procedure and an α level of 0.01 that there is insufficient evidence to support the claim that the percentage defective is less than 1%.

(b) Explain why the sample percentage is less than 1%, yet the statistical test procedure results do not support the claim.

4-118. An article in *The Engineer* ("Redesign for Suspect Wiring," June 1990) reported the results of an investigation into wiring errors on commercial transport aircraft that may produce faulty information to the flight crew. Such a wiring error may have been responsible for the crash of a British Midland Airways aircraft in January 1989 by causing the pilot to shut down the wrong engine. Of 1600 randomly selected aircraft, 8 were found to have wiring errors that could display incorrect information to the flight crew.

(a) Find a 99% two-sided CI on the proportion of aircraft that have such wiring errors.

(b) Suppose we use the information in this example to provide a preliminary estimate of p. How large a sample would be required to produce an estimate of p that we are 99% confident differs from the true value by at most 0.008?

(c) Suppose we did not have a preliminary estimate of p. How large a sample would be required if we wanted to be at least 99% confident that the sample proportion differs from the true proportion by at most 0.008 regardless of the true value of p?

(d) Comment on the usefulness of preliminary information in computing the needed sample size.

4-119. A standardized test for graduating high school seniors is designed to be completed by 75% of the students within

40 minutes. A random sample of 100 graduates showed that 64 completed the test within 40 minutes.

(a) Find a 90% two-sided CI on the proportion of such graduates completing the test within 40 minutes.

(b) Find a 95% two-sided CI on the proportion of such graduates completing the test within 40 minutes.

(c) Compare your answers to parts (a) and (b) and explain why they are the same or different.

(d) Could you use either of these CIs to determine whether the proportion is significantly different from 0.75? Explain your answer.
[*Hint:* Use the normal approximation to the binomial.]

4-120. The proportion of adults who live in Tempe, Arizona, who are college graduates is estimated to be $p = 0.4$. To test this hypothesis, a random sample of 15 Tempe adults is selected. If the number of college graduates is between 4 and 8, the hypothesis will be accepted; otherwise, we will conclude that $p \neq 0.4$.

(a) Find the type I error probability for this procedure, assuming that $p = 0.4$.

(b) Find the probability of committing a type II error if the true proportion is really $p = 0.2$.

4-121. The proportion of residents in Phoenix favoring the building of toll roads to complete the freeway system is believed to be $p = 0.3$. If a random sample of 20 residents shows that 2 or fewer favor this proposal, we will conclude that $p < 0.3$.

(a) Find the probability of type I error if the true proportion is $p = 0.3$.

(b) Find the probability of committing a type II error with this procedure if the true proportion is $p = 0.2$.

(c) What is the power of this procedure if the true proportion is $p = 0.2$?

4-122. Consider the 40 observations collected on the number of nonconforming coil springs in production batches of size 50 given in Exercise 2-57 of Chapter 2.

(a) Based on the description of the random variable and these 40 observations, is a binomial distribution an appropriate model? Perform a goodness-of-fit procedure with $\alpha = 0.05$.

(b) Calculate the P-value for this test.

4-123. Consider the 20 observations collected on the number of errors in a string of 1000 bits of a communication channel given in Exercise 2-58 of Chapter 2.

(a) Based on the description of the random variable and these 20 observations, is a binomial distribution an appropriate model? Perform a goodness-of-fit procedure with $\alpha = 0.05$.

(b) Calculate the P-value for this test.

4-124. State the null and the alternative hypotheses, and indicate the type of critical region (either two-, lower-, or upper-tailed) to test the following claims.

(a) A manufacturer of lightbulbs has a new type of lightbulb that is advertised to have a mean burning lifetime in excess of 5000 hours.

(b) A chemical engineering firm claims that its new material can be used to make automobile tires with a mean life of more than 60,000 miles.

(c) The standard deviation of breaking strength of fiber used in making drapery material does not exceed 2 psi.

(d) A safety engineer claims that more than 60% of all drivers wear safety belts for automobile trips of less than 2 miles.

(e) A biomedical device is claimed to have a mean time to failure greater than 42,000 hours.

(f) Producers of 1-inch diameter plastic pipe claim that the standard deviation of the inside diameter is less than 0.02 inch.

(g) Lightweight, handheld, laser range finders used by civil engineers are advertised to have a variance smaller than 0.05 square meters.

4-125. Consider the following Minitab output.

One-Sample T: X

Test of mu = 44.5 vs > 44.5

Variable	N	Mean	StDev	SE Mean	95% Lower Bound	T	P
X	16	45.8971	1.8273	?	45.0962	?	0.004

(a) Fill in the missing quantities.

(b) At what level of significance can the null hypothesis be rejected?

(c) If the hypotheses had been $H_0: \mu = 44$ versus $H_1: \mu > 44$, would the P-value be larger or smaller?

(d) If the hypotheses had been $H_0: \mu = 44.5$ versus $H_1: \mu \neq 44.5$, would you reject the null hypothesis at the 0.05 level?

4-126. Consider the following Minitab output.

Test and CI for One Proportion

Test of p = 0.2 vs p < 0.2

Sample	X	N	Sample p	95% Upper Bound	Z-Value	P-Value
1	146	850	0.171765	0.193044	−2.06	0.020

(a) Is this a one-sided or a two-sided test?

(b) Was this test conducted using the normal approximation to the binomial? Was that appropriate?

(c) Can the null hypothesis be rejected at the 0.05 level? At the 0.01 level?

(d) Can the null hypothesis $H_0: p = 0.2$ versus $H_0: p \neq 0.2$ be rejected at the 0.05 level?

(e) Construct an approximate 90% one-sided confidence bound for p.

TEAM EXERCISES

4-127. The thickness measurements, in millimeters, of a wall of plastic tubing used for intravenous bags were recorded as follows:

1.9976	2.0008	2.0021	1.9995
2.0004	1.9972	1.9974	1.9989
2.0017	2.0030	1.9979	2.0035
1.9997	2.0014	2.0017	2.0018

(a) Assess the assumption that the data are normally distributed.
(b) Test the hypothesis that the mean is different from 2.001 millimeters using $\alpha = 0.05$.
(c) Assume that a thickness measurement less than 1.9975 is outside the engineering specifications and considered nonconforming. Using this data set, test the hypothesis that the proportion nonconforming exceeds 0.10. Use $\alpha = 0.05$.

4-128. The following are recorded times, in hours, until failure of medical linear accelerators:

953	1037	1068	1032
988	1014	1063	1000
983	942	945	921
915	921	1090	974
997	993	997	984

(a) Assess the assumption that the data are normally distributed.
(b) Assess the claim that the mean time until failure is less than 1000 hours, at the 0.05 level of significance.
(c) If a medical linear accelerator fails in less than 925 hours, then it can be returned for warranty replacement. At the 0.05 level of significance, test the claim that the proportion returned is less than 0.20.

4-129. A detection device is used to monitor the level of CO in the air. The following data, in ppm, were collected from a single location:

7.28	6.98
8.50	6.33
5.56	7.34
3.18	5.56
4.03	4.69

(a) Assess the assumption that the data are normally distributed.
(b) There is concern that the device has significant variability in its recording. At a significance level of $\alpha = 0.05$, test the concern that the standard deviation exceeds 2.0.

4-130. Identify an example in which a standard is specified or claim is made about a population. For example, "This type of car gets an average of 30 miles per gallon in urban driving." The standard or claim may be expressed as a mean (average), variance, standard deviation, or proportion. Collect an appropriate random sample of data and perform a hypothesis test to assess the standard or claim. Report on your results. Be sure to include in your report the claim expressed as a hypothesis test, a description of the data collected, the analysis performed, and the conclusion reached.

4-131. Consider the experimental data collected to verify that the "true" speed of light is 710.5 (299,710.5 kilometers per second) in 1879 and in 1882 by the physicist A. A. Michelson. Read the story associated with the data and reported on the Web site http://lib.stat.cmu.edu/DASL/Stories/SpeedofLight.html. Use the data file to duplicate the analysis, and write a brief report summarizing your findings.

IMPORTANT TERMS AND CONCEPTS

Alternative hypothesis	Confidence coefficient	Fixed significance level hypothesis testing	One-sided alternative hypothesis
Bias in estimation	Confidence interval	Goodness of fit	One-sided confidence bounds
Chi-squared distribution	Confidence level	Hypothesis testing	Operating characteristic curves
Comparative experiment	Confidence limits	Minimum variance unbiased estimator	P-values
Confidence bound	Coverage	Null hypothesis	
	Critical region		
	Estimated standard error		

Parameter estimation
Point estimation
Power of a test
Practical significance versus statistical significance
Precision of estimation
Prediction interval

Probability of a type I error
Probability of a type II error
Procedure for hypothesis testing
Relative efficiency of an estimator

Sample size determination
Significance level
Standard error
Statistical hypothesis
Statistical inference
t-distribution

Test statistic
Tolerance interval
Two-sided alternative hypothesis
Type I error
Type II error

Decision Making
for Two Samples

CHAPTER OUTLINE

LEARNING OBJECTIVES

After careful study of this chapter, you should be able to do the following:

1. Structure comparative experiments involving two samples as hypothesis tests.
2. Perform hypothesis tests and construct confidence intervals on the difference in means of two normal distributions.
3. Perform hypothesis tests and construct confidence intervals on the ratio of the variances of two normal distributions.
4. Perform hypothesis tests and construct confidence intervals on the difference in two population proportions.
5. Compute power and type II error, and make sample size selection decisions for hypothesis tests and confidence intervals.
6. Understand how the analysis of variance can be used in an experiment to compare several means.
7. Assess the adequacy of an ANOVA model with residual plots.
8. Understand the blocking principle and how it is used to isolate the effect of nuisance factors in an experiment.
9. Design and conduct experiments using a randomized complete block design.

5-1 INTRODUCTION

The previous chapter presented hypothesis tests and confidence intervals for a single population parameter (the mean μ, the variance σ^2, or a proportion p). This chapter extends those results to the case of two independent populations.

The general situation is shown in Fig. 5-1. Population 1 has mean μ_1 and variance σ_1^2, and population 2 has mean μ_2 and variance σ_2^2. Inferences will be based on two random samples of sizes n_1 and n_2, respectively. That is, $X_{11}, X_{12}, \ldots, X_{1n_1}$ is a random sample of n_1 observations from population 1, and $X_{21}, X_{22}, \ldots, X_{2n_2}$ is a random sample of n_2 observations from population 2.

5-2 INFERENCE ON THE MEANS OF TWO POPULATIONS, VARIANCES KNOWN

In this section we consider statistical inferences on the difference in means $\mu_1 - \mu_2$ of the populations shown in Fig. 5-1, where the variances σ_1^2 and σ_2^2 are known. The assumptions for this section are summarized next.

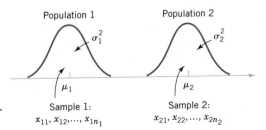

Figure 5-1   Two independent populations.

Population 1   Population 2

σ_1^2 σ_2^2

μ_1 μ_2

Sample 1:   Sample 2:
$x_{11}, x_{12}, \ldots, x_{1n_1}$ $x_{21}, x_{22}, \ldots, x_{2n_2}$

Assumptions

1. $X_{11}, X_{12}, \ldots, X_{1n_1}$ is a random sample of size n_1 from population 1.
2. $X_{21}, X_{22}, \ldots, X_{2n_2}$ is a random sample of size n_2 from population 2.
3. The two populations represented by X_1 and X_2 are independent.
4. Both populations are normal, or if they are not normal, the conditions of the central limit theorem apply.

A logical point estimator of $\mu_1 - \mu_2$ is the difference in sample means $\overline{X}_1 - \overline{X}_2$. Based on the properties of expected values in Chapter 3, we have

$$E(\overline{X}_1 - \overline{X}_2) = E(\overline{X}_1) - E(\overline{X}_2) = \mu_1 - \mu_2$$

and the variance of $\overline{X}_1 - \overline{X}_2$ is

$$V(\overline{X}_1 - \overline{X}_2) = V(\overline{X}_1) + V(\overline{X}_2) = \frac{\sigma_1^2}{n_1} + \frac{\sigma_2^2}{n_2}$$

Based on the assumptions and the preceding results, we may state the following.

Under the previous assumptions, the quantity

$$Z = \frac{\overline{X}_1 - \overline{X}_2 - (\mu_1 - \mu_2)}{\sqrt{\dfrac{\sigma_1^2}{n_1} + \dfrac{\sigma_2^2}{n_2}}} \qquad (5\text{-}1)$$

has a standard normal distribution, $N(0, 1)$.

This result will be used to form tests of hypotheses and CIs on $\mu_1 - \mu_2$. Essentially, we may think of $\mu_1 - \mu_2$ as a parameter θ, and its estimator is $\hat{\Theta} = \overline{X}_1 - \overline{X}_2$ with variance $\sigma_{\hat{\Theta}}^2 = \sigma_1^2/n_1 + \sigma_2^2/n_2$. If θ_0 is the null hypothesis value specified for θ, the test statistic will be $(\hat{\Theta} - \theta_0)/\sigma_{\hat{\Theta}}$. Note how similar this is to the test statistic for a single mean used in Chapter 4.

5-2.1   Hypothesis Testing on the Difference in Means, Variances Known

We now consider hypothesis testing on the difference in the means $\mu_1 - \mu_2$ of the two populations in Fig. 5-1. Suppose that we are interested in testing that the difference in means $\mu_1 - \mu_2$ is equal to a specified value Δ_0. Thus, the null hypothesis will be stated as H_0: $\mu_1 - \mu_2 = \Delta_0$. Obviously, in many cases, we will specify $\Delta_0 = 0$ so that we are testing the equality of two means (i.e., H_0: $\mu_1 = \mu_2$). The appropriate test statistic would be found by replacing $\mu_1 - \mu_2$ in equation 5-1 by Δ_0, and this test statistic would have a standard normal distribution under H_0. Suppose that the alternative hypothesis is H_1: $\mu_1 - \mu_2 \neq \Delta_0$. Now a sample value of $\overline{x}_1 - \overline{x}_2$ that is considerably different from Δ_0 is evidence that H_1 is true. Because Z_0 has the $N(0, 1)$ distribution when H_0 is true, we would calculate the P-value as the sum of the probabilities beyond the test statistic value z_0 and $-z_0$ in the standard normal distribution. That is, $P = 2[1 - \Phi(|z_0|)]$. This is exactly what we did in the one-sample z-test of Section 4-4.1. If we wanted to perform a fixed-significance-level test, we would take $-z_{\alpha/2}$ and $z_{\alpha/2}$ as the boundaries of the critical region just as we did in the single-sample z-test. This would give a test with level of significance α. P-values or critical regions for the one-sided alternatives would be determined similarly. Formally, we summarize these results in the following display.

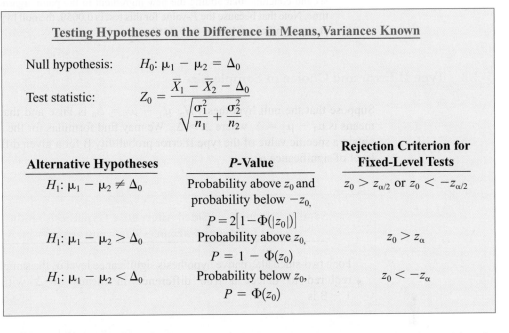

Testing Hypotheses on the Difference in Means, Variances Known

Null hypothesis: $H_0: \mu_1 - \mu_2 = \Delta_0$

Test statistic: $Z_0 = \dfrac{\overline{X}_1 - \overline{X}_2 - \Delta_0}{\sqrt{\dfrac{\sigma_1^2}{n_1} + \dfrac{\sigma_2^2}{n_2}}}$

Alternative Hypotheses	P-Value	Rejection Criterion for Fixed-Level Tests
$H_1: \mu_1 - \mu_2 \neq \Delta_0$	Probability above z_0 and probability below $-z_0$, $P = 2[1 - \Phi(|z_0|)]$	$z_0 > z_{\alpha/2}$ or $z_0 < -z_{\alpha/2}$
$H_1: \mu_1 - \mu_2 > \Delta_0$	Probability above z_0, $P = 1 - \Phi(z_0)$	$z_0 > z_\alpha$
$H_1: \mu_1 - \mu_2 < \Delta_0$	Probability below z_0, $P = \Phi(z_0)$	$z_0 < -z_\alpha$

EXAMPLE 5-1

Paint Drying Time

A product developer is interested in reducing the drying time of a primer paint. Two formulations of the paint are tested; formulation 1 is the standard chemistry, and formulation 2 has a new drying ingredient that should reduce the drying time. From experience, it is known that the standard deviation of drying time is 8 minutes, and this inherent variability should be unaffected by the addition of the new ingredient. Ten specimens are painted with formulation 1, and another 10 specimens are painted with formulation 2; the 20 specimens are painted in random order. The two sample average drying times are $\bar{x}_1 = 121$ minutes and $\bar{x}_2 = 112$ minutes, respectively. What conclusions can the product developer draw about the effectiveness of the new ingredient?

Solution. We apply the seven-step procedure to this problem as follows:

1. **Parameter of interest:** The parameter of interest is the difference in mean drying times $\mu_1 - \mu_2$ and $\Delta_0 = 0$.

2. **Null hypothesis, H_0:** $\mu_1 - \mu_2 = 0$, or H_0: $\mu_1 = \mu_2$.

3. **Alternative hypothesis, H_1:** $\mu_1 > \mu_2$. We want to reject H_0 if the new ingredient reduces mean drying time.

4. **Test statistic:** The test statistic is

$$z_0 = \frac{\bar{x}_1 - \bar{x}_2 - 0}{\sqrt{\dfrac{\sigma_1^2}{n_1} + \dfrac{\sigma_2^2}{n_2}}}$$

where $\sigma_1^2 = \sigma_2^2 = (8)^2 = 64$ and $n_1 = n_2 = 10$.

5. **Reject H_0 if:** Reject H_0: $\mu_1 = \mu_2$ if the P-value is less than 0.05.

6. **Computations:** Because $\bar{x}_1 = 121$ minutes and $\bar{x}_2 = 112$ minutes, the test statistic is

$$z_0 = \frac{121 - 112}{\sqrt{\dfrac{(8)^2}{10} + \dfrac{(8)^2}{10}}} = 2.52$$

7. **Conclusions:** Because the P-value is $P = 1 - \Phi(2.52) = 0.0059$, we reject the null hypothesis and conclude that adding the new ingredient to the paint significantly reduces the drying time. Note that because the P-value for this test is 0.0059, the null hypothesis would be rejected at **any** significance level $\alpha \geq 0.0059$. ∎

5-2.2 Type II Error and Choice of Sample Size

Suppose that the null hypothesis H_0: $\mu_1 - \mu_2 = \Delta_0$ is false and that the true difference in means is $\mu_1 - \mu_2 = \Delta$, where $\Delta > \Delta_0$. We may find formulas for the sample size required to obtain a specific value of the type II error probability β for a given difference in means Δ and level of significance α.

Sample Size for Two-Sided Alternative Hypothesis on the Difference in Means, Variances Known, when $n_1 = n_2$

For a two-sided alternative hypothesis significance level α, the sample size $n_1 = n_2 = n$ required to detect a true difference in means of Δ with power at least $1 - \beta$ is

$$n \simeq \frac{(z_{\alpha/2} + z_\beta)^2(\sigma_1^2 + \sigma_2^2)}{(\Delta - \Delta_\sigma)^2} \tag{5-2}$$

If n is not an integer, round the sample size up to the next integer.

This approximation is valid when $\Phi\left(-z_{\alpha/2} - (\Delta - \Delta_0)\sqrt{n}/\sqrt{\sigma_1^2 + \sigma_2^2}\right)$ is small compared to β.

Sample Size for One-Sided Alternative Hypothesis on the Difference in Means, Variances Known, when $n_1 = n_2$

For a one-sided alternative hypothesis significance level α, the sample size $n_1 = n_2 = n$ required to detect a true difference in means of $\Delta(\neq \Delta_0)$ with power at least $1 - \beta$ is

$$n = \frac{(z_\alpha + z_\beta)^2(\sigma_1^2 + \sigma_2^2)}{(\Delta - \Delta_0)^2} \tag{5-3}$$

The derivation of equations 5-2 and 5-3 closely follows the single-sample case in Section 4-4.2. For example, to obtain equation 5-2, we first write the expression for the β-error for the two-sided alternative, which is

$$\beta = \Phi\left(z_{\alpha/2} - \frac{\Delta - \Delta_0}{\sqrt{\dfrac{\sigma_1^2}{n_1} + \dfrac{\sigma_2^2}{n_2}}} \right) - \Phi\left(-z_{\alpha/2} - \frac{\Delta - \Delta_0}{\sqrt{\dfrac{\sigma_1^2}{n_1} + \dfrac{\sigma_2^2}{n_2}}} \right)$$

where Δ is the true difference in means of interest and Δ_0 is specified in the null hypothesis. Then by following a procedure similar to that used to obtain equation 4-24, the expression for β can be obtained for the case where $n_1 = n_2 = n$.

EXAMPLE 5-2

Paint Drying Time

To illustrate the use of these sample size equations, consider the situation described in Example 5-1, and suppose that if the true difference in drying times is as much as 10 minutes, we want to detect this with probability at least 0.90. What sample size is appropriate?

Solution. Under the null hypothesis, $\Delta_0 = 0$. We have a one-sided alternative hypothesis with $\Delta = 10$, $\alpha = 0.05$ (so $z_\alpha = z_{0.05} = 1.645$), and because the power is 0.9, $\beta = 0.10$ (so $z_\beta = z_{0.10} = 1.28$). Therefore, we may find the required sample size from equation 5-3 as follows:

$$n = \frac{(z_\alpha + z_\beta)^2(\sigma_1^2 + \sigma_2^2)}{(\Delta - \Delta_0)^2} = \frac{(1.645 + 1.28)^2[(8)^2 + (8)^2]}{(10 - 0)^2}$$

$$\approx 11 \qquad\qquad\qquad\qquad \blacksquare$$

5-2.3 Confidence Interval on the Difference in Means, Variances Known

The $100(1 - \alpha)\%$ confidence interval on the difference in two means $\mu_1 - \mu_2$ when the variances are known can be found directly from results given previously in this section. Recall that $X_{11}, X_{12}, \ldots, X_{1n_1}$ is a random sample of n_1 observations from the first population and $X_{21}, X_{22}, \ldots, X_{2n_2}$ is a random sample of n_2 observations from the second population. The difference in sample means $\overline{X}_1 - \overline{X}_2$ is a point estimator of $\mu_1 - \mu_2$, and

$$Z = \frac{\overline{X}_1 - \overline{X}_2 - (\mu_1 - \mu_2)}{\sqrt{\dfrac{\sigma_1^2}{n_1} + \dfrac{\sigma_2^2}{n_2}}}$$

has a standard normal distribution if the two populations are normal or is approximately standard normal if the conditions of the central limit theorem apply, respectively. This implies that

$$P(-z_{\alpha/2} \le Z \le z_{\alpha/2}) = 1 - \alpha$$

or

$$P\left[-z_{\alpha/2} \le \frac{\overline{X}_1 - \overline{X}_2 - (\mu_1 - \mu_2)}{\sqrt{\dfrac{\sigma_1^2}{n_1} + \dfrac{\sigma_2^2}{n_2}}} \le z_{\alpha/2} \right] = 1 - \alpha$$

This can be rearranged as

$$P\left(\overline{X}_1 - \overline{X}_2 - z_{\alpha/2}\sqrt{\frac{\sigma_1^2}{n_1} + \frac{\sigma_2^2}{n_2}} \le \mu_1 - \mu_2 \le \overline{X}_1 - \overline{X}_2 + z_{\alpha/2}\sqrt{\frac{\sigma_1^2}{n_1} + \frac{\sigma_2^2}{n_2}} \right) = 1 - \alpha$$

Therefore, the $100(1-\alpha)$% CI for $\mu_1 - \mu_2$ is defined as follows.

Confidence Interval on the Difference in Means, Variances Known

If $\bar{x}_1$ and $\bar{x}_2$ are the means of independent random samples of sizes n_1 and n_2 from populations with known variances σ_1^2 and σ_2^2, respectively, a $100(1-\alpha)$% CI for $\mu_1 - \mu_2$ is

$$\bar{x}_1 - \bar{x}_2 - z_{\alpha/2}\sqrt{\frac{\sigma_1^2}{n_1} + \frac{\sigma_2^2}{n_2}} \leq \mu_1 - \mu_2 \leq \bar{x}_1 - \bar{x}_2 + z_{\alpha/2}\sqrt{\frac{\sigma_1^2}{n_1} + \frac{\sigma_2^2}{n_2}} \quad (5\text{-}4)$$

where $z_{\alpha/2}$ is the upper $100\,\alpha/2$ percentage point and $-z_{\alpha/2}$ is the lower $100\,\alpha/2$ percentage point of the standard normal distribution in Appendix A Table I.

The confidence level $1 - \alpha$ is exact when the populations are normal. For nonnormal populations, the confidence level is approximately valid for large sample sizes.

EXAMPLE 5-3
Aircraft Spars

Tensile strength tests were performed on two different grades of aluminum spars used in manufacturing the wing of a commercial transport aircraft. From past experience with the spar manufacturing process and the testing procedure, the standard deviations of tensile strengths are assumed to be known. The data obtained are shown in Table 5-1. Find a 90% CI on the difference in means.

Solution. If μ_1 and μ_2 denote the true mean tensile strengths for the two grades of spars, then we may find a 90% CI on the difference in mean strength $\mu_1 - \mu_2$ as follows:

$$l = \bar{x}_1 - \bar{x}_2 - z_{\alpha/2}\sqrt{\frac{\sigma_1^2}{n_1} + \frac{\sigma_2^2}{n_2}} = 87.6 - 74.5 - 1.645\sqrt{\frac{(1.0)^2}{10} + \frac{(1.5)^2}{12}}$$
$$= 13.1 - 0.88 = 12.22 \text{ kg/mm}^2$$

$$u = \bar{x}_1 - \bar{x}_2 + z_{\alpha/2}\sqrt{\frac{\sigma_1^2}{n_1} + \frac{\sigma_2^2}{n_2}} = 87.6 - 74.5 + 1.645\sqrt{\frac{(1.0)^2}{10} + \frac{(1.5)^2}{12}}$$
$$= 13.1 + 0.88 = 13.98 \text{ kg/mm}^2$$

Therefore, the 90% CI on the difference in mean tensile strength is

$$12.22 \text{ kg/mm}^2 \leq \mu_1 - \mu_2 \leq 13.98 \text{ kg/mm}^2$$

Note that the CI does not include zero, which implies that the mean strength of aluminum grade 1 (μ_1) exceeds the mean strength of aluminum grade 2 (μ_2). In fact, we can state that we are 90% confident that the mean tensile strength of aluminum grade 1 exceeds that of aluminum grade 2 by between 12.22 and 13.98 kg/mm^2. ■

Table 5-1   Tensile Strength Test Result for Aluminum Spars

Spar Grade	Sample Size	Sample Mean Tensile Strength (kg/mm^2)	Standard Deviation (kg/mm^2)
1	$n_1 = 10$	$\bar{x}_1 = 87.6$	$\sigma_1 = 1.0$
2	$n_2 = 12$	$\bar{x}_2 = 74.5$	$\sigma_2 = 1.5$

One-Sided Confidence Bounds

To find a $100(1 - \alpha)\%$ lower confidence bound on $\mu_1 - \mu_2$, with known σ^2, simply replace $-z_{\alpha/2}$ with $-z_\alpha$ in the lower bound of equation 5-4 and set the upper bound to ∞. Similarly, to find a $100(1 - \alpha)\%$ upper confidence bound on μ, with known σ^2, replace $z_{\alpha/2}$ with z_α in the upper bound and set the lower bound to $-\infty$.

Choice of Sample Size

If the standard deviations σ_1 and σ_2 are known (at least approximately) and the two sample sizes n_1 and n_2 are equal ($n_1 = n_2 = n$, say), we can determine the sample size required so that the error in estimating $\mu_1 - \mu_2$ by $\bar{x}_1 - \bar{x}_2$ will be less than E at $100(1 - \alpha)\%$ confidence. The required sample size from each population is as follows.

Sample Size for a Specified E on the Difference in Means, and Variances Known when $n_1 = n_2$

If $\bar{x}_1$ and $\bar{x}_2$ are used as estimates of μ_1 and μ_2, respectively, we can be $100(1 - \alpha)\%$ confident that the error $|(\bar{x}_1 - \bar{x}_2) - (\mu_1 - \mu_2)|$ will not exceed a specified amount E when the sample size $n_1 = n_2 = n$ is

$$n = \left(\frac{z_{\alpha/2}}{E}\right)^2 (\sigma_1^2 + \sigma_2^2) \qquad (5\text{-}5)$$

Remember to round up if n is not an integer. This will ensure that the level of confidence does not drop below $100(1 - \alpha)\%$.

EXERCISES FOR SECTION 5-2

5-1. Two machines are used for filling plastic bottles with a net volume of 16.0 ounces. The fill volume can be assumed normal, with standard deviation $\sigma_1 = 0.020$ and $\sigma_2 = 0.025$ ounces. A member of the quality engineering staff suspects that both machines fill to the same mean net volume, whether or not this volume is 16.0 ounces. A random sample of 10 bottles is taken from the output of each machine.

Machine 1		Machine 2	
16.03	16.01	16.02	16.03
16.04	15.96	15.97	16.04
16.05	15.98	15.96	16.02
16.05	16.02	16.01	16.01
16.02	15.99	15.99	16.00

(a) Do you think the engineer is correct? Use the P-value approach.

(b) If $\alpha = 0.05$, what is the power of the test in part (a) for a true difference in means of 0.04?

(c) Find a 95% CI on the difference in means. Provide a practical interpretation of this interval.

(d) Assuming equal sample sizes, what sample size should be used to ensure that $\beta = 0.01$ if the true difference in means is 0.04? Assume that $\alpha = 0.05$.

5-2. Two types of plastic are suitable for use by an electronics component manufacturer. The breaking strength of this plastic is important. It is known that $\sigma_1 = \sigma_2 = 1.0$ psi. From a random sample of size $n_1 = 10$ and $n_2 = 12$, we obtain $\bar{x}_1 = 162.7$ and $\bar{x}_2 = 155.4$. The company will not adopt plastic 1 unless its mean breaking strength exceeds that of plastic 2 by at least 10 psi. Based on the sample information, should it use plastic 1? Use the P-value approach in reaching a decision.

5-3. The burning rates of two different solid-fuel propellants used in aircrew escape systems are being studied. It is known that both propellants have approximately the same standard deviation of burning rate; that is, $\sigma_1 = \sigma_2 = 3$ cm/s. Two random samples of $n_1 = 20$ and $n_2 = 20$ specimens are tested; the sample mean burning rates are $\bar{x}_1 = 18.02$ cm/s and $\bar{x}_2 = 24.37$ cm/s.

(a) Test the hypothesis that both propellants have the same mean burning rate. Use a fixed-level test with $\alpha = 0.05$.

(b) What is the P-value of the test in part (a)?

(c) What is the β-error of the test in part (a) if the true difference in mean burning rate is 2.5 cm/s?

(d) Construct a 95% CI on the difference in means $\mu_1 - \mu_2$. What is the practical meaning of this interval?

5-4. Two machines are used to fill plastic bottles with dishwashing detergent. The standard deviations of fill volume are known to be $\sigma_1 = 0.10$ and $\sigma_2 = 0.15$ fluid ounces for the two machines, respectively. Two random samples of $n_1 = 12$ bottles from machine 1 and $n_2 = 10$ bottles from machine 2 are selected, and the sample mean fill volumes are $\bar{x}_1 = 30.61$ and $\bar{x}_2 = 30.34$ fluid ounces. Assume normality.

(a) Construct a 90% two-sided CI on the mean difference in fill volume. Interpret this interval.

(b) Construct a 95% two-sided CI on the mean difference in fill volume. Compare and comment on the width of this interval to the width of the interval in part (a).

(c) Construct a 95% upper-confidence bound on the mean difference in fill volume. Interpret this interval.

5-5. Reconsider the situation described in Exercise 5-4.

(a) Test the hypothesis that both machines fill to the same mean volume. Use the P-value approach.

(b) If $\alpha = 0.05$ and the β-error of the test when the true difference in fill volume is 0.2 fluid ounces should not exceed 0.1, what sample sizes must be used? Use $\alpha = 0.05$.

5-6. Two different formulations of an oxygenated motor fuel are being tested to study their road octane numbers. The variance of road octane number for formulation 1 is $\sigma_1^2 = 1.5$, and for formulation 2 it is $\sigma_2^2 = 1.2$. Two random samples of size $n_1 = 15$ and $n_2 = 20$ are tested, and the mean road octane numbers observed are $\bar{x}_1 = 88.85$ and $\bar{x}_2 = 92.54$. Assume normality.

(a) Construct a 95% two-sided CI on the difference in mean road octane number.

(b) If formulation 2 produces a higher road octane number than formulation 1, the manufacturer would like to detect it. Formulate and test an appropriate hypothesis using the P-value approach.

5-7. Consider the situation described in Exercise 5-3. What sample size would be required in each population if we wanted the error in estimating the difference in mean burning rates to be less than 4 cm/s with 99% confidence?

5-8. Consider the road octane test situation described in Exercise 5-6. What sample size would be required in each population if we wanted to be 95% confident that the error in

estimating the difference in mean road octane number is less than 1?

5-9. A polymer is manufactured in a batch chemical process. Viscosity measurements are normally made on each batch, and long experience with the process has indicated that the variability in the process is fairly stable with $\sigma = 20$. Fifteen batch viscosity measurements are given as follows: 724, 718, 776, 760, 745, 759, 795, 756, 742, 740, 761, 749, 739, 747, 742. A process change is made that involves switching the type of catalyst used in the process. Following the process change, eight batch viscosity measurements are taken: 735, 775, 729, 755, 783, 760, 738, 780. Assume that process variability is unaffected by the catalyst change. Find a 90% CI on the difference in mean batch viscosity resulting from the process change. What is the practical meaning of this interval?

5-10. The concentration of active ingredient in a liquid laundry detergent is thought to be affected by the type of catalyst used in the process. The standard deviation of active concentration is known to be 3 grams per liter, regardless of the catalyst type. Ten observations on concentration are taken with each catalyst, and the data are shown here:

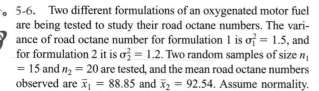

Catalyst 1:   66.1, 64.0, 64.4, 60.0, 65.3
              66.9, 61.5, 63.5, 61.6, 62.3
Catalyst 2:   66.3, 64.7, 67.6, 68.5, 68.3
              67.4, 66.1, 69.9, 70.6, 68.7

(a) Find a 95% CI on the difference in mean active concentrations for the two catalysts.

(b) Is there any evidence to indicate that the mean active concentrations depend on the choice of catalyst? Base your answer on the results of part (a).

5-11. Consider the polymer batch viscosity data in Exercise 5-9. If the difference in mean batch viscosity is 10 or less, the manufacturer would like to detect it with a high probability.

(a) Formulate and test an appropriate hypothesis using the P-value approach. What are your conclusions?

(b) Compare the results of part (a) to the 90% CI obtained in Exercise 5-9 and discuss your findings.

5-12. For the laundry detergent problem in Exercise 5-10, test the hypothesis that the mean active concentrations are the same for both types of catalyst. What is the P-value for this test? What are your conclusions? Compare your answer to that found in part (b) of Exercise 5-10, and comment on why they are the same or different.

5-3 INFERENCE ON THE MEANS OF TWO POPULATIONS, VARIANCES UNKNOWN

We now extend the results of the previous section to the difference in means of the two distributions in Fig. 5-1 when the variances of both distributions σ_1^2 and σ_2^2 are unknown. If the sample sizes n_1 and n_2 exceed 30, the normal distribution procedures in Section 5-2 could be used. However, when small samples are taken, we will assume that the populations are normally distributed and base our hypothesis tests and CIs on the t distribution. This nicely parallels the case of inference on the mean of a single sample with unknown variance.

5-3.1 Hypothesis Testing on the Difference in Means

We now consider tests of hypotheses on the difference in means $\mu_1 - \mu_2$ of two normal distributions where the variances σ_1^2 and σ_2^2 are unknown. A t-statistic will be used to test these hypotheses. As noted above and in Section 4-6, the normality assumption is required to develop the test procedure, but moderate departures from normality do not adversely affect the procedure. Two different situations must be treated. In the first case, we assume that the variances of the two normal distributions are unknown but equal; that is, $\sigma_1^2 = \sigma_2^2 = \sigma^2$. In the second, we assume that σ_1^2 and σ_2^2 are unknown and not necessarily equal.

Case 1: $\sigma_1^2 = \sigma_2^2 = \sigma^2$

Suppose we have two independent normal populations with unknown means μ_1 and μ_2, and unknown but equal variances, $\sigma_1^2 = \sigma_2^2 = \sigma^2$. We wish to test

$$H_0: \mu_1 - \mu_2 = \Delta_0$$
$$H_1: \mu_1 - \mu_2 \neq \Delta_0 \tag{5-6}$$

Let $X_{11}, X_{12}, \ldots, X_{1n_1}$ be a random sample of n_1 observations from the first population and $X_{21}, X_{22}, \ldots, X_{2n_2}$ be a random sample of n_2 observations from the second population. Let $\overline{X}_1, \overline{X}_2, S_1^2, S_2^2$ be the sample means and sample variances, respectively. Now the expected value of the difference in sample means $\overline{X}_1 - \overline{X}_2$ is $E(\overline{X}_1 - \overline{X}_2) = \mu_1 - \mu_2$, so $\overline{X}_1 - \overline{X}_2$ is an unbiased estimator of the difference in means. The variance of $\overline{X}_1 - \overline{X}_2$ is

$$V(\overline{X}_1 - \overline{X}_2) = \frac{\sigma^2}{n_1} + \frac{\sigma^2}{n_2} = \sigma^2\left(\frac{1}{n_1} + \frac{1}{n_2}\right)$$

It seems reasonable to combine the two sample variances S_1^2 and S_2^2 to form an estimator of σ^2. The **pooled estimator** of σ^2 is defined as follows.

Pooled Estimator of σ^2

> The **pooled estimator** of σ^2, denoted by S_p^2, is defined by
>
> $$S_p^2 = \frac{(n_1 - 1)S_1^2 + (n_2 - 1)S_2^2}{n_1 + n_2 - 2} \tag{5-7}$$

It is easy to see that the pooled estimator S_p^2 can be written as

$$S_p^2 = \frac{n_1 - 1}{n_1 + n_2 - 2}S_1^2 + \frac{n_2 - 1}{n_1 + n_2 - 2}S_2^2$$
$$= wS_1^2 + (1 - w)S_2^2$$

where $0 < w \leq 1$. Thus S_p^2 is a **weighted average** of the two sample variances S_1^2 and S_2^2, where the weights w and $1 - w$ depend on the two sample sizes n_1 and n_2. Obviously, if $n_1 = n_2 = n$, $w = 0.5$ and S_p^2 is just the arithmetic average of S_1^2 and S_2^2. If $n_1 = 10$ and $n_2 = 20$ (say), $w = 0.32$ and $1 - w = 0.68$. The first sample contributes $n_1 - 1$ degrees of freedom to S_p^2 and the second sample contributes $n_2 - 1$ degrees of freedom. Therefore, S_p^2 has $n_1 + n_2 - 2$ degrees of freedom.

Now we know that

$$Z = \frac{\overline{X}_1 - \overline{X}_2 - (\mu_1 - \mu_2)}{\sigma\sqrt{\dfrac{1}{n_1} + \dfrac{1}{n_2}}}$$

has an $N(0, 1)$ distribution. Replacing σ by S_p gives the following.

Given the assumptions of this section, the quantity

$$T = \frac{\overline{X}_1 - \overline{X}_2 - (\mu_1 - \mu_2)}{S_p\sqrt{\dfrac{1}{n_1} + \dfrac{1}{n_2}}} \tag{5-8}$$

has a t distribution with $n_1 + n_2 - 2$ degrees of freedom.

The use of this information to test the hypotheses in equation 5-6 is now straightforward: Simply replace $\mu_1 - \mu_2$ in equation 5-8 by Δ_0, and the resulting **test statistic** has a t distribution with $n_1 + n_2 - 2$ degrees of freedom under $H_0\colon \mu_1 - \mu_2 = \Delta_0$. The determination of P-values in the location of the critical region for fixed-level testing for both two- and one-sided alternatives parallels those in the one-sample case. This procedure is often called the **pooled t-test.**

Case 1: Testing Hypotheses on the Difference in Means of Two Normal Distributions, Variances Unknown and Equal[1]

Null hypothesis: $H_0\colon \mu_1 - \mu_2 = \Delta_0$

Test statistic:

$$T_0 = \frac{\overline{X}_1 - \overline{X}_2 - \Delta_0}{S_p\sqrt{\dfrac{1}{n_1} + \dfrac{1}{n_2}}} \tag{5-9}$$

Alternative Hypothesis	P-Value	Rejection Criterion for Fixed-Level Tests
$H_1\colon \mu_1 - \mu_2 \neq \Delta_0$	Sum of the probability above t_0 and the probability below $-t_0$	$t_0 > t_{\alpha/2, n_1+n_2-2}$ or $t_0 < -t_{\alpha/2, n_1+n_2-2}$
$H_1\colon \mu_1 - \mu_2 > \Delta_0$	Probability above t_0	$t_0 > t_{\alpha, n_1+n_2-2}$
$H_1\colon \mu_1 - \mu_2 < \Delta_0$	Probability below t_0	$t_0 < -t_{\alpha, n_1+n_2-2}$

---

[1]Although we have given the development of this procedure for the case in which the sample sizes could be different, there is an advantage to using equal sample sizes $n_1 = n_2 = n$. When the sample sizes are the same from both populations, the t-test is very robust or insensitive to the assumption of equal variances.

Table 5-2 Catalyst Yield Data (Percent) Example 5-4

Observation Number	Catalyst 1	Catalyst 2
1	91.50	89.19
2	94.18	90.95
3	92.18	90.46
4	95.39	93.21
5	91.79	97.19
6	89.07	97.04
7	94.72	91.07
8	89.21	92.75
	$\bar{x}_1 = 92.255$	$\bar{x}_2 = 92.733$
	$s_1 = 2.39$	$s_2 = 2.98$
	$n_1 = 8$	$n_2 = 8$

EXAMPLE 5-4

Chemical Process Yield

Two catalysts are being analyzed to determine how they affect the mean yield of a chemical process. Specifically, catalyst 1 is currently in use, but catalyst 2 is acceptable. Because catalyst 2 is cheaper, it should be adopted, providing it does not change the process yield. A test is run in the pilot plant and results in the data shown in Table 5-2. Is there any difference between the mean yields? Assume equal variances.

Solution. The solution using the seven-step hypothesis testing procedure is as follows:

1. **Parameter of interest:** The parameter of interest are μ_1 and μ_2, the mean process yield using catalysts 1 and 2, respectively, and we want to know whether $\mu_1 - \mu_2 = 0$.

2. **Null hypothesis, H_0:** $\mu_1 - \mu_2 = 0$, or H_0: $\mu_1 = \mu_2$

3. **Alternative hypothesis, H_1:** $\mu_1 \neq \mu_2$

4. **Test statistic:** The test statistic is

$$t_0 = \frac{\bar{x}_1 - \bar{x}_2 - 0}{s_p\sqrt{\frac{1}{n_1} + \frac{1}{n_2}}}$$

5. **Reject H_0 if:** Reject H_0 if the P-value is less than 0.05.

6. **Computations:** From Table 5-2 we have $\bar{x}_1 = 92.255$, $s_1 = 2.39$, $n_1 = 8$, $\bar{x}_2 = 92.733$, $s_2 = 2.98$, and $n_2 = 8$. Therefore,

$$s_p^2 = \frac{(n_1 - 1)s_1^2 + (n_2 - 1)s_2^2}{n_1 + n_2 - 2} = \frac{(7)(2.39)^2 + 7(2.98)^2}{8 + 8 - 2} = 7.30$$

$$s_p = \sqrt{7.30} = 2.70$$

and

$$t_0 = \frac{\bar{x}_1 - \bar{x}_2}{2.70\sqrt{\frac{1}{n_1} + \frac{1}{n_2}}} = \frac{92.255 - 92.733}{2.70\sqrt{\frac{1}{8} + \frac{1}{8}}} = -0.35$$

7. **Conclusions:** From the t-table with 14 degrees of freedom, we find that $t_{0.40,14} = 0.258$ and $t_{0.25,14} = 0.692$. Now $|t_0| = 0.35$, and $0.258 < 0.35 < 0.692$, so lower and upper bounds on the P-value are $0.5 < P < 0.8$. Therefore, since $P > 0.05$, the null hypothesis cannot be rejected. That is, at the 0.05 level of significance, we do not have strong evidence to conclude that

catalyst 2 results in a mean yield that differs from the mean yield when catalyst 1 is used. The exact *P*-value is $P = 0.73$. This was obtained from Minitab computer software.   ■

The Minitab two-sample *t*-test and confidence interval output for Example 5-4 follows:

Two-sample T for Cat 1 vs Cat 2

	N	Mean	StDev	SE Mean
Cat 1	8	92.26	2.39	0.84
Cat 2	8	92.73	2.99	1.1

Difference = mu Cat 1 − mu Cat 2
Estimate for difference: −0.48
95% CI for difference: (−3.37, 2.42)
T-Test of difference = 0 (vs not =): T-Value = −0.35, P-Value = 0.730, DF = 14
Both use Pooled StDev = 2.70

Notice that the numerical results are essentially the same as the manual computations in Example 5-4. The *P*-value is reported as $P = 0.73$. The two-sided CI on $\mu_1 - \mu_2$ is also reported. We will give the computing formula for the CI in Section 5-3.3.

Checking the Normality Assumption

Figure 5-2 shows the normal probability plot of the two samples of yield data and comparative box plots. The normal probability plots indicate that there is no problem with the normality assumption. Furthermore, both straight lines have similar slopes, providing some verification of the assumption of equal variances. The comparative box plots indicate that there is no obvious difference in the two catalysts, although catalyst 2 has slightly greater sample variability.

Case 2: $\sigma_1^2 \neq \sigma_2^2$

In some situations, we cannot reasonably assume that the unknown variances σ_1^2 and σ_2^2 are equal. There is not an exact *t*-statistic available for testing H_0: $\mu_1 - \mu_2 = \Delta_0$ in this case. However, the following test statistic is used.

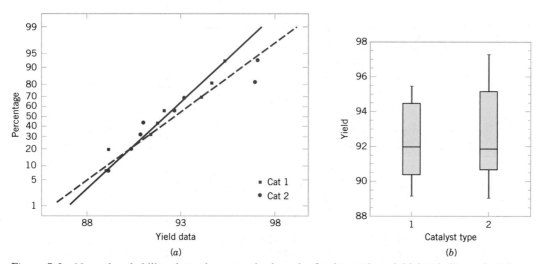

Figure 5-2   Normal probability plot and comparative box plot for the catalyst yield data in Example 5-4. (a) Normal probability plot. (b) Box plots.

> **Case 2: Test Statistic for the Difference in Means of Two Normal Distributions, Variances Unknown and Not Necessarily Equal**
>
> $$T_0^* = \frac{\overline{X}_1 - \overline{X}_2 - \Delta_0}{\sqrt{\dfrac{S_1^2}{n_1} + \dfrac{S_2^2}{n_2}}} \qquad (5\text{-}10)$$
>
> is distributed approximately as t with degrees of freedom given by
>
> $$\nu = \frac{\left(\dfrac{S_1^2}{n_1} + \dfrac{S_2^2}{n_2}\right)^2}{\dfrac{(S_1^2/n_1)^2}{n_1 - 1} + \dfrac{(S_2^2/n_2)^2}{n_2 - 1}} \qquad (5\text{-}11)$$
>
> if the null hypothesis $H_0\colon \mu_1 - \mu_2 = \Delta_0$ is true. If ν is not an integer, round down to the nearest integer.

Therefore, if $\sigma_1^2 \neq \sigma_2^2$, the hypotheses on differences in the means of two normal distributions are tested as in the equal variances case, except that T_0^* is used as the test statistic and $n_1 + n_2 - 2$ is replaced by ν in determining the degrees of freedom for the test.

EXAMPLE 5-5
Arsenic in
Drinking Water

Arsenic concentration in public drinking water supplies is a potential health risk. An article in the *Arizona Republic* (May 27, 2001) reported drinking water arsenic concentrations in parts per billion (ppb) for 10 metropolitan Phoenix communities and 10 communities in rural Arizona. The data follow:

Metro Phoenix ($\overline{x}_1 = 12.5$, $s_1 = 7.63$)	Rural Arizona ($\overline{x}_2 = 27.5$, $s_2 = 15.3$)
Phoenix, 3	Rimrock, 48
Chandler, 7	Goodyear, 44
Gilbert, 25	New River, 40
Glendale, 10	Apache Junction, 38
Mesa, 15	Buckeye, 33
Paradise Valley, 6	Nogales, 21
Peoria, 12	Black Canyon City, 20
Scottsdale, 25	Sedona, 12
Tempe, 15	Payson, 1
Sun City, 7	Casa Grande, 18

We wish to determine if there is any difference in mean arsenic concentrations between metropolitan Phoenix communities and communities in rural Arizona.

Solution. For our illustrative purposes, we are going to assume that these two data sets are representative random samples of the two communities. Figure 5-3 shows a normal probability plot for the two samples of arsenic concentration. The assumption of normality appears quite reasonable, but because the slopes of the two straight lines are very different, it is unlikely that the population variances are the same.

Applying the seven-step procedure gives the following:

1. **Parameter of interest:** The parameter of interest are the mean arsenic concentrations for the two geographic regions, say, μ_1 and μ_2, and we are interested in determining whether $\mu_1 - \mu_2 = 0$.

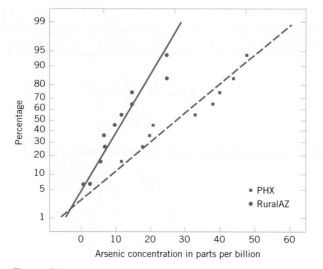

Figure 5-3 Normal probability of the arsenic concentration data from Example 5-5.

2. **Null hypothesis, H_0:** $\mu_1 - \mu_2 = 0$, or H_0: $\mu_1 = \mu_2$

3. **Alternative hypothesis, H_1:** $\mu_1 \neq \mu_2$

4. **Test statistic:** The test statistic is

$$t_0^* = \frac{\bar{x}_1 - \bar{x}_2 - 0}{\sqrt{\dfrac{s_1^2}{n_1} + \dfrac{s_2^2}{n_2}}}$$

5. **Reject H_0 if:** We will reject H_0: $\mu_1 = \mu_2$ if the P-value is less than 0.05.

6. **Computations:** The degrees of freedom on t_0^* are found from equation 5-11 as

$$\nu = \frac{\left(\dfrac{s_1^2}{n_1} + \dfrac{s_2^2}{n_2}\right)^2}{\dfrac{(s_1^2/n_1)^2}{n_1 - 1} + \dfrac{(s_2^2/n_2)^2}{n_2 - 1}} = \frac{\left[\dfrac{(7.63)^2}{10} + \dfrac{(15.3)^2}{10}\right]^2}{\dfrac{[(7.63)^2/10]^2}{9} + \dfrac{[(15.3)^2/10]^2}{9}} = 13.2 \simeq 13$$

The value of the test statistic is:

$$t_0^* = \frac{\bar{x}_1 - \bar{x}_2}{\sqrt{\dfrac{s_1^2}{n_1} + \dfrac{s_2^2}{n_2}}} = \frac{12.5 - 27.5}{\sqrt{\dfrac{(7.63)^2}{10} + \dfrac{(15.3)^2}{10}}} = -2.77$$

7. **Conclusions:** Because $t_{0.01,13} = 2.650$, $t_{0.005,13} = 3.012$, and $|t_0| = 2.77$, we find bounds on the P-value as $0.01 < P < 0.02$. Therefore, the P-value is less than 0.05, so we reject the null hypothesis. Therefore, there is evidence to conclude that mean arsenic concentration in the drinking water in rural Arizona is different from the mean arsenic concentration in metropolitan Phoenix drinking water. Furthermore, the mean arsenic concentration is higher in rural Arizona communities. ■

The Minitab output for this example follows.

**Minitab Two-Sample
t-Test and Confidence
Interval Output**

Two-sample T for PHX vs RuralAZ

	N	Mean	StDev	SE Mean
PHX	10	12.50	7.63	2.4
RuralAZ	10	27.5	15.3	4.9

Difference = mu PHX − mu RuralAZ
Estimate for difference: −15.00
95% CI for difference: (−26.71, −3.29)
T-Test of difference = 0 (vs not =): T-Value = −2.77, P-Value = 0.016, DF = 13

The numerical results from Minitab exactly match the calculations from Example 5-5. Note that a two-sided 95% CI on $\mu_1 - \mu_2$ is also reported. We will discuss its computation in Section 5-3.3; however, note that the interval does not include zero. Indeed, the upper 95% of the confidence limit is −3.29 parts per billion, well below zero, and the mean observed difference is $\bar{x}_1 - \bar{x}_2 = 12.5 - 17.5 = -15$ parts per billion.

Animation 10: Two Separate Sample Hypothesis Testing for Means

5-3.2   Type II Error and Choice of Sample Size

The operating characteristic curves in Appendix A Charts V*a*, V*b*, V*c*, and V*d* are used to evaluate the type II error for the case in which $\sigma_1^2 = \sigma_2^2 = \sigma^2$. Unfortunately, when $\sigma_1^2 \neq \sigma_2^2$, the distribution of T_0^* is unknown if the null hypothesis is false, and no operating characteristic curves are available for this case.

For the two-sided alternative $H_1: \mu_1 - \mu_2 \neq \Delta_0$, when $\sigma_1^2 = \sigma_2^2 = \sigma^2$ and $n_1 = n_2 = n$, Charts V*a* and V*b* are used with

$$d = \frac{|\Delta - \Delta_0|}{2\sigma} \tag{5-12}$$

where Δ is the true difference in means that is of interest. To use these curves, they must be entered with the sample size $n^* = 2n - 1$. For the one-sided alternative hypothesis, we use Charts V*c* and V*d* and define d and Δ as in equation 5-12. It is noted that the parameter d is a function of σ, which is unknown. As in the single-sample *t*-test, we may have to rely on a prior estimate of σ or use a subjective estimate. Alternatively, we could define the differences in the mean that we wish to detect relative to σ.

EXAMPLE 5-6

Chemical
Process Yield

Consider the catalyst experiment in Example 5-4. Suppose that if catalyst 2 produces a mean yield that differs from the mean yield of catalyst 1 by 4.0%, we would like to reject the null hypothesis with probability at least 0.85. What sample size is required?

Solution. Using $s_p = 2.70$ as a rough estimate of the common standard deviation σ, we have $d = |\Delta|/2\sigma = |4.0|/[(2)(2.70)] = 0.74$. From Appendix A Chart V*a* with $d = 0.74$ and $\beta = 0.15$, we find $n^* = 20$, approximately. Therefore, because $n^* = 2n - 1$,

$$n = \frac{n^* + 1}{2} = \frac{20 + 1}{2} = 10.5 \approx 11 \text{ (say)}$$

and we would use sample sizes of $n_1 = n_2 = n = 11$.

Minitab Will Perform Power and Sample Size Calculations for the Two-Sample *t*-Test (Equal Variances)

2-Sample *t* Test
Testing mean 1 = mean 2 (versus not =)
Calculating power for mean 1 = mean 2 + difference
Alpha = 0.05 Sigma = 2.7

Difference	Sample Size	Target Power	Actual Power
4	10	0.8500	0.8793

The results agree fairly closely with the results obtained from the OC curve.

5-3.3 Confidence Interval on the Difference in Means

Case 1: $\sigma_1^2 = \sigma_2^2 = \sigma^2$

To develop the CI for the difference in means $\mu_1 - \mu_2$ when both variances are equal, note that the distribution of the statistic

$$T = \frac{\bar{X}_1 - \bar{X}_2 - (\mu_1 - \mu_2)}{S_P \sqrt{\dfrac{1}{n_1} + \dfrac{1}{n_2}}}$$

is the *t* distribution with $n_1 + n_2 - 2$ degrees of freedom. Therefore,

$$P(-t_{\alpha/2,n_1+n_2-2} \le T \le t_{\alpha/2,n_1+n_2-2}) = 1 - \alpha$$

or

$$P\left[-t_{\alpha/2,n_1+n_2-2} \le \frac{\bar{X}_1 - \bar{X}_2 - (\mu_1 - \mu_2)}{S_P \sqrt{\dfrac{1}{n_1} + \dfrac{1}{n_2}}} \le t_{\alpha/2,n_1-n_2-2} \right] = 1 - \alpha$$

Manipulation of the quantities inside the probability statement leads to the following $100(1 - \alpha)\%$ CI on $\mu_1 - \mu_2$.

Case 1: Confidence Interval on the Difference in Means of Two Normal Distributions, Variances Unknown and Equal

If $\bar{x}_1$, $\bar{x}_2$, s_1^2, and s_2^2 are the means and variances of two random samples of sizes n_1 and n_2, respectively, from two independent normal populations with unknown but equal variances, a $100(1 - \alpha)\%$ CI on the difference in means $\mu_1 - \mu_2$ is

$$\bar{x}_1 - \bar{x}_2 - t_{\alpha/2,n_1+n_2-2}\, s_p \sqrt{\frac{1}{n_1} + \frac{1}{n_2}}$$

$$\le \mu_1 - \mu_2 \le \bar{x}_1 - \bar{x}_2 + t_{\alpha/2,n_1+n_2-2}\, s_p \sqrt{\frac{1}{n_1} + \frac{1}{n_2}} \tag{5-13}$$

where $s_p = \sqrt{[(n_1 - 1)s_1^2 + (n_2 - 1)s_2^2]/(n_1 + n_2 - 2)}$ is the pooled estimate of the common population standard deviation, and $t_{\alpha/2,n_1+n_2-2}$ is the upper $100\,\alpha/2$ percentage point of the *t* distribution with $n_1 + n_2 - 2$ degrees of freedom.

EXAMPLE 5-7

Calcium in
Doped Cement

An article in the journal *Hazardous Waste and Hazardous Materials* (Vol. 6, 1989) reported the results of an analysis of the weight of calcium in standard cement and cement doped with lead. Reduced levels of calcium would indicate that the hydration mechanism in the cement is blocked and would allow water to attack various locations in the cement structure. Ten samples of standard cement had an average weight percent calcium of $\bar{x}_1 = 90.0$, with a sample standard deviation of $s_1 = 5.0$, and 15 samples of the lead-doped cement had an average weight percent calcium of $\bar{x}_2 = 87.0$, with a sample standard deviation of $s_2 = 4.0$.

Assume that weight percent calcium is normally distributed and find a 95% CI on the difference in means, $\mu_1 - \mu_2$, for the two types of cement. Furthermore, assume that both normal populations have the same standard deviation.

Solution. The pooled estimate of the common standard deviation is found using equation 5-7 as follows:

$$s_p^2 = \frac{(n_1 - 1)s_1^2 + (n_2 - 1)s_2^2}{n_1 + n_2 - 2}$$

$$= \frac{9(5.0)^2 + 14(4.0)^2}{10 + 15 - 2}$$

$$= 19.52$$

Therefore, the pooled standard deviation estimate is $s_p = \sqrt{19.52} = 4.4$. The 95% CI is found using equation 5-13:

$$\bar{x}_1 - \bar{x}_2 - t_{0.025,23}s_p\sqrt{\frac{1}{n_1} + \frac{1}{n_2}} \leq \mu_1 - \mu_2 \leq \bar{x}_1 - \bar{x}_2 + t_{0.025,23}s_p\sqrt{\frac{1}{n_1} + \frac{1}{n_2}}$$

or, upon substituting the sample values and using $t_{0.025,23} = 2.069$,

$$90.0 - 87.0 - 2.069(4.4)\sqrt{\frac{1}{10} + \frac{1}{15}} \leq \mu_1 - \mu_2 \leq 90.0 - 87.0 + 2.069(4.4)\sqrt{\frac{1}{10} + \frac{1}{15}}$$

which reduces to

$$-0.72 \leq \mu_1 - \mu_2 \leq 6.72$$

Note that the 95% CI includes zero; therefore, at this level of confidence we cannot conclude that there is a difference in the means. Put another way, there is no evidence that doping the cement with lead affected the mean weight percent of calcium; therefore, we cannot claim that the presence of lead affects this aspect of the hydration mechanism at the 95% level of confidence. ∎

Case 2: $\sigma_1^2 \neq \sigma_2^2$

In many situations it is not reasonable to assume that $\sigma_1^2 = \sigma_2^2$. When this assumption is unwarranted, we may still find a $100(1 - \alpha)\%$ CI on $\mu_1 - \mu_2$ using the fact that

$$T^* = \frac{\bar{X}_1 - \bar{X}_2 - (\mu_1 - \mu_2)}{\sqrt{S_1^2/n_1 + S_2^2/n_2}}$$

is distributed approximately as t with degrees of freedom v given by equation 5-11. Therefore,

$$P(-t_{\alpha/2,v} \leq T^* \leq t_{\alpha/2,v}) \cong 1 - \alpha$$

and, if we substitute for T^* in this expression and isolate $\mu_1 - \mu_2$ between the inequalities, we can obtain the following CI for $\mu_1 - \mu_2$.

Case 2: Confidence Interval on the Difference in Means of Two Normal Distributions, Variances Unknown and Unequal

If $\bar{x}_1$, $\bar{x}_2$, s_1^2, and s_2^2 are the means and variances of two random samples of sizes n_1 and n_2, respectively, from two independent normal populations with unknown and unequal variances, then an approximate $100(1 - \alpha)\%$ CI on the difference in means $\mu_1 - \mu_2$ is

$$\bar{x}_1 - \bar{x}_2 - t_{\alpha/2,v}\sqrt{\frac{s_1^2}{n_1} + \frac{s_2^2}{n_2}} \leq \mu_1 - \mu_2 \leq \bar{x}_1 - \bar{x}_2 + t_{\alpha/2,v}\sqrt{\frac{s_1^2}{n_1} + \frac{s_2^2}{n_2}} \qquad (5\text{-}14)$$

where v is given by equation 5-11 and $t_{\alpha/2,v}$ is the upper $100\,\alpha/2$ percentage point of the t distribution with v degrees of freedom.

One-Sided Confidence Bounds
To find a $100(1 - \alpha)\%$ lower confidence bound on $\mu_1 - \mu_2$, with unknown σ^2 values, simply replace $-t_{\alpha/2,n_1+n_2-2}$ with $-t_{\alpha,n_1+n_2-2}$ in the lower bound of equation 5-13 for case 1 and $-t_{\alpha/2,v}$ with $-t_{\alpha,v}$ in the lower bound of equation 5-14 for case 2; the upper bound is set to ∞. Similarly, to find a $100(1 - \alpha)\%$ upper confidence bound on $\mu_1 - \mu_2$ with unknown σ^2 values, simply replace $t_{\alpha/2,n_1+n_2-2}$ with t_{α,n_1+n_2-2} in the upper bound of equation 5-13 for case 1 and $t_{\alpha/2,v}$ with $t_{\alpha,v}$ in the upper bound of equation 5-14 for case 2; the lower bound is set to $-\infty$.

EXERCISES FOR SECTION 5-3

5-13.   Consider the Minitab output shown below.

> **Two-Sample T-Test and CI: X1, X2**
>
> Two-sample T for X1 vs X2
>
> | | N | Mean | StDev | SE Mean |
> |----|----|-------|-------|---------|
> | X1 | 20 | 50.19 | 1.71 | 0.38 |
> | X2 | 20 | 52.52 | 2.48 | 0.55 |
>
> Difference = mu (X1) − mu (X2)
> Estimate for difference: −2.33341
> 95% CI for difference: (−3.69547, −0.97135)
> T-Test of difference = 0 (vs not =):
>   T-Value = −3.47, P-Value = 0.001, DF = 38
> Both use Pooled StDev = 2.1277

(a) Can the null hypothesis be rejected at the 0.05 level? Why?
(b) Is this a one-sided or a two-sided test?
(c) If the hypotheses had been H_0: $\mu_1 - \mu_2 = 2$ versus H_1: $\mu_1 - \mu_2 \neq 2$, would you reject the null hypothesis at the 0.05 level?
(d) If the hypotheses had been H_0: $\mu_1 - \mu_2 = 2$ versus H_1: $\mu_1 - \mu_2 < 2$, would you reject the null hypothesis at

the 0.05 level? Can you answer this question without doing any additional calculations? Why?
(e) Use the output and the t table to find a 95% upper confidence bound on the difference in means.
(f) What is the P-value if the alternative hypothesis is H_0: $\mu_1 - \mu_2 = 2$ versus H_1: $\mu_1 - \mu_2 \neq 2$?

5-14.   Consider the Minitab output shown below.

> **Two-Sample T-Test and CI: X1, X2**
>
> Two-sample T for X1 vs X2
>
> | | N | Mean | StDev | SE Mean |
> |----|----|-------|-------|---------|
> | X1 | 15 | 75.47 | 1.63 | ? |
> | X2 | 25 | 76.06 | 1.99 | 0.40 |
>
> Difference = mu (X1) − mu (X2)
> Estimate for difference: −0.590171
> 95% upper bound for difference: ?
> T-Test of difference = 0 (vs <): T-Value = −0.97,
>   P-Value = 0.170, DF = ?
> Both use Pooled StDev = ?

(a) Fill in the missing values in the output. Can the null hypothesis be rejected at the 0.05 level? Why?

(b) Is this a one-sided or a two-sided test?

(c) Use the output and the t table to find a 99% one-sided upper confidence bound on the difference in means.

(d) What is the P-value if the alternative hypothesis is $H_0: \mu_1 - \mu_2 = 1$ versus $H_1: \mu_1 - \mu_2 < 1$?

5-15. An article in *Electronic Components and Technology Conference* (Vol. 52, 2001, pp. 1167–1171) describes a study comparing single versus dual spindle saw processes for copper metallized wafers. A total of 15 devices of each type were measured for the width of the backside chipouts, $\bar{x}_{single} = 66.385$, $s_{single} = 7.895$ and $\bar{x}_{double} = 45.278$, $s_{double} = 8.612$.

(a) Do the sample data support the claim that both processes have the same chip outputs? Assume that both populations are normally distributed and have the same variance. Answer this question by finding and interpreting the P-value for the test.

(b) Construct a 95% two-sided confidence interval on the mean difference in spindle saw process. Compare this interval to the results in part (a).

(c) If the β-error of the test when the true difference in chip outputs is 15 should not exceed 0.1 when $\alpha = 0.05$, what sample sizes must be used?

5-16. An article in *IEEE International Symposium on Electromagnetic Compatibility* (Vol. 2, 2002, pp. 667–670) describes the quantification of the absorption of electromagnetic energy and the resulting thermal effect from cellular phones. The experimental results were obtained from in vivo experiments conducted on rats. The arterial blood pressure values (mmHg) for the control group (8 rats) during the experiment are $\bar{x}_1 = 90$, $s_1 = 5$ and for the test group (9 rats) are $\bar{x}_2 = 115$, $s_2 = 10$.

(a) Is there evidence to support the claim that the test group has higher mean blood pressure? Assume that both populations are normally distributed but the variances are not equal. Answer this question by finding the P-value for this test.

(b) Calculate a confidence interval to answer the question in part (a).

5-17. The diameter of steel rods manufactured on two different extrusion machines is being investigated. Two random samples of sizes $n_1 = 15$ and $n_2 = 17$ are selected, and the sample means and sample variances are $\bar{x} = 8.73$, $s_1^2 = 0.35$, $\bar{x}_2 = 8.68$, and $s_2^2 = 0.40$, respectively. Assume that $\sigma_1^2 = \sigma_2^2$ and that the data are drawn from a normal distribution.

(a) Is there evidence to support the claim that the two machines produce rods with different mean diameters? Use a P-value in arriving at this conclusion.

(b) Construct a 95% CI for the difference in mean rod diameter. Interpret this interval.

5-18. An article in *Fire Technology* describes the investigation of two different foam expanding agents that can be used in the nozzles of firefighting spray equipment. A random sample of five observations with an aqueous film-forming foam (AFFF)

had a sample mean of 4.340 and a standard deviation of 0.508. A random sample of five observations with alcohol-type concentrates (ATC) had a sample mean of 7.091 and a standard deviation of 0.430. Assume that both populations are well represented by normal distributions with the same standard deviations.

(a) Is there evidence to support the claim that there is no difference in mean foam expansion of these two agents? Use a fixed-level test with $\alpha = 0.10$.

(b) Calculate the P-value for this test.

(c) Construct a 90% CI for the difference in mean foam expansion. Explain how this interval confirms your finding in part (a).

5-19. A consumer organization collected data on two types of automobile batteries, A and B. The summary statistics for 12 observations of each type are $\bar{x}_A = 36.51$, $\bar{x}_B = 34.21$, $s_A = 1.43$, and $s_B = 0.93$. Assume that the data are normally distributed with $\sigma_A = \sigma_B$.

(a) Is there evidence to support the claim that type A battery mean life exceeds that of type B? Use the P-value in answering this question.

(b) Construct a one-sided 99% confidence bound for the difference in mean battery life. Explain how this interval confirms your finding in part (a).

(c) Suppose that if the mean life of type A batteries exceeds that of type B by as much as 2 months, it is important to detect this difference with probability at least 0.95 when $\alpha = 0.01$. Is the choice of $n_1 = n_2 = 12$ of this problem adequate?

5-20. The deflection temperature under load for two different types of plastic pipe is being investigated. Two random samples of 15 pipe specimens are tested, and the deflection temperatures observed are reported here (in °F):

Type 1			Type 2		
206	193	192	177	176	198
188	207	210	197	185	188
205	185	194	206	200	189
187	189	178	201	197	203
194	213	205	180	192	192

(a) Construct box plots and normal probability plots for the two samples. Do these plots provide support of the assumptions of normality and equal variances? Write a practical interpretation for these plots.

(b) Do the data support the claim that the deflection temperature under load for type 1 pipe exceeds that of type 2? In reaching your conclusions, use a P-value in answering this question.

(c) Suppose that if the mean deflection temperature for type 2 pipe exceeds that of type 1 by as much as 5°F, it is important to detect this difference with probability at least 0.90 when $\alpha = 0.05$. Is the choice of $n_1 = n_2 = 15$ in part (b) of this problem adequate?

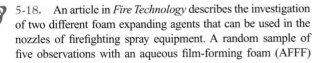

5-21. In semiconductor manufacturing, wet chemical etching is often used to remove silicon from the backs of wafers prior to metalization. The etch rate is an important characteristic in this process and known to follow a normal distribution. Two different etching solutions have been compared, using two random samples of 10 wafers for etch solution. The observed etch rates are as follows (in mils/min):

Solution 1		Solution 2	
9.9	10.6	10.2	10.0
9.4	10.3	10.6	10.2
9.3	10.0	10.7	10.7
9.6	10.3	10.4	10.4
10.2	10.1	10.5	10.3

(a) Do the data support the claim that the mean etch rate is the same for both solutions? In reaching your conclusions, use a fixed-level test with $\alpha = 0.05$ and assume that both population variances are equal.
(b) Calculate a P-value for the test in part (a).
(c) Find a 95% CI on the difference in mean etch rates.
(d) Construct normal probability plots for the two samples. Do these plots provide support for the assumptions of normality and equal variances? Write a practical interpretation for these plots.

5-22. Two suppliers manufacture a plastic gear used in a laser printer. The impact strength of these gears measured in foot-pounds is an important characteristic. A random sample of 10 gears from supplier 1 results in $\bar{x}_1 = 289.30$ and $s_1 = 22.5$, and another random sample of 16 gears from the second supplier results in $\bar{x}_2 = 321.50$ and $s_2 = 21$.
(a) Is there evidence to support the claim that supplier 2 provides gears with higher mean impact strength? Use the P-value approach, and assume that both populations are normally distributed but the variances are not equal.
(b) Do the data support the claim that the mean impact strength of gears from supplier 2 is at least 25 foot-pounds higher than that of supplier 1? Make the same assumptions as in part (a).
(c) Construct an appropriate 95% CI on the difference in mean impact strength. Use this interval to answer the question posed in part (b).

5-23. A photoconductor film is manufactured at a nominal thickness of 25 mils. The product engineer wishes to decrease the energy absorption of the film, and he believes this can be achieved by reducing the thickness of the film to 20 mils. Eight samples of each film thickness are manufactured in a pilot production process, and the film absorption (in $\mu J/in^2$) is measured. For the 25-mil film, the sample data result is $\bar{x}_1 = 1.179$ and $s_1 = 0.088$, and for the 20-mil film, the data yield $\bar{x}_2 = 1.036$ and $s_2 = 0.093$.
(a) Do the data support the claim that reducing the film thickness decreases the energy absorption of the film? Use a fixed-level test with $\alpha = 0.10$ and assume that the two population variances are equal and the underlying population is normally distributed.
(b) What is the P-value for this test?
(c) Find a 95% CI on the difference in the two means.

5-24. The melting points of two alloys used in formulating solder were investigated by melting 21 samples of each material. The sample mean and standard deviation for alloy 1 was $\bar{x}_1 = 420.48°F$ and $s_1 = 2.34°F$, and for alloy 2 they were $\bar{x}_2 = 425°F$ and $s_2 = 2.5°F$.
(a) Do the sample data support the claim that both alloys have the same melting point? Use a fixed-level test with $\alpha = 0.05$ and assume that both populations are normally distributed and have the same standard deviation.
(b) Find the P-value for this test.

5-25. Referring to the melting point experiment in Exercise 5-24, suppose that the true mean difference in melting points is 3°F. How large a sample would be required to detect this difference using an $\alpha = 0.05$ level test with probability at least 0.9? Use $\sigma_1 = \sigma_2 = 4$ as an initial estimate of the common standard deviation.

5-26. Two companies manufacture a rubber material intended for use in an automotive application. The part will be subjected to abrasive wear in the field application, so we decide to compare the material produced by each company in a test. Twenty-five samples of material from each company are tested in an abrasion test, and the amount of wear after 1000 cycles is observed. For company 1, the sample mean and standard deviation of wear are $\bar{x}_1 = 20.12$ mg/1000 cycles and $s_1 = 1.9$ mg/1000 cycles, and for company 2, we obtain $\bar{x}_2 = 11.64$ mg/1000 cycles and $s_2 = 7.9$ mg/1000 cycles.
(a) Do the data support the claim that the two companies produce material with different mean wear? Use the P-value approach and assume each population is normally distributed but that their variances are not equal.
(b) Do the data support a claim that the material from company 1 has higher mean wear than the material from company 2? Use the same assumptions as in part (a).

5-27. The thickness of a plastic film (in mils) on a substrate material is thought to be influenced by the temperature at which the coating is applied. A completely randomized experiment is carried out. Eleven substrates are coated at 125°F, resulting in a sample mean coating thickness of $\bar{x}_1 = 101.28$ and a sample standard deviation of $s_1 = 5.08$. Another 13 substrates are coated at 150°F, for which $\bar{x}_2 = 101.70$ and $s_2 = 20.15$ are observed. It was originally suspected that raising the process temperature would reduce mean coating thickness. Do the data support this claim? Use the P-value approach and assume that the two population standard deviations are not equal.

5-28. Reconsider the abrasive wear test in Exercise 5-26. Construct a CI that will address the questions in parts (a) and (b) in that exercise.

5-29. Reconsider the coating thickness experiment in Exercise 5-27. Answer the question posed regarding the effect of temperature on coating thickness by using a CI. Explain your answer.

5-4   THE PAIRED *t*-TEST

A special case of the two-sample *t*-tests of Section 5-3 occurs when the observations on the two populations of interest are collected in **pairs.** Each pair of observations—say, (X_{1j}, X_{2j})—are taken under homogeneous conditions, but these conditions may change from one pair to another. For example, suppose that we are interested in comparing two different types of tips for a hardness-testing machine. This machine presses the tip into a metal specimen with a known force. By measuring the depth of the depression caused by the tip, the hardness of the specimen can be determined. If several specimens were selected at random, half tested with tip 1, half tested with tip 2, and the pooled or independent *t*-test in Section 5-3 was applied, the results of the test could be erroneous. The metal specimens could have been cut from bar stock that was produced in different heats, or they might not be homogeneous in some other way that might affect hardness. Then the observed difference between mean hardness readings for the two tip types also includes hardness differences between specimens.

A more powerful experimental procedure is to collect the data in pairs—that is, to make two hardness readings on each specimen, one with each tip. The test procedure would then consist of analyzing the *differences* between hardness readings on each specimen. If there is no difference between tips, the mean of the differences should be zero. This test procedure is called the **paired *t*-test.**

Let $(X_{11}, X_{21}), (X_{12}, X_{22}), \ldots, (X_{1n}, X_{2n})$ be a set of n paired observations where we assume that the mean and variance of the population represented by X_1 are μ_1 and σ_1^2, and the mean and variance of the population represented by X_2 are μ_2 and σ_2^2. Define the differences between each pair of observations as $D_j = X_{1j} - X_{2j}, j = 1,2,\ldots, n$. The D_j's are assumed to be normally distributed with mean

$$\mu_D = E(X_1 - X_2) = E(X_1) - E(X_2) = \mu_1 - \mu_2$$

and variance σ_D^2, so testing hypotheses about the difference between μ_1 and μ_2 can be accomplished by performing a one-sample *t*-test on μ_D. Specifically, testing H_0: $\mu_1 - \mu_2 = \Delta_0$ against H_1: $\mu_1 - \mu_2 \neq \Delta_0$ is equivalent to testing

$$H_0\text{: } \mu_D = \Delta_0$$
$$H_1\text{: } \mu_D \neq \Delta_0 \tag{5-15}$$

The test statistic is given next.

The Paired *t*-Test

Null hypothesis: H_0: $\mu_D = \Delta_0$

Test statistic: $T_0 = \dfrac{\overline{D} - \Delta_0}{S_D/\sqrt{n}}$ (5-16)

Alternative Hypothesis	*P*-Value	Rejection Region for Fixed-Level Tests
H_1: $\mu_D \neq \Delta_0$	Sum of the probability above t_0 and the probability below $-t_0$,	$t_0 > t_{\alpha/2,n-1}$ or $t_0 < -t_{\alpha/2,n-1}$
H_1: $\mu_D > \Delta_0$	Probability above t_0	$t_0 > t_{\alpha,n-1}$
H_1: $\mu_D < \Delta_0$	Probability below t_0	$t_0 < -t_{\alpha,n-1}$

In equation 5-16, $\overline{D}$ is the sample average of the n differences $D_1, D_2, \ldots, D_n$ and S_D is the sample standard deviation of these differences.

EXAMPLE 5-8

Shear Strength of Steel Girders

An article in the *Journal of Strain Analysis* (Vol. 18, No. 2,1983) reports a comparison of several methods for predicting the shear strength for steel plate girders. Data for two of these methods, the Karlsruhe and Lehigh procedures, when applied to nine specific girders, are shown in Table 5-3. We wish to determine whether there is any difference (on the average) between the two methods.

Solution. The seven-step procedure is applied as follows:

1. **Parameter of interest:** The parameter of interest is the difference in mean shear strength between the two methods—say, $\mu_D = \mu_1 - \mu_2 = 0$.

2. **Null hypothesis, H_0:** $\mu_D = 0$

3. **Alternative hypothesis, H_1:** $\mu_D \neq 0$

4. **Test statistic:** The test statistic is

$$t_0 = \frac{\overline{d}}{s_d/\sqrt{n}}$$

5. **Reject H_0 if:** Reject H_0 if the *P*-value is <0.05.

6. **Computations:** The sample average and standard deviation of the differences d_j are $\overline{d} = 0.2769$ and $s_d = 0.1350$, and so the test statistic is

$$t_0 = \frac{\overline{d}}{s_d/\sqrt{n}} = \frac{0.2769}{0.1350/\sqrt{9}} = 6.15$$

7. **Conclusions:** Because $t_{0.0005,8} = 5.041$ and the value of the test statistic $t_0 = 6.15$ exceeds this value, the *P*-value is less than $2(0.0005) = 0.001$. Therefore, we conclude that the strength prediction methods yield different results. Specifically, the data indicate that the Karlsruhe method produces, on the average, higher strength predictions than does the Lehigh method. ∎

Minitab Paired *t*-Test and Confidence Interval Output for Example 5-8

Paired T-Test and CI: Karlsruhe, Lehigh

Paired T for Karlsruhe–Lehigh

	N	Mean	StDev	SE Mean
Karlsruhe	9	1.34011	0.14603	0.04868
Lehigh	9	1.06322	0.05041	0.01680
Difference	9	0.276889	0.135027	0.045009

95% CI for mean difference: (0.173098, 0.380680)
T-Test of mean difference = 0 (vs not = 0): T-Value = 6.15, P-Value = 0.000

The results essentially agree with the manual calculations. In addition to the hypothesis test results, Minitab reports a two-sided CI on the difference in means. This CI was found by constructing a single-sample CI on μ_D. We will give the details below.

Table 5-3   Strength Predictions for Nine Steel Plate Girders (Predicted Load/Observed Load)

Girder	Karlsruhe Method	Lehigh Method	Difference d_j
S1/1	1.186	1.061	0.125
S2/1	1.151	0.992	0.159
S3/1	1.322	1.063	0.259
S4/1	1.339	1.062	0.277
S5/1	1.200	1.065	0.135
S2/1	1.402	1.178	0.224
S2/2	1.365	1.037	0.328
S2/3	1.537	1.086	0.451
S2/4	1.559	1.052	0.507

Paired versus Unpaired Comparisons

In performing a comparative experiment, the investigator can sometimes choose between the paired experiment and the two-sample (or unpaired) experiment. If n measurements are to be made on each population, the two-sample t-statistic is

$$T_0 = \frac{\bar{X}_1 - \bar{X}_2 - \Delta_0}{S_p\sqrt{\frac{1}{n} + \frac{1}{n}}}$$

which would be compared to t_{2n-2}, and of course, the paired t-statistic is

$$T_0 = \frac{\bar{D} - \Delta_0}{S_D/\sqrt{n}}$$

which is compared to t_{n-1}. Note that because

$$\bar{D} = \sum_{j=1}^{n} \frac{D_j}{n} = \sum_{j=1}^{n} \frac{(X_{1j} - X_{2j})}{n} = \sum_{j=1}^{n} \frac{X_{1j}}{n} - \sum_{j=1}^{n} \frac{X_{2j}}{n} = \bar{X}_1 - \bar{X}_2$$

the numerators of both statistics are identical. However, the denominator of the two-sample t-test is based on the assumption that X_1 and X_2 are *independent*. In many paired experiments, a strong positive correlation ρ exists between X_1 and X_2. Then it can be shown that

$$V(\bar{D}) = V(\bar{X}_1 - \bar{X}_2 - \Delta_0) = V(\bar{X}_1) + V(\bar{X}_2) - 2\,\mathrm{cov}(\bar{X}_1, \bar{X}_2) = \frac{2\sigma^2(1 - \rho)}{n}$$

assuming that both populations X_1 and X_2 have identical variances σ^2. Furthermore, S_D^2/n estimates the variance of $\bar{D}$. Whenever there is positive correlation within the pairs, the denominator for the paired t-test will be smaller than the denominator of the two-sample t-test. This can cause the two-sample t-test to considerably understate the significance of the data if it is incorrectly applied to paired samples.

   Although pairing will often lead to a smaller value of the variance of $\bar{X}_1 - \bar{X}_2$, it does have a disadvantage—namely, the paired t-test leads to a loss of $n - 1$ degrees of freedom in comparison to the two-sample t-test. Generally, we know that increasing the degrees of freedom of a test increases the power against any fixed alternative values of the parameter.

So how do we decide to conduct the experiment? Should we pair the observations or not? Although there is no general answer to this question, we can give some guidelines based on the previous discussion:

1. If the experimental units are relatively homogeneous (small σ) and the correlation within pairs is small, the gain in precision attributable to pairing will be offset by the loss of degrees of freedom, so an independent-sample experiment should be used.

2. If the experimental units are relatively heterogeneous (large σ) and there is large positive correlation within pairs, the paired experiment should be used. Typically, this case occurs when the experimental units are the *same* for both treatments; as in Example 5-8, the same girders were used to test the two methods.

Implementing the rules still requires judgment because σ and ρ are never known precisely. Furthermore, if the number of degrees of freedom is large (say, 40 or 50), the loss of $n - 1$ of them for pairing may not be serious. However, if the number of degrees of freedom is small (say, 10 or 20), losing half of them is potentially serious if not compensated for by increased precision from pairing.

Confidence Interval for μ_D

To construct the confidence interval for μ_D, note that

$$T = \frac{\overline{D} - \mu_D}{S_D/\sqrt{n}}$$

follows a t distribution with $n - 1$ degrees of freedom. Then, because

$$P(-t_{\alpha/2,n-1} \leq T \leq t_{\alpha/2,n-1}) = 1 - \alpha$$

we can substitute for T in the above expression and perform the necessary steps to isolate $\mu_D = \mu_1 - \mu_2$ between the inequalities. This leads to the following $100(1 - \alpha)\%$ CI on $\mu_D = \mu_1 - \mu_2$.

Confidence Interval on μ_D for Paired Observations

If $\overline{d}$ and s_d are the sample mean and standard deviation, respectively, of the normally distributed difference of n random pairs of measurements, a $100(1 - \alpha)\%$ CI on the difference in means $\mu_D = \mu_1 - \mu_2$ is

$$\overline{d} - t_{\alpha/2,n-1}s_d/\sqrt{n} \leq \mu_D \leq \overline{d} + t_{\alpha/2,n-1}s_d/\sqrt{n} \qquad (5\text{-}17)$$

where $t_{\alpha/2,n-1}$ is the upper $100\,\alpha/2$ percentage point of the t-distribution with $n - 1$ degrees of freedom.

This CI is also valid for the case where $\sigma_1^2 \neq \sigma_2^2$ because s_D^2 estimates $\sigma_D^2 = V(X_1 - X_2)$. Also, for large samples (say, $n \geq 30$ pairs), the explicit assumption of normality is unnecessary because of the central limit theorem. Equation 5-17 was used to calculate the CI in the shear strength experiment in Example 5-8.

Table 5-4   Time in Seconds to Parallel Park Two Automobiles

	Automobile		Difference
Subject	1 (x_{1j})	2 (x_{2j})	(d_j)
1	37.0	17.8	19.2
2	25.8	20.2	5.6
3	16.2	16.8	−0.6
4	24.2	41.4	−17.2
5	22.0	21.4	0.6
6	33.4	38.4	−5.0
7	23.8	16.8	7.0
8	58.2	32.2	26.0
9	33.6	27.8	5.8
10	24.4	23.2	1.2
11	23.4	29.6	−6.2
12	21.2	20.6	0.6
13	36.2	32.2	4.0
14	29.8	53.8	−24.0

EXAMPLE 5-9
Parallel Parking

The journal *Human Factors* (1962, pp. 375–380) reports a study in which $n = 14$ subjects were asked to parallel park two cars having very different wheel bases and turning radii. The time in seconds for each subject was recorded and is given in Table 5-4. Find a 90% CI on the mean difference in times.

Solution. From the column of observed differences we calculate $\bar{d} = 1.21$ and $s_d = 12.68$. The 90% CI for $\mu_D = \mu_1 - \mu_2$ is found from equation 5-17 as follows:

$$\bar{d} - t_{0.05,13}s_d/\sqrt{n} \le \mu_D \le \bar{d} + t_{0.05,13}s_d/\sqrt{n}$$

$$1.21 - 1.771(12.68)/\sqrt{14} \le \mu_D \le 1.21 + 1.771(12.68)/\sqrt{14}$$

$$-4.79 \le \mu_D \le 7.21$$

Note that the CI on μ_D includes zero. This implies that, at the 90% level of confidence, the data do not support the claim that the two cars have different mean parking times μ_1 and μ_2. That is, the value $\mu_D = \mu_1 - \mu_2 = 0$ is not inconsistent with the observed data. ∎

EXERCISES FOR SECTION 5-4

5-30. Consider the Minitab output shown on the right.
(a) Fill in the missing values in the output, including a bound on the P-value. Can the null hypothesis be rejected at the 0.05 level? Why?
(b) Is this a one-sided or a two-sided test?
(c) Use the output and the t table to find a 99% two-sided CI on the difference in means.
(d) How can you tell that there is insufficient evidence to reject the null hypothesis just by looking at the computer output and not making any additional computations?

Paired T-Test and CI: X1, X2

Paired T for X1 − X2

	N	Mean	StDev	SE Mean
X1	12	74.2430	1.8603	0.5370
X2	12	73.4797	1.9040	0.5496
Difference	12	?	2.905560	0.838763

95% CI for mean difference: $(-1.082805, 2.609404)$
T-Test of mean difference = 0 (vs not = 0):
T-Value = ?, P-Value = ?

5-31. Consider the Minitab output shown below.

> **Paired T-Test and CI: X1, X2**
>
> Paired T for X1 − X2
>
> | | N | Mean | StDev | SE Mean |
> |---|---|------|-------|---------|
> | X1 | 10 | 100.642 | ? | 0.488 |
> | X2 | 10 | 105.574 | 2.867 | 0.907 |
> | Difference | 10 | −4.93262 | 3.66736 | ? |
>
> 95% CI for mean difference: (−7.55610, −2.30915)
> T-Test of mean difference = 0 (vs not = 0):
> T-Value = ? P-Value = 0.002

(a) Fill in the missing values in the output. Can the null hypothesis be rejected at the 0.05 level? Why?
(b) Is this a one-sided or a two-sided test?
(c) Use the output and the t table to find a 99% two-sided CI on the difference in means.
(d) What is the P-value for the test statistic if the objective of this experiment was to demonstrate that the mean of population 1 is smaller than the mean of population 2?
(e) What is the P-value for the test statistic if the objective of this experiment was to demonstrate that the difference in means is equal to 4?

5-32. Consider the shear strength experiment described in Example 5-8. Construct a 95% CI on the difference in mean shear strength for the two methods. Is the result you obtained consistent with the findings in Example 5-8? Explain why.

5-33. Reconsider the shear strength experiment described in Example 5-8. Does each of the individual shear strengths have to be normally distributed for the paired t-test to be appropriate, or is it only the difference in shear strengths that must be normal? Use a normal probability plot to investigate the normality assumption.

5-34. Consider the parking data in Example 5-9. Use the paired t-test to investigate the claim that the two types of cars have different levels of difficulty to parallel park. Use $\alpha = 0.10$. Compare your results with the confidence interval constructed in Example 5-9 and comment on why they are the same or different.

5-35. Reconsider the parking data in Example 5-9. Investigate the assumption that the differences in parking times are normally distributed.

5-36. The manager of a fleet of automobiles is testing two brands of radial tires. She assigns one tire of each brand at random to the two rear wheels of eight cars and runs the cars until the tires wear out. The data are shown here (in kilometers). Find a 99% CI on the difference in mean life. Which brand would you prefer, based on this calculation?

Car	Brand 1	Brand 2	Car	Brand 1	Brand 2
1	36,925	34,318	5	37,210	38,015
2	45,300	42,280	6	48,360	47,800
3	36,240	35,500	7	38,200	37,810
4	32,100	31,950	8	33,500	33,215

5-37. A computer scientist is investigating the usefulness of two different design languages in improving programming tasks. Twelve expert programmers, familiar with both languages, are asked to code a standard function in both languages, and the time in minutes is recorded. The data are shown here:

Programmer	Time Design Language 1	Time Design Language 2
1	17	18
2	16	14
3	21	19
4	14	11
5	18	23
6	24	21
7	16	10
8	14	13
9	21	19
10	23	24
11	13	15
12	18	20

(a) Find a 95% CI on the difference in mean coding times. Is there any indication that one design language is preferable?
(b) Is it reasonable to assume that the difference in coding time is normally distributed? Show evidence to support your answer.

5-38. Fifteen adult males between the ages of 35 and 50 participated in a study to evaluate the effect of diet and exercise on blood cholesterol levels. The total cholesterol was measured in each subject initially, and then 3 months after participating in an aerobic exercise program and switching to a low-fat diet, as shown in the following table.

(a) Do the data support the claim that low-fat diet and aerobic exercise are of value in producing a mean reduction in blood cholesterol levels? Use the P-value approach.
(b) Find a 95% CI on the mean reduction in blood cholesterol level.

Blood Cholesterol Level					
Subject	Before	After	Subject	Before	After
1	265	229	9	260	247
2	240	231	10	279	239
3	258	227	11	283	246
4	295	240	12	240	218
5	251	238	13	238	219
6	245	241	14	225	226
7	287	234	15	247	233
8	314	256			

to 158°F for 168 hours. The tensile strength of the specimens of this material is measured before and after the aging process. The following data (in psi) are recorded:

Specimen	Original	Aged	Specimen	Original	Aged
1	215	203	6	231	218
2	226	216	7	234	224
3	226	217	8	219	210
4	219	211	9	209	201
5	222	215	10	216	207

5-39. An article in the *Journal of Aircraft* (Vol. 23, 1986, pp. 859–864) describes a new equivalent plate analysis method formulation that is capable of modeling aircraft structures such as cranked wing boxes and that produces results similar to the more computationally intensive finite element analysis method. Natural vibration frequencies for the cranked wing box structure are calculated using both methods, and results for the first seven natural frequencies are shown here.

	Finite Element, Cycle/s	Equivalent Plate, Cycle/s		Finite Element, Cycle/s	Equivalent Plate, Cycle/s
Car			Car		
1	14.58	14.76	5	174.73	181.22
2	48.52	49.10	6	212.72	220.14
3	97.22	99.99	7	277.38	294.80
4	113.99	117.53			

(a) Do the data suggest that the two methods provide the same mean value for natural vibration frequency? Use the *P*-value approach.
(b) Find a 95% CI on the mean difference between the two methods and use it to answer the question in part (a).

5-40. The Federal Aviation Administration requires material used to make evacuation systems retain their strength over the life of the aircraft. In an accelerated life test, the principal material, polymer-coated nylon weave, is aged by exposing it

(a) Is there evidence to support the claim that the nylon weave tensile strength is the same before and after the aging process? Use the *P*-value approach.
(b) Find a 99% CI on the mean difference in tensile strength and use it to answer the question from part (a).

5-41. Two different analytical tests can be used to determine the impurity level in steel alloys. Eight specimens are tested using both procedures, and the results are shown in the following tabulation. Is there sufficient evidence to conclude that both tests give the same mean impurity level, using $\alpha = 0.01$?

Specimen	Test 1	Test 2	Specimen	Test 1	Test 2
1	1.2	1.4	5	1.7	2.0
2	1.3	1.7	6	1.8	2.1
3	1.5	1.5	7	1.4	1.7
4	1.4	1.3	8	1.3	1.6

5-42. Consider the tensile strength data in Exercise 5-40. Is there evidence to support the claim that the accelerated life test will result in a mean loss of at least 5 psi? Use $\alpha = 0.05$.

5-43. Consider the impurity level data in Exercise 5-41. Construct a 99% CI on the mean difference between the two testing procedures. Use it to answer the question posed in Exercise 5-41.

5-5 INFERENCE ON THE RATIO OF VARIANCES OF TWO NORMAL POPULATIONS

We now introduce tests and CIs for the two population variances shown in Fig. 5-1. We will assume that both populations are normal. Both the hypothesis testing and confidence interval procedures are relatively sensitive to the normality assumption.

5-5.1   Hypothesis Testing on the Ratio of Two Variances

Suppose that two independent normal populations are of interest, where the population means and variances—say, μ_1, σ_1^2, μ_2, and σ_2^2—are unknown. We wish to test hypotheses about the equality of the two variances—say, H_0: $\sigma_1^2 = \sigma_2^2$. Assume that two random samples of size n_1 from population 1 and of size n_2 from population 2 are available, and let S_1^2 and S_2^2 be the sample variances. We wish to test the hypotheses

$$H_0: \sigma_1^2 = \sigma_2^2$$
$$H_1: \sigma_1^2 \neq \sigma_2^2$$

The development of a test procedure for these hypotheses requires a new probability distribution.

The F Distribution

One of the most useful distributions in statistics is the F distribution. The random variable F is defined to be the ratio of two independent chi-square random variables, each divided by its number of degrees of freedom. That is,

$$F = \frac{W/u}{Y/v}$$

where W and Y are independent chi-square random variables with u and v degrees of freedom, respectively. We now formally state the sampling distribution of F.

The F Distribution

Let W and Y be independent chi-square random variables with u and v degrees of freedom, respectively. Then the ratio

$$F = \frac{W/u}{Y/v} \tag{5-18}$$

has the probability density function

$$f(x) = \frac{\Gamma\left(\dfrac{u+v}{2}\right)\left(\dfrac{u}{v}\right)^{u/2} x^{(u/2)-1}}{\Gamma\left(\dfrac{u}{2}\right)\Gamma\left(\dfrac{v}{2}\right)\left[\left(\dfrac{u}{v}\right)x + 1\right]^{(u+v)/2}}, 0 < x < \infty \tag{5-19}$$

and is said to follow the F distribution with u degrees of freedom in the numerator and v degrees of freedom in the denominator. It is usually abbreviated as $F_{u,v}$.

The mean and variance of the F distribution are $\mu = v/(v-2)$ for $v > 2$, and

$$\sigma^2 = \frac{2v^2(u+v-2)}{u(v-2)^2(v-4)}, \qquad v > 4$$

Two F distributions are shown in Fig. 5-4. The F random variable is nonnegative, and the distribution is skewed to the right. The F distribution looks very similar to the chi-square distribution in Fig. 4-21; however, the two parameters u and v provide extra flexibility regarding shape.

The percentage points of the F distribution are given in Table IV of Appendix A. Let $f_{\alpha,u,v}$ be the percentage point of the F distribution, with numerator degrees of freedom u and denominator degrees of freedom v such that the probability that the random variable F exceeds this value is

$$P(F > f_{\alpha,u,v}) = \int_{f_{\alpha,u,v}}^{\infty} f(x)\, dx = \alpha$$

This is illustrated in Fig. 5-5. For example, if $u = 5$ and $v = 10$, we find from Table IV of Appendix A that

$$P(F > f_{0.05,5,10}) = P(F_{5,10} > 3.33) = 0.05$$

That is, the upper 5 percentage point of $F_{5,10}$ is $f_{0.05,5,10} = 3.33$.

Table IV contains only upper-tail percentage points (for selected values of $f_{\alpha,u,v}$ for $\alpha \le 0.25$) of the F distribution. The lower-tail percentage points $f_{1-\alpha,u,v}$ can be found as follows:

$$f_{1-\alpha,u,v} = \frac{1}{f_{\alpha,v,u}} \tag{5-20}$$

Find a Lower-Tail Percentage Point of the F-Distribution

For example, to find the lower-tail percentage point $f_{0.95,5,10}$, note that

$$f_{0.95,5,10} = \frac{1}{f_{0.05,10,5}} = \frac{1}{4.74} = 0.211$$

The Test Procedure

A hypothesis testing procedure for the equality of two variances is based on the following result.

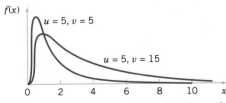

Figure 5-4   Probability density functions of two F distributions.

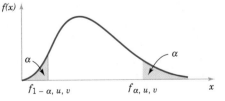

Figure 5-5   Upper and lower percentage points of the F distribution.

Let $X_{11}, X_{12}, \ldots, X_{1n_1}$ be a random sample from a normal population with mean μ_1 and variance σ_1^2, and let $X_{21}, X_{22}, \ldots, X_{2n_2}$ be a random sample from a second normal population with mean μ_2 and variance σ_2^2. Assume that both normal populations are independent. Let S_1^2 and S_2^2 be the sample variances. Then the ratio

$$F = \frac{S_1^2/\sigma_1^2}{S_2^2/\sigma_2^2}$$

has an F distribution with $n_1 - 1$ numerator degrees of freedom and $n_2 - 1$ denominator degrees of freedom.

This result is based on the fact that $(n_1 - 1)S_1^2/\sigma_1^2$ is a chi-square random variable with $n_1 - 1$ degrees of freedom, that $(n_2 - 1)S_2^2/\sigma_2^2$ is a chi-square random variable with $n_2 - 1$ degrees of freedom, and that the two normal populations are independent. Clearly under the null hypothesis $H_0:\sigma_1^2 = \sigma_2^2$ the ratio $F_0 = S_1^2/S_2^2$ has an F_{n_1-1,n_2-1} distribution. This is the basis of the following test procedure.

Summary

Testing Hypotheses on the Equality of Variances of Two Normal Distributions

Null hypothesis: $H_0: \sigma_1^2 = \sigma_2^2$

Test statistic: $F_0 = \dfrac{S_1^2}{S_2^2}$ (5-21)

Alternative Hypotheses	Rejection Criterion
$H_1: \sigma_1^2 \neq \sigma_2^2$	$f_0 > f_{\alpha/2,n_1-1,n_2-1}$ or $f_0 < f_{1-\alpha/2,n_1-1,n_2-1}$
$H_1: \sigma_1^2 > \sigma_2^2$	$f_0 > f_{\alpha,n_1-1,n_2-1}$
$H_1: \sigma_1^2 < \sigma_2^2$	$f_0 < f_{1-\alpha,n_1-1,n_2-1}$

The critical regions are shown in Fig. 5-6.

EXAMPLE 5-10
Etching Wafers

Oxide layers on semiconductor wafers are etched in a mixture of gases to achieve the proper thickness. The variability in the thickness of these oxide layers is a critical characteristic of the wafer, and low variability is desirable for subsequent processing steps. Two different mixtures of gases are being studied to

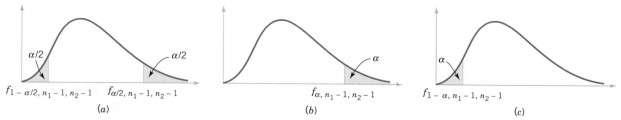

Figure 5-6 The F distribution for the test of $H_0: \sigma_1^2 = \sigma_2^2$ with critical region values for (a) $H_1: \sigma_1^2 \neq \sigma_2^2$, (b) $H_1: \sigma_1^2 > \sigma_2^2$, and (c) $H_1: \sigma_1^2 < \sigma_2^2$.

determine whether one is superior in reducing the variability of the oxide thickness. Sixteen wafers are etched in each gas. The sample standard deviations of oxide thickness are $s_1 = 1.96$ angstroms and $s_2 = 2.13$ angstroms, respectively. Is there any evidence to indicate that either gas is preferable? Use a fixed-level test with $\alpha = 0.05$.

Solution. The seven-step hypothesis testing procedure may be applied to this problem as follows:

1. **Parameter of interest:** The parameter of interest are the variances of oxide thickness σ_1^2 and σ_2^2. We will assume that oxide thickness is a normal random variable for both gas mixtures.

2. **Null hypothesis, H_0:** $\sigma_1^2 = \sigma_2^2$

3. **Alternative hypothesis, H_1:** $\sigma_1^2 \neq \sigma_2^2$

4. **Test statistic:** The test statistic is given by equation 5-21:

$$f_0 = \frac{s_1^2}{s_2^2}$$

5. **Reject H_0 if:** Because $n_1 = n_2 = 16$ and $\alpha = 0.05$, we will reject $H_0: \sigma_1^2 = \sigma_2^2$ if $f_0 > f_{0.025,15,15} = 2.86$ or if $f_0 < f_{0.975,15,15} = 1/f_{0.025,15,15} = 1/2.86 = 0.35$. Refer to Figure 5-6a.

6. **Computations:** Because $s_1^2 = (1.96)^2 = 3.84$ and $s_2^2 = (2.13)^2 = 4.54$, the test statistic is

$$f_0 = \frac{s_1^2}{s_2^2} = \frac{3.84}{4.54} = 0.85$$

7. **Conclusions:** Because $f_{0.975,15,15} = 0.35 < 0.85 < f_{0.025,15,15} = 2.86$, we cannot reject the null hypothesis $H_0: \sigma_1^2 = \sigma_2^2$ at the 0.05 level of significance. Therefore, there is no strong evidence to indicate that either gas results in a smaller variance of oxide thickness. ■

P-Values for the F-Test

The P-value approach can also be used with F-tests. To show how to do this, consider the upper-tailed one-tailed test. The P-value is the area (probability) under the F distribution with $n_1 - 1$ and $n_2 - 1$ degrees of freedom that lies beyond the computed value of the test statistic f_0. Appendix A Table IV can be used to obtain upper and lower bounds on the P-value. For example, consider an F-test with 9 numerator and 14 denominator degrees of freedom for which $f_0 = 3.05$. From Appendix A Table IV we find that $f_{0.05,9,14} = 2.65$ and $f_{0.025,9,14} = 3.21$, so because $f_0 = 3.05$ lies between these two values, the P-value is between 0.05 and 0.025; that is, $0.025 < P < 0.05$. The P-value for a lower-tailed test would be found similarly, although since Appendix A Table IV contains only upper-tail points of the F distribution, equation 5-20 would have to be used to find the necessary lower-tail points. For a two-tailed test, the bounds obtained from a one-tail test would be doubled to obtain the P-value.

To illustrate calculating bounds on the P-value for a two-tailed F-test, reconsider Example 5-10. The computed value of the test statistic in this example is $f_0 = 0.85$. This value falls in the lower tail of the $F_{15,15}$ distribution. The lower-tail point that has 0.25 probability to the left of it is $f_{0.75,15,15} = 1/f_{0.25,15,15} = 1/1.43 = 0.70$, and since $0.70 < 0.85$, the probability that lies to the left of 0.85 exceeds 0.25. Therefore, we would conclude that the P-value for $f_0 = 0.85$ is greater than $2(0.25) = 0.5$, so there is insufficient evidence to reject the null hypothesis. This is consistent with the original conclusions from Example 5-10. The actual P-value is 0.7570. This value was obtained from a calculator from which we found that $P(F_{15,15} \leq 0.85) = 0.3785$ and $2(0.3785) = 0.7570$. Minitab can also be used to calculate the required probabilities.

Finding the P-Value for
Example 5-10

5-5.2   Confidence Interval on the Ratio of Two Variances

To find the CI, recall that the sampling distribution of

$$F = \frac{S_2^2/\sigma_2^2}{S_1^2/\sigma_1^2}$$

is an F with $n_2 - 1$ and $n_1 - 1$ degrees of freedom. *Note:* We start with S_2^2 in the numerator and S_1^2 in the denominator to simplify the algebra used to obtain an interval for σ_1^2/σ_2^2. Therefore,

$$P(f_{1-\alpha/2,n_2-1,n_1-1} \le F \le f_{\alpha/2,n_2-1,n_1-1}) = 1 - \alpha$$

Substitution for F and manipulation of the inequalities will lead to the following $100(1 - \alpha)\%$ CI for σ_1^2/σ_2^2.

Definition

Confidence Interval on the Ratio of Variances of Two Normal Distributions

If s_1^2 and s_2^2 are the sample variances of random samples of sizes n_1 and n_2, respectively, from two independent normal populations with unknown variances σ_1^2 and σ_2^2, a $100(1 - \alpha)\%$ CI on the ratio σ_1^2/σ_2^2 is

$$\frac{s_1^2}{s_2^2} f_{1-\alpha/2,n_2-1,n_1-1} \le \frac{\sigma_1^2}{\sigma_2^2} \le \frac{s_1^2}{s_2^2} f_{\alpha/2,n_2-1,n_1-1} \qquad (5\text{-}22)$$

where $f_{\alpha/2,n_2-1,n_1-1}$ and $f_{1-\alpha/2,n_2-1,n_1-1}$ are the upper and lower $100\,\alpha/2$ percentage points of the F distribution with $n_2 - 1$ numerator and $n_1 - 1$ denominator degrees of freedom, respectively.

EXAMPLE 5-11

Surface Finish

A company manufactures impellers for use in jet-turbine engines. One of the operations involves grinding a particular surface finish on a titanium alloy component. Two different grinding processes can be used, and both processes can produce parts at identical mean surface roughness. The manufacturing engineer would like to select the process having the least variability in surface roughness. A random sample of $n_1 = 11$ parts from the first process results in a sample standard deviation $s_1 = 5.1$ microinches, and a random sample of $n_2 = 16$ parts from the second process results in a sample standard deviation of $s_2 = 4.7$ microinches. We need to find a 90% CI on the ratio of the two variances σ_1^2/σ_2^2.

Solution. Assuming that the two processes are independent and that surface roughness is normally distributed, we can use equation 5-22 as follows:

$$\frac{s_1^2}{s_2^2} f_{0.95,15,10} \le \frac{\sigma_1^2}{\sigma_2^2} \le \frac{s_1^2}{s_2^2} f_{0.05,15,10}$$

$$\frac{(5.1)^2}{(4.7)^2} 0.39 \le \frac{\sigma_1^2}{\sigma_2^2} \le \frac{(5.1)^2}{(4.7)^2} 2.85$$

or

$$0.46 \leq \frac{\sigma_1^2}{\sigma_2^2} \leq 3.36$$

Note that we have used equation 5-20 to find $f_{0.95,15,10} = 1/f_{0.05,10,15} = 1/2.54 = 0.39$. Because this CI includes unity, we cannot claim that the standard deviations of surface roughness for the two processes are different at the 90% level of confidence. ∎

One-Sided Confidence Bounds

To find a $100(1 - \alpha)\%$ lower confidence bound on σ_1^2/σ_2^2, simply replace $f_{1-\alpha/2,n_2-1,n_1-1}$ with $f_{1-\alpha,n_2-1,n_1-1}$ in the lower bound of equation 5-22; the upper bound is set to ∞. Similarly, to find a $100(1 - \alpha)\%$ upper confidence bound on σ_1^2/σ_2^2, simply replace $f_{\alpha/2,n_2-1,n_1-1}$ with f_{α,n_2-1,n_1-1} in the upper bound of equation 5-22; the lower bound is set to 0. To find the CI or confidence bounds of σ_1/σ_2, simply take the square root of the ends of the interval or bounds.

EXERCISES FOR SECTION 5-5

5-44. For an F distribution, find the following:
(a) $f_{0.25,5,10}$ (b) $f_{0.10,24,9}$
(c) $f_{0.05,8,15}$ (d) $f_{0.75,5,10}$
(e) $f_{0.90,24,9}$ (f) $f_{0.95,8,15}$

5-45. For an F distribution, find the following:
(a) $f_{0.25,7,15}$ (b) $f_{0.10,10,12}$
(c) $f_{0.01,20,10}$ (d) $f_{0.75,7,15}$
(e) $f_{0.90,10,12}$ (f) $f_{0.99,20,10}$

5-46. An experiment was conducted to compare the variances of two independent normal populations. The sample sizes from both populations was 10 and the computed value of the F-statistic was $f_0 = 4.45$. Find a bound on the P-value for this test statistic.

5-47. An experiment was conducted to compare the variances of two independent normal populations. The null hypothesis was $H_0: \sigma_1^2 = \sigma_2^2$ versus $H_1: \sigma_1^2 > \sigma_2^2$. The sample sizes from both populations were 16, and the computed value of the F-statistic was $f_0 = 2.75$. Find a bound on the P-value for this test statistic.

5-48. Eleven resilient modulus observations of a ceramic mixture of type A are measured and found to have a sample average of 18.42 psi and sample standard deviation of 2.77 psi. Ten resilient modulus observations of a ceramic mixture of type B are measured and found to have a sample average of 19.28 psi and sample standard deviation of 2.41 psi. Is there sufficient evidence to support the investigator's claim that type A ceramic has larger variability than type B? Use $\alpha = 0.05$.

5-49. Consider the etch rate data in Exercise 5-21. Test the hypothesis $H_0: \sigma_1^2 = \sigma_2^2$ against $H_1: \sigma_1^2 \neq \sigma_2^2$ using $\alpha = 0.05$, and draw conclusions.

5-50. Consider the diameter data in Exercise 5-17. Construct the following:
(a) A 90% two-sided CI on σ_1/σ_2.
(b) A 95% two-sided CI on σ_1/σ_2. Comment on the comparison of the width of this interval with the width of the interval in part (a).
(c) A 90% lower-confidence bound on σ_1/σ_2.

5-51. Consider the foam data in Exercise 5-18. Construct the following:
(a) A 90% two-sided CI on σ_1^2/σ_2^2.
(b) A 95% two-sided CI on σ_1^2/σ_2^2. Comment on the comparison of the width of this interval with the width of the interval in part (a).
(c) A 90% lower-confidence bound on σ_1/σ_2.

5-52. Consider the film data in Exercise 5-23. Test $H_0: \sigma_1^2 = \sigma_2^2$ versus $H_1: \sigma_1^2 \neq \sigma_2^2$ using $\alpha = 0.02$.

5-53. Consider the gear impact strength data in Exercise 5-22. Is there sufficient evidence to conclude that the variance of impact strength is different for the two suppliers? Use $\alpha = 0.05$.

5-54. Consider the melting point data in Exercise 5-24. Do the sample data support a claim that both alloys have the same variance of melting point? Use $\alpha = 0.05$ in reaching your conclusion.

5-55. Exercise 5-27 presented measurements of plastic coating thickness at two different application temperatures. Test the appropriate hypothesis to demonstrate that the variance of the thickness is less for the 125°F process than the 150°F process, using $\alpha = 0.10$.

5-56. A study was performed to determine whether men and women differ in their repeatability in assembling components on printed circuit boards. Two samples of 25 men and

21 women were selected, and each subject assembled the units. The two sample standard deviations of assembly time were $s_{\text{men}} = 0.914$ min and $s_{\text{women}} = 1.093$ min. Is there evidence to support the claim that men have less repeatability than women for this assembly task? Use $\alpha = 0.01$ and state

any necessary assumptions about the underlying distribution of the data.

5-57. Reconsider the assembly repeatability experiment described in Exercise 5-56 . Find a 99% lower bound on the ratio of the two variances. Provide an interpretation of the interval.

5-6 INFERENCE ON TWO POPULATION PROPORTIONS

We now consider the case in which there are two binomial parameters of interest—say, p_1 and p_2—and we wish to draw inferences about these proportions. We will present large-sample hypothesis testing and CI procedures based on the normal approximation to the binomial.

5-6.1 Hypothesis Testing on the Equality of Two Binomial Proportions

Suppose that the two independent random samples of sizes n_1 and n_2 are taken from two populations, and let X_1 and X_2 represent the number of observations that belong to the class of interest in samples 1 and 2, respectively. Furthermore, suppose that the normal approximation to the binomial is applied to each population so that the estimators of the population proportions $\hat{P}_1 = X_1/n_1$ and $\hat{P}_2 = X_2/n_2$ have approximately normal distributions. We are interested in testing the hypotheses

$$H_0: p_1 = p_2$$
$$H_1: p_1 \neq p_2$$

The quantity

$$Z = \frac{\hat{P}_1 - \hat{P}_2 - (p_1 - p_2)}{\sqrt{\dfrac{p_1(1 - p_1)}{n_1} + \dfrac{p_2(1 - p_2)}{n_2}}} \tag{5-23}$$

has approximately a standard normal distribution, $N(0, 1)$.

This result is the basis of a test for $H_0: p_1 = p_2$. Specifically, if the null hypothesis $H_0: p_1 = p_2$ is true, using the fact that $p_1 = p_2 = p$, the random variable

$$Z = \frac{\hat{P}_1 - \hat{P}_2}{\sqrt{p(1 - p)\left(\dfrac{1}{n_1} + \dfrac{1}{n_2}\right)}}$$

is distributed approximately $N(0, 1)$. An estimator of the common parameter p is

$$\hat{P} = \frac{X_1 + X_2}{n_1 + n_2}$$

The test statistic for $H_0: p_1 = p_2$ is then

$$Z_0 = \frac{\hat{P}_1 - \hat{P}_2}{\sqrt{\hat{P}(1 - \hat{P})\left(\dfrac{1}{n_1} + \dfrac{1}{n_2}\right)}}$$

This leads to the following test procedures.

Testing Hypotheses on the Equality of Two Binomial Proportions

Null hypothesis: $H_0: p_1 = p_2$

Test statistic: $Z_0 = \dfrac{\hat{P}_1 - \hat{P}_2}{\sqrt{\hat{P}(1 - \hat{P})\left(\dfrac{1}{n_1} + \dfrac{1}{n_2}\right)}}$ (5-24)

Alternative Hypotheses	P-Value	Rejection Criterion for Fixed-Level Tests
$H_1: p_1 \neq p_2$	Probability above z_0 and probability below $-z_0$, $P = 2[1 - \Phi(|z_0|)]$	$z_0 > z_{\alpha/2}$ or $z_0 < -z_{\alpha/2}$
$H_1: p_1 > p_2$	Probability above z_0, $P = 1 - \Phi(z_0)$	$z_0 > z_{\alpha}$
$H_1: p_1 < p_2$	Probability below z_0, $P = \Phi(z_0)$	$z_0 < -z_{\alpha}$

EXAMPLE 5-12

Interocular Lenses

Two different types of polishing solution are being evaluated for possible use in a tumble-polish operation for manufacturing interocular lenses used in the human eye following cataract surgery. Three hundred lenses were tumble-polished using the first polishing solution, and of this number 253 had no polishing-induced defects. Another 300 lenses were tumble-polished using the second polishing solution, and 196 lenses were satisfactory on completion. Is there any reason to believe that the two polishing solutions differ?

Solution. The seven-step hypothesis procedure leads to the following results:

1. **Parameter of interest:** The parameter of interest are p_1 and p_2, the proportion of lenses that are satisfactory following tumble-polishing with polishing fluid 1 or 2.

2. **Null hypothesis, H_0:** $p_1 = p_2$
3. **Alternative hypothesis, H_1:** $p_1 \neq p_2$
4. **Test statistic:** The test statistic is

$$z_0 = \frac{\hat{p}_1 - \hat{p}_2}{\sqrt{\hat{p}(1 - \hat{p})\left(\dfrac{1}{n_1} + \dfrac{1}{n_2}\right)}}$$

where $\hat{p}_1 = 253/300 = 0.8433$, $\hat{p}_2 = 196/300 = 0.6533$, $n_1 = n_2 = 300$, and

$$\hat{p} = \frac{x_1 + x_2}{n_1 + n_2} = \frac{253 + 196}{300 + 300} = 0.7483$$

5. **Reject H_0 if:** Reject H_0: $p_1 = p_2$ if the P-value is less than 0.05.
6. **Computations:** The value of the test statistic is

$$z_0 = \frac{0.8433 - 0.6533}{\sqrt{0.7483(0.2517)\left(\dfrac{1}{300} + \dfrac{1}{300}\right)}} = 5.36$$

7. **Conclusions:** Because $z_0 = 5.36$, the P-value is $P = 2[1 - \Phi(5.36)] \approx 0$, we reject the null hypothesis. This is the closest we can get to the exact P-value using Appendix A Table I. Using a calculator, we can find a better approximate P-value as $P \approx 8.32 \times 10^{-8}$. There is strong evidence to support the claim that the two polishing fluids are different. Fluid 1 produces a higher fraction of nondefective lenses. ∎

Minitab Test of Two Proportions and Confidence Interval Output for Example 5-12

Test and CI for Two Proportions

Sample	X	N	Sample p
1	253	300	0.843333
2	196	300	0.653333

Difference = p (1) − p (2)
Estimate for difference: 0.19
95% CI for difference: (0.122236, 0.257764)
Test for difference = 0 (vs not = 0): Z = 5.36, P-Value = 0.000

The results agree with the manual calculations. In addition to the hypothesis test results, Minitab reports a two-sided CI on the difference in the two proportions. We will give the equation for constructing the CI in Section 5-6.3.

5-6.2 Type II Error and Choice of Sample Size

The computation of the β-error for the foregoing test is somewhat more involved than in the single-sample case. The problem is that the denominator of Z_0 is an estimate of the standard deviation of $\hat{P}_1 - \hat{P}_2$ under the assumption that $p_1 = p_2 = p$. When H_0: $p_1 = p_2$ is false, the standard deviation of $\hat{P}_1 - \hat{P}_2$ is

$$\sigma_{\hat{P}_1 - \hat{P}_2} = \sqrt{\frac{p_1(1 - p_1)}{n_1} + \frac{p_2(1 - p_2)}{n_2}} \tag{5-25}$$

If the alternative hypothesis is two-sided, the β-error is

$$
\beta = \Phi\left[\frac{z_{\alpha/2}\sqrt{\overline{pq}(1/n_1 + 1/n_2)} - (p_1 - p_2)}{\sigma_{\hat{P}_1 - \hat{P}_2}}\right]
$$
$$
- \Phi\left[\frac{-z_{\alpha/2}\sqrt{\overline{pq}(1/n_1 + 1/n_2)} - (p_1 - p_2)}{\sigma_{\hat{P}_1 - \hat{P}_2}}\right] \tag{5-26}
$$

where

$$
\overline{p} = \frac{n_1 p_1 + n_2 p_2}{n_1 + n_2}
$$

$$
\overline{q} = \frac{n_1(1 - p_1) + n_2(1 - p_2)}{n_1 + n_2} = 1 - \overline{p}
$$

and $\sigma_{\hat{P}_1 - \hat{P}_2}$ is given by equation 5-25.

If the alternative hypothesis is $H_1: p_1 > p_2$, then

$$
\beta = \Phi\left[\frac{z_{\alpha}\sqrt{\overline{pq}(1/n_1 + 1/n_2)} - (p_1 - p_2)}{\sigma_{\hat{P}_1 - \hat{P}_2}}\right] \tag{5-27}
$$

and if the alternative hypothesis is $H_1: p_1 < p_2$,

$$
\beta = 1 - \Phi\left[\frac{-z_{\alpha}\sqrt{\overline{pq}(1/n_1 + 1/n_2)} - (p_1 - p_2)}{\sigma_{\hat{P}_1 - \hat{P}_2}}\right] \tag{5-28}
$$

For a specified pair of values p_1 and p_2, we can find the sample sizes $n_1 = n_2 = n$ required to give the test of size α that has specified type II error β. The formula is as follows.

Sample Size for a Two-Sided Hypothesis Test on the Difference in Two Binomial Proportions

For the two-sided alternative, the common sample size is

$$
n = \frac{\left(z_{\alpha/2}\sqrt{(p_1 + p_2)(q_1 + q_2)/2} + z_{\beta}\sqrt{p_1 q_1 + p_2 q_2}\right)^2}{(p_1 - p_2)^2} \tag{5-29}
$$

where $q_1 = 1 - p_1$ and $q_2 = 1 - p_2$.

For a one-sided alternative, replace $z_{\alpha/2}$ in equation 5-29 by z_α.

5-6.3   Confidence Interval on the Difference in Binomial Proportions

The CI for $p_1 - p_2$ can be found directly because we know that

$$z = \frac{\hat{P}_1 - \hat{P}_2 - (p_1 - p_2)}{\sqrt{\dfrac{p_1(1 - p_1)}{n_1} + \dfrac{p_2(1 - p_2)}{n_2}}}$$

is a standard normal random variable. Thus

$$P(-z_{\alpha/2} \le Z \le z_{\alpha/2}) \simeq 1 - \alpha$$

so we can substitute for Z in this last expression and use an approach similar to the one employed previously to find the following approximate $100(1 - \alpha)\%$ CI for $p_1 - p_2$.

Confidence Interval on the Difference in Binomial Proportions

If $\hat{p}_1$ and $\hat{p}_2$ are the sample proportions of observation in two independent random samples of sizes n_1 and n_2 that belong to a class of interest, an approximate $100(1 - \alpha)\%$ CI on the difference in the true proportions $p_1 - p_2$ is

$$\hat{p}_1 - \hat{p}_2 - z_{\alpha/2}\sqrt{\frac{\hat{p}_1(1 - \hat{p}_1)}{n_1} + \frac{\hat{p}_2(1 - \hat{p}_2)}{n_2}}$$

$$\le p_1 - p_2 \le \hat{p}_1 - \hat{p}_2 + z_{\alpha/2}\sqrt{\frac{\hat{p}_1(1 - \hat{p}_1)}{n_1} + \frac{\hat{p}_2(1 - \hat{p}_2)}{n_2}} \tag{5-30}$$

where $z_{\alpha/2}$ is the upper $\alpha/2$ percentage point of the standard normal distribution.

EXAMPLE 5-13
Crankshaft
Bearings

Consider the process manufacturing crankshaft bearings described in Example 4-14. Suppose that a modification is made in the surface finishing process and that, subsequently, a second random sample of 85 axle shafts is obtained. The number of defective shafts in this second sample is 8. Is there evidence to support a claim that the process change has led to an improvement in the surface finish of the bearings?

Solution. We will answer this question by finding a CI on the difference in the proportion of defective bearings before and after the process change. Because $n_1 = 85, \hat{p}_1 = 0.12, n_2 = 85$, and $\hat{p}_2 = 8/85 = 0.09$, we can obtain an approximate 95% CI on the difference in the proportion of defective bearings produced under the two processes from equation 5-30 as follows:

$$\hat{p}_1 - \hat{p}_2 - z_{0.025}\sqrt{\frac{\hat{p}_1(1 - \hat{p}_1)}{n_1} + \frac{\hat{p}_2(1 - \hat{p}_2)}{n_2}}$$

$$\le p_1 - p_2 \le \hat{p}_1 - \hat{p}_2 + z_{0.025}\sqrt{\frac{\hat{p}_1(1 - \hat{p}_1)}{n_1} + \frac{\hat{p}_2(1 - \hat{p}_2)}{n_2}}$$

or

$$0.12 - 0.09 - 1.96\sqrt{\frac{0.12(0.88)}{85} + \frac{0.09(0.91)}{85}}$$

$$\leq p_1 - p_2 \leq 0.12 - 0.09 + 1.96\sqrt{\frac{0.12(0.88)}{85} + \frac{0.09(0.91)}{85}}$$

This simplifies to

$$-0.06 \leq p_1 - p_2 \leq 0.12$$

This CI includes zero, so, based on the sample data, it seems unlikely that the changes made in the surface finish process have reduced the proportion of defective crankshaft bearings being produced. ■

EXERCISES FOR SECTION 5-6

5-58. Fill in the blanks in the Minitab output shown below.

Test and CI for Two Proportions

Sample	X	N	Sample p
1	285	500	0.570000
2	521	?	0.651250

Difference = p (1) − p (2)
Estimate for difference: ?
95% CI for difference: (−0.135782, −0.0267185)
Test for difference = 0 (vs not = 0): Z = ?
P-Value = 0.003

(a) Is this a one-sided or a two-sided test?
(b) Can the null hypothesis be rejected at the 0.05 level?
(c) Can the null hypothesis H_0: $p_1 = p_2$ versus H_0: $p_1 < p_2$ be rejected at the 0.05 level? How can you do this without performing any additional calculations?
(d) Can the null hypothesis H_0: $p_1 - p_2 = -0.02$ versus H_0: $p_1 - p_2 \neq -0.02$ be rejected at the 0.05 level? Can you do this without performing any additional calculations?
(e) Construct an approximate 90% CI for p.

5-59. Fill in the blanks in the Minitab output shown below.

Test and CI for Two Proportions

Sample	X	N	Sample p
1	190	250	0.760000
2	240	350	0.685714

Difference = p (1) − p (2)
Estimate for difference: ?
95% lower bound for difference: 0.0139543
Test for difference = 0 (vs > 0): Z = ? P-Value = ?

(a) Is this a one-sided or a two-sided test?
(b) Can the null hypothesis be rejected at the 0.05 level?
(c) Can the null hypothesis H_0: $p_1 = p_2$ versus H_0: $p_1 > p_2$ be rejected at the 0.05 level? How can you do this without performing any additional calculations?
(d) Construct an approximate 95% two-sided CI for p.

5-60. Two different types of injection-molding machines are used to form plastic parts. A part is considered defective if it has excessive shrinkage or is discolored. Two random samples, each of size 300, are selected, and 15 defective parts are found in the sample from machine 1 whereas 8 defective parts are found in the sample from machine 2. Is it reasonable to conclude that both machines produce the same fraction of defective parts, using $\alpha = 0.05$? Find the P-value for this test.

5-61. Consider the situation described in Exercise 5-60. Suppose that $p_1 = 0.05$ and $p_2 = 0.01$.
(a) With the sample sizes given here, what is the power of the test for this two-sided alternative?
(b) Determine the sample size needed to detect this difference with a probability of at least 0.9. Use $\alpha = 0.05$.

5-62. Consider the situation described in Exercise 5-60. Suppose that $p_1 = 0.05$ and $p_2 = 0.02$.
(a) With the sample sizes given here, what is the power of the test for this two-sided alternative?
(b) Determine the sample size needed to detect this difference with a probability of at least 0.9. Use $\alpha = 0.05$.

5-63. The rollover rate of sport utility vehicles is a transportation safety issue. Safety advocates claim that manufacturer A's vehicle has a higher rollover rate than that of manufacturer B. One hundred crashes for each of these vehicles were examined. The rollover rates were $p_A = 0.35$ and $p_B = 0.25$.
(a) Does manufacturer A's vehicle have a higher rollover rate than manufacturer B's? Use the P-value approach.

(b) What is the power of this test, assuming $\alpha = 0.05$?

(c) Assume that the rollover rate of manufacturer A's vehicle is 0.15 higher than B's. Is the sample size sufficient for detecting this difference with probability level at least 0.90, if $\alpha = 0.053$?

5-64. Construct a 95% CI on the difference in the two fractions defective for Exercise 5-60.

5-65. Construct a 95% lower bound on the difference in the two rollover rates for Exercise 5-63. Provide a practical interpretation of this interval.

5-7 SUMMARY TABLES FOR INFERENCE PROCEDURES FOR TWO SAMPLES

The tables on the inside back cover of the book summarize all of the two-sample inference procedures given in this chapter. The tables contain the null hypothesis statements, the test statistics, the criteria for rejection of the various alternative hypotheses, and the formulas for constructing the $100(1 - \alpha)\%$ CIs.

5-8 WHAT IF WE HAVE MORE THAN TWO SAMPLES?

As this chapter and Chapter 4 have illustrated, testing and experimentation are a natural part of the engineering analysis and decision-making process. Suppose, for example, that a civil engineer is investigating the effect of different curing methods on the mean compressive strength of concrete. The experiment would consist of making up several test specimens of concrete using each of the proposed curing methods and then testing the compressive strength of each specimen. The data from this experiment could be used to determine which curing method should be used to provide maximum mean compressive strength.

If there are only two curing methods of interest, this experiment could be designed and analyzed using the two-sample t-test presented in this chapter. That is, the experimenter has a **single factor** of interest—curing methods—and there are only two **levels** of the factor.

Many single-factor experiments require that more than two levels of the factor be considered. For example, the civil engineer may want to investigate five different curing methods. In this chapter we show how the **analysis of variance** (ANOVA) can be used for comparing means when there are more than two levels of a single factor. We will also discuss **randomization** of the experimental runs and the important role this concept plays in the overall experimentation strategy. In Chapter 7, we will show how to design and analyze experiments with several factors.

5-8.1 Completely Randomized Experiment and Analysis of Variance

A manufacturer of paper used for making grocery bags is interested in improving the tensile strength of the product. Product engineering thinks that tensile strength is a function of the hardwood concentration in the pulp and that the range of hardwood concentrations of practical interest is between 5 and 20%. A team of engineers responsible for the study decides to investigate four levels of hardwood concentration: 5, 10, 15, and 20%. They decide to make up six test specimens at each concentration level, using a pilot plant. All 24 specimens are tested on a laboratory tensile tester, in random order. The data from this experiment are shown in Table 5-5.

This is an example of a completely randomized single-factor experiment with four levels of the factor. The levels of the factor are sometimes called **treatments**, and each treatment has six observations or **replicates.** The role of **randomization** in this experiment is extremely important. By randomizing the order of the 24 runs, the effect of any nuisance variable that may influence the observed tensile strength is approximately balanced out. For example, suppose

Table 5-5   Tensile Strength of Paper (psi)

Hardwood Concentration (%)	Observations						Totals	Averages
	1	2	3	4	5	6		
5	7	8	15	11	9	10	60	10.00
10	12	17	13	18	19	15	94	15.67
15	14	18	19	17	16	18	102	17.00
20	19	25	22	23	18	20	127	21.17
							383	15.96

that there is a warm-up effect on the tensile testing machine; that is, the longer the machine is on, the greater the observed tensile strength. If all 24 runs are made in order of increasing hardwood concentration (that is, all six 5% concentration specimens are tested first, followed by all six 10% concentration specimens, etc.), any observed differences in tensile strength could also be due to the warm-up effect.

It is important to graphically analyze the data from a designed experiment. Figure 5-7a presents box plots of tensile strength at the four hardwood concentration levels. This figure indicates that changing the hardwood concentration has an effect on tensile strength; specifically, higher hardwood concentrations produce higher observed tensile strength. Furthermore, the distribution of tensile strength at a particular hardwood level is reasonably symmetric, and the variability in tensile strength does not change dramatically as the hardwood concentration changes.

Graphical interpretation of the data is always a good idea. Box plots show the variability of the observations *within* a treatment (factor level) and the variability *between* treatments. We now discuss how the data from a single-factor randomized experiment can be analyzed statistically.

Analysis of Variance

Suppose we have a different levels of a single factor that we wish to compare. Sometimes each factor level is called a treatment, a very general term that can be traced to the early applications

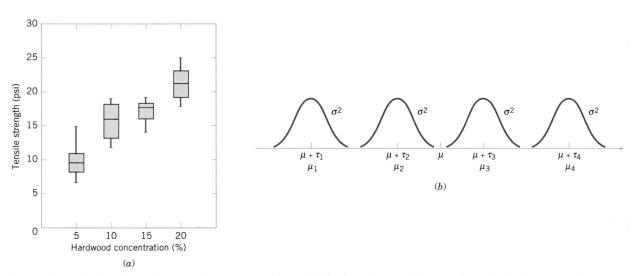

Figure 5-7   (*a*) Box plots of hardwood concentration data. (*b*) Display of the model in equation 5-31 for the completely randomized single-factor experiment.

Table 5-6   Typical Data for a Single-Factor Experiment

Treatment	Observations				Totals	Averages
1	y_{11}	y_{12}	. . .	y_{1n}	$y_{1\cdot}$	$\bar{y}_{1\cdot}$
2	y_{21}	y_{22}	. . .	y_{2n}	$y_{2\cdot}$	$\bar{y}_{2\cdot}$
.	.	.	. . .	.	.	.
.	.	.	. . .	.	.	.
.	.	.	. . .	.	.	.
a	y_{a1}	y_{a2}	. . .	y_{an}	$y_{a\cdot}$	$\bar{y}_{a\cdot}$
					$y_{\cdot\cdot}$	$\bar{y}_{\cdot\cdot}$

of experimental design methodology in the agricultural sciences. The response for each of the a treatments is a random variable. The observed data would appear as shown in Table 5-6. An entry in Table 5-6—say, y_{ij}—represents the jth observation taken under treatment i. We initially consider the case in which there are an equal number of observations, n, on each treatment.

We may describe the observations in Table 5-6 by the **linear statistical model**

$$Y_{ij} = \mu + \tau_i + \epsilon_{ij} \begin{cases} i = 1, 2, \ldots, a \\ j = 1, 2, \ldots, n \end{cases} \tag{5-31}$$

where Y_{ij} is a random variable denoting the (ij)th observation, μ is a parameter common to all treatments called the **overall mean,** τ_i is a parameter associated with the ith treatment called the ith **treatment effect,** and ϵ_{ij} is a random error component. Notice that the model could have been written as

$$Y_{ij} = \mu_i + \epsilon_{ij} \begin{cases} i = 1, 2, \ldots, a \\ j = 1, 2, \ldots, n \end{cases}$$

where $\mu_i = \mu + \tau_i$ is the mean of the ith treatment. In this form of the model, we see that each treatment defines a population that has mean μ_i, consisting of the overall mean μ plus an effect τ_i that is due to that particular treatment. We will assume that the errors ϵ_{ij} are normally and independently distributed with mean zero and variance σ^2. Therefore, each treatment can be thought of as a normal population with mean μ_i and variance σ^2. See Fig. 5-7b.

Equation 5-31 is the underlying model for a single-factor experiment. Furthermore, because we require that the observations are taken in random order and that the environment (often called the experimental units) in which the treatments are used is as uniform as possible, this design is called a **completely randomized experiment.**

We now present the analysis of variance for testing the equality of a population means. However, the ANOVA is a far more useful and general technique; it will be used extensively in the next two chapters. In this section we show how it can be used to test for equality of treatment effects. In our application the treatment effects τ_i are usually defined as deviations from the overall mean μ, so

$$\sum_{i=1}^{a} \tau_i = 0 \tag{5-32}$$

Let $y_{i\cdot}$ represent the total of the observations under the ith treatment and $\bar{y}_{i\cdot}$ represent the average of the observations under the ith treatment. Similarly, let $y_{\cdot\cdot}$ represent the grand total of all observations and $\bar{y}..$ represent the grand mean of all observations. Expressed mathematically,

$$y_{i\cdot} = \sum_{j=1}^{n} y_{ij} \qquad \bar{y}_{i\cdot} = y_{i\cdot}/n \qquad i = 1, 2, \ldots, a$$

$$y_{\cdot\cdot} = \sum_{i=1}^{a} \sum_{j=1}^{n} y_{ij} \qquad \bar{y}_{\cdot\cdot} = y_{\cdot\cdot}/N \tag{5-33}$$

where $N = an$ is the total number of observations. Thus the "dot" subscript notation implies summation over the subscript that it replaces.

We are interested in testing the equality of the a treatment means $\mu_1, \mu_2, \ldots, \mu_a$. Using equation 5-32, we find that this is equivalent to testing the hypotheses

$$H_0: \tau_1 = \tau_2 = \cdots = \tau_a = 0$$
$$H_1: \tau_i \neq 0 \quad \text{for at least one } i \tag{5-34}$$

Thus, if the null hypothesis is true, each observation consists of the overall mean μ plus a realization of the random error component ϵ_{ij}. This is equivalent to saying that all N observations are taken from a normal distribution with mean μ and variance σ^2. Therefore, if the null hypothesis is true, changing the levels of the factor has no effect on the mean response.

The analysis of variance partitions the total variability in the sample data into two component parts. Then the test of the hypothesis in equation 5-34 is based on a comparison of two independent estimates of the population variance. The total variability in the data is described by the **total sum of squares**

$$SS_T = \sum_{i=1}^{a} \sum_{j=1}^{n} (y_{ij} - \bar{y}..)^2$$

The partition of the total sum of squares is given in the following definition.

The **ANOVA sum of squares identity** is

$$\sum_{i=1}^{a} \sum_{j=1}^{n} (y_{ij} - \bar{y}..)^2 = n \sum_{i=1}^{a} (\bar{y}_{i\cdot} - \bar{y}..)^2 + \sum_{i=1}^{a} \sum_{j=1}^{n} (y_{ij} - \bar{y}_{i\cdot})^2 \tag{5-35}$$

The proof of this identity is straightforward, and is provided in Montgomery and Runger (2006).

The identity in equation 5-35 shows that the total variability in the data, measured by the total sum of squares, can be partitioned into a sum of squares of differences between treatment means and the grand mean and a sum of squares of differences of observations within a treatment from the treatment mean. Differences between observed treatment means and the grand mean measure the differences between treatments, whereas differences of observations within a treatment from the treatment mean can be due only to random error. Therefore, we write equation 5-35 symbolically as

$$SS_T = SS_{\text{Treatments}} + SS_E \qquad (5\text{-}36)$$

where

$$SS_T = \sum_{i=1}^{a} \sum_{j=1}^{n} (y_{ij} - \bar{y}..)^2 = \text{total sum of squares}$$

$$SS_{\text{Treatments}} = n \sum_{i=1}^{a} (\bar{y}_{i\cdot} - \bar{y}..)^2 = \text{treatment sum of squares}$$

and

$$SS_E = \sum_{i=1}^{a} \sum_{j=1}^{n} (y_{ij} - \bar{y}_{i\cdot})^2 = \text{error sum of squares}$$

We can gain considerable insight into how the analysis of variance works by examining the expected values of $SS_{\text{Treatments}}$ and SS_E. This will lead us to an appropriate statistic for testing the hypothesis of no differences among treatment means (or $\tau_i = 0$).

It can be shown that

$$E\left(\frac{SS_{\text{Treatments}}}{a-1}\right) = \sigma^2 + \frac{n \sum_{i=1}^{a} \tau_i^2}{a-1} \qquad (5\text{-}37)$$

The ratio

$$MS_{\text{Treatments}} = SS_{\text{Treatments}}/(a-1)$$

is called the **mean square for treatments.** Thus, if H_0 is true, $MS_{\text{Treatments}}$ is an unbiased estimator of σ^2 because under H_0 each $\tau_i = 0$. If H_1 is true, $MS_{\text{Treatments}}$ estimates σ^2 plus a positive term that incorporates variation due to the systematic difference in treatment means.

We can also show that the expected value of the error sum of squares is $E(SS_E) = a(n-1)\sigma^2$. Therefore, the **error mean square**

$$MS_E = SS_E/[a(n-1)]$$

is an unbiased estimator of σ^2 regardless of whether or not H_0 is true. There is also a partition of the number of degrees of freedom that corresponds to the sum of squares identity in equation 5-35. That is, there are $an = N$ observations; thus, SS_T has $an - 1$ degrees of freedom. There are a levels of the factor, so $SS_{\text{Treatments}}$ has $a - 1$ degrees of freedom. Finally, within any treatment there are n replicates providing $n - 1$ degrees of freedom with which to estimate the experimental error. Because there are a treatments, we have $a(n-1)$ degrees of freedom for error. Therefore, the degrees of freedom partition is

$$an - 1 = a - 1 + a(n-1)$$

Now assume that each of the a populations can be modeled as a normal distribution. Using this assumption we can show that if the null hypothesis H_0 is true, the ratio

$$F_0 = \frac{SS_{\text{Treatments}}/(a-1)}{SS_E/[a(n-1)]} = \frac{MS_{\text{Treatments}}}{MS_E} \qquad (5\text{-}38)$$

has an F distribution with $a - 1$ and $a(n - 1)$ degrees of freedom. Furthermore, from the expected mean squares, we know that MS_E is an unbiased estimator of σ^2. Also, under the null hypothesis, $MS_{\text{Treatments}}$ is an unbiased estimator of σ^2. However, if the null hypothesis is false, the expected value of $MS_{\text{Treatments}}$ is greater than σ^2. Therefore, under the alternative hypothesis, the expected value of the numerator of the test statistic (equation 5-38) is greater than the expected value of the denominator. Consequently, we should reject H_0 if the statistic is large. This implies an upper-tail, one-tail test procedure. Therefore, the P-value would be the probability to the right of the value of the test statistic in the $F_{a-1,\,a(n-1)}$ distribution. For a fixed-level test, we would reject H_0 if $f_0 > F_{\alpha,\,a-1,\,a(n-1)}$ where f_0 is computed from equation 5-38. These results are summarized as follows.

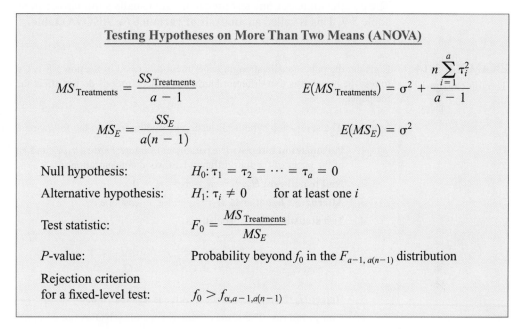

Testing Hypotheses on More Than Two Means (ANOVA)

$$MS_{\text{Treatments}} = \frac{SS_{\text{Treatments}}}{a - 1} \qquad\qquad E(MS_{\text{Treatments}}) = \sigma^2 + \frac{n \sum_{i=1}^{a} \tau_i^2}{a - 1}$$

$$MS_E = \frac{SS_E}{a(n - 1)} \qquad\qquad E(MS_E) = \sigma^2$$

Null hypothesis: $H_0: \tau_1 = \tau_2 = \cdots = \tau_a = 0$

Alternative hypothesis: $H_1: \tau_i \neq 0 \qquad$ for at least one i

Test statistic: $F_0 = \dfrac{MS_{\text{Treatments}}}{MS_E}$

P-value: Probability beyond f_0 in the $F_{a-1,\,a(n-1)}$ distribution

Rejection criterion
for a fixed-level test: $f_0 > f_{\alpha,\,a-1,a(n-1)}$

Efficient computational formulas for the sums of squares may be obtained by expanding and simplifying the definitions of $SS_{\text{Treatments}}$ and SS_T. This yields the following results.

Completely Randomized Experiment with Equal Sample Sizes

The computing formulas for the sums of squares in the analysis of variance for a completely randomized experiment with equal sample sizes in each treatment are

$$SS_T = \sum_{i=1}^{a} \sum_{j=1}^{n} y_{ij}^2 - \frac{y_{..}^2}{N}$$

and

$$SS_{\text{Treatments}} = \sum_{i=1}^{a} \frac{y_{i.}^2}{n} - \frac{y_{..}^2}{N}$$

The error sum of squares is usually obtained by subtraction as

$$SS_E = SS_T - SS_{\text{Treatments}}$$

Table 5-7 The Analysis of Variance for a Single-Factor Experiment

Source of Variation	Sum of Squares	Degrees of Freedom	Mean Square	F_0
Treatments	$SS_{\text{Treatments}}$	$a - 1$	$MS_{\text{Treatments}}$	$\dfrac{MS_{\text{Treatments}}}{MS_E}$
Error	SS_E	$a(n - 1)$	MS_E	
Total	SS_T	$an - 1$		

The computations for this test procedure are usually summarized in tabular form as shown in Table 5-7. This is called an **analysis of variance** (or ANOVA) **table.**

EXAMPLE 5-14

Tensile Strength

Consider the paper tensile strength experiment described in Section 5-8.1. Use the analysis of variance to test the hypothesis that different hardwood concentrations do not affect the mean tensile strength of the paper.

Solution. The seven-step hypothesis testing procedure leads to the following results:

1. **Parameter of interest:** The parameter of interest are τ_1, τ_2, τ_3, and τ_4, the mean tensile strength of the paper of the four different hardwood concentrations.

2. **Null hypothesis, H_0:** $\tau_1 = \tau_2 = \tau_3 = \tau_4 = 0$

3. **Alternative hypothesis, H_1:** $\tau_i \neq 0$ for at least one i

4. **Test statistic:** The test statistic is

$$f_0 = \frac{MS_{\text{Treatments}}}{MS_E}$$

5. **Reject H_0 if:** Reject H_0 if the P-value is less than 0.05.

6. **Computations:** The sums of squares for the ANOVA are computed from equation 5-36 as follows:

$$SS_T = \sum_{i=1}^{4} \sum_{j=1}^{6} y_{ij}^2 - \frac{y_{..}^2}{N}$$

$$= (7)^2 + (8)^2 + \cdots + (20)^2 - \frac{(383)^2}{24} = 512.96$$

$$SS_{\text{Treatments}} = \sum_{i=1}^{4} \frac{y_{i.}^2}{n} - \frac{y_{..}^2}{N}$$

$$= \frac{(60)^2 + (94)^2 + (102)^2 + (127)^2}{6} - \frac{(383)^2}{24} = 382.79$$

$$SS_E = SS_T - SS_{\text{Treatments}}$$

$$= 512.96 - 382.79 = 130.17$$

We usually do not perform these calculations by hand. The ANOVA computed by Minitab is presented in Table 5-8.

7. **Conclusions:** From Table 5-8, we note that the computed value of the test statistic is $f_0 = 19.61$ and the P-value is reported as $P = 0.000$. Because the P-value is considerably smaller than

Table 5-8 Minitab Analysis of Variance Output for the Paper Tensile Strength Experiment

One-Way Analysis of Variance					
Analysis of Variance					
Source	DF	SS	MS	F	P
Factor	3	382.79	127.60	19.61	0.000
Error	20	130.17	6.51		
Total	23	512.96			

				Individual 95% Cls For Mean Based on Pooled StDev	
Level	N	Mean	StDev	----- + ---------- + ---------- + ---------- +-	
5	6	10.000	2.828	(---*---)	
10	6	15.667	2.805	(---*---)	
15	6	17.000	1.789	(---*---)	
20	6	21.167	2.639	(---*---)	
				----- + ---------- + ---------- + ---------- +-	
Pooled StDev = 2.551				10.0 15.0 20.0 25.0	

$\alpha = 0.05$, we have strong evidence to conclude that H_0 is not true. That is, the hardwood concentration in the pulp affects the tensile strength of the paper. Since this is an upper-tailed F-test, we could bound the P-value by using the F-table in Appendix A Table IV. From this table, we find that $f_{0.01,3,20} = 4.94$, and because $f_0 = 19.61$ exceeds this value, we know that the P-value is less than 0.01. The actual P-value (found from a calculator) is 3.59×10^{-6}. Note that Minitab also provides some summary information about each level of hardwood concentration, including the confidence interval on each mean. ∎

In some single-factor experiments, the number of observations taken under each treatment may be different. We then say that the design is **unbalanced.** The analysis of variance described earlier is still valid, but slight modifications must be made in the sums of squares formulas. Let n_i observations be taken under treatment $i (i = 1, 2, \ldots, a)$, and let the total number of observations $N = \sum_{i=1}^{a} n_i$. The computational formulas for SS_T and $SS_{\text{Treatments}}$ are as shown in the following definition.

Completely Randomized Experiment with Unequal Sample Sizes

The computing formulas for the sums of squares in the analysis of variance for a completely randomized experiment with unequal sample sizes n_i in each treatment are

$$SS_T = \sum_{i=1}^{a} \sum_{j=1}^{n_i} y_{ij}^2 - \frac{y_{..}^2}{N}$$

$$SS_{\text{Treatments}} = \sum_{i=1}^{a} \frac{y_{i.}^2}{n_i} - \frac{y_{..}^2}{N}$$

and

$$SS_E = SS_T - SS_{\text{Treatments}}$$

Which Means Differ?

Finally, note that the analysis of variance tells us whether there is a difference among means. It does not tell us which means differ. If the analysis of variance indicates that there is a statistically significant difference among means, there is a simple graphical procedure that can be used to isolate the specific differences. Suppose that $\bar{y}_1., \bar{y}_2., \dots, \bar{y}_a.$ are the observed averages for these factor levels. Each treatment average has standard deviation $\sigma/\sqrt{n}$, where σ is the standard deviation of an individual observation. If all treatment means are equal, the observed means $\bar{y}_i.$ would behave as if they were a set of observations drawn at random from a normal distribution with mean μ and standard deviation $\sigma/\sqrt{n}$.

Visualize this normal distribution capable of being slid along an axis below which the treatment means $\bar{y}_1., \bar{y}_2., \dots, \bar{y}_a.$ are plotted. If all treatment means are equal, there should be some position for this distribution that makes it obvious that the $\bar{y}_i.$ values were drawn from the same distribution. If this is not the case, the $\bar{y}_i.$ values that do not appear to have been drawn from this distribution are associated with treatments that produce different mean responses.

The only flaw in this logic is that σ is unknown. However, we can use $\sqrt{MS_E}$ from the analysis of variance to estimate σ. This implies that a t distribution should be used instead of the normal in making the plot, but because the t looks so much like the normal, sketching a normal curve that is approximately $6\sqrt{MS_E/n}$ units wide will usually work very well.

Examining Differences Among Means

Figure 5-8 shows this arrangement for the hardwood concentration experiment. The standard deviation of this normal distribution is

$$\sqrt{MS_E/n} = \sqrt{6.51/6} = 1.04$$

If we visualize sliding this distribution along the horizontal axis, we note that there is no location for the distribution that would suggest that all four observations (the plotted means) are typical, randomly selected values from that distribution. This, of course, should be expected because the analysis of variance has indicated that the means differ, and the display in Fig. 5-8 is only a graphical representation of the analysis of variance results. The figure does indicate that treatment 4 (20% hardwood) produces paper with higher mean tensile strength than do the other treatments and that treatment 1 (5% hardwood) results in lower mean tensile strength than do the other treatments. The means of treatments 2 and 3 (10 and 15% hardwood, respectively) do not differ.

This simple procedure is a rough but very useful and effective technique for comparing means following an analysis of variance. There are more quantitative techniques, called **multiple comparison procedures,** for testing for differences between specific means following an analysis of variance. Because these procedures typically involve a series of tests, the type I error compounds to produce an **experiment-wise** or **family error rate.** For more details on these procedures, see Montgomery (2005).

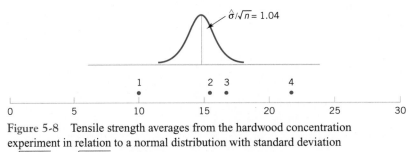

Figure 5-8 Tensile strength averages from the hardwood concentration experiment in relation to a normal distribution with standard deviation $\sqrt{MS_E/n} = \sqrt{6.51/6} = 1.04$.

Table 5-9 Residuals for the Tensile Strength Experiment

Hardwood Concentration (%)	Residuals					
5	−3.00	−2.00	5.00	1.00	−1.00	0.00
10	−3.67	1.33	−2.67	2.33	3.33	−0.67
15	−3.00	1.00	2.00	0.00	−1.00	1.00
20	−2.17	3.83	0.83	1.83	−3.17	−1.17

Residual Analysis and Model Checking

The one-way model analysis of variance assumes that the observations are normally and independently distributed with the same variance for each treatment or factor level. These assumptions should be checked by examining the residuals. A residual is the difference between an observation y_{ij} and its estimated (or fitted) value from the statistical model being studied, denoted as $\hat{y}_{ij}$. For the completely randomized design $\hat{y}_{ij} = \bar{y}_{i\cdot}$ and each residual is $e_{ij} = y_{ij} - \bar{y}_{i\cdot}$—that is, the difference between an observation and the corresponding observed treatment mean. The residuals for the hardwood percentage experiment are shown in Table 5-9. Using $\bar{y}_{i\cdot}$ to calculate each residual essentially removes the effect of hardwood concentration from those data; consequently, the residuals contain information about unexplained variability.

The normality assumption can be checked by constructing a normal probability plot of the residuals. To check the assumption of equal variances at each factor level, plot the residuals against the factor levels and compare the spread in the residuals. It is also useful to plot the residuals against $\bar{y}_{i\cdot}$ (sometimes called the fitted value); the variability in the residuals should not depend in any way on the value of $\bar{y}_{i\cdot}$. Most statistics software packages will construct these plots on request. When a pattern appears in these plots, it usually suggests the need for a transformation—that is, analyzing the data in a different metric. For example, if the variability in the residuals increases with $\bar{y}_{i\cdot}$, a transformation such as log y or $\sqrt{y}$ should be considered. In some problems, the dependency of residual scatter on the observed mean $\bar{y}_{i\cdot}$ is very important information. It may be desirable to select the factor level that results in maximum response; however, this level may also cause more variation in response from run to run.

The independence assumption can be checked by plotting the residuals against the time or run order in which the experiment was performed. A pattern in this plot, such as sequences of positive and negative residuals, may indicate that the observations are not independent. This suggests that time or run order is important or that variables that change over time are important and have not been included in the experimental design.

Normal Probability Plot Interpretation

A normal probability plot of the residuals from the paper tensile strength experiment is shown in Fig. 5-9. Figures 5-10 and 5-11 present the residuals plotted against the factor levels and the fitted value $\bar{y}_{i\cdot}$, respectively. These plots do not reveal any model inadequacy or unusual problem with the assumptions.

Animation 13: ANOVA

5-8.2 Randomized Complete Block Experiment

In many experimental design problems, it is necessary to design the experiment so that the variability arising from a nuisance factor can be controlled. For example, consider the situation of Example 5-8, where two different methods were used to predict the shear strength of steel plate girders. Because each girder has different strength (potentially) and this variability in strength was not of direct interest, we designed the experiment by using the two test

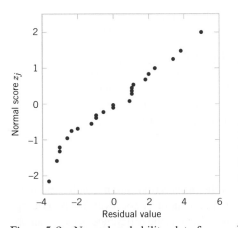

Figure 5-9 Normal probability plot of residuals from the hardwood concentration experiment.

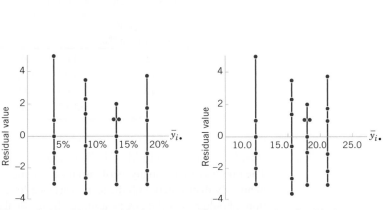

Figure 5-10 Plot of residuals versus factor levels (hardwood concentration).

Figure 5-11 Plot of residuals versus $\bar{y}_{i\cdot}$.

methods on each girder and then comparing the average difference in strength readings on each girder to zero using the paired t-test. The paired t-test is a procedure for comparing two treatment means when all experimental runs cannot be made under homogeneous conditions. Alternatively, we can view the paired t-test as a method for reducing the background noise in the experiment by blocking out a **nuisance factor** effect. The block is the nuisance factor, and in this case, the nuisance factor is the actual **experimental unit**—the steel girder specimens used in the experiment.

The randomized block design is an extension of the paired t-test to situations where the factor of interest has more than two levels; that is, more than two treatments must be compared. For example, suppose that three methods could be used to evaluate the strength readings on steel plate girders. We may think of these as three treatments—say, t_1, t_2, and t_3. If we use four girders as the experimental units, a **randomized complete block design** would appear as shown in Fig. 5-12. The design is called a randomized complete block design because each block is large enough to hold all the treatments and because the actual assignment of each of the three treatments within each block is done randomly. Once the experiment has been conducted, the data are recorded in a table, such as is shown in Table 5-10. The observations in this table—say, y_{ij}—represent the response obtained when method i is used on girder j.

The general procedure for a randomized complete block design consists of selecting b blocks and running a complete replicate of the experiment in each block. The data that result from running a randomized complete block design for investigating a single factor with a levels and b blocks are shown in Table 5-11. There will be a observations (one per factor level)

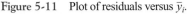

Figure 5-12 A randomized complete block design.

Table 5-10   A Randomized Complete Block Design

Treatment (Method)	Block (Girder)			
	1	2	3	4
1	y_{11}	y_{12}	y_{13}	y_{14}
2	y_{21}	y_{22}	y_{23}	y_{24}
3	y_{31}	y_{32}	y_{33}	y_{34}

Table 5-11 A Randomized Complete Block Design with a Treatments and b Blocks

Treatments	Blocks 1	2	$\cdots$	b	Totals	Averages
1	y_{11}	y_{12}	$\cdots$	y_{1b}	$y_{1\cdot}$	$\bar{y}_{1\cdot}$
2	y_{21}	y_{22}	$\cdots$	y_{2b}	$y_{2\cdot}$	$\bar{y}_{2\cdot}$
$\vdots$	$\vdots$	$\vdots$		$\vdots$	$\vdots$	$\vdots$
a	y_{a1}	y_{a2}	$\cdots$	y_{ab}	$y_{a\cdot}$	$\bar{y}_{a\cdot}$
Totals	$y_{\cdot 1}$	$y_{\cdot 2}$	$\cdots$	$y_{\cdot b}$	$y_{\cdot\cdot}$	
Averages	$\bar{y}_{\cdot 1}$	$\bar{y}_{\cdot 2}$	$\cdots$	$\bar{y}_{\cdot b}$		$\bar{y}_{\cdot\cdot}$

in each block, and the order in which these observations are run is randomly assigned within the block.

We will now describe the ANOVA for a randomized complete block design. Suppose that a single factor with a levels is of interest and that the experiment is run in b blocks. The observations may be represented by the linear statistical model.

$$Y_{ij} = \mu + \tau_i + \beta_j + \epsilon_{ij} \begin{cases} i = 1, 2, \ldots, a \\ j = 1, 2, \ldots, b \end{cases} \tag{5-39}$$

where μ is an overall mean, τ_i is the effect of the ith treatment, β_j is the effect of the jth block, and ϵ_{ij} is the random error term, which is assumed to be normally and independently distributed with mean zero and variance σ^2. For our purpose, the treatments and blocks will be considered as fixed factors. Furthermore, the treatment and block effects are defined as deviations from the overall mean, so $\sum_{i=1}^{a} \tau_i = 0$ and $\sum_{j=1}^{b} \beta_j = 0$. We also assume that treatments and blocks do not interact; that is, the effect of treatment i is the same regardless of which block (or blocks) it is tested in. We are interested in testing the equality of the treatment effects; that is,

$$H_0: \tau_1 = \tau_2 = \cdots = \tau_a = 0 \tag{5-40}$$
$$H_1: \tau_i \neq 0 \text{ at least one } i$$

As in the completely randomized experiment, testing the hypothesis that all the treatment effects τ_i are equal to zero is equivalent to testing the hypothesis that the treatment means are equal.

The ANOVA procedure for a randomized complete block design uses a sum of squares identity that partitions the total sum of squares into three components.

The **sum of squares identity for the randomized complete block design** is

$$\sum_{i=1}^{a} \sum_{j=1}^{b} (y_{ij} - \bar{y}_{\cdot\cdot})^2 = b \sum_{i=1}^{a} (\bar{y}_{i\cdot} - \bar{y}_{\cdot\cdot})^2 + a \sum_{j=1}^{b} (\bar{y}_{\cdot j} - \bar{y}_{\cdot\cdot})^2$$

$$+ \sum_{i=1}^{a} \sum_{j=1}^{b} (y_{ij} - \bar{y}_{\cdot j} - \bar{y}_{i\cdot} + \bar{y}_{\cdot\cdot})^2 \tag{5-41}$$

The sum of squares identity may be represented symbolically as

$$SS_T = SS_{\text{Treatments}} + SS_{\text{Blocks}} + SS_E$$

where

$$SS_T = \sum_{i=1}^{a} \sum_{j=1}^{b} (y_{ij} - \bar{y}..)^2 = \text{total sum of squares}$$

$$SS_{\text{Treatments}} = b \sum_{i=1}^{a} (\bar{y}_{i\cdot} - \bar{y}..)^2 = \text{treatment sum of squares}$$

$$SS_{\text{Blocks}} = a \sum_{j=1}^{b} (\bar{y}_{\cdot j} - \bar{y}..)^2 = \text{block sum of squares}$$

$$SS_E = \sum_{i=1}^{a} \sum_{j=1}^{b} (\bar{y}_{ij} - \bar{y}_{\cdot j} - \bar{y}_{i\cdot} + \bar{y}..)^2 = \text{error sum of squares}$$

Furthermore, the degree-of-freedom breakdown corresponding to these sums of squares is

$$ab - 1 = (a - 1) + (b - 1) + (a - 1)(b - 1)$$

For the randomized block design, the relevant mean squares are

$$MS_{\text{Treatments}} = \frac{SS_{\text{Treatments}}}{a - 1} \qquad MS_{\text{Blocks}} = \frac{SS_{\text{Blocks}}}{b - 1} \qquad MS_E = \frac{SS_E}{(a - 1)(b - 1)} \qquad (5\text{-}42)$$

The expected values of these mean squares can be shown to be as follows:

$$E(MS_{\text{Treatments}}) = \sigma^2 + \frac{b \sum_{i=1}^{a} \tau_i^2}{a - 1}$$

$$E(MS_{\text{Blocks}}) = \sigma^2 + \frac{a \sum_{j=1}^{b} \beta_j^2}{b - 1}$$

$$E(MS_E) = \sigma^2$$

Therefore, if the null hypothesis H_0 is true so that all treatment effects $\tau_i = 0$, $MS_{\text{Treatments}}$ is an unbiased estimator of σ^2, whereas if H_0 is false, $MS_{\text{Treatments}}$ overestimates σ^2. The mean square for error is always an unbiased estimate of σ^2. To test the null hypothesis that the treatment effects are all zero, we compute the ratio

$$F_0 = \frac{MS_{\text{Treatments}}}{MS_E} \qquad (5\text{-}43)$$

which has an F distribution with $a - 1$ and $(a - 1)(b - 1)$ degrees of freedom if the null hypothesis is true. The P-value would be computed as in any upper-tailed F-test. We would reject the null hypothesis for small P-values. For fixed-level testing at the α level of significance, we would reject H_0 if the computed value of the test statistic in equation 5-43

$$f_0 > f_{\alpha,\, a-1,\, (a-1)(b-1)}.$$

In practice, we compute SS_T, $SS_{\text{Treatments}}$, and SS_{Blocks} and then obtain the error sum of squares SS_E by subtraction. The appropriate computing formulas are as follows.

Randomized Complete Block Experiment

The computing formulas for the sums of squares in the analysis of variance for a randomized complete block design are

$$SS_T = \sum_{i=1}^{a} \sum_{j=1}^{b} y_{ij}^2 - \frac{y_{..}^2}{ab}$$

$$SS_{\text{Treatments}} = \frac{1}{b} \sum_{i=1}^{a} y_{i\cdot}^2 - \frac{y_{..}^2}{ab}$$

$$SS_{\text{Blocks}} = \frac{1}{a} \sum_{j=1}^{b} y_{\cdot j}^2 - \frac{y_{..}^2}{ab}$$

and

$$SS_E = SS_T - SS_{\text{Treatments}} - SS_{\text{Blocks}}$$

The computations are usually arranged in an analysis of variance table, such as is shown in Table 5-12. Generally, a computer software package will be used to perform the ANOVA for the randomized complete block design. Here, we give the computational details rather than explicitly listing the seven-step procedure.

EXAMPLE 5-15
Fabric Strength

An experiment was performed to determine the effect of four different chemicals on the strength of a fabric. These chemicals are used as part of the permanent press finishing process. Five fabric samples were selected, and a randomized complete block design was run by testing each chemical type once in

Table 5-12   Analysis of Variance for a Randomized Complete Block Design

Source of Variation	Sum of Squares	Degrees of Freedom	Mean Square	F_0
Treatments	$SS_{\text{Treatments}}$	$a - 1$	$\dfrac{SS_{\text{Treatments}}}{a - 1}$	$\dfrac{MS_{\text{Treatments}}}{MS_E}$
Blocks	SS_{Blocks}	$b - 1$	$\dfrac{SS_{\text{Blocks}}}{b - 1}$	
Error	SS_E (by subtraction)	$(a - 1)(b - 1)$	$\dfrac{SS_E}{(a - 1)(b - 1)}$	
Total	SS_T	$ab - 1$		

Table 5-13   Fabric Strength Data—Randomized Complete Block Design

Chemical Type	Fabric Sample					Treatment Totals $y_{i\cdot}$	Treatment Averages $\bar{y}_{i\cdot}$
	1	2	3	4	5		
1	1.3	1.6	0.5	1.2	1.1	5.7	1.14
2	2.2	2.4	0.4	2.0	1.8	8.8	1.76
3	1.8	1.7	0.6	1.5	1.3	6.9	1.38
4	3.9	4.4	2.0	4.1	3.4	17.8	3.56
Block totals $y_{\cdot j}$	9.2	10.1	3.5	8.8	7.6	39.2$(y..)$	
Block averages $\bar{y}_{\cdot j}$	2.30	2.53	0.88	2.20	1.90		1.96$(\bar{y}..)$

random order on each fabric sample. The data are shown in Table 5-13. We will test for differences in means using the analysis of variance with $\alpha = 0.01$.

The sums of squares for the ANOVA are computed as follows:

$$SS_T = \sum_{i=1}^{4} \sum_{j=1}^{5} y_{ij}^2 - \frac{y_{..}^2}{ab}$$

$$= (1.3)^2 + (1.6)^2 + \cdots + (3.4)^2 - \frac{(39.2)^2}{20} = 25.69$$

$$SS_{\text{Treatments}} = \sum_{i=1}^{4} \frac{y_{i\cdot}^2}{b} - \frac{y_{..}^2}{ab}$$

$$= \frac{(5.7)^2 + (8.8)^2 + (6.9)^2 + (17.8)^2}{5} - \frac{(39.2)^2}{20} = 18.04$$

$$SS_{\text{Blocks}} = \sum_{j=1}^{5} \frac{y_{\cdot j}^2}{a} - \frac{y_{..}^2}{ab}$$

$$= \frac{(9.2)^2 + (10.1)^2 + (3.5)^2 + (8.8)^2 + (7.6)^2}{4} - \frac{(39.2)^2}{20} = 6.69$$

$$SS_E = SS_T - SS_{\text{Blocks}} - SS_{\text{Treatments}}$$

$$= 25.69 - 6.69 - 18.04 = 0.96$$

The ANOVA is summarized in Table 5-14. Because $f_0 = 75.13 > f_{0.01,3,12} = 5.95$, the P-value is less than 0.01, so we conclude that there is a significant difference in the chemical types so far as their effect on mean fabric strength is concerned. The actual P-value (found from a calculator) is 4.79×10^{-8}. ■

Table 5-14   Analysis of Variance for the Randomized Complete Block Experiment

Source of Variation	Sum of Squares	Degrees of Freedom	Mean Square	f_0	P-value
Chemical types (treatments)	18.04	3	6.01	75.13	4.79 E-8
Fabric samples (blocks)	6.69	4	1.67		
Error	0.96	12	0.08		
Total	25.69	19			

When Is Blocking Necessary?

Suppose an experiment was conducted as a randomized block design, and blocking was not really necessary. There are ab observations and $(a - 1)(b - 1)$ degrees of freedom for error. If the experiment had been run as a completely randomized single-factor design with b replicates, we would have had $a(b - 1)$ degrees of freedom for error. Therefore, blocking has cost $a(b - 1) - (a - 1)(b - 1) = b - 1$ degrees of freedom for error. Thus, because the loss in error degrees of freedom is usually small, if there is a reasonable chance that block effects may be important, the experimenter should use the randomized block design.

Computer Solution

Table 5-15 presents the computer output from Minitab for the randomized complete block design example. We used the analysis of variance menu for balanced designs to solve this problem. The results agree closely with the hand calculations from Table 5-14. Note that Minitab computes an F-statistic for the blocks (the fabric samples). The validity of this ratio as a test statistic for the null hypothesis of no block effects is doubtful, because the blocks represent a **restriction on randomization**; that is, we have only randomized within the blocks. If the blocks are not chosen at random, or if they are not run in random order, the F-ratio for blocks may not provide reliable information about block effects. For more discussion see Montgomery (2005, Chapter 4).

Which Means Differ?

When the ANOVA indicates that a difference exists between the treatment means, we may need to perform some follow-up tests to isolate the specific differences. The graphical method previously described can be used for this purpose. The four chemical type averages are

$$\bar{y}_{1.} = 1.14 \quad \bar{y}_{2.} = 1.76 \quad \bar{y}_{3.} = 1.38 \quad \bar{y}_{4.} = 3.56$$

Each treatment average uses $b = 5$ observations (one from each block). Therefore, the standard deviation of a treatment average is $\sigma/\sqrt{b}$. The estimate of σ is $\sqrt{MS_E}$. Thus, the standard deviation used for the normal distribution is

$$\sqrt{MS_E/b} = \sqrt{0.0792/5} = 0.126$$

Table 5-15 Minitab Analysis of Variance for the Randomized Complete Block Design in Example 5-15

Analysis of Variance (Balanced Designs)						
Factor	Type	Levels	Values			
Chemical	fixed	4	1	2	3	4
Fabric S	fixed	5	1	2	3	4	5

Analysis of Variance for strength

Source	DF	SS	MS	F	P
Chemical	3	18.0440	6.0147	75.89	0.000
Fabric S	4	6.6930	1.6733	21.11	0.000
Error	12	0.9510	0.0792		
Total	19	25.6880			

F-test with denominator: Error
Denominator MS = 0.079250 with 12 degrees of freedom

Numerator	DF	MS	F	P
Chemical	3	6.015	75.89	0.000
Fabric S	4	1.673	21.11	0.000

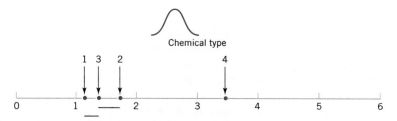

Figure 5-13   Strength averages from the fabric experiment in relation to a normal distribution with standard deviation $\sqrt{MS_E/b} = \sqrt{0.0792/5} = 0.126$.

Examining Differences among Means

A sketch of a normal distribution that is $6\sqrt{MS_E/b} = 0.755$ units wide is shown in Fig. 5-13. If we visualize sliding this distribution along the horizontal axis, we note that there is no location for the distribution that would suggest that all four means are typical, randomly selected values from that distribution. This should be expected because the analysis of variance has indicated that the means differ. The underlined pairs of means are not different. Chemical type 4 results in significantly different strengths than the other three types. Chemical types 2 and 3 do not differ, and types 1 and 3 do not differ. There may be a small difference in strength between types 1 and 2.

Residual Analysis and Model Checking

In any designed experiment, it is always important to examine the residuals and to check for violation of basic assumptions that could invalidate the results. As usual, the residuals for the randomized complete block design are only the difference between the observed and estimated (or fitted) values from the statistical model—say,

$$e_{ij} = y_{ij} - \hat{y}_{ij}$$

and the fitted values are

$$\hat{y}_{ij} = \bar{y}_{i\cdot} + \bar{y}_{\cdot j} - \bar{y}_{\cdot\cdot} \tag{5-44}$$

The fitted value represents the estimate of the mean response when the ith treatment is run in the jth block. The residuals from the chemical type experiment are shown in Table 5-16.

Figures 5-14, 5-15, 5-16, and 5-17 present the important residual plots for the experiment. These residual plots are usually constructed by computer software packages. There is some indication that fabric sample (block) 3 has greater variability in strength when treated with the four chemicals than the other samples. Chemical type 4, which provides the greatest strength, also has somewhat more variability in strength. Follow-up experiments may be necessary to confirm these findings, if they are potentially important.

Table 5-16   Residuals from the Randomized Complete Block Design

Chemical Type	Fabric Sample				
	1	2	3	4	5
1	−0.18	−0.10	0.44	−0.18	0.02
2	0.10	0.08	−0.28	0.00	0.10
3	0.08	−0.24	0.30	−0.12	−0.02
4	0.00	0.28	−0.48	0.30	−0.10

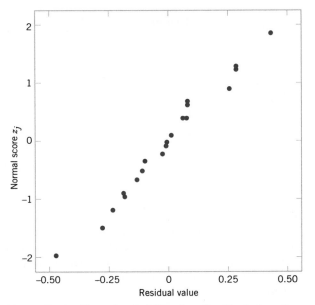

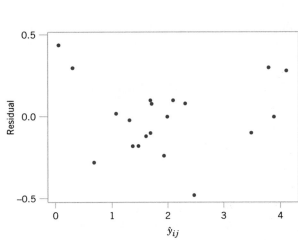

Figure 5-14   Normal probability plot of residuals from the randomized complete block design.

Figure 5-15   Residuals versus $\hat{y}_{ij}$.

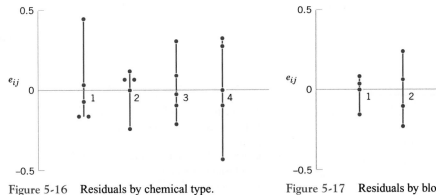

Figure 5-16   Residuals by chemical type.

Figure 5-17   Residuals by block.

EXERCISES FOR SECTION 5-8

5-66. A Minitab ANOVA output is shown below. Fill in the blanks. You may give bounds on the *P*-value.

One-way ANOVA:

Source	DF	SS	MS	F	P
Factor	3	36.15	?	?	?
Error	?	?	?		
Total	19	196.04			

5-67. A Minitab ANOVA output is shown below. Fill in the blanks. You may give bounds on the *P*-value.

One-way ANOVA:

Source	DF	SS	MS	F	P
Factor	?	?	246.93	?	?
Error	25	186.53	?		
Total	29	1174.24			

5-68. In "Orthogonal Design for Process Optimization and Its Application to Plasma Etching" (*Solid State Technology,* May 1987), G. Z. Yin and D. W. Jillie describe an experiment to determine the effect of C_2F_6 flow rate on the uniformity of the etch on a silicon wafer used in integrated circuit manufacturing. Three flow rates are used in the experiment, and the resulting uniformity (in percent) for six replicates is shown here.

C_2F_6 Flow (SCCM)	Observations					
	1	2	3	4	5	6
125	2.7	4.6	2.6	3.0	3.2	3.8
160	4.9	4.6	5.0	4.2	3.6	4.2
200	4.6	3.4	2.9	3.5	4.1	5.1

(a) Does C_2F_6 flow rate affect etch uniformity? Construct box plots to compare the factor levels and perform the analysis of variance. What is the approximate *P*-value? What are your conclusions?
(b) Which gas flow rates produce different mean etch uniformities?

5-69. In *Design and Analysis of Experiments*, 6th edition (John Wiley & Sons, 2005), D. C. Montgomery describes an experiment in which the tensile strength of a synthetic fiber is of interest to the manufacturer. It is suspected that strength is related to the percentage of cotton in the fiber. Five levels of cotton percentage are used, and five replicates are run in random order, resulting in the data that follow.

Cotton Percentage	Observations				
	1	2	3	4	5
15	7	7	15	11	9
20	12	17	12	18	18
25	14	18	18	19	19
30	19	25	22	19	23
35	7	10	11	15	11

(a) Does cotton percentage affect breaking strength? Draw comparative box plots and perform an analysis of variance. Use a *P*-value approach.
(b) Plot average tensile strength against cotton percentage and interpret the results.
(c) Which specific means are different?
(d) Perform residual analysis and model checking.

5-70. An experiment was run to determine whether four specific firing temperatures affect the density of a certain type of brick. The experiment led to the following data.
(a) Does the firing temperature affect the density of the bricks? Use a *P*-value.
(b) Find the *P*-value for the *F*-statistic computed in part (a).

Temperature (°F)			
100	125	150	175
Density			
21.8	21.7	21.9	21.9
21.9	21.4	21.8	21.7
21.7	21.5	21.8	21.8
21.6	21.5	21.6	21.7
21.7	—	21.5	21.6
21.5	—	—	21.8
21.8	—	—	—

5-71. A study was carried out to determine if curing temperature significantly affects the tensile strength of silicone rubber. An axially controlled automated hydraulic force applicator was used to measure the tensile strength (in megapascals, MPa) of each of the specimens. The results are given below.

Temperature, Celsius		
25	40	55
2.09	2.22	2.03
2.14	2.09	2.22
2.18	2.10	2.10
2.05	2.02	2.07
2.18	2.05	2.03
2.11	2.01	2.15

(a) Test the hypothesis that the curing temperatures affect the tensile strength of the silicone rubber. Use a *P*-value approach.
(b) Construct box plots of the data. Do these support your conclusions? Explain.
(c) Perform residual analysis and model checking.

5-72. An electronics engineer is interested in the effect on tube conductivity of five different types of coating for cathode ray tubes in a telecommunications system display device. The following conductivity data are obtained. If $\alpha = 0.05$, can you isolate any differences in mean conductivity due to the coating type?

Coating Type	Conductivity			
1	143	141	150	146
2	152	149	137	143
3	134	133	132	127
4	129	127	132	129
5	147	148	144	142

5-73. A Minitab ANOVA from a randomized complete block experiment output is shown below.

Two-way ANOVA: Y versus Treatment, Block

Source	DF	SS	MS	F	P
Treatment	4	1010.56	?	29.84	?
Block	?	?	64.765	?	?
Error	20	169.33	?		
Total	29	1503.71			

(a) Fill in the blanks. You may give bounds on the *P*-value.
(b) How many blocks were used in this experiment?
(c) What conclusions can you draw?

5-74. Consider the Minitab ANOVA output from the completely randomized single-factor experiment shown in Exercise 5-67. Suppose that this experiment had been conducted in a randomized complete block design, and that the sum of squares for blocks was 80.00. Modify the ANOVA table to show the correct analysis for the randomized complete block experiment.

5-75. In "The Effect of Nozzle Design on the Stability and Performance of Turbulent Water Jets" (*Fire Safety Journal*, Vol. 4, August 1981), C. Theobald describes an experiment in which a shape measurement was determined for several different nozzle types at different levels of jet efflux velocity. Interest in this experiment focuses primarily on nozzle type, and velocity is a nuisance factor. The data are as follows.

Nozzle	Jet Efflux Velocity (m/s)					
Type	11.73	14.37	16.59	20.43	23.46	28.74
1	0.78	0.80	0.81	0.75	0.77	0.78
2	0.85	0.85	0.92	0.86	0.81	0.83
3	0.93	0.92	0.95	0.89	0.89	0.83
4	1.14	0.97	0.98	0.88	0.86	0.83
5	0.97	0.86	0.78	0.76	0.76	0.75

(a) Does nozzle type affect shape measurement? Compare the nozzles with box plots and the analysis of variance.
(b) Use the graphical method from Section 5-8.1 to determine specific differences between the nozzles. Does a graph of the average (or standard deviation) of the shape measurements versus nozzle type assist with the conclusions?
(c) Analyze the residuals from this experiment.

5-76. In *Design and Analysis of Experiments*, 6th edition (John Wiley & Sons, 2005), D. C. Montgomery describes an experiment that determined the effect of four different types of tips in a hardness tester on the observed hardness of a metal alloy. Four specimens of the alloy were obtained, and each tip was tested once on each specimen, producing the following data:

Type of	Specimen			
Tip	1	2	3	4
1	9.3	9.4	9.6	10.0
2	9.4	9.3	9.8	9.9
3	9.2	9.4	9.5	9.7
4	9.7	9.6	10.0	10.2

(a) Is there any difference in hardness measurements between the tips?
(b) Use the graphical method from Section 5-8.1 to investigate specific differences between the tips.
(c) Analyze the residuals from this experiment.

5-77. An article in the *American Industrial Hygiene Association Journal* (Vol. 37, 1976, pp. 418–422) describes a field test for detecting the presence of arsenic in urine samples. The test has been proposed for use among forestry workers because of the increasing use of organic arsenics in that industry. The experiment compared the test as performed by both a trainee and an experienced trainer to an analysis at a remote laboratory. Four subjects were selected for testing and are considered as blocks. The response variable is arsenic content (in ppm) in the subject's urine. The data are as follows.

	Subject			
Test	1	2	3	4
Trainee	0.05	0.05	0.04	0.15
Trainer	0.05	0.05	0.04	0.17
Lab	0.04	0.04	0.03	0.10

(a) Is there any difference in the arsenic test procedure?
(b) Analyze the residuals from this experiment.

5-78. An article in the *Food Technology Journal* (Vol. 10, 1956, pp. 39–42) describes a study on the protopectin content of tomatoes during storage. Four storage times were selected, and samples from nine lots of tomatoes were analyzed. The protopectin content (expressed as hydrochloric acid-soluble fraction mg/kg) is in the table on the next page.
(a) The researchers in this study hypothesized that mean protopectin content would be different at different storage times. Can you confirm this hypothesis with a statistical test using $\alpha = 0.05$?
(b) Find the *P*-value for the test in part (a).
(c) Which specific storage times are different? Would you agree with the statement that protopectin content decreases as storage time increases?
(d) Analyze the residuals from this experiment.

Storage Time (days)	Lot								
	1	2	3	4	5	6	7	8	9
0	1694.0	989.0	917.3	346.1	1260.0	965.6	1123.0	1106.0	1116.0
7	1802.0	1074.0	278.8	1375.0	544.0	672.2	818.0	406.8	461.6
14	1568.0	646.2	1820.0	1150.0	983.7	395.3	422.3	420.0	409.5
21	415.5	845.4	377.6	279.4	447.8	272.1	394.1	356.4	351.2

5-79. An experiment was conducted to investigate leaking current in a near-micrometer SOS MOSFETS device. The purpose of the experiment was to investigate how leakage current varies as the channel length changes. Four channel lengths were selected. For each channel length, five different widths were also used, and width is to be considered a nuisance factor. The data are as follows.

Channel Length	Width				
	1	2	3	4	5
1	0.7	0.8	0.8	0.9	1.0
2	0.8	0.8	0.9	0.9	1.0
3	0.9	1.0	1.7	2.0	4.0
4	1.0	1.5	2.0	3.0	20.0

(a) Test the hypothesis that mean leakage voltage does not depend on the channel length, using $\alpha = 0.05$.
(b) Analyze the residuals from this experiment. Comment on the residual plots.

5-80. Consider the leakage voltage experiment described in Exercise 5-79. The observed leakage voltage for channel length 4 and width 5 was erroneously recorded. The correct observation is 4.0. Analyze the corrected data from this experiment. Is there evidence to conclude that mean leakage voltage increases with channel length?

SUPPLEMENTAL EXERCISES

5-81. A procurement specialist has purchased 25 resistors from vendor 1 and 35 resistors from vendor 2. Each resistor's resistance is measured with the following results.

Vendor 1

96.8	100.0	100.3	98.5	98.3	98.2	99.6
99.4	99.9	101.1	103.7	97.7	99.7	101.1
97.7	98.6	101.9	101.0	99.4	99.8	99.1
99.6	101.2	98.2	98.6			

Vendor 2

106.8	106.8	104.7	104.7	108.0	102.2
103.2	103.7	106.8	105.1	104.0	106.2
102.6	100.3	104.0	107.0	104.3	105.8
104.0	106.3	102.2	102.8	104.2	103.4
104.6	103.5	106.3	109.2	107.2	105.4
106.4	106.8	104.1	107.1	107.7	

(a) What distributional assumption is needed to test the claim that the variance of resistance of product from vendor 1 is not significantly different from the variance of resistance of product from vendor 2? Perform a graphical procedure to check this assumption.

(b) Perform an appropriate statistical hypothesis testing procedure to determine whether the procurement specialist can claim that the variance of resistance of product from vendor 1 is significantly different from the variance of resistance of product from vendor 2.

5-82. An article in the *Journal of Materials Engineering* (Vol 11, No. 4, 1989, pp. 275–282) reported the results of an experiment to determine failure mechanisms for plasma-sprayed thermal barrier coatings. The failure stress for one particular coating (NiCrAlZr) under two different test conditions is as follows:

Failure stress ($\times 10^6$ Pa) after nine 1-hr cycles: 19.8, 18.5, 17.6, 16.7, 16.7, 14.8, 15.4, 14.1, 13.6

Failure stress ($\times 10^6$ Pa) after six 1-hr cycles: 14.9, 12.7, 11.9, 11.4, 10.1, 7.9

(a) What assumptions are needed to construct confidence intervals for the difference in mean failure stress under the two different test conditions? Use normal probability plots of the data to check these assumptions.

(b) Perform a hypothesis test to determine if the mean failure stress of the two different test conditions is the same at the 0.05 significance level.

(c) Confirm that the *P*-value of this test is 0.001.

(d) Construct a 99.9% CI of this difference. Use this CI to again test the hypothesis that the mean failure stress of the two different test conditions is the same. Explain why your results are the same or different from those found in part (a).

5-83. A manufacturing company uses a screen printing process to deposit ink on thin plastic substrates. The thickness of the deposit is a critical quality characteristic. A new automated ink depositing system has been added to reduce the variability in the thickness of the deposit. Weight measurements in grams, used to characterize the thickness, are taken using the old manual and new automated processes. The recorded sample standard deviations are $s_{old} = 0.094$ grams based on 21 observations and $s_{new} = 0.047$ grams based on 16 observations.

(a) Determine if the new system results in a variance that is significantly less than the old at $\alpha = 0.1$. State any necessary assumptions of your analysis.

(b) Find the *P*-value of this test.

(c) Construct a 90% CI on the ratio of the variances.

(d) Use the CI found in part (c) to determine if the new system results in a variance that is significantly less than the old. Explain why your answer is the same or different.

5-84. A liquid dietary product implies in its advertising that use of the product for 1 month results in an average weight loss of at least 3 pounds. Eight subjects use the product for 1 month, and the resulting weight loss data are reported here. Use hypothesis testing procedures to answer the following questions.

Subject	Initial Weight (lb)	Final Weight (lb)	Subject	Initial Weight (lb)	Final Weight (lb)
1	165	161	5	155	150
2	201	195	6	143	141
3	195	192	7	150	146
4	198	193	8	187	183

(a) Do the data support the claim of the producer of the dietary product with the probability of a type I error of 0.05?

(b) Do the data support the claim of the producer of the dietary product with the probability of a type I error set to 0.01?

(c) In an effort to improve sales, the producer is considering changing its claim from "at least 3 pounds" to "at least 5 pounds." Repeat parts (a) and (b) to test this new claim.

5-85. The breaking strength of yarn supplied by two manufacturers is being investigated. We know from experience with the manufacturers' processes that $\sigma_1 = 5$ and $\sigma_2 = 4$ psi. A random sample of 20 test specimens from each manufacturer results in $\bar{x}_1 = 88$ and $\bar{x}_2 = 91$ psi, respectively.

(a) Using a 90% CI on the difference in mean breaking strength, comment on whether or not there is evidence to support the claim that manufacturer 2 produces yarn with higher mean breaking strength.

(b) Using a 98% CI on the difference in mean breaking strength, comment on whether or not there is evidence to support the claim that manufacturer 2 produces yarn with higher mean breaking strength.

(c) Comment on why the results from parts (a) and (b) are different or the same. Which would you choose to make your decision and why?

5-86. Consider the previous exercise. Suppose that prior to collecting the data, you decide that you want the error in estimating $\mu_1 - \mu_2$ by $\bar{x}_1 - \bar{x}_2$ to be less than 1.5 psi. Specify the sample size for the following percentage confidence:

(a) 90%

(b) 98%

(c) Comment on the effect of increasing the percentage confidence on the sample size needed.

(d) Repeat parts (a)–(c) with an error of less than 0.75 psi instead of 1.5 psi.

(e) Comment on the effect of decreasing the error on the sample size needed.

5-87. The Salk polio vaccine experiment in 1954 focused on the effectiveness of the vaccine in combating paralytic polio. Because it was felt that without a control group of children there would be no sound basis for evaluating the efficacy of the Salk vaccine, the vaccine was administered to one group, and a placebo (visually identical to the vaccine but known to have no effect) was administered to a second group. For ethical reasons, and because it was suspected that knowledge of vaccine administration would affect subsequent diagnosis, the experiment was conducted in a double-blind fashion. That is, neither the subjects nor the administrators knew who received the vaccine and who received the placebo. The actual data for this experiment are as follows:

Placebo group: $n = 201,299$: 110 cases of polio observed

Vaccine group: $n = 200,745$: 33 cases of polio observed

(a) Use a hypothesis testing procedure to determine whether the proportion of children in the two groups who contracted paralytic polio is statistically different. Use a probability of a type I error equal to 0.05.

(b) Repeat part (a) using a probability of a type I error equal to 0.01.

(c) Compare your conclusions from parts (a) and (b) and explain why they are the same or different.

5-88. A study was carried out to determine the accuracy of Medicaid claims. In a sample of 1095 physician-filed claims, 942 claims exactly matched information in medical records. In a sample of 1042 hospital-filed claims, the corresponding number was 850.

(a) Is there a difference in the accuracy between these two sources? What is the P-value of the test? What are your conclusions using $\alpha = 0.05$?

(b) Suppose a second study was conducted. Of the 550 physician-filed claims examined, 473 were accurate, whereas of the 550 hospital-filed claims examined, 451 were accurate. Is there a statistically significant difference in the accuracy in the second study's data? Again, calculate the P-value and make your decision using $\alpha = 0.05$.

(c) Note that the estimated accuracy percentages are nearly identical for the first and second studies; however, the results of the hypothesis tests in parts (a) and (b) are different. Explain why this occurs.

(d) Construct a 95% CI on the difference of the two proportions for part (a). Then construct a 95% CI on the difference of the two proportions for part (b). Explain why the estimated accuracy percentages for the two studies are nearly identical but the lengths of the confidence intervals are different.

5-89. In a random sample of 200 Phoenix residents who drive a domestic car, 165 reported wearing their seat belt regularly, whereas another sample of 250 Phoenix residents who drive a foreign car revealed 198 who regularly wore their seat belt.

(a) Perform a hypothesis testing procedure to determine whether there is a statistically significant difference in seat belt usage between domestic and foreign car drivers. Set your probability of a type I error to 0.05.

(b) Perform a hypothesis testing procedure to determine whether there is a statistically significant difference in seat belt usage between domestic and foreign car drivers. Set your probability of a type I error to 0.1.

(c) Compare your answers for parts (a) and (b) and explain why they are the same or different.

(d) Suppose that all the numbers in the problem description were doubled. That is, in a random sample of 400 Phoenix residents who drive a domestic car, 330 reported wearing their seat belt regularly, whereas another sample of 500 Phoenix residents who drive a foreign car revealed 396 who regularly wore their seat belt. Repeat parts (a) and (b) and comment on the effect of increasing the sample size without changing the proportions on your results.

5-90. Consider the previous exercise, which summarized data collected from drivers about their seat belt usage.

(a) Do you think there is a reason not to believe these data? Explain your answer.

(b) Is it reasonable to use the hypothesis testing results from the previous problem to draw an inference about the difference in proportion of seat belt usage

    (i) of the spouses of these drivers of domestic and foreign cars? Explain your answer.

    (ii) of the children of these drivers of domestic and foreign cars? Explain your answer.

    (iii) of all drivers of domestic and foreign cars? Explain your answer.

    (iv) of all drivers of domestic and foreign trucks? Explain your answer.

5-91. Consider Example 5-12 in the text.

(a) Redefine the parameters of interest to be the proportion of lenses that are unsatisfactory following tumble polishing with polishing fluids 1 or 2. Test the hypothesis that the two polishing solutions give different results using $\alpha = 0.01$.

(b) Compare your answer in part (a) with that in the example. Explain why they are the same or different.

5-92. A manufacturer of a new pain relief tablet would like to demonstrate that its product works twice as fast as the competior's product. Specifically, it would like to test

$$H_0: \mu_1 = 2\mu_2$$
$$H_1: \mu_1 > 2\mu_2$$

where μ_1 is the mean absorption time of the competitive product and μ_2 is the mean absorption time of the new product. Assuming that the variances σ_1^2 and σ_2^2 are known, develop a procedure for testing this hypothesis.

5-93. Suppose that we are testing $H_0: \mu_1 = \mu_2$ versus $H_1: \mu_1 \neq \mu_2$, and we plan to use equal sample sizes from the two populations. Both populations are assumed to be normal with unknown but equal variances. If we use $\alpha = 0.05$ and if the true mean $\mu_1 = \mu_2 + \sigma$, what sample size must be used for the power of this test to be at least 0.90?

5-94. A fuel-economy study was conducted for two German automobiles, Mercedes and Volkswagen. One vehicle of each brand was selected, and the mileage performance was observed for 10 tanks of fuel in each car. The data are as follows (in mpg):

Mercedes		Volkswagen	
24.7	24.9	41.7	42.8
24.8	24.6	42.3	42.4
24.9	23.9	41.6	39.9
24.7	24.9	39.5	40.8
24.5	24.8	41.9	29.6

(a) Construct a normal probability plot of each of the data sets. Based on these plots, is it reasonable to assume that they are each drawn from a normal population?

(b) Suppose it was determined that the lowest observation of the Mercedes data was erroneously recorded and should be 24.6. Furthermore, the lowest observation of the Volkswagen data was also mistaken and should be 39.6. Again construct normal probability plots of each of the data sets with the corrected values. Based on these new plots, is it reasonable to assume that they are each drawn from a normal population?

(c) Compare your answers from parts (a) and (b) and comment on the effect of these mistaken observations on the normality assumption.

(d) Using the corrected data from part (b) and a 95% CI, is there evidence to support the claim that the variability in mileage performance is greater for a Volkswagen than for a Mercedes?

5-95. An article in *Neurology* (Vol. 50, 1998, pp. 1246–1252) describes a finding that monozygotic twins share numerous physical, psychological, and pathological traits. The investigators measured an intelligence score of 10 pairs of twins, and the data are as follows:

Pair	Birth Order: 1	Birth Order: 2
1	6.08	5.73
2	6.22	5.80
3	7.99	8.42
4	7.44	6.84
5	6.48	6.43
6	7.99	8.76
7	6.32	6.32
8	7.60	7.62
9	6.03	6.59
10	7.52	7.67

(a) Is the assumption that the difference in score in normally distributed reasonable? Show results to support your answer.

(b) Find a 95% confidence interval on the difference in mean score. Is there any evidence that mean score depends on birth order?

(c) It is important to detect a mean difference in score of one point, with a probability of at least 0.90. Was the use of 10 pairs an adequate sample size? If not, how many pairs should have been used?

5-96. An article in the *Journal of the Environmental Engineering Division* ("Distribution of Toxic Substances in Rivers," Vol. 108, 1982, pp. 639–649) describes a study of the concentration of several hydrophobic organic substances in the Wolf River in Tennessee. Measurements of hexachlorobenzene (HCB) in nanograms per liter were taken at different depths downstream of an abandoned dump site. Data for two depths follow:

Surface: 3.74, 4.61, 4.00, 4.67, 4.87, 5.12, 4.52, 5.29, 5.74, 5.48
Bottom: 5.44, 6.88, 5.37, 5.44, 5.03, 6.48, 3.89, 5.85, 6.85, 7.16

(a) What assumptions are required to test the claim that the mean HCB concentration is the same at both depths?

Check those assumptions for which you have the information.

(b) Apply an appropriate procedure to determine if the data support the claim in part (a).

(c) Suppose that the true difference in mean concentration is 2.0 nanograms per liter. For $\alpha = 0.05$, what is the power of a statistical test for $H_0: \mu_1 = \mu_2$ versus $H_1: \mu_1 \neq \mu_2$?

(d) What sample size would be required to detect a difference of 1.0 nanograms per liter at $\alpha = 0.05$ if the power must be at least 0.9?

5-97. Consider the fire-fighting foam expanding agents investigated in Exercise 5-18, in which five observations of each agent were recorded. Suppose that, if agent 1 produces a mean expansion that differs from the mean expansion of agent 2 by 1.5, we would like to reject the null hypothesis with probability at least 0.95.

(a) What sample size is required?

(b) Do you think that the original sample size in Exercise 5-18 was appropriate to detect this difference? Explain your answer.

5-98. A manufacturer of heart pacemakers is investigating changing the casing material to decrease the weight of the device. Three different alloys are being considered. Eight prototype parts are made from each alloy material and weighed, in grams. The data are compiled into the following partially complete analysis of variance table:

Source of Variation	Sums of Squares	Degrees of Freedom	Mean Square	F	P
Factor	4.1408				
Error		21			
Total	4.8596				

(a) Complete the analysis of variance table.

(b) Use the analysis of variance table to test the hypothesis that the weight differs among the alloy types. Use $\alpha = 0.10$.

5-99. A materials engineer performs an experiment to investigate whether there is a difference among five types of foam pads used under carpeting. A mechanical device is constructed to simulate "walkers" on the pad, and four samples of each pad are randomly tested on the simulator. After a certain amount of time, the foam pad is removed from the simulator, examined, and scored for wear quality. The data are compiled into the following partially complete analysis of variance table.

Source of Variation	Sum of Squares	Degrees of Freedom	Mean Square	F_0
Treatments	95.129			
Error	86.752			
Total		19		

(a) Complete the analysis of variance table.
(b) Use the analysis of variance table to test the hypothesis that wear quality differs among the types of foam pads. Use $\alpha = 0.05$.

5-100.   A Rockwell hardness-testing machine presses a tip into a test coupon and uses the depth of the resulting depression to indicate hardness. Two different tips are being compared to determine whether they provide the same Rockwell C-scale hardness readings. Nine coupons are tested, with both tips being tested on each coupon. The data are as follows.

Coupon	Tip 1	Tip 2	Coupon	Tip 1	Tip 2
1	47	46	6	41	41
2	42	40	7	45	46
3	43	45	8	45	46
4	40	41	9	49	48
5	42	43			

(a) State any assumptions necessary to test the claim that both tips produce the same Rockwell C-scale hardness readings. Check those assumptions for which you have the data.
(b) Apply an appropriate statistical method to determine whether the data support the claim that the difference in Rockwell C-scale hardness readings of the two tips is significantly different from zero.
(c) Suppose that, if the two tips differ in mean hardness readings by as much as 1.0, we want the power of the test to be at least 0.9. For an $\alpha = 0.01$, how many coupons should have been used in the test?

5-101.   Two different gauges can be used to measure the depth of bath material in a Hall cell used in smelting aluminum. Each gauge is used once in 15 cells by the same operator. Depth measurements from both gauges for 15 cells are shown below.

Cell	Gauge 1	Gauge 2	Cell	Gauge 1	Gauge 2
1	46 in.	47 in.	9	52 in.	51 in.
2	50	53	10	47	45
3	47	45	11	49	51
4	53	50	12	45	45
5	49	51	13	47	49
6	48	48	14	46	43
7	53	54	15	50	51
8	56	53			

(a) State any assumptions necessary to test the claim that both gauges produce the same mean bath depth readings. Check those assumptions for which you have the data.

(b) Apply an appropriate statistical method to determine whether the data support the claim that the two gauges produce different bath depth readings.
(c) Suppose that if the two gauges differ in mean bath depth readings by as much as 1.65 inch, we want the power of the test to be at least 0.8. For $\alpha = 0.01$, how many cells should have been used?

5-102.   An article in the *Materials Research Bulletin* (Vol. 26, No. 11, 1991) reported a study of four different methods of preparing the superconducting compound $PbMo_6S_8$. The authors contend that the presence of oxygen during the preparation process affects the material's superconducting transition temperature T_c. Preparation methods 1 and 2 use techniques that are designed to eliminate the presence of oxygen, whereas methods 3 and 4 allow oxygen to be present. Five observations on T_c (in kelvins, K) were made for each method, and the results are as follows.

Preparation Method	Transition Temperature T_c (K)				
1	14.8	14.8	14.7	14.8	14.9
2	14.6	15.0	14.9	14.8	14.7
3	12.7	11.6	12.4	12.7	12.1
4	14.2	14.4	14.4	12.2	11.7

(a) Is there evidence to support the claim that the presence of oxygen during preparation affects the mean transition temperature? Use $\alpha = 0.05$.
(b) What is the P-value for the F-test in part (a)?

5-103.   A paper in the *Journal of the Association of Asphalt Paving Technologists* (Vol. 59, 1990) describes an experiment to determine the effect of air voids on percentage retained strength of asphalt. For purposes of the experiment, air voids are controlled at three levels: low (2–4%), medium (4–6%), and high (6–8%). The data are shown in the following table.

Air Voids	Retained Strength (%)							
Low	106	90	103	90	79	88	92	95
Medium	80	69	94	91	70	83	87	83
High	78	80	62	69	76	85	69	85

(a) Do the different levels of air voids significantly affect mean retained strength? Use $\alpha = 0.01$.
(b) Find the P-value for the F-statistic in part (a).

5-104.   An article in *Environment International* (Vol. 18, No. 4, 1992) describes an experiment in which the amount of radon released in showers was investigated. Radon-enriched water was used in the experiment, and six different orifice diameters were tested in shower heads. The data from the experiment are shown in the following table.

Orifice Diameter	Radon Released (%)			
0.37	80	83	83	85
0.51	75	75	79	79
0.71	74	73	76	77
1.02	67	72	74	74
1.40	62	62	67	69
1.99	60	61	64	66

(a) Does the size of the orifice affect the mean percentage of radon released? Use $\alpha = 0.05$.

(b) Find the P-value for the F-statistic in part (a).

5-105. A team of computer engineers is interested in determining if the percentage of disk space that is used for virtual memory will significantly improve the response time of a server. The team measured the response time in milliseconds when 2, 4, 6, and 8% of the server's disk space is used for virtual memory. The results of the experiment are as follows:

2%	4%	6%	8%
2.2	2.0	1.8	1.9
2.1	2.0	2.0	2.0
1.9	1.9	1.7	2.0
1.9	1.8	1.8	1.9
2.2	2.0	2.0	1.8
1.8	2.0	1.9	2.0

(a) Does the percentage of the server allocated to memory change the response time of the server at the 0.05 level of significance?

(b) Find the P-value of the F-statistic in part (a).

TEAM EXERCISES

5-106. Construct a data set for which the paired t-test statistic is very large, indicating that when this analysis is used, the two population means are different; however, t_0 for the two-sample t-test is very small, so the incorrect analysis would indicate that there is no significant difference between the means.

5-107. Identify an example in which a comparative standard or claim is made about two independent populations. For example, Car Type A gets more average miles per gallon in urban driving than Car Type B. The standard or claim may be expressed as a mean (average), variance, standard deviation, or proportion. Collect two appropriate random samples of data and perform a hypothesis test. Report on your results. Be sure to include in your report the comparison expressed as a hypothesis test, a description of the data collected, the analysis performed, and the conclusion reached.

IMPORTANT TERMS AND CONCEPTS

Alternative hypothesis
Analysis of variance (ANOVA)
Blocking
Chi-square distribution
Completely randomized design
Confidence intervals
Confidence level
Connection between hypothesis tests and confidence intervals

Critical region for a test statistic
F distribution
Null hypothesis
One- and two-sided alternative hypotheses
One-sided confidence bounds
Operating characteristic curves
Paired t-test
Pooled t-test

Power of a test
P-value
Randomized complete block design
Sample size determination for confidence intervals
Sample size determination for hypothesis tests

Statistical hypotheses
Statistical versus practical significance
t distribution
Test statistic
Two-sample t-test
Type I error
Type II error

6

Building Empirical Models

CHAPTER OUTLINE

LEARNING OBJECTIVES

After careful study of this chapter, you should be able to do the following:

1. Use simple linear or multiple linear regression for building empirical models of engineering and scientific data.

2. Analyze residuals to determine if the regression model is an adequate fit to the data or to see if any underlying assumptions are violated.

3. Test statistical hypotheses and construct confidence intervals on regression model parameters.

4. Use the regression model either to compute a mean or to make a prediction of a future observation.

5. Use confidence intervals or prediction intervals to describe the error in estimation from a regression model.

6. Comment on the strengths and weaknesses of your empirical model.

6-1 INTRODUCTION TO EMPIRICAL MODELS

Engineers frequently use **models** in problem formulation and solution. Sometimes these models are based on our physical, chemical, or engineering science knowledge of the phenomenon, and in such cases we call these models **mechanistic models.** Examples of mechanistic models include Ohm's law, the gas laws, and Kirchhoff's laws. However, there are many situations in which two or more variables of interest are related, and the mechanistic model relating these variables is unknown. In these cases it is necessary to build a model relating the variables based on observed data. This type of model is called an **empirical model.** An empirical model can be manipulated and analyzed just as a mechanistic model can.

As an illustration, consider the data in Table 6-1. In this table, y is the salt concentration (milligrams/liter) found in surface streams in a particular watershed and x is the percentage of the watershed area consisting of paved roads. The data are consistent with those found in an article in the *Journal of Environmental Engineering* (Vol. 115, No. 3, 1989). A scatter diagram of the data (with dot diagrams of the individual variables) is shown in Fig. 6-1. There is no obvious physical mechanism that relates the salt concentration to the roadway area, but the scatter diagram indicates that some relationship, possibly linear, does exist. A linear relationship will not pass exactly through all of the points in Fig. 6-1, but there is an indication that the points are scattered randomly about a straight line. Therefore, it is probably reasonable to assume that the mean of the random variable Y (the salt concentration) is related to roadway area x by the following straight-line relationship:

$$E(Y|x) = \mu_{Y|x} = \beta_0 + \beta_1 x$$

where the slope and intercept of the line are unknown parameters. The notation $E(Y|x)$ represents the expected value of the **response variable** Y at a particular value of the **regressor variable** x. Although the mean of Y is a linear function of x, the actual observed value y does not fall exactly

Table 6-1 Salt Concentration in Surface Streams and Roadway Area

Observation	Salt concentration (y)	Roadway area (x)
1	3.8	0.19
2	5.9	0.15
3	14.1	0.57
4	10.4	0.40
5	14.6	0.70
6	14.5	0.67
7	15.1	0.63
8	11.9	0.47
9	15.5	0.75
10	9.3	0.60
11	15.6	0.78
12	20.8	0.81
13	14.6	0.78
14	16.6	0.69
15	25.6	1.30
16	20.9	1.05
17	29.9	1.52
18	19.6	1.06
19	31.3	1.74
20	32.7	1.62

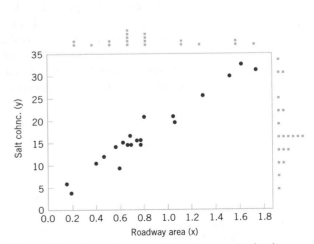

Figure 6-1 Scatter diagram of the salt concentration in surface streams and roadway area data in Table 6-1.

on a straight line. The appropriate way to generalize this to a **probabilistic linear model** is to assume that the expected value of Y is a linear function of x, but that for a fixed value of x the actual value of Y is determined by the mean value function (the linear model) plus a random error term ϵ.

Simple Linear Regression Model

In the **simple linear regression model** the dependent variable, or **response,** is related to one independent, or **regressor variable,** as

$$Y = \beta_0 + \beta_1 x + \epsilon \qquad (6\text{-}1)$$

where ϵ is the random error term. The parameters β_0 and β_1 are called **regression coefficients.**

To gain more insight into this model, suppose that we can fix the value of x and observe the value of the random variable Y. Now if x is fixed, the random component ϵ on the right-hand side of the model in equation 6-1 determines the properties of Y. Suppose that the mean and variance of ϵ are 0 and σ^2, respectively. Then

$$E(Y|x) = E(\beta_0 + \beta_1 x + \epsilon) = \beta_0 + \beta_1 x + E(\epsilon) = \beta_0 + \beta_1 x$$

Note that this is the same relationship we initially wrote down empirically from inspection of the scatter diagram in Fig. 6-1. The variance of Y given x is

$$V(Y|x) = V(\beta_0 + \beta_1 x + \epsilon) = V(\beta_0 + \beta_1 x) + V(\epsilon) = 0 + \sigma^2 = \sigma^2$$

Thus the true regression model $\mu_{Y|x} = \beta_0 + \beta_1 x$ is a line of mean values; that is, the height of the regression line at any value of x is simply the expected value of Y for that x. The slope, β_1, can be interpreted as the change in the mean of Y for a unit change in x. Furthermore, the variability of Y at a particular value of x is determined by the error variance σ^2. This implies that there is a distribution of Y values at each x and that the variance of this distribution is the same at each x.

For example, suppose that the true regression model relating salt concentration to roadway area is $\mu_{Y|x} = 3 + 15x$, and suppose that the variance is $\sigma^2 = 2$. Figure 6-2 illustrates this situation. Note that we have used a normal distribution to describe the random variation in ϵ. Because Y is the sum of a constant $\beta_0 + \beta_1 x$ (the mean) and a normally distributed random variable, Y is a normally distributed random variable. The variance σ^2 determines the variability in the observations Y on salt concentration. Thus, when σ^2 is small, the observed values of Y

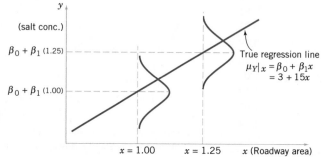

Figure 6-2 The distribution of Y for a given value of x for the salt concentration-roadway area data.

will fall close to the line, and when σ^2 is large, the observed values of Y may deviate considerably from the line. Because σ^2 is constant, the variability in Y at any value of x is the same.

The regression model describes the relationship between salt concentration Y and roadway area x. Thus, for any value of roadway area, salt concentration has a normal distribution with mean $3 + 15x$ and variance 2. For example, if $x = 1.25$, then Y has mean value $\mu_{Y|x} = 3 + 15(1.25) = 21.75$ and variance 2.

There are many empirical model building situations in which there is more than one regressor variable. Once again, a regression model can be used to describe the relationship. A regression model that contains more than one regressor variable is called a **multiple regression model.**

As an example, suppose that the effective life of a cutting tool depends on the cutting speed and the tool angle. A multiple regression model that might describe this relationship is

$$Y = \beta_0 + \beta_1 x_1 + \beta_2 x_2 + \epsilon \qquad (6\text{-}2)$$

where Y represents the tool life, x_1 represents the cutting speed, x_2 represents the tool angle, and ϵ is a random error term. This is a **multiple linear regression model** with two regressors. The term *linear* is used because equation 6-2 is a linear function of the unknown parameters β_0, β_1, and β_2.

The regression model in equation 6-2 describes a plane in the three-dimensional space of Y, x_1, and x_2. Figure 6-3*a* shows this plane for the regression model

$$E(Y) = 50 + 10x_1 + 7x_2$$

where we have assumed that the expected value of the error term is zero; that is, $E(\epsilon) = 0$. The parameter β_0 is the **intercept** of the plane. We sometimes call β_1 and β_2 **partial regression coefficients** because β_1 measures the expected change in Y per unit change in x_1 when x_2 is held constant, and β_2 measures the expected change in Y per unit change in x_2 when x_1 is held constant. Figure 6-3*b* shows a **contour plot** of the regression model—that is, lines of constant $E(Y)$ as a function of x_1 and x_2. Note that the contour lines in this plot are straight lines.

Multiple Linear Regression Model

In a **multiple linear regression model,** the **dependent variable** or **response** is related to k **independent** or **regressor variables.** The model is

$$Y = \beta_0 + \beta_1 x_1 + \beta_2 x_2 + \cdots + \beta_k x_k + \epsilon \qquad (6\text{-}3)$$

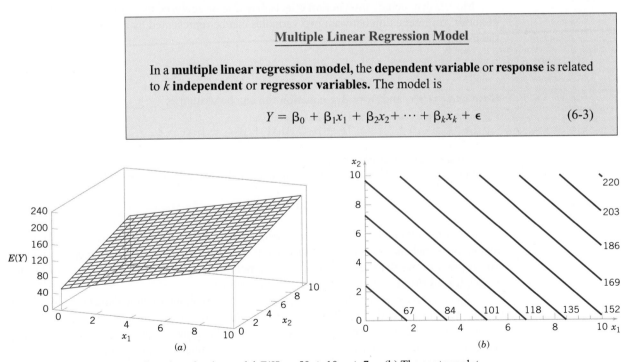

Figure 6-3   (a) The regression plane for the model $E(Y) = 50 + 10x_1 + 7x_2$. (b) The contour plot.

The parameters β_j, $j = 0, 1, \ldots, k$, are called regression coefficients. This model describes a **hyperplane** in the space of the regressor variables $\{x_j\}$ and Y. The parameter β_j represents the expected change in response Y per unit change in x_j when all the remaining regressors x_i $(i \neq j)$ are held constant.

Multiple linear regression models are often used as **empirical models**. That is, the mechanistic model that relates Y and $x_1, x_2, \ldots, x_k$ is unknown, but over certain ranges of the independent variables the linear regression model is an adequate approximation.

These empirical models are related to the important, well-known Taylor series approximation of a complex function, as discussed in Chapter 3. For example, the first-order Taylor series approximation of the unknown function $f(x)$ about the mean μ_x

$$f(x) \cong f(\mu_x) + \left.\frac{df(x)}{dx}\right|_{x=\mu_x} (x - \mu_x) + R$$

$$\cong \beta_0 + \beta_1(x - \mu_x)$$

which, when the remainder term is ignored, is a simple linear model about the mean without the error term. Furthermore, a higher-order Taylor series of $f(x)$ or function $f(x_1, x_2, \ldots, x_k)$ with k independent variables can be captured in a multiple linear regression model. That is, complex functional relationships can often be analyzed using equation 6-3 and multiple linear regression techniques.

For example, consider the cubic polynomial model in one regressor variable.

$$Y = \beta_0 + \beta_1 x + \beta_2 x^2 + \beta_3 x^3 + \epsilon \tag{6-4}$$

If we let $x_1 = x$, $x_2 = x^2$, $x_3 = x^3$, equation 6-4 can be written as

$$Y = \beta_0 + \beta_1 x_1 + \beta_2 x_2 + \beta_3 x_3 + \epsilon \tag{6-5}$$

which is a multiple linear regression model with three regressor variables.

Models that include **interaction** effects may also be analyzed by multiple linear regression methods. An interaction between two variables can be represented by a cross-product term in the model, such as

$$Y = \beta_0 + \beta_1 x_1 + \beta_2 x_2 + \beta_{12} x_1 x_2 + \epsilon \tag{6-6}$$

If we let $x_3 = x_1 x_2$ and $\beta_3 = \beta_{12}$, equation 6-6 can be written as

$$Y = \beta_0 + \beta_1 x_1 + \beta_2 x_2 + \beta_3 x_3 + \epsilon$$

which is a linear regression model.

Figure 6-4a and b shows the three-dimensional plot of the regression model

$$Y = 50 + 10x_1 + 7x_2 + 5x_1 x_2$$

and the corresponding two-dimensional contour plot. Note that, although this model is a linear regression model, the shape of the surface that is generated by the model is not linear. In general, **any regression model that is linear in parameters** (the βs) **is a linear regression model, regardless of the shape of the surface that it generates.**

Figure 6-4 provides a nice graphical interpretation of an interaction. Generally, interaction implies that the effect produced by changing one variable (x_1, say) depends on the level of the other variable (x_2). For example, Fig. 6-4 shows that changing x_1 from 2 to 8 produces a much smaller change in $E(Y)$ when $x_2 = 2$ than when $x_2 = 10$. Interaction effects occur frequently in

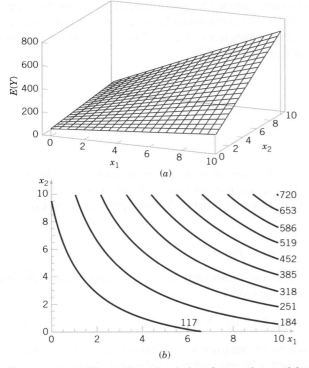

Figure 6-4   (a) Three-dimensional plot of regression model $E(Y) = 50 + 10x_1 + 7x_2 + 5x_1x_2$. (b) The contour plot.

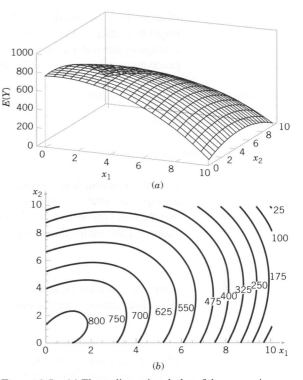

Figure 6-5   (a) Three-dimensional plot of the regression model $E(Y) = 800 + 10x_1 + 7x_2 - 8.5x_1^2 - 5x_2^2 + 4x_1x_2$. (b) The contour plot.

the product and process design, process optimization, and other engineering activities, and regression methods are one of the techniques that we can use to describe them.

As a final example, consider the second-order model with interaction

$$Y = \beta_0 + \beta_1 x_1 + \beta_2 x_2 + \beta_{11} x_1^2 + \beta_{22} x_2^2 + \beta_{12} x_1 x_2 + \epsilon \qquad (6\text{-}7)$$

If we let $x_3 = x_1^2$, $x_4 = x_2^2$, $x_5 = x_1 x_2$, $\beta_3 = \beta_{11}$, $\beta_4 = \beta_{22}$, and $\beta_5 = \beta_{12}$, equation 6-7 can be written as a multiple linear regression model as follows:

$$Y = \beta_0 + \beta_1 x_1 + \beta_2 x_2 + \beta_3 x_3 + \beta_4 x_4 + \beta_5 x_5 + \epsilon$$

Figure 6-5*a* and *b* shows the three-dimensional plot and the corresponding contour plot for

$$E(Y) = 800 + 10x_1 + 7x_2 - 8.5x_1^2 - 5x_2^2 + 4x_1x_2$$

These plots indicate that the expected change in Y when x_1 is changed by one unit (say) is a function of *both* x_1 and x_2. The quadratic and interaction terms in this model produce a mound-shaped function. Depending on the values of the regression coefficients, the second-order model with interaction is capable of assuming a wide variety of shapes; thus it is a very flexible regression model.

In most real-world problems, the values of the parameters (the regression coefficients β_j) and the error variance σ^2 will not be known, and they must be estimated from sample data. **Regression analysis** is a collection of statistical tools for finding estimates of the parameters in the regression model. Then this fitted regression equation or model is typically used in prediction of future observations of Y or for estimating the mean response at a particular level of x. To illustrate with the simple linear regression model example, an environmental engineer might be interested in estimating the mean salt concentration in surface streams when the percentage of the watershed area that is paved roads is $x = 1.25\%$. This chapter discusses these procedures and applications for linear regression models.

Animation 15: Drawing Regression Lines

6-2   SIMPLE LINEAR REGRESSION

6-2.1   Least Squares Estimation

The case of **simple linear regression** considers a *single regressor* or *predictor x* and a dependent or *response* variable Y. Suppose that the true relationship between Y and x is a straight line and that the observation Y at each level of x is a random variable. As noted previously, the expected value of Y for each value of x is

$$E(Y|x) = \beta_0 + \beta_1 x$$

where the intercept β_0 and the slope β_1 are unknown regression coefficients. We assume that each observation, Y, can be described by the model

$$Y = \beta_0 + \beta_1 x + \epsilon \qquad (6\text{-}8)$$

where ϵ is a random error with mean zero and variance σ^2. The random errors corresponding to different observations are also assumed to be uncorrelated random variables.

Suppose that we have n pairs of observations $(x_1, y_1), (x_2, y_2), \ldots (x_n, y_n)$. Figure 6-6 shows a typical scatter plot of observed data and a candidate for the estimated regression line. The estimates of β_0 and β_1 should result in a line that is (in some sense) a "best fit" to the data. The German scientist Karl Gauss (1777–1855) proposed estimating the parameters β_0 and β_1 in equation 6-8 to minimize the sum of the squares of the vertical deviations in Fig. 6-6.

We call this approach to estimating the regression coefficients the **method of least squares.** Using equation 6-8, we may express the n observations in the sample as

$$y_i = \beta_0 + \beta_1 x_i + \epsilon_i, \qquad i = 1, 2, \ldots, n \qquad (6\text{-}9)$$

and the sum of the squares of the deviations of the observations from the true regression line is

$$L = \sum_{i=1}^{n} \epsilon_i^2 = \sum_{i=1}^{n} (y_i - \beta_0 - \beta_1 x_i)^2 \qquad (6\text{-}10)$$

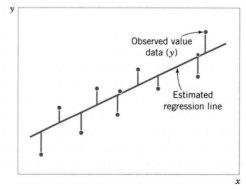

Figure 6-6   Deviations of the data from the estimated regression model.

The least squares estimators of β_0 and β_1, say, $\hat{\beta}_0$ and $\hat{\beta}_1$, must satisfy

$$\left.\frac{\partial L}{\partial \beta_0}\right|_{\hat{\beta}_0,\hat{\beta}_1} = -2\sum_{i=1}^{n}(y_i - \hat{\beta}_0 - \hat{\beta}_1 x_i) = 0 \tag{6-11}$$

$$\left.\frac{\partial L}{\partial \beta_1}\right|_{\hat{\beta}_0,\hat{\beta}_1} = -2\sum_{i=1}^{n}(y_i - \hat{\beta}_0 - \hat{\beta}_1 x_i)x_i = 0$$

Simplifying these two equations yields

$$n\hat{\beta}_0 + \hat{\beta}_1\sum_{i=1}^{n}x_i = \sum_{i=1}^{n}y_i \tag{6-12}$$

$$\hat{\beta}_0\sum_{i=1}^{n}x_i + \hat{\beta}_1\sum_{i=1}^{n}x_i^2 = \sum_{i=1}^{n}y_i x_i$$

Equations 6-12 are called the **least squares normal equations.** The solution to the normal equations results in the least squares estimates $\hat{\beta}_0$ and $\hat{\beta}_1$.

Computing Formulas for Simple Linear Regression

The **least squares estimates** of the intercept and slope in the simple linear regression model are

$$\hat{\beta}_0 = \bar{y} - \hat{\beta}_1\bar{x} \tag{6-13}$$

$$\hat{\beta}_1 = \frac{\displaystyle\sum_{i=1}^{n}y_i x_i - \frac{\left(\sum_{i=1}^{n}y_i\right)\left(\sum_{i=1}^{n}x_i\right)}{n}}{\displaystyle\sum_{i=1}^{n}x_i^2 - \frac{\left(\sum_{i=1}^{n}x_i\right)^2}{n}} \tag{6-14}$$

$$= \frac{\displaystyle\sum_{i=1}^{n}(x_i - \bar{x})(y_i - \bar{y})}{\displaystyle\sum_{i=1}^{n}(x_i - \bar{x})^2} = \frac{S_{xy}}{S_{xx}}$$

where $\bar{y} = (1/n)\sum_{i=1}^{n}y_i$ and $\bar{x} = (1/n)\sum_{i=1}^{n}x_i$.

The **fitted** or **estimated regression line** is therefore

$$\hat{y} = \hat{\beta}_0 + \hat{\beta}_1 x \tag{6-15}$$

Note that each pair of observations satisfies the relationship

$$y_i = \hat{\beta}_0 + \hat{\beta}_1 x_i + e_i, \quad i = 1, 2, \ldots, n$$

where $e_i = y_i - \hat{y}_i$ is called the **residual.** The residual describes the error in the fit of the model to the ith observation y_i. Subsequently we will use the residuals to provide information about the **adequacy** of the fitted model.

EXAMPLE 6-1

**Salt Concentration
and Roadway Data**

Fit a simple linear regression model to the data on salt concentration and roadway area in Table 6-1.

Solution. To build the regression model, the following quantities are computed:

$$n = 20 \quad \sum_{i=1}^{20} x_i = 16.480 \quad \sum_{i=1}^{20} y_i = 342.70 \quad \bar{x} = 0.824 \quad \bar{y} = 17.135$$

$$\sum_{i=1}^{20} y_i^2 = 7060.00 \quad \sum_{i=1}^{20} x_i^2 = 17.2502 \quad \sum_{i=1}^{20} x_i y_i = 346.793$$

$$S_{xx} = \sum_{i=1}^{20} x_i^2 - \frac{\left(\sum_{i=1}^{20} x_i\right)^2}{20} = 17.2502 - \frac{(16.486)^2}{20} = 3.67068$$

and

$$S_{xy} = \sum_{i=1}^{20} x_i y_i - \frac{\left(\sum_{i=1}^{20} x_i\right)\left(\sum_{i=1}^{20} y_i\right)}{20} = 346.793 - \frac{(16.480)(342.70)}{20} = 64.4082$$

Therefore, the least squares estimates of the slope and intercept are

$$\hat{\beta}_1 = \frac{S_{xy}}{S_{xx}} = \frac{64.4082}{3.67068} = 17.5467$$

and

$$\hat{\beta}_0 = \bar{y} - \hat{\beta}_1 \bar{x} = 17.135 - (17.5467)0.824 = 2.6765$$

The fitted simple linear regression model is

$$\hat{y} = 2.6765 + 17.5467x$$

$$\uparrow \qquad\qquad \uparrow$$

$$\hat{\beta}_0 \qquad\quad \hat{\beta}_1$$

This model is plotted in Fig. 6-7, along with the sample data.

Using the linear regression model from Example 6-1, we would predict that the salt concentration in surface streams, where the percentage of paved roads in the watershed is 1.25%, is $\hat{y} = 2.6765 + 17.5467(1.25) = 24.61$ milligrams/liter. The predicted value can be interpreted either as an estimate of the mean salt concentration when roadway area $x = 1.25\%$ or as an estimate of a *new* observation when $x = 1.25\%$. These estimates are, of course, subject to error; that is, it is unlikely that either the true mean salt concentration or a future observation would be *exactly* 24.61 milligrams/liter

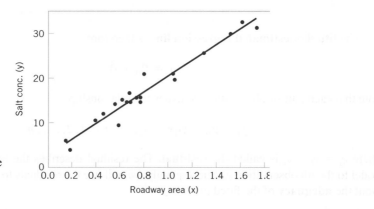

Figure 6-7 Scatter diagram of salt concentration y versus roadway area x and the fitted regression model.

when the roadway area is 1.25%. Subsequently, we will see how to use confidence intervals and prediction intervals to describe the error in estimation from a regression model. ∎

Computer software is widely used to fit regression models. Some output from Minitab for the regression model on salt concentration and roadway area is shown in Table 6-2.

Table 6-2   Minitab Regression Analysis Output for Salt Concentration and Roadway Data

Regression Analysis: Salt conc (y) versus Roadway area (x)

The regression equation is
Salt conc (y) = 2.68 + 17.5 Roadway area (x)

Predictor	Coef	SE Coef	T	P
Constant	2.6765 ◄─ $\hat{\beta}_0$	0.8680	3.08	0.006
Roadway area	17.5467 ◄─ $\hat{\beta}_1$	0.9346	18.77	0.000

S = 1.791   R-Sq = 95.1%   R-Sq(adj) = 94.9%

Analysis of Variance

Source	DF	SS	MS	F	P
Regression	1	1130.1	1130.1	352.46	0.000
Residual Error	18	57.7 ◄─ SS_E	3.2 ◄─ $\hat{\sigma}^2$		
Total	19	1187.9 ◄─ SS_T			

Obs	Roadway area	Salt conc	Fit	SE Fit	Residual
1	0.19	3.800	6.010	0.715	−2.210
2	0.15	5.900	5.309	0.746	0.591
3	0.57	14.100	12.678	0.465	1.422
4	0.40	10.400	9.695	0.563	0.705
5	0.70	14.600	14.959	0.417	−0.359
6	0.67	14.500	14.433	0.425	0.067
7	0.63	15.100	13.731	0.440	1.369
8	0.47	11.900	10.923	0.519	0.977
9	0.75	15.500	15.837	0.406	−0.337
10	0.60	9.300	13.205	0.452	−3.905
11	0.78	15.600	16.363	0.403	−0.763
12	0.81	20.800	16.889	0.401	3.911
13	0.78	14.600	16.363	0.403	−1.763
14	0.69	16.600	14.784	0.420	1.816
15	1.30	25.600	25.487	0.599	0.113
16	1.05	20.900	21.101	0.453	−0.201
17	1.52	29.900	29.347	0.764	0.553
18	1.06	19.600	21.276	0.457	−1.676
19	1.74	31.300	33.208	0.945	−1.908
20	1.62	32.700	31.102	0.845	1.598

Predicted Values for New Observations

New Obs	Fit	SE Fit	95.0% CI	95.0% PI
1	24.610	0.565	(23.424, 25.796)	(20.665, 28.555)

Values of Predictors for New Observations

New Obs	Roadway area
1	1.25

We have highlighted several entries in the Minitab output including the estimates of β_0 and β_1 (in the column labeled "Coef" in the upper portion of Table 6-2). Notice that Minitab computes the model **residuals**; that is, it successively substitutes each value of x_i ($i = 1, 2, \ldots, n$) in the sample into the fitted regression model, calculates the fitted values $\hat{y}_i = \hat{\beta}_0 + \hat{\beta}_1 x_i$, and then finds the residuals as $e_i = y_i - \hat{y}_i$, $i = 1, 2, \ldots, n$. For example, the ninth observation has $x_9 = 0.75$ and $y_9 = 15.5$, and the regression model predicts that $\hat{y}_9 = 15.837$, so the corresponding residual is $e_9 = 15.5 - 15.837 = -0.337$. The residuals for all 20 observations are listed toward the bottom of the output.

The residuals from the fitted regression model are used to estimate the variance of the model errors σ^2. Recall that σ^2 determines the amount of variability in the observations on the response y at a given value of the regressor variable x. The sum of the squared residuals is used to obtain the estimate of σ^2.

Definition

> The **residual sum of squares** (sometimes called the **error sum of squares**) is defined as
>
> $$SS_E = \sum_{i=1}^{n} (y_i - \hat{y}_i)^2 = \sum_{i=1}^{n} e_i^2 \qquad (6\text{-}16)$$
>
> and the estimate of σ^2 is
>
> $$\hat{\sigma}^2 = \frac{SS_E}{n - 2} \qquad (6\text{-}17)$$

EXAMPLE 6-1 (continued)

Both $SS_E = 57.7$ and the estimate of the variance $\hat{\sigma}^2 = 3.2$ for the salt concentration–roadway area regression model are highlighted in Table 6-2. The reported quantity $s = 1.791$ is an estimate of the standard deviation of the model errors. (Note that s is not exactly equal to $\sqrt{\hat{\sigma}^2} = \sqrt{3.2}$ due to the way that Minitab rounds the numerical output quantities.)

Regression Assumptions and Model Properties

In linear regression, we customarily assume that the model errors ϵ_i, $i = 1, 2, \ldots, n$ are normally and independently distributed with mean zero and variance σ^2. The values of the regressor variables x_i are assumed to be fixed before the data is collected, so the response variable Y_i has a normal distribution with mean $\beta_0 + \beta_1 x_i$ and variance σ^2. Furthermore, both $\hat{\beta}_0$ and $\hat{\beta}_1$ can be written as linear combinations of the Y_i's. The properties of linear functions of normally and independently distributed random variables leads to the following results.

> 1. Both $\hat{\beta}_0$ and $\hat{\beta}_1$ are **unbiased estimators** of the intercept and slope, respectively. That is, the distribution of $\hat{\beta}_1$ (and $\hat{\beta}_0$) is centered at the true value of β_1 (and β_0).
>
> 2. The variances of $\hat{\beta}_0$ and $\hat{\beta}_1$ are
>
> $$V(\hat{\beta}_0) = \sigma^2 \left(\frac{1}{n} + \frac{\bar{x}^2}{S_{xx}} \right) \qquad \text{and} \qquad V(\hat{\beta}_1) = \frac{\sigma^2}{S_{xx}}$$
>
> 3. The distributions of $\hat{\beta}_0$ and $\hat{\beta}_1$ are **normal.**

If we replace σ^2 in the expressions for the variances of the slope and intercept by $\hat{\sigma}^2$ from equation 6-17 and take the square root, we obtain the **standard errors** of the slope and intercept.

Standard Error of the Slope and Intercept

The standard errors of the slope and intercept in simple linear regression are

$$se(\hat{\beta}_1) = \sqrt{\frac{\hat{\sigma}^2}{S_{xx}}} \qquad (6\text{-}18)$$

and

$$se(\hat{\beta}_0) = \sqrt{\hat{\sigma}^2 \left(\frac{1}{n} + \frac{\bar{x}^2}{S_{xx}} \right)} \qquad (6\text{-}19)$$

respectively.

EXAMPLE 6-1
(continued)

Minitab calculates the standard errors of the slope and intercept and reports them in the computer output (see Table 6-2) immediately adjacent to the coefficient estimates $\hat{\beta}_0$ and $\hat{\beta}_1$ in the column headed "SE Coef." We find from the Minitab display that $se(\hat{\beta}_0) = 0.8680$ and $se(\hat{\beta}_1) = 0.9346$. These standard errors will be used to find confidence intervals and test hypotheses about the slope and intercept.

Regression and Analysis of Variance
The **total sum of squares** of the observed y values

$$SS_T = S_{yy} = \sum_{i=1}^{n} (y_i - \bar{y})^2 = \sum_{i=1}^{n} y_i^2 - \frac{\left(\sum_{i=1}^{n} y_i \right)^2}{n} \qquad (6\text{-}20)$$

is a measure of the total variability in the response. The total sum of squares (SS_T) can be written as

---

$$SS_T = \sum_{i=1}^{n} (y_i - \bar{y})^2 = \sum_{i=1}^{n} (\hat{y}_i - \bar{y})^2 + \sum_{i=1}^{n} (y_i - \hat{y}_i)^2 \qquad (6\text{-}21)$$

This is an **analysis of variance** (ANOVA), similar to the ANOVA we encountered in Section 5-8. It partitions the total variability in the response into two components. One of these is the **error** or **residual sum of squares** from equation 6-16, which is a measure of unexplained variability in the y's, and the other

---

$$SS_R = \sum_{i=1}^{n} (\hat{y}_i - \bar{y})^2$$

---

measures variability explained by the regression model. SS_R is usually called the **regression sum of squares**, or the **model** sum of squares. These sums of squares are listed in Table 6-2 in the section of output headed "Analysis of Variance." We usually think of the ratio SS_E/SS_T as the proportion of variability in the response variable that cannot be accounted for by the regression model. Consequently, $1 - SS_E/SS_T$ is the proportion of variability in the response that is accounted for by the model.

Coefficient of Determination (R^2)

The coefficient of determination is defined as

$$R^2 = 1 - \frac{SS_E}{SS_T} \qquad (6\text{-}22)$$

It is interpreted as the proportion of variability in the observed response variable that is explained by the linear regression model. Sometimes the quantity reported is $100R^2$, and it is referred to as the percentage of variability explained by the model.

Minitab calculates and reports the R^2 statistic. For example, for the regression model for the salt concentration–roadway area data in Table 6-2, Minitab reports the quantity $100R^2$ as 95.1%, implying that the regression model accounts for 95.1% of the observed variability in the salt concentration data.

The ANOVA partition forces $0 \leq R^2 \leq 1$. A large value of R^2 suggests that the model has been successful in explaining the variability in the response. When R^2 is small, it may be an indication that we need to find an alternative model, such as a multiple regression model, that can account for more of the variability in y.

Other Aspects of Regression

Regression models are used primarily for **interpolation**. That is, when predicting a new observation on the response (or estimating the mean response) at a particular value of the regressor x, we should only use values of x that are within the range of the x's used to fit the model. For example, in the salt concentration–roadway area problem of Example 6-1, values of roadway area between 0.15 and 1.74% are appropriate, but a value of $x = 2.5$ would not be reasonable because it is far outside of the original range of the regressors. Basically, as one moves outside the range of the original data, the reliability of the linear approximation as an empirical model of the true relationship will deteriorate.

We have assumed throughout this section that the regressor variable x is controllable and is set to levels chosen by the analyst and that the response variable Y is a random variable. There are many situations where this would not be the case. In fact, in the salt concentration–roadway area data, the roadway areas were not controlled. The analyst selected a group of 20 watersheds, and both the salt concentration *and* the roadway area were random variables. The regression methods that we described in this chapter can be employed both when the regressor values are fixed in advance and when they are random, but the fixed-regressor case is somewhat easier to describe, so we concentrate on it. When Y and X are both random, we can also use **correlation** as a measure of the association between the two variables. We will discuss this briefly in Section 6-2.6.

The term "regression analysis" was first used in the late nineteenth century by Sir Francis Galton, who studied the relationship between the heights of fathers and sons. Galton fit a model to predict the height of a son from the height of the father. He discovered that if the father were of above-average height, the son would tend to also be of above-average height, but not by as much as the father. Thus, height of the son regressed toward the mean.

Animation 15: Drawing Regression Lines

6-2.2   Testing Hypotheses in Simple Linear Regression

It is often useful to test hypotheses about the slope and intercept in a linear regression model. The normality assumption on the model errors and hence on the response variable that we introduced in Section 6-2.1 continues to apply.

Use of t-Tests

Suppose we wish to test the hypothesis that the slope equals a constant, say, $\beta_{1,0}$. The appropriate hypotheses are

$$H_0: \beta_1 = \beta_{1,0}$$
$$H_1: \beta_1 \neq \beta_{1,0} \tag{6-23}$$

Because the responses Y_i are normally and independently distributed random variables, $\hat{\beta}_1$ is $N(\beta_1, \sigma^2/S_{xx})$. As a result, the **test statistic**

$$T_0 = \frac{\hat{\beta}_1 - \beta_{1,0}}{\sqrt{\hat{\sigma}^2/S_{xx}}} = \frac{\hat{\beta}_1 - \beta_{1,0}}{se(\hat{\beta}_1)} \tag{6-24}$$

follows the t distribution with $n - 2$ degrees of freedom under $H_0: \beta_1 = \beta_{1,0}$. A P-value would be calculated as in any t-test. For a fixed-level test, we would reject $H_0: \beta_1 = \beta_{1,0}$ if the computed value of the test statistic

$$|t_0| > t_{\alpha/2, n-2} \tag{6-25}$$

where t_0 is computed from equation 6-24. A similar procedure can be used to test hypotheses about the intercept. To test

$$H_0: \beta_0 = \beta_{0,0}$$
$$H_1: \beta_0 \neq \beta_{0,0} \tag{6-26}$$

we would use the **test statistic**

$$T_0 = \frac{\hat{\beta}_0 - \beta_{0,0}}{\sqrt{\hat{\sigma}^2 \left[\dfrac{1}{n} + \dfrac{\bar{x}^2}{S_{xx}} \right]}} = \frac{\hat{\beta}_0 - \beta_{0,0}}{se(\hat{\beta}_0)} \tag{6-27}$$

The P-value would be computed as in any t-test. For a fixed-level test, we would reject the null hypothesis if the computed value of the test statistic, t_0, is such that $|t_0| > t_{\alpha/2, n-2}$.

A very important special case of the hypotheses of equation 6-23 is

$$H_0: \beta_1 = 0$$
$$H_1: \beta_1 \neq 0 \tag{6-28}$$

Figure 6-8   The hypothesis H_0: $\beta_1 = 0$ is not rejected.

These hypotheses relate to the **significance of regression.** Failure to reject H_0: $\beta_1 = 0$ is equivalent to concluding that there is no linear relationship between x and Y. This situation is illustrated in Fig. 6-8. Note that this may imply either that x is of little value in explaining the variation in Y and that the best estimator of Y for any x is $\hat{y} = \overline{Y}$ (Fig. 6-8a) or that the true relationship between x and Y is not linear (Fig. 6-8b). Alternatively, if H_0: $\beta_1 = 0$ is rejected, this implies that x is of value in explaining the variability in Y (see Fig. 6-9). Rejecting H_0: $\beta_1 = 0$ could mean either that the straight-line model is adequate (Fig. 6-9a) or that, although there is a linear effect of x, better results could be obtained with the addition of higher-order polynomial terms in x (Fig. 6-9b).

EXAMPLE 6-2
Salt Concentration and Roadway Data

Test for significance of regression using the model for the salt concentration–roadway area data from Example 6-1.

Solution. The hypotheses are

$$H_0: \beta_1 = 0$$
$$H_1: \beta_1 \neq 0$$

and we will use $\alpha = 0.01$. From Example 6-1 and the Minitab output in Table 6-2, we have

$$\hat{\beta}_1 = 17.5467 \quad n = 20, \quad S_{xx} = 3.67068, \quad \hat{\sigma}^2 = 3.2$$

so the t-statistic in equation 6-24 is

$$t_0 = \frac{\hat{\beta}_1}{\sqrt{\hat{\sigma}^2/S_{xx}}} = \frac{\hat{\beta}_1}{se(\hat{\beta}_1)} = \frac{17.5467}{\sqrt{3.2/3.67068}} = 18.77$$

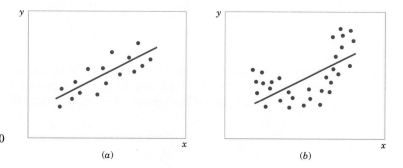

Figure 6-9   The hypothesis H_0: $\beta_1 = 0$ is rejected.

Because the critical value of t is $t_{0.005,18} = 2.88$, the value of the test statistic 18.77 is very far into the critical region, implying that $H_0: \beta_1 = 0$ should be rejected. The P-value for this test is $P \simeq 2.87 \times 10^{-13}$. This was obtained manually with a calculator and strongly indicates significance of β_1.

The Minitab output for this problem is shown in Table 6-2. Notice that the t-statistic value for the slope is computed as 18.77 and that the reported P-value is $P = 0.000$. Minitab also reports the t-statistic for testing the hypothesis $H_0: \beta_0 = 0$. This statistic is computed from equation 6-27, with $\beta_{0,0} = 0$, as $t_0 = 3.08$. Because the P-value is 0.006, the hypothesis that the intercept is zero is rejected. ∎

Analysis of Variance Approach

The analysis of variance can also be used to test for the significance of regression. The ANOVA identity in equation 6-21 can be written symbolically as

$$SS_T = SS_R + SS_E \qquad (6\text{-}29)$$

where SS_T is the **total sum of squares,** SS_R is the **regression** or **model sum of squares,** and SS_E is the **error** or **residual sum of squares.** If the null hypothesis for significance of regression, $H_0: \beta_1 = 0$, is true, SS_R/σ^2 is a chi-squared random variable with 1 degree of freedom. Note that the number of degrees of freedom for this chi-squared random variable is equal to the number of regressor variables in the model. We can also show that SS_E/σ^2 is a chi-squared random variable with $n - 2$ degrees of freedom, and that SS_E and SS_R are independent.

Testing for Significance of Regression in Simple Linear Regression

$$MS_R = \frac{SS_R}{1} \qquad MS_E = \frac{SS_E}{n - p}$$

Null hypothesis: $H_0: \beta_1 = 0$
Alternative hypothesis: $H_1: \beta_1 \neq 0$

Test statistic: $F_0 = \dfrac{MS_R}{MS_E}$ $\qquad (6\text{-}30)$

Rejection criterion for a fixed-level test: $f_0 > f_{\alpha,1,n-2}$

P-value: Probability beyond f_0 in the $F_{1,n-2}$ distribution

---

The ANOVA test for significance of regression is usually summarized in a table, such as shown in Table 6-3.

Table 6-3   Analysis of Variance for Testing Significance of Regression

Source of Variation	Sum of Squares	Degrees of Freedom	Mean Square	F_0
Regression	SS_R	1	MS_R	MS_R/MS_E
Error or residual	SS_E	$n - 2$	MS_E	
Total	SS_T	$n - 1$		

**EXAMPLE 6-1
(continued)**

The Minitab output displayed in Table 6-2 (page 291) contains the analysis of variance test for significance of regression. The computed value of the F statistic for significance of regression is $f_0 = MS_R/MS_E = 1130.1/3.2 = 352.46$. Minitab reports the P-value for this test as 0.000 (the actual P-value is 2.87×10^{-13}). Therefore, we reject the null hypothesis that the slope of the regression line is zero and conclude that there is a linear relationship between salt concentration and roadway area. ■

The t-test for significance of regression is closely related to the ANOVA F-test. In fact, the two tests produce identical results. This should not be surprising because both procedures are testing the same hypotheses. It turns out that the square of the computed value of the test statistic t_0 is equal to the computed value of the ANOVA test statistic f_0 (however, rounding may affect the results). To see this, refer to the Minitab output in Table 6-2 and note that $t_0^2 = 18.77^2 = 352.3$, which apart from rounding in the numbers reported by Minitab, is equal to the ANOVA F-statistic. In general, the square of a t random variable with r degrees of freedom is equal to an F random variable with one numerator degree of freedom and r denominator degrees of freedom.

6-2.3 Confidence Intervals in Simple Linear Regression

Confidence Intervals on the Slope and Intercept
In addition to point estimates of the slope and intercept, it is possible to obtain confidence interval estimates of these parameters. The width of these CIs is a measure of the overall quality of the regression line. If the error terms, ϵ_i, in the regression model are normally and independently distributed,

$$(\hat{\beta}_1 - \beta_1)/se(\hat{\beta}_1) \quad \text{and} \quad (\hat{\beta}_0 - \beta_0)/se(\hat{\beta}_0)$$

are both distributed as t random variables with $n - 2$ degrees of freedom. This leads to the following definition of $100(1 - \alpha)\%$ CIs on the slope and intercept.

Confidence Intervals on the Model Parameters, Simple Linear Regression

Under the assumption that the observations are normally and independently distributed, a $100(1 - \alpha)\%$ **confidence interval on the slope** β_1 in a simple linear regression is

$$\hat{\beta}_1 - t_{\alpha/2, n-2}\, se(\hat{\beta}_1) \leq \beta_1 \leq \hat{\beta}_1 + t_{\alpha/2, n-2}\, se(\hat{\beta}_1) \tag{6-31}$$

Similarly, a $100(1 - \alpha)\%$ **CI on the intercept** β_0 is

$$\hat{\beta}_0 - t_{\alpha/2, n-2}\, se(\hat{\beta}_0) \leq \beta_0 \leq \hat{\beta}_0 + t_{\alpha/2, n-2}\, se(\hat{\beta}_0) \tag{6-32}$$

where $se(\hat{\beta}_1)$ and $se(\hat{\beta}_0)$ are defined in equations 6-18 and 6-19, respectively.

EXAMPLE 6-3

Salt Concentration
and Roadway Data

Find a 95% CI on the slope of the regression line using the data in Example 6-1.

Solution. Recall that $\hat{\beta}_1 = 17.5467$ and that $se(\hat{\beta}_1) = 0.9346$ (see Table 6-2, page 291). Then, from equation 6-31 we find

$$\hat{\beta}_1 - t_{0.025, 18}\, se(\hat{\beta}_1) \leq \beta_1 \leq \hat{\beta}_1 + t_{0.025, 18}\, se(\hat{\beta}_1)$$

or

$$17.5467 - 2.101(0.9346) \leq \beta_1 \leq 17.5467 + 2.101(0.9346)$$

This simplifies to

$$15.5831 \leq \beta_1 \leq 19.5103$$

■

Confidence Interval on the Mean Response

A CI may be constructed on the mean response at a specified value of x, say, x_0. This is a CI about $E(Y|x_0) = \mu_{Y|x_0}$ and is often called a CI about the regression line. Because $E(Y|x_0) = \mu_{Y|x_0} = \beta_0 + \beta_1 x_0$, we may obtain a point estimate of the mean of Y at $x = x_0$ (or $\mu_{Y|x_0}$) from the fitted model as

$$\hat{\mu}_{Y|x_0} = \hat{y}_0 = \hat{\beta}_0 + \hat{\beta}_1 x_0$$

Now $\hat{\mu}_{Y|x_0}$ is an unbiased point estimator of $\mu_{Y|x_0}$ because $\hat{\beta}_0$ and $\hat{\beta}_1$ are unbiased estimators of β_0 and β_1. The variance of $\hat{\mu}_{Y|x_0}$ is

$$V(\hat{\mu}_{Y|x_0}) = \sigma^2 \left[\frac{1}{n} + \frac{(x_0 - \bar{x})^2}{S_{xx}} \right] \tag{6-33}$$

Also, $\hat{\mu}_{Y|x_0}$ is normally distributed because $\hat{\beta}_1$ and $\hat{\beta}_0$ are normally distributed, and if $\hat{\sigma}^2$ is used as an estimate of σ^2, it is easy to show that

$$\frac{\hat{\mu}_{Y|x_0} - \mu_{Y|x_0}}{\sqrt{\hat{\sigma}^2 \left[\frac{1}{n} + \frac{(x_0 - \bar{x})^2}{S_{xx}} \right]}} = \frac{\hat{\mu}_{Y|x_0} - \mu_{Y|x_0}}{se(\hat{\mu}_{Y|x_0})}$$

has a t distribution with $n - 2$ degrees of freedom. The quantity $se(\hat{\mu}_{Y|x_0})$ is sometimes called the standard error of the fitted value. This leads to the following confidence interval definition.

Confidence Interval on the Mean Response, Simple Linear Regression

A $100(1 - \alpha)\%$ **CI about the mean response** at the value of $x = x_0$, say $\mu_{Y|x_0}$, is given by

$$\hat{\mu}_{Y|x_0} - t_{\alpha/2,n-2}\, se(\hat{\mu}_{Y|x_0}) \leq \mu_{Y|x_0} \leq \hat{\mu}_{Y|x_0} + t_{\alpha/2,n-2}\, se(\hat{\mu}_{Y|x_0}) \tag{6-34}$$

where $\hat{\mu}_{Y|x_0} = \hat{\beta}_0 + \hat{\beta}_1 x_0$ is computed from the fitted regression model.

Note that the width of the CI for $\mu_{Y|x_0}$ is a function of the value specified for x_0. The interval width is a minimum for $x_0 = \bar{x}$ and widens as $|x_0 - \bar{x}|$ increases.

EXAMPLE 6-4

Salt Concentration
and Roadway Data

Construct a 95% CI about the mean response for the data in Example 6-1.

Solution. The fitted model is $\hat{\mu}_{Y|x_0} = 2.6765 + 17.5467x_0$, and the 95% CI on $\mu_{Y|x_0}$ is found from equation 6-33 as

$$\hat{\mu}_{Y|x_0} \pm 2.101 \; se(\hat{\mu}_{Y|x_0}) \qquad se(\hat{\mu}_{Y|x_0}) = \sqrt{3.2\left[\frac{1}{20} + \frac{(x_0 - 0.824)^2}{3.67068}\right]}$$

Suppose that we are interested in predicting mean salt concentration when roadway area $x_0 = 1.25\%$. Then

$$\hat{\mu}_{Y|1.25} = 2.6765 + 17.5467(1.25) = 24.61$$

and the 95% CI is

$$\left\{24.61 \pm 2.101\sqrt{3.2\left[\frac{1}{20} + \frac{(1.25 - 0.824)^2}{3.67068}\right]}\right\}$$

or

$$24.61 \pm 2.101(0.564)$$

Therefore, the 95% CI on $\mu_{Y|1.25}$ is

$$23.425 \le \mu_{Y|1.25} \le 24.795$$

Minitab will also perform these calculations. Refer to Table 6-2. The predicted value of y at $x = 1.25$ is shown along with $se(\hat{\mu}_{Y|1.25})$ and the 95% CI on the mean of y at this level of x. Minitab labels the standard error $se(\hat{\mu}_{Y|1.25})$ as "SE Fit."

By repeating these calculations for several different values for x_0 we can obtain confidence limits for each corresponding value of $\mu_{Y|x_0}$. Minitab calculated the standard error $se(\hat{\mu}_{Y|x_0})$ at each of the x-values in the sample. In Table 6-2, these standard errors are in the column labeled "SE Fit." Figure 6-10 displays the scatter diagram from Minitab with the fitted model and the corresponding 95% confidence limits plotted as the upper and lower lines. The 95% confidence level applies only to the interval obtained at one value of x and not to the entire set of x-levels. Notice that the width of the confidence interval on $\mu_{Y|x_0}$ increases as $|x_0 - \bar{x}|$ increases. ∎

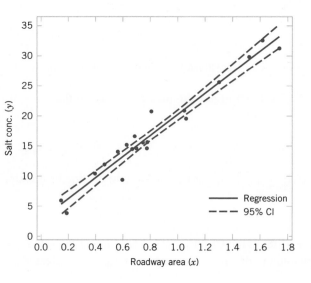

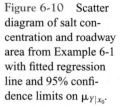

Figure 6-10   Scatter diagram of salt concentration and roadway area from Example 6-1 with fitted regression line and 95% confidence limits on $\mu_{Y|x_0}$.

6-2.4   Prediction of New Observations

An important application of a regression model is predicting new or future observations Y corresponding to a specified level of the regressor variable x. If x_0 is the value of the regressor variable of interest,

$$\hat{Y}_0 = \hat{\mu}_{Y|x_0} = \hat{\beta}_0 + \hat{\beta}_1 x_0 \tag{6-35}$$

is the point estimator of the new or future value of the response Y_0.

Now consider obtaining an interval estimate for this future observation Y_0. This new observation is independent of the observations used to develop the regression model. Therefore, the confidence interval for $\mu_{Y|x_0}$ in equation 6-34 is inappropriate because it is based only on the data used to fit the regression model. The CI about $\mu_{Y|x_0}$ refers to the true mean response at $x = x_0$ (that is, a population parameter), not to future observations.

Let Y_0 be the future observation at $x = x_0$, and let $\hat{Y}_0$ given by equation 6-35 be the estimator of Y_0. Note that the error in prediction $Y_0 - \hat{Y}_0$ is a normally distributed random variable with mean zero and variance

$$V(Y_0 - \hat{Y}_0) = \sigma^2 \left[1 + \frac{1}{n} + \frac{(x_0 - \bar{x})^2}{S_{xx}} \right]$$

because Y_0 is independent of $\hat{Y}_0$. If we use $\hat{\sigma}^2$ to estimate σ^2, we can show that

$$\frac{Y_0 - \hat{Y}_0}{\sqrt{\hat{\sigma}^2 \left[1 + \frac{1}{n} + \frac{(x_0 - \bar{x})^2}{S_{xx}} \right]}}$$

has a t distribution with $n - 2$ degrees of freedom. From this we can develop the following **prediction interval** (PI) definition.

Prediction Interval on a Future Observation, Simple Linear Regression

A $100(1 - \alpha)\%$ **PI on a future observation** Y_0 at the value x_0 is given by

$$\hat{y}_0 - t_{\alpha/2, n-2} \sqrt{\hat{\sigma}^2 \left[1 + \frac{1}{n} + \frac{(x_0 - \bar{x})^2}{S_{xx}} \right]}$$

$$\leq Y_0 \leq \hat{y}_0 + t_{\alpha/2, n-2} \sqrt{\hat{\sigma}^2 \left[1 + \frac{1}{n} + \frac{(x_0 - \bar{x})^2}{S_{xx}} \right]} \tag{6-36}$$

where the value $\hat{y}_0$ is computed from the regression model $\hat{y}_0 = \hat{\beta}_0 + \hat{\beta}_1 x_0$.

Notice that the PI is of minimum width at $x_0 = \bar{x}$ and widens as $|x_0 - \bar{x}|$ increases. By comparing equation 6-36 with equation 6-34, we observe that the PI at the point x_0 is always wider than the CI at x_0. This results because the PI depends on both the error from the fitted model and the error associated with future observations. The PI in equation 6-36 is similar to the PI on a future observation drawn from a normal distribution introduced in Section 4-8.1, except that now a regressor variable is involved in determining the future value.

EXAMPLE 6-5
Salt Concentration and Roadway Data

Using the data in Example 6-1, find a 95% PI on a future observation of salt concentration when roadway area $x_0 = 1.25\%$.

Solution. Using equation 6-36 and recalling from Example 6-4 that $\hat{y}_0 = 24.61$, we find that the PI is

$$24.61 - 2.101\sqrt{3.2\left[1 + \frac{1}{20} + \frac{(1.25 - 0.824)^2}{3.67068}\right]}$$

$$\leq Y_0 \leq 24.61 + 2.101\sqrt{3.2\left[1 + \frac{1}{20} + \frac{(1.25 - 0.824)^2}{3.67068}\right]}$$

which simplifies to

$$20.66 \leq y_0 \leq 28.55$$

Minitab will also calculate PIs. Refer to the output in Table 6-2. The 95% PI on the future observation at $x_0 = 1.25\%$ is shown in the table.

By repeating the foregoing calculations at different levels of x_0, we may obtain the 95% PIs shown graphically as the lower and upper lines about the fitted regression model in Fig. 6-11. Notice that this graph also shows the 95% confidence limits on $\mu_{Y|x_0}$ calculated in Example 6-4. It illustrates that the prediction limits are always wider than the confidence limits. ■

6-2.5 Checking Model Adequacy

Fitting a regression model requires several assumptions. Estimation of the model parameters requires the assumption that the errors are uncorrelated random variables with mean zero and constant variance. Tests of hypotheses and interval estimation require that the errors be normally distributed. In addition, we assume that the order of the model is correct; that is, if we fit a simple linear regression model, we are assuming that the phenomenon actually behaves in a linear or first-order manner.

The analyst should always consider the validity of these assumptions to be doubtful and conduct analyses to examine the adequacy of the model that has been tentatively entertained. The residuals from the regression model, defined as $e_i = y_i - \hat{y}_i$, $i = 1, 2, \ldots, n$, are useful

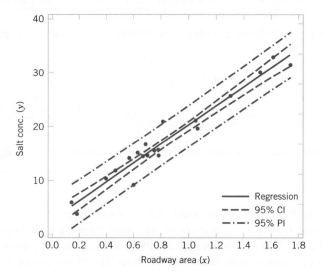

Figure 6-11 Scatter diagram of salt concentration–roadway area data from Example 6-1 with fitted regression line, 95% prediction limits (outer lines), and 95% confidence limits on $\mu_{Y|x_0}$.

in checking the assumptions of normality and constant variance, and in determining whether additional terms in the model would be useful.

As an approximate check of normality, the experimenter can construct a frequency histogram of the residuals or a **normal probability plot of residuals.** Many computer programs will produce a normal probability plot of residuals, and because the sample sizes in regression are often too small for a histogram to be meaningful, the normal probability plotting method is preferred. It requires judgment to assess the abnormality of such plots. (Refer to the discussion of the "fat pencil" method in Chapter 3.)

We may also **standardize** the residuals by computing $d_i = e_i/\sqrt{\hat{\sigma}^2}$, $i = 1, 2, \ldots, n$. If the errors are normally distributed, approximately 95% of the standardized residuals should fall in the interval $(-2, +2)$. Residuals that are far outside this interval may indicate the presence of an **outlier**; that is, an observation that is not typical of the rest of the data. Various rules have been proposed for discarding outliers. However, outliers sometimes provide important information about unusual circumstances of interest to experimenters and should not be discarded. For further discussion of outliers, see Montgomery, Peck, and Vining (2006).

It is frequently helpful to plot the residuals (1) in time sequence (if known), (2) against the $\hat{y}_i$, and (3) against the independent variable x. These graphs will usually look like one of the four general patterns shown in Fig. 6-12. Pattern (a) in Fig. 6-12 represents the ideal situation, whereas patterns (b), (c), and (d) represent anomalies. If the residuals appear as in (b), the variance of the observations may be increasing with time or with the magnitude of y_i or x_i. Data transformation on the response y is often used to eliminate this problem. Widely used variance-stabilizing transformations include the use of $\sqrt{y}$, $\ln y$, or $1/y$ as the response. See Montgomery, Peck, and Vining (2006) for more details regarding methods for selecting an appropriate transformation. If a plot of the residuals against time has the appearance of (b), the variance of the observations is increasing with time. Plots of residuals against $\hat{y}_i$ and x_i that look like (c) also indicate inequality of variance. Residual plots that look like (d) indicate model inadequacy; that is, higher-order terms should be added to the model, a transformation on the x-variable or the y-variable (or both) should be considered, or other regressors should be considered. Outliers can have a dramatic impact on a regression model. As noted below, a large residual is often evidence that an outlier is present.

EXAMPLE 6-6	The residuals for the regression model for the salt concentration–roadway area data are shown in Table 6-2. Analyze the residuals to determine if the regression model provides an adequate fit to the data or if any underlying assumptions are violated.
Salt Concentration and Roadway Data	

Solution. A normal probability plot of these residuals is shown in Fig. 6-13. No severe deviations from normality are obviously apparent, although the two largest residuals do not fall extremely close to a straight line drawn through the remaining residuals. The residuals are plotted against $\hat{y}$ in Fig. 6-14. There is no indication of a problem with the assumption of constant variance. ∎

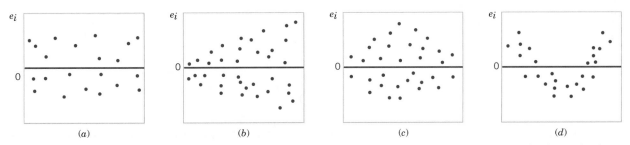

Figure 6-12   Patterns for residual plots: (a) satisfactory, (b) funnel, (c) double bow, (d) nonlinear. Horizontal axis may be time, $\hat{y}_i$, or x_i.

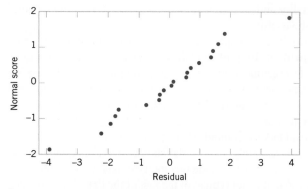

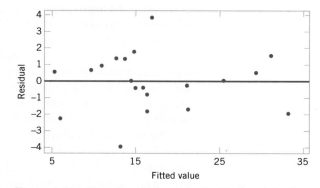

Figure 6-13   Normal probability plot of residuals from the salt concentration–roadway area regression model.

Figure 6-14   Plot of residuals versus fitted values $\hat{y}$ for the salt concentration–roadway area regression model.

The two largest residuals are $e_{10} = -3.905$ and $e_{12} = 3.911$ (see Table 6-2). The standardized residuals are $d_{10} = e_{10}/\sqrt{\hat{\sigma}^2} = -3.905/\sqrt{3.2} = -2.183$ and $d_{12} = e_{12}/\sqrt{\hat{\sigma}^2} = 3.911/\sqrt{3.2} = 2.186$, and these are not far enough outside the nominal $-2, +2$ range where we would expect most of the standardized residuals to fall to cause any alarm.

It is easy to demonstrate the impact of an outlier. Suppose that the salt concentration for observation 12 was $y_{12} = 28.8$ (instead of 20.8). Figure 6-15 shows a scatter plot of this modified data set with the resulting least squares fit. Using Minitab, you can easily verify that the fitted value corresponding to observation 12 is now $\hat{y}_{12} = 17.29$, and the corresponding residual is $y_{12} - \hat{y}_{12} = 28.8 - 17.29 = 11.51$. The standardized value of this residual is $d_{12} = e_{12}/\sqrt{\hat{\sigma}^2} = 11.51/\sqrt{10.1} = 3.62$ (MS_E or $\hat{\sigma}^2 = 10.1$ in the new regression model), which is far enough outside the nominal $-2, +2$ range for us to classify observation 12 as an outlier. The actual impact of this outlier on the regression line, however, seems fairly modest. Comparing Fig. 6-15 and 6-7 (the least squares fit to the original data) reveals that the slope of the regression model has not been seriously affected by the outlier (17.516 versus 17.5467) but that the intercept has increased by a greater amount, proportionally, from 2.6765 to 3.102. The outlier has basically raised the average height of the fitted line.

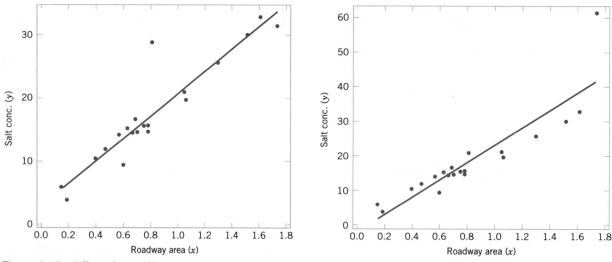

Figure 6-15   Effect of an outlier.

Figure 6-16   Effect of an influential observation.

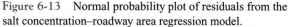

Now suppose that the response for observation 19 was 61.3 instead of 31.3. The scatter plot and fitted line are shown in Fig. 6-16. This outlier has had a more dramatic impact on the least squares fit and has actually begun to pull the fitted line away from the rest of the data. This is due both to the size of the outlier and to its location along the x-axis. Sample points near the ends of the range in x-space have potentially more impact on the least squares fit than points that are located near the middle of the x-space. Points that are remote in the x-space *and* that have large residuals are often called **influential observations**. A scatter plot is useful in identifying these influential observations in simple linear regression. However, in multiple regression, the dimensionality of the problem can make detecting them difficult. We will say more about influential observations in Section 6-3.3.

6-2.6   Correlation and Regression

We commented in Section 6-2.1 that our development of regression assumed that the regressor variable x was fixed or chosen in advance and that the response variable Y was a random variable, but that our results for parameter estimation and model inference still apply even if Y and X both are random variables. In this section we discuss this point further and show some of the connections between regression and correlation.

Suppose that Y and X are jointly normally distributed random variables with correlation coefficient ρ (joint distributions were introduced in Section 3-11). We refer to ρ as the **population correlation coefficient**, which is a measure of the strength of the linear relationship between Y and X in the population or joint distribution. We also have a collection of sample pairs (x_i, y_i), $i = 1, 2, \ldots, n$, and the **sample correlation coefficient** between Y and X is given by

$$r = \frac{\sum_{i=1}^{n} (y_i - \bar{y})(x_i - \bar{x})}{\sqrt{\sum_{i=1}^{n} (y_i - \bar{y})^2}\sqrt{\sum_{i=1}^{n} (x_i - \bar{x})^2}} \tag{6-37}$$

We illustrated the computation of r in Section 2-6 and discussed its interpretation. The sample correlation coefficient for the salt concentration–roadway area data is $r = 0.975$, a strong positive correlation. Note that r is just the square root of the coefficient of determination, R^2, with the sign taken to be equal to the sign of the slope.

The sample correlation coefficient is also closely related to the slope in a linear regression model; in fact,

$$r = \hat{\beta}_1 \left(\frac{S_{xx}}{SS_T}\right)^{1/2} \tag{6-38}$$

so testing the hypothesis that the slope equals zero (significance of regression) is really equivalent to testing that the population correlation coefficient $\rho = 0$. We can also conduct this test directly; that is, to test

$$H_0: \rho = 0$$
$$H_1: \rho \neq 0 \tag{6-39}$$

the computed value of the **test statistic** is

$$t_0 = \frac{r\sqrt{n-3}}{\sqrt{1-r^2}} \qquad (6\text{-}40)$$

and if $|t_0| > t_{\alpha/2,n-2}$, the null hypothesis in equation 6-38 is rejected.

EXAMPLE 6-6 (continued)

Using the salt concentration and roadway data and Minitab output

$$r = 17.5467\left(\frac{3.67068}{1187.9}\right)^{1/2} = 0.9754$$

Therefore, the **test statistic** is

$$t_0 = \frac{0.9754\sqrt{20-3}}{\sqrt{1-(0.9754)^2}}$$

$$= 18.24$$

With a critical value $t_{-0.005,\,18} = 2.88$, the value of the test statistic 18.24 exceeds 2.88 and has a P-value of 0. Note that this test statistic value is the same (except for rounding) as the test statistic for β_1 at 18.77. Both tests yield the same conclusion: the correlation between Y and X is significant or, equivalently, the regression model is significant. ∎

There are also other hypothesis testing and CI procedures associated with the sample correlation coefficient ρ. For details of these procedures and examples, refer to Montgomery and Runger (2006).

EXERCISES FOR SECTION 6-2

For Exercises 6-1 through 6-5, perform the following.
(a) Estimate the intercept β_0 and slope β_1 regression coefficients. Write the estimated regression line.
(b) Compute the residuals.
(c) Compute SS_E and estimate the variance.
(d) Find the standard error of the slope and intercept coefficients.
(e) Show that $SS_T = SS_R + SS_E$.
(f) Compute the coefficient of determination, R^2. Comment on the value.
(g) Use a t-test to test for significance of the intercept and slope coefficients at $\alpha = 0.05$. Give the P-values of each and comment on your results.
(h) Construct the ANOVA table and test for significance of regression using the P-value. Comment on your results and their relationship to your results in part (g).
(i) Construct 95% CIs on the intercept and slope. Comment on the relationship of these CIs and your findings in parts (g) and (h).
(j) Perform model adequacy checks. Do you believe the model provides an adequate fit?
(k) Compute the sample correlation coefficient and test for its significance at $\alpha = 0.05$. Give the P-value and comment on your results and their relationship to your results in parts (g) and (h).

6-1. Establishing the properties of materials is an important problem in identifying a suitable substitute for biodegradable materials in the fast-food packaging industry. Consider the following data on product density (g/cm^3) and thermal conductivity K-factor (W/mK) published in *Materials Research and Innovation* (1999, pp. 2–8).

Thermal Conductivity y	Product Density x
0.0480	0.1750
0.0525	0.2200
0.0540	0.2250
0.0535	0.2260
0.0570	0.2500
0.0610	0.2765

6-2 The number of pounds of steam used per month by a chemical plant is thought to be related to the average ambient temperature (in °F) for that month. The past year's usage and temperature are shown in the following table.

Month	Temp. x	Usage/1000 y	Month	Temp. x	Usage/1000 y
Jan.	21	185.79	July	68	621.55
Feb.	24	214.47	Aug.	74	675.06
Mar.	32	288.03	Sept.	62	562.03
Apr.	47	424.84	Oct.	50	452.93
May	50	454.58	Nov.	41	369.95
June	59	539.03	Dec.	30	273.98

6-3. Regression methods were used to analyze the data from a study investigating the relationship between roadway surface temperature (x) and pavement deflection (y). The data follow.

Temperature x	Deflection y	Temperature x	Deflection y
70.0	0.621	72.7	0.637
77.0	0.657	67.8	0.627
72.1	0.640	76.6	0.652
72.8	0.623	73.4	0.630
78.3	0.661	70.5	0.627
74.5	0.641	72.1	0.631
74.0	0.637	71.2	0.641
72.4	0.630	73.0	0.631
75.2	0.644	72.7	0.634
76.0	0.639	71.4	0.638

6-4. Turbidity is a measure of the cloudiness of the water and is used to indicate water quality levels. Higher turbidity levels are usually associated with higher levels of disease-causing microbes like viruses, bacteria, and parasites. The turbidity units of measure are reported as formazin suspension units, or FAUs. Data were collected on the Rio Grande River during the late spring and summer months in order to study the relationship between temperature and turbidity. The data follow.

Temperature x	Turbidity y	Temperature x	Turbidity y
22.9	125	26.1	100
24.0	118	26.9	105
22.9	103	22.8	55
23.0	105	27.0	267
20.5	26	26.1	286
26.2	90	26.2	235
25.8	99	26.6	265

6-5. An article in *Concrete Research* ("Near Surface Characteristics of Concrete: Intrinsic Permeability," Vol. 41,

1989) presented data on compressive strength x and intrinsic permeability y of various concrete mixes and cures. The following data are consistent with those reported.

Strength x	Permeability y	Strength x	Permeability y
3.1	33.0	2.4	35.7
4.5	31.0	3.5	31.9
3.4	34.9	1.3	37.3
2.5	35.6	3.0	33.8
2.2	36.1	3.3	32.8
1.2	39.0	3.2	31.6
5.3	30.1	1.8	37.7
4.8	31.2		

6-6. An article in the *Journal of Sound and Vibration* (Vol. 151, 1991, pp. 383–394) described a study investigating the relationship between noise exposure and hypertension. The following data are representative of those reported in the article.

y	1	0	1	2	5	1	4	6	2	3
x	60	63	65	70	70	70	80	90	80	80

y	5	4	6	8	4	5	7	9	7	6
x	85	89	90	90	90	90	94	100	100	100

(a) Draw a scatter diagram of y (blood pressure rise in millimeters of mercury) versus x (sound pressure level in decibels). Does a simple linear regression model seem reasonable in this situation?

(b) Fit the simple linear regression model using least squares. Find an estimate of σ^2.

(c) Find the predicted mean rise in blood pressure level associated with a sound pressure level of 85 decibels.

6-7. Consider the data and simple linear regression in Exercise 6-1.

(a) Find the mean thermal conductivity given that the product density is 0.2350.

(b) Compute a 95% CI on this mean response.

(c) Compute a 95% PI on a future observation when product density is equal to 0.2350.

(d) What do you notice about the relative size of these two intervals? Which is bigger and why?

6-8. Consider the data and simple linear regression model in Exercise 6-2.

(a) Find the mean pounds of steam given that the ambient temperature is 52 degrees.

(b) Compute a 99% CI on this mean response.

(c) Compute a 99% PI on a future observation when the ambient temperature is equal to 52 degrees.

(d) What do you notice about the relative size of these two intervals? Which is bigger and why?

6-9. Consider the data and simple linear regression model in Exercise 6-5.

(a) Find the mean permeability given that the strength is 2.1.

(b) Compute a 99% CI on this mean response.

(c) Compute a 99% PI on a future observation when the strength is equal to 2.1.

(d) What do you notice about the relative size of these two intervals? Which is bigger and why?

6-10. Consider the data and simple linear regression model in Exercise 6-6.

(a) Find the mean rise in blood pressure given that the sound pressure level is 85 decibels.

(b) Compute a 99% CI on this mean response.

(c) Compute a 99% PI on a future observation when the sound pressure level is 85 decibels.

(d) What do you notice about the relative size of these two intervals? Which is bigger and why?

6-11. An article in *Wood Science and Technology* ("Creep in Chipboard, Part 3: Initial Assessment of the Influence of Moisture Content and Level of Stressing on Rate of Creep and Time to Failure," Vol. 15, 1981, pp. 125–144) describes the results of a study of the deflection (mm) of particleboard from stress levels of relative humidity. Assume that the two variables are related according to the simple linear regression model. The data are shown below:

x = Stress level (%): 54   54   61   61   68
y = Deflection (mm): 16.473 18.693 14.305 15.121 13.505
x = Stress level (%): 68   75   75   75
y = Deflection (mm): 11.640 11.168 12.534 11.224

(a) Calculate the least square estimates of the slope and intercept. What is the estimate of σ^2? Graph the regression model and the data.

(b) Find the estimate of the mean deflection if the stress level can be limited to 65%.

(c) Estimate the change in the mean deflection associated with a 5% increment in stress level.

(d) To decrease the mean deflection by one millimeter, how much increase in stress level must be generated?

(e) Given that the stress level is 68%, find the fitted value of deflection and the corresponding residual.

6-12. An article in the *Journal of the Environmental Engineering Division* ("Least Squares Estimates of BOD Parameters," Vol. 106, 1980, pp. 1197–1202) describes the results of testing a sample from the Holston River below Kingport, TN, during August 1977. The biochemical oxygen demand (BOD) test is conducted over a period of time in days. The resulting data is shown below:

Time (days):   1   2   4   6   8   10   12   14   16
           18   20
BOD (mg/liter):   0.6   0.7   1.5   1.9   2.1   2.6   2.9   3.7   3.5
           3.7   3.8

(a) Assuming that a simple linear regression model is appropriate, fit the regression model relating BOD (y) to the time (x). What is the estimate of σ^2?

(b) What is the estimate of expected BOD level when the time is 15 days?

(c) What change in mean BOD is expected when the time increases by 3 days?

(d) Suppose the time used is 6 days. Calculate the fitted value of y and the corresponding residual.

(e) Calculate the fitted $\hat{y}_i$ for each value of x_i used to fit the model. Then construct a graph of $\hat{y}_i$ versus the corresponding observed values y_i and comment on what this plot would look like if the relationship between y and x was a deterministic (no random error) straight line. Does the plot actually obtained indicate that time is an effective regressor variable in predicting BOD?

6-13. Use the partially complete Minitab output below to answer the following questions.

(a) Find all of the missing values.

(b) Find the estimate of σ^2.

(c) Test for significance of regression. Test for significance of β_1. Comment on the two results. Use $\alpha = 0.05$.

(d) Construct a 95% CI on β_1. Use this CI to test for significance of regression.

(e) Comment on results found in parts (c) and (d).

(f) Write the regression model and use it to compute the residual when $x = 2.18$ and $y = 2.8$.

(g) Use the regression model to compute the mean and predicted future response when $x = 1.5$. Given that $\bar{x} = 1.76$ and $S_{xx} = 5.326191$, construct a 95% CI on the mean response and a 95% PI on the future response. Which interval in larger? Why?

Predictor	Coef	SE Coef	T	P
Constant	0.6649	0.1594	4.17	0.001
X	0.83075	0.08552	?	?

S = ?   R − Sq = 88.7%   R − Sq(adj) = ?

Analysis of Variance

Source	DF	SS	MS	F	P
Regression	1	3.6631	3.6631	?	?
Residual Error	12	0.4658	?		
Total	13	?			

6-14. Use the partially complete Minitab output on the next page to answer the following questions.

(a) Find all of the missing values.

(b) Find the estimate of σ^2.

(c) Test for significance of regression. Test for significance of β_1. Comment on the two results. Use $\alpha = 0.05$.

(d) Construct a 95% CI on β_1. Use this CI to test for significance of regression.
(e) Comment on results found in parts (c) and (d).
(f) Write the regression model and use it to compute the residual when $x = 0.58$ and $y = -3.30$.
(g) Use the regression model to compute the mean and predicted future response when $x = 0.6$. Given that $\bar{x} = 0.52$ and $S_{xx} = 1.218294$, construct a 95% CI on the mean response and a 95% PI on the future response. Which interval in larger? Why?

Predictor	Coef	SE Coef	T	P
Constant	0.9798	0.3367	2.91	0.011
X	-8.3088	0.5725	?	?

$S = ?$ $R - Sq = 93.8\%$ $R - Sq(adj) = ?$

Analysis of Variance

Source	DF	SS	MS	F	P
Regression	1	84.106	84.106	?	?
Residual Error	14	5.590	?		
Total	15	?			

6-3   MULTIPLE REGRESSION

We now consider the multiple linear regression model introduced in Section 6-1. As we did for simple linear regression, we will show how to estimate the model parameters using the method of least squares, test hypotheses and construct CIs on the model parameters, predict future observations, and investigate model adequacy.

6-3.1   Estimation of Parameters in Multiple Regression

The method of least squares may be used to estimate the regression coefficents in the multiple regression model, equation 6-3. Suppose that $n > k$ observations are available, and let x_{ij} denote the ith observation or level of variable x_j. The observations are

$$(x_{i1}, x_{i2}, \ldots, x_{ik}, y_i) \quad i = 1, 2, \ldots, n > k$$

It is customary to present the data for multiple regression in a table such as Table 6-4.
Each observation $(x_{i1}, x_{i2}, \ldots, x_{ik}, y_i)$, satisfies the model in equation 6-3, or

$$y_i = \beta_0 + \beta_1 x_{i1} + \beta_2 x_{i2} + \cdots + \beta_k x_{ik} + \epsilon_i$$

$$= \beta_0 + \sum_{j=1}^{k} \beta_j x_{ij} + \epsilon_i \quad i = 1, 2, \ldots, n \tag{6-41}$$

The least squares function is

$$L = \sum_{i=1}^{n} \epsilon_i^2 = \sum_{i=1}^{n} \left(y_i - \beta_0 - \sum_{j=1}^{k} \beta_j x_{ij} \right)^2 \tag{6-42}$$

We want to minimize L with respect to $\beta_0, \beta_1, \ldots, \beta_k$. The **least squares estimates** of β_0, $\beta_1, \ldots, \beta_k$ must satisfy

Table 6-4   Data for Multiple Linear Regression

y	x_1	x_2	$\cdots$	x_k
y_1	x_{11}	x_{12}	$\cdots$	x_{1k}
y_2	x_{21}	x_{22}	$\cdots$	x_{2k}
$\vdots$	$\vdots$	$\vdots$		$\vdots$
y_n	x_{n1}	x_{n2}	$\cdots$	x_{nk}

$$\left.\frac{\partial L}{\partial \beta_0}\right|_{\hat\beta_0, \hat\beta_1, \ldots, \hat\beta_k} = -2\sum_{i=1}^{n}\left(y_i - \hat\beta_0 - \sum_{j=1}^{k}\hat\beta_j x_{ij}\right) = 0 \tag{6-43a}$$

and

$$\left.\frac{\partial L}{\partial \beta_j}\right|_{\hat\beta_0, \hat\beta_1, \ldots, \hat\beta_k} = -2\sum_{i=1}^{n}\left(y_i - \hat\beta_0 - \sum_{j=1}^{k}\hat\beta_j x_{ij}\right) x_{ij} = 0 \quad j = 1, 2, \ldots, k \tag{6-43b}$$

Simplifying equations 6-43a and 6-43b, we obtain the **least squares normal equations**

$$n\hat\beta_0 + \hat\beta_1 \sum_{i=1}^{n} x_{i1} + \hat\beta_2 \sum_{i=1}^{n} x_{i2} + \cdots + \hat\beta_k \sum_{i=1}^{n} x_{ik} = \sum_{i=1}^{n} y_i$$

$$\hat\beta_0 \sum_{i=1}^{n} x_{i1} + \hat\beta_1 \sum_{i=1}^{n} x_{i1}^2 + \hat\beta_2 \sum_{i=1}^{n} x_{i1}x_{i2} + \cdots + \hat\beta_k \sum_{i=1}^{n} x_{i1}x_{ik} = \sum_{i=1}^{n} x_{i1}y_i$$

$$\vdots \qquad \vdots \qquad \vdots \qquad \vdots \qquad \vdots$$

$$\hat\beta_0 \sum_{i=1}^{n} x_{ik} + \hat\beta_1 \sum_{i=1}^{n} x_{ik}x_{i1} + \hat\beta_2 \sum_{i=1}^{n} x_{ik}x_{i2} + \cdots + \hat\beta_k \sum_{i=1}^{n} x_{ik}^2 = \sum_{i=1}^{n} x_{ik}y_i \tag{6-44}$$

Note that there are $p = k + 1$ normal equations, one for each of the unknown regression coefficients. The solution to the normal equations will be the **least squares estimators** of the regression coefficients, $\hat\beta_0, \hat\beta_1, \ldots, \hat\beta_k$. The normal equations can be solved by any method appropriate for solving a system of linear equations.

EXAMPLE 6-7

Wire Bond Pull Strength

In Chapter 1, we used data on pull strength of a wire bond in a semiconductor manufacturing process, wire length, and die height to illustrate building an empirical model. We will use the same data, repeated for convenience in Table 6-5, and show the details of estimating the model parameters. Scatter plots of the data are presented in Figs. 1-11a and 1-11b. Figure 6-17 shows a matrix of two-dimensional scatter plots of the data. These displays can be helpful in visualizing the relationships among variables in a multivariable data set.

Fit the multiple linear regression model

$$Y = \beta_0 + \beta_1 x_1 + \beta_2 x_2 + \epsilon$$

Table 6-5 Wire Bond Pull Strenght Data for Example 6-7

Observation Number	Pull Strength y	Wire Length x_1	Die Height x_2	Observation Number	Pull Strength y	Wire Length x_1	Die Height x_2
1	9.95	2	50	14	11.66	2	360
2	24.45	8	110	15	21.65	4	205
3	31.75	11	120	16	17.89	4	400
4	35.00	10	550	17	69.00	20	600
5	25.02	8	295	18	10.30	1	585
6	16.86	4	200	19	34.93	10	540
7	14.38	2	375	20	46.59	15	250
8	9.60	2	52	21	44.88	15	290
9	24.35	9	100	22	54.12	16	510
10	27.50	8	300	23	56.63	17	590
11	17.08	4	412	24	22.13	6	100
12	37.00	11	400	25	21.15	5	400
13	41.95	12	500				

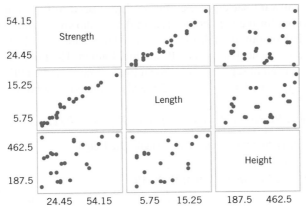

Figure 6-17 Matrix of scatter plots (from Minitab) for the wire bond pull strength data in Table 6-5.

where Y = pull strength, x_1 = wire length, and x_2 = die height.

Solution. From the data in Table 6-5 we calculate

$$n = 25, \quad \sum_{i=1}^{25} y_i = 725.82, \quad \sum_{i=1}^{25} x_{i1} = 206, \quad \sum_{i=1}^{25} x_{i2} = 8{,}294$$

$$\sum_{i=1}^{25} x_{i1}^2 = 2{,}396, \quad \sum_{i=1}^{25} x_{i2}^2 = 3{,}531{,}848$$

$$\sum_{i=1}^{25} x_{i1} x_{i2} = 77{,}177, \quad \sum_{i=1}^{25} x_{i1} y_i = 8{,}008.47, \quad \sum_{i=1}^{25} x_{i2} y_i = 274{,}816.71$$

For the model $Y = \beta_0 + \beta_1 x_1 + \beta_2 x_2 + \epsilon$, the normal equations 6-43 are

$$n\hat{\beta}_0 + \hat{\beta}_1 \sum_{i=1}^{n} x_{i1} + \hat{\beta}_2 \sum_{i=1}^{n} x_{i2} = \sum_{i=1}^{n} y_i$$

$$\hat{\beta}_0 \sum_{i=1}^{n} x_{i1} + \hat{\beta}_1 \sum_{i=1}^{n} x_{i1}^2 + \hat{\beta}_2 \sum_{i=1}^{n} x_{i1} x_{i2} = \sum_{i=1}^{n} x_{i1} y_i$$

$$\hat{\beta}_0 \sum_{i=1}^{n} x_{i2} + \hat{\beta}_1 \sum_{i=1}^{n} x_{i1} x_{i2} + \hat{\beta}_2 \sum_{i=1}^{n} x_{i2}^2 = \sum_{i=1}^{n} x_{i2} y_i$$

Inserting the computed summations into the normal equations, we obtain

$$25\hat{\beta}_0 + 206\hat{\beta}_1 + 8{,}294\hat{\beta}_2 = 725.82$$

$$206\hat{\beta}_0 + 2{,}396\hat{\beta}_1 + 77{,}177\hat{\beta}_2 = 8{,}008.47$$

$$8{,}294\hat{\beta}_0 + 77{,}177\hat{\beta}_1 + 3{,}531{,}848\hat{\beta}_2 = 274{,}816.71$$

The solution to this set of equations is

$$\hat{\beta}_0 = 2.26379, \quad \hat{\beta}_1 = 2.74427, \quad \hat{\beta}_2 = 0.01253$$

Using these estimated model parameters, the fitted regression equation is

$$\hat{y} = 2.26379 + 2.74427 x_1 + 0.01253 x_2.$$

This equation can be used to predict pull strength for pairs of values of the regressor variables wire length (x_1) and die height (x_2). This is essentially the same regression model given in equation 1-6, Section 1-3. Figure 1-13 shows a three-dimentional plot of the plane of predicted values $\hat{y}$ generated from this equation. ■

Computer software packages are almost always used to fit multiple regression models. The Minitab output for the wire bond strength data is shown in Table 6-6.

In the Minitab output of Table 6-6, x_1 = wire length and x_2 = die height. The estimates of the regression coefficients $\hat{\beta}_0$, $\hat{\beta}_1$, and $\hat{\beta}_2$ are highlighted. The fitted values $\hat{y}_i$ from the model corresponding to each observation are shown in the column labeled "Fit." The residuals, computed from $e_i = y_i - \hat{y}_i$, $i = 1, 2, \ldots, 25$ are also shown in the bottom section of the output.

Table 6-6   Minitab Regression Analysis Output for Wire Bond Pull Strength Data

Regression Analysis: Strength versus Wire Length, Die Height

The regression equation is

Strength = 2.26 + 2.74 Wire Ln + 0.0125 Die Ht

Predictor	Coef	SE Coef	T	P
Constant	2.264 ← $\hat{\beta}_0$	1.060	2.14	0.044
Wire Ln	2.74427 ← $\hat{\beta}_1$	0.09352	29.34	0.000
Die Ht	0.012528 ← $\hat{\beta}_2$	0.002798	4.48	0.000

S = 2.288   R-Sq = 98.1%   R-Sq (adj) = 97.9%

Analysis of Variance

Source	DF	SS	MS	F	P
Regression	2	5990.8	2995.4	572.17	0.000
Residual Error	22	115.2 ← SS_E	5.2 ← $\hat{\sigma}^2$		
Total	24	6105.9 ← SS_T			

Obs	Strength	Fit	SE Fit	Residual	St Resid
1	9.950	8.379	0.907	1.571	0.75
2	24.450	25.596	0.765	−1.146	−0.53
3	31.750	33.954	0.862	−2.204	−1.04
4	35.000	36.597	0.730	−1.597	−0.74
5	25.020	27.914	0.468	−2.894	−1.29
6	16.860	15.746	0.626	1.114	0.51
7	14.380	12.450	0.786	1.930	0.90
8	9.600	8.404	0.904	1.196	0.57
9	24.350	28.215	0.819	−3.865	−1.81
10	27.500	27.976	0.465	−0.476	−0.21
11	17.080	18.402	0.696	−1.322	−0.61
12	37.000	37.462	0.525	−0.462	−0.21
13	41.950	41.459	0.655	0.491	0.22
14	11.660	12.262	0.769	−0.602	−0.28
15	21.650	15.809	0.621	5.841	2.65
16	17.890	18.252	0.679	−0.362	−0.17
17	69.000	64.666	1.165	4.334	2.20
18	10.300	12.337	1.238	−2.037	−1.06
19	34.930	36.472	0.710	−1.542	−0.71
20	46.590	46.560	0.878	0.030	0.01
21	44.880	47.061	0.824	−2.181	−1.02
22	54.120	52.561	0.843	1.559	0.73
23	56.630	56.308	0.977	0.322	0.16
24	22.130	19.982	0.756	2.148	0.99
25	21.150	20.996	0.618	0.154	0.07

Many of the computations and analysis procedures that we introduced for simple linear regression carry over to the multiple regression case. For example, the sum of the squared residuals is used to estimate the error variance σ^2. The residual (or error) sum of squares is $SS_E = \sum_{i=1}^{n}(y_i - \hat{y}_i)^2$, and the estimate of σ^2 for a multiple linear regression model with p parameters is

$$\hat{\sigma}^2 = \frac{\sum_{i=1}^{n}(y_i - \hat{y}_i)^2}{n-p} = \frac{SS_E}{n-p} \tag{6-45}$$

EXAMPLE 6-7
(continued)

Compute Estimate of
Error Variance $\hat{\sigma}^2$

The Minitab output in Table 6-6 shows that the residual sum of squares for the bond strength regression model is $SS_E = 115.2$, there are $n = 25$ observations, and the model has $p = 3$ parameters (β_0, β_1, and β_2), so the estimate of the error variance is computed from equation 6-45 as $\hat{\sigma}^2 = SS_E/(n-p) = 115.2/(25 - 3) = 5.2$, as reported by Minitab in Table 6-6.

The analysis of variance partitioning of the total sum of squares given in equation 6-21 is also valid for multiple regression. The Minitab output in Table 6-6 contains the ANOVA results. The total sum of squares has $n - 1$ degrees of freedom, the model or regression sum of squares has $k = p - 1$ degrees of freedom (recall that k is the number of regressor variables), and the error or residual sum of squares has $n - p$ degrees of freedom.

Compute and Interpret R^2

The coefficient of determination or R^2 in multiple regression is computed exactly as it is in simple linear regression; that is, $R^2 = SS_R/SS_T = 1 - (SS_E/SS_T)$. In a multiple linear regression model, it is customary to refer to R^2 as the coefficient of *multiple* determination. For the bond strength regression model, Minitab calculates $R^2 = 1 - (115.2/6105.9) = 0.981$, and the output reports $R^2 \times 100\% = 98.1\%$. This can be interpreted as indicating that the model containing wire length and die height accounts for approximately 98.1% of the observed variability in bond strength.

The numerical value of R^2 cannot decrease as variables are added to a regression model. For instance, if we were to use only the regressor variable $x_1 =$ wire length for the wire bond strength data, the value of $R^2 = 0.964$. Adding the second regressor $x_2 =$ die height increases the value to $R^2 = 0.981$. To more accurately reflect the value of adding another regressor to the model, an **adjusted R^2** statistic can be used.

The Adjusted Coefficient of Multiple Determination (R^2_{Adjusted})

The **adjusted coefficient of multiple determination** for a multiple regression model with k regressors is

$$R^2_{\text{Adjusted}} = 1 - \frac{SS_E/(n-p)}{SS_T/(n-1)} = \frac{(n-1)R^2 - k}{n-p} \tag{6-46}$$

The adjusted R^2 statistic essentially penalizes the usual R^2 statistic by taking the number of regressor variables in the model into account. In general, the adjusted R^2 statistic will not always increase when a variable is added to the model. The adjusted R^2 will only increase if the addition of a new variable produces a large enough reduction in the residual sum of squares to compensate for the loss of one residual degree of freedom.

EXAMPLE 6-7
(continued)

To illustrate this point, consider the regression model for the wire bond strength data with one regressor variable x_1 = wire length. The value of the residual sum of squares for this model is $SS_E = 220.09$. From equation 6-46, the adjusted R^2 statistic is

Compute and Interpret
R^2_{Adjusted}

$$R^2_{\text{Adjusted}} = 1 - \frac{SS_E/(n-p)}{SS_T/(n-1)} = 1 - \frac{220.09/(25-2)}{6105.9/(25-1)} = 0.962$$

When both regressors are included in the model, the value of $R^2_{\text{Adjusted}} = 0.979$ (refer to Table 6-6). Because the adjusted R^2 statistic has increased with the addition of the second regressor variable, we would conclude that adding this new variable to the model was probably a good idea because it resulted in slightly more explained variability in the response.

Contribution of Second Regressor

Another way to view the contribution of adding a regressor to the model is to examine the change in the residual mean square. For the wire bond strength data with only x_1 = wire length as the regressor the residual sum of squares is $SS_E = 220.09$, and the residual mean square is $SS_E/(n-p) = 220.09/(25-2)$ = 9.57. When both regressors are in the model, the residual mean square is 5.2. Because this residual mean square estimates σ^2, the variance of the unexplained variability in the response, as much smaller, we conclude that the model with two regressors is superior. Note that using the residual mean square as an estimate of σ^2 results in a **model-dependent** estimate. However, a regression model with a small value of the residual mean square is almost always superior to a model with a large residual mean square.

6-3.2   Inferences in Multiple Regression

Just as in simple linear regression, it is important to test hypotheses and construct CIs in multiple regression. In this section we will describe these procedures. In most cases, they are straightforward modifications of the procedures we used in simple linear regression.

Test for Significance of Regression
The test for significance of regression in simple linear regression investigated whether there was a useful linear relationship between the response y and the single regressor x. In multiple regression, this is a test of the hypothesis that there is no useful linear relationship between the response y and *any* of the regressors $x_1, x_2, \ldots, x_k$. The hypotheses are

$$H_0: \beta_1 = \beta_2 = \cdots = \beta_k = 0$$

$$H_1: \text{At least one } \beta_j \neq 0$$

Testing for Significance of Regression in Multiple Regression

$$MS_R = \frac{SS_R}{k} \qquad MS_E = \frac{SS_E}{n-p}$$

Null hypothesis: $H_0: \beta_1 = \beta_2 = \cdots = \beta_k = 0$

Alternative hypothesis: $H_1: \text{At least one } \beta_j \neq 0$

Test statistic: $F_0 = \dfrac{MS_R}{MS_E}$ (6-47)

P-value: Probability above f_0 in the $F_{k,n-p}$ distribution

Rejection criterion for a
fixed-level test: $f_0 > f_{\alpha,k,n-p}$

Thus if the null hypothesis is rejected, at least one of the regressor variables in the model is linearly related to the response. The ANOVA partition of the total variability in the response y (equation 6-21) is used to test these hypotheses.

Interpret ANOVA Table

The test procedure is usually summarized in an ANOVA table, and it is also included in multiple regression computer output. For the wire bond strength regression model, refer to Table 6-6, page 312. The hypotheses for this model are

$$H_0: \beta_1 = \beta_2 = 0$$

$$H_1: \text{At least one } \beta_j \neq 0$$

In the Minitab output under the section headed "Analysis of Variance," we find the values of the mean squares for regression and residual, and the computed value of the test statistic in equation 6-47 is $f_0 = 572.17$. Because the P-value is very small, we would conclude that at least one of the regressors is related to the response y.

Inference on Individual Regression Coefficients

Because the estimates of the regression coefficients $\hat{\beta}_j, j = 0, 1, \ldots, k$ in a multiple regression model are just linear combinations of the y's, and the y's are assumed to have a normal distribution, the $\hat{\beta}_j$'s are normally distributed. Furthermore, the $\hat{\beta}_j$'s are unbiased estimators of the true model coefficients and their standard errors, $se(\hat{\beta}_j), j = 0, 1, \ldots, k$, can be computed as the product of $\hat{\sigma}$ and a function of the x's. The standard error is a rather complicated expression, but it is computed and reported by all multiple regression computer programs. In the Minitab output in Table 6-6 the standard errors of the model regression coefficients are listed in the column labeled "SE Coef." Inferences on an individual regression coefficient are based on the quantity

$$T = \frac{\hat{\beta}_j - \beta_j}{se(\hat{\beta}_j)}$$

which has the t distribution with $n - p$ degrees of freedom. This leads to the following hypothesis testing and CI results for an individual regression coefficient β_j.

Inferences on the Model Parameters in Multiple Regression

1. The test for $H_0: \beta_j = \beta_{j,0}$ versus $H_1: \beta_j \neq \beta_{j,0}$ employs the **test statistic**

$$T_0 = \frac{\hat{\beta}_j - \beta_{j,0}}{se(\hat{\beta}_j)} \quad (6\text{-}48)$$

and the null hypothesis is rejected if $|t_0| > t_{\alpha/2,n-p}$. A P-value approach can also be used. One-sided alternative hypotheses can also be tested.

2. A $100(1 - \alpha)\%$ CI for an individual regression coefficient is given by

$$\hat{\beta}_j - t_{\alpha/2,n-p}se(\hat{\beta}_j) \leq \beta_j \leq \hat{\beta}_j + t_{\alpha/2,n-p}se(\hat{\beta}_j) \quad (6\text{-}49)$$

A very important special case of the test on an individual regression coefficient is a hypothesis of the form $H_0: \beta_j = 0$ versus $H_1: \beta_j \neq 0$. Most regression packages calculate the test statistic for this hypothesis for each variable that is in the model. This is a measure of the contribution of each *individual* regressor variable to the overall model. The test statistic is

$$T_0 = \frac{\hat{\beta}_j}{se(\hat{\beta}_j)} \tag{6-50}$$

This test is often called a **partial** or **marginal** test because it is evaluating the contribution of each regressor variable to the model *given* that all of the *other* regressors are also included.

EXAMPLE 6-7 (continued)

Interpret t-Values

The Minitab output in Table 6-6 shows the values of the test statistic computed from equation 6-50 for each of the regressor variables, wire length and die height. The *t*-value for wire length $t_0 = 29.34$ has a *P*-value of 0.0, which indicates that the regressor wire length contributes significantly to the model, given that the other regressor die height is also in the model. Also, the *t*-value for die height $t_0 = 4.48$ has a *P*-value of 0.0, which indicates that die height contributes significantly to the model, given that the other regressor wire length is also included.

Generally, if the *t*-statistic for any individual regression coefficient is insignificant, so the hypothesis H_0: $\beta_j = 0$ is not rejected, this is an indication that the regressor x_j should be removed from the model. In some situations, these *t*-tests may indicate that more than one regressor is not important. The correct approach in this situation is to remove the least significant regressor and refit the model. Then *t*-tests are conducted for the regressors in this new model to determine which, if any, regressors are still not significant, and continue in this manner until all of the regressors in the model are significant.

Confidence Intervals on the Mean Response and Prediction Intervals

A multiple regression model is often used to obtain a **point estimate** of the mean response at a particular set of *x*'s; that is, when $x_1 = x_{10}, x_2 = x_{20}, \ldots, x_k = x_{k0}$. The true mean response at this point is $\mu_{Y|x_{10},x_{20},\ldots,x_{k0}} = \beta_0 + \beta_1 x_{10} + \beta_2 x_{20} + \cdots + \beta_k x_{k0}$, and the corresponding point estimate is

$$\hat{\mu}_{Y|x_{10},x_{20},\ldots,x_{k0}} = \hat{\beta}_0 + \hat{\beta}_1 x_{10} + \hat{\beta}_2 x_{20} + \cdots + \hat{\beta}_k x_{k0} \tag{6-51}$$

The standard error of this point estimate is a complicated function that depends on the *x*'s used to fit the regression model and the coordinates at which the point estimate is computed, but it is provided by many regression software packages and is denoted by $se(\hat{\mu}_{Y|x_{10},x_{20},\ldots,x_{k0}})$. The confidence interval on the mean response at the point $x_1 = x_{10}, x_2 = x_{20}, \ldots, x_k = x_{k0}$ is given by the following expression.

Confidence Interval on the Mean Response in Multiple Regression

A $100(1 - \alpha)\%$ CI on the mean response at the point $x_1 = x_{10}, x_2 = x_{20}, \ldots, x_k = x_{k0}$ in a multiple regression model is given by

$$\hat{\mu}_{Y|x_{10},x_{20},\ldots,x_{k0}} - t_{\alpha/2,n-p} se(\hat{\mu}_{Y|x_{10},x_{20},\ldots,x_{k0}}) \leq \mu_{Y|x_{10},x_{20},\ldots,x_{k0}}$$
$$\leq \hat{\mu}_{Y|x_{10},x_{20},\ldots,x_{k0}} + t_{\alpha/2,n-p} se(\hat{\mu}_{Y|x_{10},x_{20},\ldots,x_{k0}}) \tag{6-52}$$

where $\hat{\mu}_{Y|x_{10},x_{20},\ldots,x_{k0}}$ is computed from equation 6-51.

Interpret Mean Response and 95% CI

Minitab will calculate the confidence interval in equation 6-52 for a point of interest. For example, suppose that we are interested in finding an estimate of the mean pull strength at two points: (1) when wire length $x_1 = 11$ and die height $x_2 = 35$ and (2) when wire length $x_1 = 5$ and die height $x_2 = 20$. The point estimate is found by first substituting $x_1 = 11$ and $x_2 = 35$ and then $x_1 = 5$ and $x_2 = 20$ into the

Table 6-7 Minitab Output

Predicted Values for New Observations

New Obs	Fit	SE Fit	95.0% CI	95.0% PI
1	32.889	1.062	(30.687, 35.092)	(27.658, 38.121)

New Obs	Fit	SE Fit	95.0% CI	95.0% PI
2	16.236	0.929	(14.310, 18.161)	(11.115, 21.357)

Values of Predictors for New Observations

New Obs	Wire Ln	Die Ht
1	11.0	35.0

New Obs	Wire Ln	Die Ht
2	5.00	20.0

fitted regression model and calculating the fitted values of the responses at the two points. Minitab reports the point estimates and their associated 95% CIs in Table 6-7.

The estimated mean response $\mu_{y|11,35} = 32.899$ has a 95% CI of (30.687, 35.092), where as the estimated mean response $\mu_{y|5,20} = 16.236$ has a 95% CI of (14.310, 18.161). Notice that the 95% CI for the second point is narrower than the 95% CI for the first point. As in simple linear regression, the width of the CI on the mean response increases as the point moves away from the center of the region of x-variable space, and point 1 is farther away from the center of the of x-space than is point 2.

Interpret Mean
Response and 95% CI

A $100(1 - \alpha)\%$ PI on a future observation at the point $x_1 = x_{10}, x_2 = x_{20}, \ldots, x_k = x_{k0}$ in a multiple regression model can also be determined. The response at the point of interest is

$$Y_0 = \beta_0 + \beta_1 x_{10} + \beta_2 x_{20} + \cdots + \beta_k x_{k0} + \epsilon$$

and the corresponding predicted value is

$$\hat{Y}_0 = \hat{\mu}_{Y|x_{10}, x_{20}, \ldots, x_{k0}} = \hat{\beta}_0 + \hat{\beta}_1 x_{10} + \hat{\beta}_2 x_{20} + \cdots + \hat{\beta}_k x_{k0} \tag{6-53}$$

The prediction error is $Y_0 - \hat{Y}_0$, and the standard deviation of this prediction error is

$$\sqrt{\hat{\sigma}^2 + \left[se(\hat{\mu}_{Y|x_{10}, x_{20}, \ldots, x_{k0}}) \right]^2}$$

Therefore, the PI on a future observation is as follows.

Prediction Interval on a Future Observation

A $100(1 - \alpha)\%$ PI on a future observation at the point $x_1 = x_{10}, x_2 = x_{20}, \ldots, x_k = x_{k0}$ in a multiple regression model is given by

$$\hat{y}_0 - t_{\alpha/2, n-p} \sqrt{\hat{\sigma}^2 + \left[se(\hat{\mu}_{Y|x_{10}, x_{20}, \ldots, x_{k0}}) \right]^2} \le Y_0$$

$$\le \hat{y}_0 + t_{\alpha/2, n-p} \sqrt{\hat{\sigma}^2 + \left[se(\hat{\mu}_{Y|x_{10}, x_{20}, \ldots, x_{k0}}) \right]^2} \tag{6-54}$$

where $\hat{y}_0 = \hat{\mu}_{Y|x_{10}, x_{20}, \ldots, x_{k0}}$ is computed from equation 6-53.

EXAMPLE 6-7
(continued)

Interpret Future
Observation and 95% PI

The Minitab output in Table 6-7 shows the 95% PIs for the pull strength at two new points where wire length $x_1 = 11$ and die height $x_2 = 35$ and where $x_1 = 5$ and die height $x_2 = 20$. The future observation $\hat{y}_0 = \hat{\mu}_{y|11,35} = 32.889$ has a 95% PI of $(27.658, 38.121)$, whereas the future observation $\hat{y}_0 = \hat{\mu}_{y|5,20} = 16.236$ has a 95% PI of $(11.115, 21.357)$.

Notice that these PIs are wider than the corresponding CIs and that they increase in width as the point where the prediction is made moves farther away from the center of the x-space. ∎

A Test for the Significance of a Group of Regressors
There are some situations in regression model building where interest centers on a subset of the regressors in the full model. For example, suppose that we are considering fitting a second-order model

$$Y = \beta_0 + \beta_1 x_1 + \beta_2 x_2 + \beta_{12} x_1 x_2 + \beta_{11} x_1^2 + \beta_{22} x_2^2 + \epsilon$$

but we are uncertain about the contribution of the second-order terms to the model. Therefore, we are interested in testing the hypothesis

$$H_0: \beta_{12} = \beta_{11} = \beta_{22} = 0$$
$$H_1: \text{At least one of the } \beta\text{'s} \neq 0$$

We can use an F-test for these hypotheses.

In general, suppose that the **full model** has k regressors, and we are interested in testing whether the last $k - r$ of them can be deleted from the model. This smaller model is called the **reduced model.** That is, the full model is

$$Y = \beta_0 + \beta_1 x_1 + \beta_2 x_2 + \cdots + \beta_r x_r + \beta_{r+1} x_{r+1} + \cdots + \beta_k x_k + \epsilon$$

and the reduced model has $\beta_{r+1} = \beta_{r+2} = \cdots = \beta_k = 0$, so the reduced model is

$$Y = \beta_0 + \beta_1 x_1 + \beta_2 x_2 + \cdots + \beta_r x_r + \epsilon$$

The test is performed by fitting both the full and the reduced models and comparing the residual sums of squares for the two models. Let $SS_E(FM)$ be the residual sum of squares for the full model and let $SS_E(RM)$ be the residual sum of squares for the reduced model. Then to test the hypotheses

$$H_0: \beta_{r+1} = \beta_{r+2} = \cdots = \beta_k = 0$$
$$H_1: \text{At least one of the } \beta\text{'s} \neq 0 \tag{6-55}$$

we would use the test statistic

$$F_0 = \frac{[SS_E(RM) - SS_E(FM)]/(k - r)}{SS_E(FM)/(n - p)} \tag{6-56}$$

The null hypothesis in equation 6-55 is rejected if $f_0 > f_{a,k-r,n-p}$. A P-value approach can also be used.

6-3.3 Checking Model Adequacy

Residual Analysis
The residuals $e_i = y_i - \hat{y}_i, i = 1, 2, \ldots, n$ should be graphically analyzed to check the adequacy of a multiple linear regression model. A normal probability plot of the residuals is used

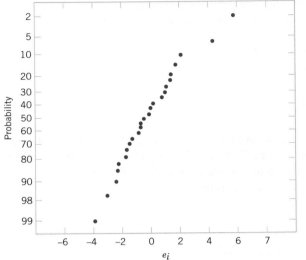

Figure 6-18   Normal probability plot of residuals for wire bond empirical model.

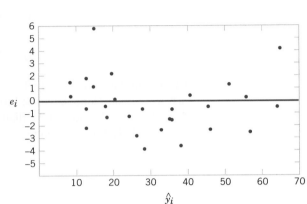

Figure 6-19   Plot of residuals against $\hat{y}$ for wire bond empirical model.

to check the normality assumption, and plots of the residuals versus the fitted values and possibly versus each of the individual regressors can reveal other model inadequacies, such as inequality of variance and the possible need for additional regressors in the model.

Interpret Residual Plots

Figures 6-18, 6-19, 6-20, and 6-21 present the normal probability plot of the residuals and the plots of the residuals versus the fitted values $\hat{y}$ and both of the regressors x_1 and x_2 for the wire bond strength regression model. The normal probability plot is satisfactory, but the plots of the residuals versus $\hat{y}$ and x_1 reveal slight curvature. Possibly another regressor variable is needed in the model. In general, though, none of the plots suggests any dramatic problems with the model.

In multiple regression, we often examine scaled residuals. A common residual scaling is the **standardized residual,** $d_i = e_i / \sqrt{\hat{\sigma}^2}, i = 1, 2, \ldots, n$. We discussed the standardized residual in simple linear regression and observed that it can be useful in looking for outliers. Another scaled residual, the **studentized residual,** is very useful in multiple regression. The studentized residual scales the usual least squares residual by dividing it by its exact standard error. We now show how the studentized residuals are computed. The regression coefficients $\hat{\beta}$ are linear combinations of the observations y. Because the predicted values $\hat{y}_i$ are linear

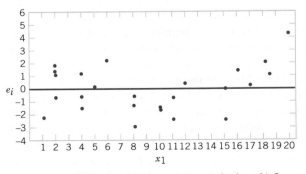

Figure 6-20   Plot of residuals against x_1 (wire length) for wire bond empirical model.

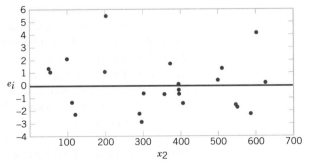

Figure 6-21   Plot of residuals against x_2 (die height) for wire bond empirical model.

combinations of the regression coefficients, they are also linear combinations of the observations y_i. We can write the relationship between the $\hat{y}_i$ and the y_i values as

$$\hat{y}_1 = h_{11}y_1 + h_{12}y_2 + \cdots + h_{1n}y_n$$
$$\hat{y}_2 = h_{21}y_1 + h_{22}y_2 + \cdots + h_{2n}y_n$$
$$\vdots$$
$$\hat{y}_n = h_{n1}y_1 + h_{n2}y_2 + \cdots + h_{nn}y_n \qquad (6\text{-}57)$$

The h_{ij}'s are functions of only the x's that are used to fit the model, and they are actually fairly easy to compute (for details, see Montgomery, Peck, and Vining [2006]). Furthermore, we can show that $h_{ij} = h_{ji}$ and that the diagonal values in this system of equations take on the values $0 < h_{ii} \leq 1$. The studentized residuals are defined as follows.

Studentized Residuals

The studentized residuals are defined as

$$r_i = \frac{e_i}{se(e_i)} = \frac{e_i}{\sqrt{\hat{\sigma}^2(1 - h_{ii})}}, i = 1, 2, \ldots, n \qquad (6\text{-}58)$$

Because the h_{ii}'s are always between zero and unity, a studentized residual is always larger than the corresponding standardized residual. Consequently, studentized residuals are a more sensitive diagnostic when looking for outliers.

Interpret Studentized Residuals

The Minitab output in Table 6-6 lists the studentized residuals (in the column labeled "St Resid") for the wire bond pull strength regression model. None of these studentized residuals is large enough to indicate that outliers may be present.

Influential Observations

When using multiple regression, we occasionally find that some subset of the observations is unusually influential. Sometimes these influential observations are relatively far away from the vicinity where the rest of the data were collected. A hypothetical situation for two variables is depicted in Fig. 6-22, where one observation in x-space is remote from the rest of the data. The disposition of points in the x-space is important in determining the properties of the model. For example, point (x_{i1}, x_{i2}) in Fig. 6-22 may be very influential in determining R^2, the estimates of the regression coefficients, and the magnitude of the error mean square.

We would like to examine the influential points to determine whether they control many model properties. If these influential points are "bad" points, or erroneous in any way, they should be eliminated. On the other hand, there may be nothing wrong with these points, but at least we would like to determine whether or not they produce results consistent with the rest of the data. In any event, even if an influential point is a valid one, if it controls important model properties, we would like to know this because it could have an impact on the use of the model.

One useful method is to inspect the h_{ii}'s, defined in equation 6-57. The value of h_{ii} can be interpreted as a measure of the distance of the point $(x_{i1}, x_{i2}, \ldots x_{ik})$ from the average of all of the points in the data set. The value of h_{ii} is not the usual distance measure, but it has similar properties. Consequently, a large value for h_{ii} implies that the ith data point in x-space is remote from the center of the data (as in Fig. 6-22). A rule of thumb is that h_{ii}'s greater than $2p/n$ should be investigated.

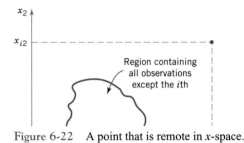

Figure 6-22   A point that is remote in x-space.

A data point for which h_{ii} exceeds this value is considered a **leverage point.** Because it is remote, it has substantial leverage, or potential to change the regression analysis. The average value of h_{ii} in any data set is p/n. Therefore, the rule flags values greater than twice the average.

Montgomery, Peck, and Vining (2006) and Myers (1990) describe several other methods for detecting influential observations. An excellent diagnostic is the distance measure developed by Professor Dennis R. Cook. This is a measure of the squared distance between the usual least squares estimate (the $\hat{\beta}$'s) based on all n observations and the estimate obtained when the ith point is deleted.

Cook's Distance Measure

Cook's distance measure is defined as

$$D_i = \frac{r_i^2}{p} \frac{h_{ii}}{(1 - h_{ii})} \quad i = 1, 2, \ldots, n \qquad (6\text{-}59)$$

Clearly, if the ith point is influential, its removal will result in some of the regression coefficients changing considerably from the value obtained using all n observations. Thus, a large value of D_i implies that the ith point is influential. We see that D_i consists of the squared studentized residual, which reflects how well the model fits the ith observation y_i. (Recall that $r_i = e_i/\sqrt{\hat{\sigma}^2(1 - h_{ii})}$ in equation 6-58, and a component that measures how far that point is from the rest of the data $[h_{ii}/(1 - h_{ii})]$ is a measure of the distance of the ith point in x-space from the centroid of the remaining $n - 1$ points.) A value of $D_i > 1$ would indicate that the point is influential. Either component of D_i (or both) may contribute to a large value.

EXAMPLE 6-8

Wire Bond Pull
Strength

Table 6-8 lists the values of h_{ii} and Cook's distance measure D_i for the wire bond pull strength data. To illustrate the calculations, consider the first observation

$$D_1 = \frac{r_1^2}{p} \cdot \frac{h_{11}}{(1 - h_{11})} = \frac{[e_1/\sqrt{\hat{\sigma}^2(1 - h_{11})}]^2}{p} \cdot \frac{h_{11}}{(1 - h_{11})}$$

$$= \frac{[1.571/\sqrt{5.2(1 - 0.1573)}]^2}{3} \cdot \frac{0.1573}{(1 - 0.1573)} = 0.035$$

Table 6-8   Influence Diagnostics for the Wire Bond Pull Strength Data

Observations i	h_{ii}	Cook's Distance Measure D_i	Observations i	h_{ii}	Cook's Distance Measure D_i
1	0.1573	0.035	14	0.1129	0.003
2	0.1116	0.012	15	0.0737	0.187
3	0.1419	0.060	16	0.0879	0.001
4	0.1019	0.021	17	0.2593	0.565
5	0.0418	0.024	18	0.2929	0.155
6	0.0749	0.007	19	0.0962	0.018
7	0.1181	0.036	20	0.1473	0.000
8	0.1561	0.020	21	0.1296	0.052
9	0.1280	0.160	22	0.1358	0.028
10	0.0413	0.001	23	0.1824	0.002
11	0.0925	0.013	24	0.1091	0.040
12	0.0526	0.001	25	0.0729	0.000
13	0.0820	0.001			

Interpret Influence Diagnostics

Twice the average of the h_{ii} values is $2p/n = 2(3)/25 = 0.2400$. Two data points (17 and 18) have values of h_{ii} that exceed this cutoff, so they could be classified as leverage points. However, because the D_{17} and D_{18} values are less than unity, these points are not unusually influential. Note that none of the D_i values is large enough to cause concern. ∎

Multicollinearity

In multiple regression problems we expect to find dependencies between the regressor variables and the response. However, in many regression problems, we also find that there are dependencies among the regressors. In situations where these dependencies are strong, we say that **multicollinearity** exists. Multicollinearity can have serious effects on the estimates of the parameters in a regression model, resulting in parameters that are poorly estimated (large variance or standard error) and that are unstable in the sense that a different sample from the same process or system can produce very different estimates of the β's. Models with strong multicollinearity are often not reliable prediction equations.

There are several diagnostics that can be used to determine whether multicollinearity is present. The simplest of these is the **variance inflation factor,** or VIF.

Variance Inflation Factors

The VIFs for a multiple linear regression model are

$$VIF(\beta_j) = \frac{1}{1 - R_j^2}, \quad j = 1, 2, \ldots, k \tag{6-60}$$

where R_j^2 is the coefficient of multiple determination that results from regressing x_j on the other $k - 1$ regressors.

It is easy to see why the VIF defined in equation 6-60 is a good measure of multicollinearity. If regressor x_j is strongly linearly dependent on some subset of the other regressor variables in the model, R_j^2 will be large, say, close to unity, and the corresponding VIF will be large. Alternatively, if regressor x_j is not nearly linearly dependent on the other regressors, the value of R_j^2 will be small and the corresponding VIF will be small. Generally, if the VIF associated with any regressor exceeds 10, we would suspect that multicollinearity is present.

Interpret VIF

Many regression computer programs will calculate the VIFs. The following display shows the results from Minitab for the wire bond pull strength regression model.

Predictor	Coef	SE Coef	T	P	VIF
Constant	2.264	1.060	2.14	0.044	
Wire Ln	2.74427	0.09352	29.34	0.000	1.2
Die Ht	0.01258	0.002798	4.48	0.000	1.2

Because the VIFs here are quite small, there is no apparent problem with multicollinearity in the wire bond pull strength data. If strong multicollinearity is present, many regression analysts will recommend investigating one of several possible remedial measures, including deleting some of the regressor variables from the model or using a technique other than the method of least squares to estimate the model parameters. A comprehensive discussion of multicollinearity is in Montgomery, Peck, and Vining (2006).

EXERCISES FOR SECTION 6-3

For Exercises 6-15 through 6-18, use Minitab to assist you in answering the following.

(a) Estimate the regression coefficients. Write the multiple linear regression model. Comment on the relationship found between the set of independent variables and the dependent variable.

(b) Compute the residuals.

(c) Compute SS_E and estimate the variance.

(d) Compute the coefficient of determination, R^2, and adjusted coefficient of multiple determination, R_{Adjusted}^2. Comment on their values.

(e) Construct the ANOVA table and test for significance of regression. Comment on your results.

(f) Find the standard error of the individual coefficients.

(g) Use a t-test to test for significance of the individual coefficients at $\alpha = 0.05$. Comment on your results.

(h) Construct 95% CIs on the individual coefficients. Compare your results with those found in part (g) and comment.

(i) Perform a model adequacy check, including computing studentized residuals and Cook's distance measure for each of the observations. Comment on your results.

(j) Compute the variance inflation factor and comment on the presence of multicollinearity.

6-15. Consider the bearing wear data in Exercise 2-47.

6-16. Consider the MPG data in Exercise 2-48.

6-17. The electric power consumed each month by a chemical plant is thought to be related to the average ambient temperature (x_1), the number of days in the month (x_2), the average product purity (x_3), and the tons of product produced (x_4). The past year's historical data are available and are presented in the following table:

y	x_1	x_2	x_3	x_4
240	25	24	91	100
236	31	21	90	95
290	45	24	88	110
274	60	25	87	88
301	65	25	91	94
316	72	26	94	99
300	80	25	87	97
296	84	25	86	96
267	75	24	88	110
276	60	25	91	105
288	50	25	90	100
261	38	23	89	98

6-18. An engineer at a semiconductor company wants to model the relationship between the device HFE (y) and three parameters: Emitter-RS (x_1), Base-RS (x_2), and Emitter-to-Base RS (x_3). The data are shown in the following table.

x_1 Emitter-RS	x_2 Base-RS	x_3 Emitter-to-Base RS	y HFE
14.620	226.00	7.000	128.40
15.630	220.00	3.375	52.62
14.620	217.40	6.375	113.90
15.000	220.00	6.000	98.01
14.500	226.50	7.625	139.90
15.250	224.10	6.000	102.60
16.120	220.50	3.375	48.14
15.130	223.50	6.125	109.60
15.500	217.60	5.000	82.68
15.130	228.50	6.625	112.60
15.500	230.20	5.750	97.52
16.120	226.50	3.750	59.06
15.130	226.60	6.125	111.80
15.630	225.60	5.375	89.09
15.380	229.70	5.875	101.00
14.380	234.00	8.875	171.90
15.500	230.00	4.000	66.80
14.250	224.30	8.000	157.10
14.500	240.50	10.870	208.40
14.620	223.70	7.375	133.40

6-19. Consider the bearing wear data and multiple linear regression model in Exercise 6-15.

(a) Find the mean bearing wear given that the oil viscosity is 25.0 and the load is 1100.
(b) Compute a 99% CI on this mean response.
(c) Compute a 99% PI on a future observation when the oil viscosity is 25.0 and the load is 1100.
(d) What do you notice about the relative size of these two intervals? Which is bigger and why?

6-20. Consider the MPG data and multiple linear regression model in Exercise 6-16.

(a) Find the mean MPG given that the weight is 2650 and the horsepower is 120.
(b) Compute a 95% CI on this mean response.
(c) Compute a 95% PI on a future observation when the weight is 2650 and the horsepower is 120.
(d) What do you notice about the relative size of these two intervals? Which is bigger and why?

6-21. Consider the power consumption data and multiple linear regression model in Exercise 6-17.

(a) Find the mean power consumption given that $x_1 = 75°F$, $x_2 = 24$ days, $x_3 = 90\%$, and $x_4 = 98$ tons.
(b) Compute a 95% CI on this mean response.
(c) Compute a 95% PI on a future observation when $x_1 = 75°F$, $x_2 = 24$ days, $x_3 = 90\%$, and $x_4 = 98$ tons.
(d) What do you notice about the relative size of these two intervals? Which is bigger and why?

6-22. Consider the HFE data and multiple linear regression model in Exercise 6-18.

(a) Find the mean HFE given that $x_1 = 14.5$, $x_2 = 220$, and $x_3 = 5.0$.
(b) Compute a 90% CI on this mean response.
(c) Compute a 90% PI on a future observation when $x_1 = 14.5$, $x_2 = 220$, and $x_3 = 5.0$.
(d) What do you notice about the relative size of these two intervals? Which is bigger and why?

6-23. An article in *Optical Engineering* ("Operating Curve Extraction of a Correlator's Filter," Vol. 43, 2004, pp. 2775–2779) reported the use of an optical correlator to perform an experiment by varying brightness and contrast. The resulting modulation is characterized by the useful range of gray levels. The data are shown below:

Brightness (%):	54	61	65	100	100	100	50	57	54
Contrast (%):	56	80	70	50	65	80	25	35	26
Useful range (ng):	96	50	50	112	96	80	155	144	255

(a) Fit a multiple linear regression model to these data.
(b) Estimate σ^2 and the standard errors of the regression coefficients.
(c) Test for significance of β_1 and β_2.
(d) Predict the useful range when brightness = 80 and contrast = 75. Construct a 95% PI.
(e) Compute the mean response of the useful range when brightness = 80 and contrast = 75. Compute a 95% CI.
(f) Interpret parts (d) and (e) and comment on the comparison between the 95% PI and 95% CI.

6-24. An article in *Biotechnology Progress* (Vol. 17, 2001, pp. 366–368) reported an experiment to investigate and optimize nisin extraction in aqueous two-phase systems (ATPS). The nisin recovery was the dependent variable (y). The two regressor variables were concentration (%) of PEG 4000 (denoted as x_1) and concentration (%) of Na_2SO_4 (denoted as x_2).

x_1	x_2	y
13	11	62.8739
15	11	76.1328
13	13	87.4667
15	13	102.3236
14	12	76.1872
14	12	77.5287
14	12	76.7824
14	12	77.4381
14	12	78.7417

(a) Fit a multiple linear regression model to these data.
(b) Estimate σ^2 and the standard errors of the regression coefficients.
(c) Test for significance of β_1 and β_2.
(d) Use the model to predict the nisin recovery when $x_1 = 14.5$ and $x_2 = 12.5$. Construct a 95% PI.

(e) Compute the mean response of the nisin recovery when $x_1 = 14.5$ and $x_2 = 12.5$. Construct a 95% CI.

(f) Interpret parts (d) and (e) and comment on the comparison between the 95% PI and 95% CI.

6-25. Use the partially complete Minitab output below to answer the following questions.

(a) Find all of the missing values.

(b) Find the estimate of σ^2.

(c) Test for significance of regression. Use $\alpha = 0.05$

(d) Test for significance of β_1 and β_2 using a t-test with $\alpha = 0.05$. Comment on the two results.

(e) Construct a 95% CI on β_1. Use this CI to test for significance of β_1.

(f) Construct a 95% CI on β_2. Use this CI to test for significance of β_2.

(g) Comment on results found in parts (c)–(f). Is this regression model appropriate? What is your recommended next step in the analysis?

6-26. Use the partially complete Minitab output below to answer the following questions.

(a) Find all of the missing values.

(b) Find the estimate of σ^2.

(c) Test for significance of regression. Use $\alpha = 0.05$.

(d) Test for significance of β_1, β_2, and β_3 using a t-test with $\alpha = 0.05$. Comment on these results.

(e) Construct a 95% CI on β_1. Use this CI to test for significance.

(f) Construct a 95% CI on β_2. Use this CI to test for significance.

(g) Construct a 95% CI on β_3. Use this CI to test for significance.

(h) Comment on results found in parts (c)–(g). Is this regression model appropriate? What is your recommended next step in the analysis?

Predictor	Coef	SE Coef	T	P
Constant	3.318	1.007	3.29	0.003
x1	0.7417	0.5768	?	?
x2	9.1142	0.6571	?	?

S = ? R − Sq = ? R − Sq (adj) = 87.6%

Analysis of Variance

Source	DF	SS	MS	F	P
Regression	2	133.366	66.683	?	?
Residual Error	?	17.332	?		
Total	27	150.698			

Predictor	Coef	SE Coef	T	P
Constant	6.188	2.704	2.29	0.027
x1	9.6864	0.4989	?	?
x2	−0.3796	0.2339	?	?
x3	2.9447	0.2354	?	?

S = ? R − Sq = ? R − Sq (adj) = 90.2%

Analysis of Variance

Source	DF	SS	MS	F	P
Regression	3	363.01	121.00	?	?
Residual Error	44	36.62	?		
Total	47	399.63			

6-4 OTHER ASPECTS OF REGRESSION

In this section we briefly present three other aspects of using multiple regression: building models with polynomial terms, categorical or qualitative variables as regressors, and selection of variables for a regression model. For more discussion of these (and other) topics, consult Montgomery and Runger (2006) or Montgomery, Peck, and Vining (2006).

6-4.1 Polynomial Models

In Section 6-1 we observed that models with polynomial terms in the regressors, such as the second-order model

$$Y = \beta_0 + \beta_1 x_1 + \beta_{11} x_1^2 + \epsilon$$

are really linear regression models and can be fit and analyzed using the methods discussed in Section 6-3. Polynomial models arise frequently in engineering and the sciences, and this contributes greatly to the widespread use of linear regression in these fields.

EXAMPLE 6-9

Acetylene Yield

To illustrate fitting a polynomial regression model, consider the data on yield of acetylene and two process variables, reactor temperature, and ratio of H_2 to n-heptane [for more discussion and analysis of this data and references to the original sources, see Montgomery, Peck, and Vining (2006)] shown in Table 6-9. Engineers often consider fitting a second-order model to this type of chemical process data.

Solution. The second-order model in two regressors is

$$Y = \beta_0 + \beta_1 x_1 + \beta_2 x_2 + \beta_{12} x_1 x_2 + \beta_{11} x_1^2 + \beta_{22} x_2^2 + \epsilon$$

When fitting a polynomial model, it is usually a good idea to **center** the regressors (by subtracting the average $\bar{x}_j$) from each observation and to use these centered regressors to obtain the cross-product and squared terms in the model. This reduces the **multicollinearity** in the data and often results in a regression model that is more reliable in the sense that the model coefficients are estimated with better precision. For the acetylene data, this would involve subtracting 1212.5 from each observation on x_1 = temperature and 12.444 from each observation on x_2 = ratio. Therefore, the regression model that we will fit is

Computation Using
Centered Data

$$Y = \beta_0 + \beta_1(T - 1212.5) + \beta_2(R - 12.444)$$
$$+ \beta_{12}(T - 1212.5)(R - 12.444) + \beta_{11}(T - 1212.5)^2 + \beta_{22}(R - 12.444)^2 + \epsilon$$

A portion of the Minitab output for this model is shown below.

The regression equation is

Yield = 36.4 + 0.130 Temp + 0.480 Ratio − 0.00733 T × R + 0.000178 T^2
        − 0.0237 R^2

Predictor	Coef	SE Coef	T	P	VIF
Constant	36.4339	0.5529	65.90	0.000	
Temp	0.130476	0.003642	35.83	0.000	1.1
Ratio	0.48005	0.05860	8.19	0.000	1.5
T × R	−0.0073346	0.0007993	−9.18	0.000	1.4
T^2	0.00017820	0.00005854	3.04	0.012	1.2
R^2	−0.02367	0.01019	−2.32	0.043	1.7

S = 1.066   R-Sq = 99.5%   R-Sq (adj) = 99.2%

Analysis of Variance

Source	DF	SS	MS	F	P
Regression	5	2112.34	422.47	371.49	0.000
Residual Error	10	11.37	1.14		
Total	15	2123.71			

Table 6-9   The Acetylene Data

Observation	Yield, Y	Temp, T	Ratio, R	Observation	Yield, Y	Temp, T	Ratio, R
1	49.0	1300	7.5	9	34.5	1200	11.0
2	50.2	1300	9.0	10	35.0	1200	13.5
3	50.5	1300	11.0	11	38.0	1200	17.0
4	48.5	1300	13.5	12	38.5	1200	23.0
5	47.5	1300	17.0	13	15.0	1100	5.3
6	44.5	1300	23.0	14	17.0	1100	7.5
7	28.0	1200	5.3	15	20.5	1100	11.0
8	31.5	1200	7.5	16	29.5	1100	17.0

Remember that the regression coefficients in this display refer to the *centered* regressors in the model shown above. Note that the ANOVA test for significance of regression suggests that at least some of the variables in the model are important and that the *t*-tests on the individual variables indicates that all of the terms are necessary in the model. The VIFs are all small, so there is no apparent problem with multicollinearity. ∎

Suppose that we wanted to test the contribution of the second-order terms to this model. In other words, what is the value of expanding the model to include the additional terms? The hypotheses that need to be tested are

$$H_0: \beta_{r+1} = \beta_{r+2} = \cdots = \beta_k = 0$$
$$H_1: \text{At least one of the } \beta\text{'s} \neq 0$$

We showed how to test these hypotheses in Section 6-3.2. Recall that the procedure involves considering the quadratic model as the **full model** and then fitting a **reduced model** that in this case would be the first-order model

$$Y = \beta_0 + \beta_1(T - 1212.5) + \beta_2(R - 12.444) + \epsilon$$

Computation of Reduced Model

The Minitab regression output for this reduced model is as follows.

The regression equation is

Yield = 36.1 + 0.134 Temp + 0.351 Ratio

Predictor	Coef	SE Coef	T	P	VIF
Constant	36.1063	0.9060	39.85	0.000	
Temp	0.13396	0.01191	11.25	0.000	1.1
Ratio	0.3511	0.1696	2.07	0.059	1.1

S = 3.624   R-Sq = 92.0%   R-Sq(adj) = 90.7%

Analysis of Variance

Source	DF	SS	MS	F	P
Regression	2	1952.98	976.49	74.35	0.000
Residual Error	13	170.73	13.13		
Total	15	2123.71			

The test statistic for the above hypotheses was originally given in equation 6-56, repeated below for convenience:

$$F_0 = \frac{[SS_E(RM) - SS_E(FM)]/(k - r)}{SS_E(FM)/(n - p)}$$

Interpret Reduced Model Fit

In the test statistic, $SS_E(RM) = 170.73$ is the residual sum of squares for the reduced model, $SS_E(FM) = 11.37$ is the residual sum of squares for the full model, $n = 16$ is the number of observations, $p = 6$ is the number of parameters in the full model, $k = 5$ is the number of regressors in the full model, and $r = 2$ is the number of regressors in the reduced model. Therefore, the computed value of the test statistic is

$$f_0 = \frac{[SS_E(RM) - SS_E(FM)]/(k - r)}{SS_E(FM)/(n - p)} = \frac{(170.73 - 11.37)/(5 - 2)}{11.37/(16 - 6)} = 46.72$$

This value would be compared to $f_{\alpha,3,10}$. Alternatively, the P-value is 3.49×10^{-6}. Because the P-value is very small, we would reject the null hypothesis H_0: $\beta_{12} = \beta_{11} = \beta_{22} = 0$ and conclude that at least one of the second-order terms contributes significantly to the model. Actually, we know from the t-tests in the Minitab output that all three of the second-order terms are important. ■

6-4.2 Categorical Regressors

In the regression models studied previously, all of the regressor variables have been **quantitative** variables; that is, they have either been numerical variables, or they have been measurable on a well-defined scale. Sometimes we encounter qualitative or categorical variables that need to be included in a regression model.

EXAMPLE 6-10

Gas Mileage

Suppose that we are studying gasoline mileage on a fleet of automobiles. The response variable is Y = gas mileage, and two regressors of interest are x_1 = engine displacement (in^3) and x_2 = horsepower. Most of the cars in the fleet have an automatic transmission, but some of them have a manual transmission.

It is easy to incorporate categorical information like this into a regression model. Let x_3 = type of transmission and define

$$x_3 = \begin{cases} 0 \text{ if the car has an automatic transmission} \\ 1 \text{ if the car has a manual transmission} \end{cases}$$

Sometimes a variable with this 0,1 coding is called an **indicator variable.** Build the regression model.

Solution. The regression model for the gas mileage analysis is then

$$Y = \beta_0 + \beta_1 x_1 + \beta_2 x_2 + \beta_3 x_3 + \epsilon$$

This model actually describes two different regression models. When $x_3 = 0$ and the car has an automatic transmission, the model for gas mileage is

$$Y = \beta_0 + \beta_1 x_1 + \beta_2 x_2 + \epsilon$$

but when the car has a manual transmission ($x_3 = 1$), the model is

$$Y = \beta_0 + \beta_1 x_1 + \beta_2 x_2 + \beta_3(1) + \epsilon$$
$$= (\beta_0 + \beta_3) + \beta_1 x_1 + \beta_2 x_2 + \epsilon$$

Notice that the two models have different intercepts, but the model parameters that convey the impacts of engine displacement and horsepower are not affected by the type of transmission in the cars. This might be unreasonable. In fact, we might expect an **interaction** between the regressor variables, engine displacement and type of transmission, and between horsepower and type of transmission.

Use Indicator Variables to Model Interaction

Including interaction terms such as this is easy. The appropriate model is

$$Y = \beta_0 + \beta_1 x_1 + \beta_2 x_2 + \beta_3 x_3 + \beta_{13} x_1 x_3 + \beta_{23} x_2 x_3 + \epsilon$$

Now when the car has an automatic transmission ($x_3 = 0$), the model for gas mileage is

$$Y = \beta_0 + \beta_1 x_1 + \beta_2 x_2 + \epsilon$$

but when the car has a manual transmission ($x_3 = 1$) the model becomes

$$Y = \beta_0 + \beta_1 x_1 + \beta_2 x_2 + \beta_3(1) + \beta_{13} x_1(1) + \beta_{23} x_2(1) + \epsilon$$
$$= (\beta_0 + \beta_3) + (\beta_1 + \beta_{13}) x_1 + (\beta_2 + \beta_{23}) x_2 + \epsilon$$

Thus all three of the model coefficients are affected by whether the car has a manual transmission. This could have a dramatic impact on the shape of the regression function.

EXAMPLE 6-11
Shampoo Data

As an illustration, reconsider the shampoo data in Table 2-10. One of the variables, Region (Eastern, Western), is categorical, and it can be incorporated into the regression model just as we did with the gas mileage problem. If the shampoo is manufactured in the East, we will let Region = 0, and if it is made in the West, we will let Region = 1. Build the regression model.

Solution. The Minitab output for a linear regression model using Foam, Residue, and Region as the regressors follows.

The regression equation is

Quality = 89.8 + 1.82 Foam − 3.38 Residue − 3.41 Region

Predictor	Coef	SE Coef	T	P
Constant	89.806	2.990	30.03	0.000
Foam	1.8192	0.3260	5.58	0.000
Residue	−3.3795	0.6858	−4.93	0.000
Region	−3.4062	0.9194	−3.70	0.001

S = 2.21643   R-Sq = 77.6%   R-Sq (adj) = 74.2%

Analysis of Variance

Source	DF	SS	MS	F	P
Regression	3	339.75	113.25	23.05	0.000
Residual Error	20	98.25	4.91		
Total	23	438.00			

Interpret Output and
Add Interaction

Notice that all three regressors are important. In this model, the effect of the Region regressor is to shift the intercept by an amount equal to −3.41 units when predicting quality for shampoo manufactured in the West.

Potential interaction effects in this data can be investigated by including the interaction terms Foam × Region and Residue × Region in the model. The Minitab output for this model follows.

The regression equation is

Quality = 88.3 + 1.98 Foam − 3.22 Residue − 1.71 Region − 0.642 F × R + 0.43 R × Res

Predictor	Coef	SE Coef	T	P
Constant	88.257	4.840	18.24	0.000
Foam	1.9825	0.4292	4.62	0.000
Residue	−3.2153	0.9525	−3.38	0.003
Region	−1.707	6.572	−0.26	0.798
F × R	−0.6419	0.9434	−0.68	0.505
R × Res	0.430	1.894	0.23	0.823

S = 2.30499   R-Sq = 78.2%   R-Sq (adj) = 72.1%

Analysis of Variance

Source	DF	SS	MS	F	P
Regression	5	342.366	68.473	12.89	0.000
Residual Error	18	95.634	5.313		
Total	23	438.000			

Clearly, the interaction terms are not necessary in this model. ■

Indicator variables can be used when there are more than two levels of a categorical variable. For example, suppose that the shampoo had been produced in three regions, the East, the Midwest, and the West. Two indicator variables (say x_1 and x_2) would be defined as follows:

Region	x_1	x_2
East	0	0
Midwest	1	0
West	0	1

In general, if there are r levels of a categorical variable, we will need $r - 1$ indicator variables to incorporate the categorical variable in the regression model.

6-4.3 Variable Selection Techniques

Many applications of regression involve a data set with a relatively large number of regressors, and we wish to build a model with (perhaps) a smaller number of these regressor variables. Employing a smaller number of regressors will make the model easier to use operationally and could result in a model that is both easier to interpret and produces more reliable predictions than a model containing all of the regressors.

If the number of regressors is not too large, one way to select the subset of regressors for the model is to fit all possible subset models and evaluate these candidate models with respect to appropriate criteria to make the final choice. This sounds awkward, but it is actually very practical and easy to do in many problems. The practical limitation in Minitab is about 20 candidate regressors; the actual limitation will always depend on the specific software used and how the particular package is implemented.

Two criteria that are often used in evaluating a subset regression model are R^2 and the residual mean square MS_E. The objective is to find a model for which R^2 is large and MS_E is small. Now, R^2 cannot decrease as variables are added to the model, so the objective is to find a subset model for which the value of R^2 is nearly as large as the R^2 when all of the variables are in the model. A model with the minimum value of MS_E is desirable because it implies that the model has explained as much of the variability in the response as possible.

The third criterion is based on the standardized total squared estimation error

$$\Gamma_p = \frac{E\left\{ \sum_{i=1}^{n} [\hat{Y}_i - E(Y_i)]^2 \right\}}{\sigma^2} = \frac{E[SS_E(p)]}{\sigma^2} - n + 2p$$

where $\hat{Y}_i$ is the predicted response from the subset model with p parameters, $SS_E(p)$ is the residual sum of squares from this model, and $E(Y_i)$ is the expected response from the "true" model, that is, the model with the correct subset of regressors. Now, the quantities $E[SS_E(p)]$ and σ^2 are unknown, but they can be estimated by the observed value of $SS_E(p)$ and by the estimate $\hat{\sigma}^2$ obtained from the full model, the model containing all of the candidate regressors denoted $\hat{\sigma}^2(FM)$. The criterion then becomes

$$C_p = \frac{SS_E(p)}{\hat{\sigma}^2(FM)} - n + 2p$$

A model with a small value of C_p is considered desirable.

EXAMPLE 6-10 (continued)

To illustrate the "all possible regressions" approach, we will apply the technique to the shampoo data in Table 2-10. Minitab will provide the best subset regression models of size m ($1 \le m \le 10$) for up to 20 candidate regressors, where "best" is the model with maximum R^2 or minimum MS_E. The Minitab output with $m = 5$ is shown in Table 6-10.

In the Minitab output, "S" is the square root of the residual mean square. The model with the smallest value of the residual mean square is the four-variable model containing Foam, Scent, Residue, and Region (with east = 0 and west = 1). This model also has the smallest value of C_p, so assuming that the residual analysis is satisfactory, it would be a good candidate for the best regression equation that describes the relationships in this data set. The answer for the best model with smallest value of the residual mean square wasn't change! ∎

Table 6-10   Minitab Best Subsets Regression for Shampoo Data

Response is Quality

Vars	R-Sq	R-Sq(adj)	Mallows C-p	S	F o a m	S c e n t	C o l o r	R e s i d u e	R e g i o n
1	26.2	22.9	46.4	3.8321	X				
1	25.7	22.3	46.9	3.8455					X
1	23.9	20.5	48.5	3.8915			X		
1	6.3	2.1	64.3	4.3184		X			
1	3.8	0.0	66.7	4.3773			X		
2	62.2	58.6	16.1	2.8088	X		X		
2	50.3	45.6	26.7	3.2185	X				X
2	42.6	37.2	33.6	3.4589				X	X
2	40.9	35.3	35.2	3.5098	X		X		
2	32.6	26.2	42.7	3.7486	X	X			
3	77.6	74.2	4.2	2.2164	X			X	X
3	63.1	57.6	17.2	2.8411	X		X	X	
3	62.5	56.9	17.7	2.8641	X	X		X	
3	52.9	45.9	26.4	3.2107	X		X		X
3	51.8	44.6	27.4	3.2491	X	X			X
4	79.9	75.7	4.1	2.1532	X	X		X	X
4	78.6	74.1	5.3	2.2205	X		X	X	X
4	64.8	57.4	17.7	2.8487	X	X	X	X	
4	53.0	43.1	28.3	3.2907	X	X	X		X
4	51.4	41.2	29.7	3.3460		X	X	X	X
5	80.0	74.5	6.0	2.2056	X	X	X	X	X

Another approach to selecting subset regression models is **stepwise regression.** This is actually a collection of related methods that are designed to work effectively with large data sets. A widely used stepwise procedure is **backward elimination.** This method starts with all of the regressors in the model and successively eliminates them based on the value of the t-test statistics $\hat{\beta}_j/se(\hat{\beta}_j)$. If the smallest absolute value of this t-ratio is less than a cutoff value t_{out}, the regressor associated with this t-ratio is removed from the model. The model is then refit and the backward elimination process continues until no more regressors can be eliminated. Minitab uses a cutoff value for t_{out} with a significance level of 0.1.

The Minitab backward elimination procedure applied to the shampoo data is in Table 6-11. The final model contains Foam, Residue, and Region. Notice that this model is slightly different from the one that we found using all possible regressions.

Another variation of stepwise regression is **forward selection.** This procedure begins with no variables in the model and adds them one at a time. The variable that results in the largest t-statistic value is inserted at each step, as long as the value of the test statistic exceeds the threshold value t_{in}. Minitab uses a significance level of 0.25 to determine the threshold value t_{in}. The procedure terminates when no remaining candidate variables meet the criterion for variable entry.

Table 6-11   Stepwise Regression Backward Elimination for Shampoo Data: Quality versus Foam, Scent, Color, Residue, Region

Backward elimination. Alpha-to-Remove: 0.1

Response is Quality on 5 predictors, with N = 24

Step	1	2	3
Constant	86.33	86.14	89.81
Foam	1.82	1.87	1.82
T-Value	5.07	5.86	5.58
P-Value	0.000	0.000	0.000
Scent	1.03	1.18	
T-Value	1.12	1.48	
P-Value	0.277	0.155	
Color	0.23		
T-Value	0.33		
P-Value	0.746		
Residue	−4.00	−3.93	−3.38
T-Value	−4.93	−5.15	−4.93
P-Value	0.000	0.000	0.000
Region	−3.86	−3.71	−3.41
T-Value	−3.70	−4.05	−3.70
P-Value	0.002	0.001	0.001
S	2.21	2.15	2.22
R-Sq	80.01	79.89	77.57
R-Sq (adj)	74.45	75.65	74.20
Mallows C-p	6.0	4.1	4.2

Table 6-12   Stepwise Regression Forward Selection for Shampoo Data: Quality versus Foam, Scent, Color, Residue, Region

Forward selection. Alpha-to-Enter: 0.25

Response is Quality on 5 predictors, with N = 24

Step	1	2	3	4
Constant	76.00	89.45	89.81	86.14
Foam	1.54	1.90	1.82	1.87
T-Value	2.80	4.61	5.58	5.86
P-Value	0.010	0.000	0.000	0.000
Residue		−3.82	−3.38	−3.93
T-Value		−4.47	−4.93	−5.15
P-Value		0.000	0.000	0.000
Region			−3.41	−3.71
T-Value			−3.70	−4.05
P-Value			0.001	0.001
Scent				1.18
T-Value				1.48
P-Value				0.155
S	3.83	2.81	2.22	2.15
R-Sq	26.24	62.17	77.57	79.89
R-Sq (adj)	22.89	58.57	74.20	75.65
Mallows C-p	46.4	16.1	4.2	4.1

The Minitab stepwise regression output for the shampoo data is given in Table 6-12. The final model contains Foam, Scent, Residue, and Region, and this is the same model found by the all possible regression method.

The most common variation of stepwise regression uses a combination of forward and backward stepping and is usually called stepwise regression. This procedure begins with a forward step, but immediately after inserting a variable, a backward elimination step is conducted to determine if any variables that were added at previous steps can now be removed. Two cutoffs values t_{in} and t_{out} must be selected, and usually we set $t_{in} = t_{out}$. Minitab uses a significance level of 0.15 for both t_{in} and t_{out}.

The Minitab stepwise regression output for the shampoo data is in Table 6-13. The stepwise regression procedure selects, Scent, Residue, and Region as the regressors for the final model. This is the same model found by the backwards elimination procedure.

Most regression analysts consider the all possible regressions approach the best of the available methods because it is possible to implicitly evaluate all of the candidate equations. Consequently, one can be assured of finding the model that minimizes the residual mean square or minimizes C_p. Stepwise methods are myopic because they only change one variable at a time in each successive equation. They are not assured to produce a final equation that optimizes any particular criterion. However, a lot of practical experience with stepwise methods indicates that the equation that results is usually a very good one.

Table 6-13 Stepwise Regression Combined Forward and Backward Elimination: Quality versus Foam, Scent, Color, Residue, Region

Alpha-to-Enter: 0.15 Alpha-to-Remove: 0.15

Response is Quality on 5 predictors, with N = 24

Step	1	2	3
Constant	76.00	89.45	89.81
Foam	1.54	1.90	1.82
T-Value	2.80	4.61	5.58
P-Value	0.010	0.000	0.000
Residue		−3.82	−3.38
T-Value		−4.47	−4.93
P-Value		0.000	0.000
Region			−3.41
T-Value			−3.70
P-Value			0.001
S	3.83	2.81	2.22
R-Sq	26.24	62.17	77.57
R-Sq (adj)	22.89	58.57	74.20
Mallows C-p	46.4	16.1	4.2

EXERCISES FOR SECTION 6-4

For Exercises 6-27 and 6-28, use Minitab to assist you in answering the following.

(a) Fit second-order polynomial models.

(b) Check for multicolliearity in the data for each of the polynomial models. Comment on your results.

(c) Test the contribution of the second-order terms in the models when compared to the reduced first-order model. Comment on your results.

6-27. Consider the bearing wear data and multiple linear regression problem in Exercises 6-15 and 6-19.

6-28. Consider the MPG data and multiple linear regression problem in Exercises 6-16 and 6-20.

For Exercises 6-29 and 6-30, using only first-order terms, build regression models using the following techniques:

(a) All possible regression. Find the C_p and S values.

(b) Forward selection

(c) Backward elimination

(d) Comment on the models obtained. Which model would you prefer?

6-29. Consider the power consumption data in Exercise 6-17.

6-30. Consider the HFE data in Exercise 6-18.

6-31. A mechanical engineer is investigating the surface finish of metal parts produced on a lathe and its relationship to the speed (in revolutions per minute) and type of cutting tool of the lathe. The data are shown in Table 6-14 on the next page.

(a) Build a regression model using indicator variables and comment on the significance of regression.

(b) Build a separate model for each tool type and comment on the significance of regression of each model.

Table 6-14   Surface Finish Data for Exercise 6-31

Observation Number, i	Surface Finish y_i	RPM	Type of Cutting Tool	Observation Number, i	Surface Finish y_i	RPM	Type of Cutting Tool
1	45.44	225	302	11	33.50	224	416
2	42.03	200	302	12	31.23	212	416
3	50.10	250	302	13	37.52	248	416
4	48.75	245	302	14	37.13	260	416
5	47.92	235	302	15	34.70	243	416
6	47.79	237	302	16	33.92	238	416
7	52.26	265	302	17	32.13	224	416
8	50.52	259	302	18	35.47	251	416
9	45.58	221	302	19	33.49	232	416
10	44.78	218	302	20	32.29	216	416

SUPPLEMENTAL EXERCISES

6-32. An industrial engineer at a furniture manufacturing plant wishes to investigate how the plant's electricity usage depends on the amount of the plant's production. He suspects that there is a simple linear relationship between production measured as the value in million-dollar units of furniture produced in that month (x) and the electrical usage in units of kWh (kilowatt-hours, y). The following data were collected:

Dollars	kWh	Dollars	kWh
4.70	2.59	4.01	2.65
4.00	2.61	4.31	2.64
4.59	2.66	4.51	2.38
4.70	2.58	4.46	2.41
4.65	2.32	4.55	2.35
4.19	2.31	4.14	2.55
4.69	2.52	4.25	2.34
3.95	2.32		

(a) Draw a scatter diagram of these data. Does a straight-line relationship seem plausible?
(b) Fit a simple linear regression model to these data.
(c) Test for significance of regression using $\alpha = 0.05$. What is the P-value for this test?
(d) Find a 95% CI estimate on the slope.
(e) Test the hypothesis H_0: $\beta_1 = 0$ versus H_1: $\beta_1 \neq 0$ using $\alpha = 0.05$. What conclusion can you draw about the slope coefficient?
(f) Test the hypothesis H_0: $\beta_0 = 0$ versus H_1: $\beta_0 \neq 0$ using $\alpha = 0.05$. What conclusions can you draw about the best model?

6-33. Show that an equivalent way to define the test for significance of regression in simple linear regression is to base the test on R^2 as follows: To test H_0: $\beta_1 = 0$ versus H_1: $\beta_1 \neq 0$, calculate

$$F_0 = \frac{R^2(n-2)}{1-R^2}$$

and reject H_0: $\beta_1 = 0$ if the computed value $f_0 > f_{\alpha,1,n-2}$.

6-34. Suppose that the simple linear regression model has been fit to $n = 25$ observations and $R^2 = 0.90$.

(a) Test for significance of regression at $\alpha = 0.05$. Use the results of Exercise 6-33.
(b) What is the smallest value of R^2 that would lead to the conclusion of a significant regression if $\alpha = 0.05$?

6-35. **Studentized Residuals.** Show that in a simple linear regression model the variance of the ith residual is

$$V(e_i) = \sigma^2 \left[1 - \left(\frac{1}{n} + \frac{(x_i - \bar{x})^2}{S_{xx}} \right) \right]$$

Hint:

$$\text{cov}(Y_i, \hat{Y}_i) = \sigma^2 \left[\frac{1}{n} + \frac{(x_i - \bar{x})^2}{S_{xx}} \right]$$

The ith studentized residual for this model is defined as

$$r_i = \frac{e_i}{\sqrt{\hat{\sigma}^2 \left[1 - \left(\frac{1}{n} + \frac{(x_i - \bar{x})^2}{S_{xx}} \right) \right]}}$$

(a) Explain why r_i has unit standard deviation (for σ known).
(b) Do the standardized residuals have unit standard deviation?
(c) Discuss the behavior of the studentized residual when the sample value x_i is very close to the middle of the range of x.
(d) Discuss the behavior of the studentized residual when the sample value x_i is very near one end of the range of x.

6-36. The data on the next page are DC output from a windmill (y) and wind velocity (x).

(a) Draw a scatter diagram of these data. What type of relationship seems appropriate in relating y to x?
(b) Fit a simple linear regression model to these data.
(c) Test for significance of regression using $\alpha = 0.05$. What conclusions can you draw?

(d) Plot the residuals from the simple linear regression model versus $\hat{y}_i$ and versus wind velocity x. What do you conclude about model adequacy?

(e) Based on the analysis, propose another model relating y to x. Justify why this model seems reasonable.

(f) Fit the regression model you have proposed in part (e). Test for significance of regression (use $\alpha = 0.05$), and graphically analyze the residuals from this model. What can you conclude about model adequacy?

Observation Number	Wind Velocity (MPH), x_i	DC Output y_i
1	5.00	1.582
2	6.00	1.822
3	3.40	1.057
4	2.70	0.500
5	10.00	2.236
6	9.70	2.386
7	9.55	2.294
8	3.05	0.558
9	8.15	2.166
10	6.20	1.866
11	2.90	0.653
12	6.35	1.930
13	4.60	1.562
14	5.80	1.737
15	7.40	2.088
16	3.60	1.137
17	7.85	2.179
18	8.80	2.112
19	7.00	1.800
20	5.45	1.501
21	9.10	2.303
22	10.20	2.310
23	4.10	1.194
24	3.95	1.144
25	2.45	0.123

6-37. The h_{ii} are often used to denote **leverage**—that is, a point that is unusual in its location in the x-space and that may be influential. Generally, the ith point is called a **leverage point** if h_{ii} exceeds $2p/n$, which is twice the average size of all the hat diagonals. Recall that $p = k + 1$.

(a) Table 6-8 contains the hat matrix diagonal for the wire bond pull strength data used in Example 6-7. Find the average size of these elements.

(b) Based on the criterion given, are there any observations that are leverage points in the data set?

6-38. The data shown in Table 6-15 on the next page represent the thrust of a jet-turbine engine (y) and six candidate regressors: $x_1 = $ primary speed of rotation, $x_2 = $ secondary speed of rotation, $x_3 = $ fuel flow rate, $x_4 = $ pressure, $x_5 = $ exhaust temperature, and $x_6 = $ ambient temperature at time of test.

(a) Fit a multiple linear regression model using $x_3 = $ fuel flow rate, $x_4 = $ pressure, and $x_5 = $ exhaust temperature as the regressors.

(b) Test for significance of regression using $\alpha = 0.01$. Find the P-value for this test. What are your conclusions?

(c) Find the t-test statistic for each regressor. Using $\alpha = 0.01$, explain carefully the conclusion you can draw from these statistics.

(d) Find R^2 and the adjusted statistic for this model. Comment on the meaning of each value and its usefulness in assessing the model.

(e) Construct a normal probability plot of the residuals and interpret this graph.

(f) Plot the residuals versus $\hat{y}$. Are there any indications of inequality of variance or nonlinearity?

(g) Plot the residuals versus x_3. Is there any indication of nonlinearity?

(h) Predict the thrust for an engine for which $x_3 = 20000$, $x_4 = 170$, and $x_5 = 1589$.

6-39. Consider the engine thrust data in Exercise 6-38. Refit the model using $y^* = \ln y$ as the response variable and $x_3^* = \ln x_3$ as the regressor (along with x_4 and x_5).

(a) Test for significance of regression using $\alpha = 0.01$. Find the P-value for this test and state your conclusions.

(b) Use the t-statistic to test $H_0: \beta_j = 0$ versus $H_1: \beta_j \neq 0$ for each variable in the model. If $\alpha = 0.01$, what conclusions can you draw?

(c) Plot the residuals versus $\hat{y}^*$ and versus x_3^*. Comment on these plots. How do they compare with their counterparts obtained in Exercise 6-38 parts (f) and (g)?

6-40. Following are data on $y = $ green liquor (g/l) and $x = $ paper machine speed (ft/min) from a Kraft paper machine. (The data were read from a graph in an article in the *Tappi Journal*, March 1986.)

y	16.0	15.8	15.6	15.5	14.8
x	1700	1720	1730	1740	1750

y	14.0	13.5	13.0	12.0	11.0
x	1760	1770	1780	1790	1795

(a) Fit the model $Y = \beta_0 + \beta_1 x + \beta_2 x^2 + \epsilon$ using least squares.

(b) Test for significance of regression using $\alpha = 0.05$. What are your conclusions?

(c) Test the contribution of the quadratic term to the model, over the contribution of the linear term, using a t-test. If $\alpha = 0.05$, what conclusion can you draw?

(d) Plot the residuals from the model in part (a) versus $\hat{y}$. Does the plot reveal any inadequacies?

(e) Construct a normal probability plot of the residuals. Comment on the normality assumption.

Table 6-15   Jet-Turbine Engine Data for Exercise 6-38

Observation Number	y	x_1	x_2	x_3	x_4	x_5	x_6
1	4540	2140	20640	30250	205	1732	99
2	4315	2016	20280	30010	195	1697	100
3	4095	1905	19860	29780	184	1662	97
4	3650	1675	18980	29330	164	1598	97
5	3200	1474	18100	28960	144	1541	97
6	4833	2239	20740	30083	216	1709	87
7	4617	2120	20305	29831	206	1669	87
8	4340	1990	19961	29604	196	1640	87
9	3820	1702	18916	29088	171	1572	85
10	3368	1487	18012	28675	149	1522	85
11	4445	2107	20520	30120	195	1740	101
12	4188	1973	20130	29920	190	1711	100
13	3981	1864	19780	29720	180	1682	100
14	3622	1674	19020	29370	161	1630	100
15	3125	1440	18030	28940	139	1572	101
16	4560	2165	20680	30160	208	1704	98
17	4340	2048	20340	29960	199	1679	96
18	4115	1916	19860	29710	187	1642	94
19	3630	1658	18950	29250	164	1576	94
20	3210	1489	18700	28890	145	1528	94
21	4330	2062	20500	30190	193	1748	101
22	4119	1929	20050	29960	183	1713	100
23	3891	1815	19680	29770	173	1684	100
24	3467	1595	18890	29360	153	1624	99
25	3045	1400	17870	28960	134	1569	100
26	4411	2047	20540	30160	193	1746	99
27	4203	1935	20160	29940	184	1714	99
28	3968	1807	19750	29760	173	1679	99
29	3531	1591	18890	29350	153	1621	99
30	3074	1388	17870	28910	133	1561	99
31	4350	2071	20460	30180	198	1729	102
32	4128	1944	20010	29940	186	1692	101
33	3940	1831	19640	29750	178	1667	101
34	3480	1612	18710	29360	156	1609	101
35	3064	1410	17780	28900	136	1552	101
36	4402	2066	20520	30170	197	1758	100
37	4180	1954	20150	29950	188	1729	99
38	3973	1835	19750	29740	178	1690	99
39	3530	1616	18850	29320	156	1616	99
40	3080	1407	17910	28910	137	1569	100

6-41. An article in the *Journal of Environmental Engineering* (Vol. 115, No. 3, 1989, pp. 608–619) reported the results of a study on the occurrence of sodium and chloride in surface streams in central Rhode Island. The data shown are chloride concentration y (in mg/1) and roadway area in the watershed x (in %).

y	4.4	6.6	9.7	10.6	10.8
x	0.19	0.15	0.57	0.70	0.67

y	10.9	11.8	12.1	14.3	14.7
x	0.63	0.47	0.70	0.60	0.78

y	15.0	17.3	19.2	23.1	27.4
x	0.81	0.78	0.69	1.30	1.05

y	27.7	31.8	39.5
x	1.06	1.74	1.62

(a) Draw a scatter diagram of the data. Does a simple linear regression model seem appropriate here?

(b) Fit the simple linear regression model using the method of least squares.

(c) Estimate the mean chloride concentration for a watershed that has 1% roadway area.

(d) Find the fitted value corresponding to $x = 0.47$ and the associated residual.

(e) Suppose we wish to fit a regression model for which the true regression line passes through the point (0, 0). The appropriate model is $Y = \beta x + \epsilon$. Assume that we have n pairs of data $(x_1, y_1), (x_2, y_2), \ldots, (x_n, y_n)$. Show that the least squares estimate of β is $\Sigma\, y_i x_i / \Sigma\, x_i^2$.

(f) Use the results of part (e) to fit the model $Y = \beta x + \epsilon$ to the chloride concentration–roadway area data in this exercise. Plot the fitted model on a scatter diagram of the data and comment on the appropriateness of the model.

6-42. Consider the no-intercept model $Y = \beta x + \epsilon$ with the ϵ's NID$(0, \sigma^2)$. The estimate of σ^2 is $s^2 = \Sigma_{i=1}^{n}(y_i - \hat{\beta}x_i)^2/(n-1)$ and $V(\hat{\beta}) = \sigma^2/\Sigma_{i=1}^{n}x_i^2$.

(a) Devise a test statistic for H_0: $\beta = 0$ versus H_1: $\beta \neq 0$.

(b) Apply the test in part (a) to the model from Exercise 6-41 part (f).

6-43. A rocket motor is manufactured by bonding together two types of propellants, an igniter and a sustainer. The shear strength of the bond y is thought to be a linear function of the age of the propellant x when the motor is cast. Twenty observations are shown in the table on the right.

(a) Draw a scatter diagram of the data. Does the straight-line regression model seem to be plausible?

(b) Find the least squares estimates of the slope and intercept in the simple linear regression model.

(c) Estimate the mean shear strength of a motor made from propellant that is 20 weeks old.

(d) Obtain the fitted values $\hat{y}_i$ that correspond to each observed value y_i. Plot $\hat{y}_i$ versus y_i, and comment on what this plot would look like if the linear relationship between shear strength and age were perfectly deterministic (no error). Does this plot indicate that age is a reasonable choice of regressor variable in this model?

Observation Number	Strength y (psi)	Age x (weeks)
1	2158.70	15.50
2	1678.15	23.75
3	2316.00	8.00
4	2061.30	17.00
5	2207.50	5.00
6	1708.30	19.00
7	1784.70	24.00
8	2575.00	2.50
9	2357.90	7.50
10	2277.70	11.00
11	2165.20	13.00
12	2399.55	3.75
13	1779.80	25.00
14	2336.75	9.75
15	1765.30	22.00
16	2053.50	18.00
17	2414.40	6.00
18	2200.50	12.50
19	2654.20	2.00
20	1753.70	21.50

6-44. Consider the simple linear regression model $Y = \beta_0 + \beta_1 x + \epsilon$. Suppose that the analyst wants to use $z = x - \bar{x}$ as the regressor variable.

(a) Using the data in Exercise 6-43, construct one scatter plot of the (x_i, y_i) points and then another of the $(z_i = x_i - \bar{x}, y_i)$ points. Use the two plots to explain intuitively how the two models, $Y = \beta_0 + \beta_1 x + \epsilon$ and $Y = \beta_0^* + \beta_1^* z + \epsilon$, are related.

(b) Find the least squares estimates of β_0^* and β_1^* in the model $Y = \beta_0^* + \beta_1^* z + \epsilon$. How do they relate to the least squares estimates $\hat{\beta}_0$ and $\hat{\beta}_1$?

6-45. Suppose that each value of x_i is multiplied by a positive constant a, and each value of y_i is multiplied by another positive constant b. Show that the t-statistic for testing H_0: $\beta_1 = 0$ versus H_1: $\beta_1 \neq 0$ is unchanged in value.

6-46. Test the hypothesis H_0: $\beta_1 = 10$ versus H_1: $\beta_1 \neq 10$ (using $\alpha = 0.01$) for the steam usage data in Exercise 6-2 using a new test statistic. (*Hint:* In the usual hypothesis test, H_0: $\beta_1 = 0.0$ versus $\beta_1 \neq 0.0$. In this exercise, the hypothesized value is 10.0.) Also, find the P-value for this test.

6-47. Consider the engine data as transformed in Exercise 6-39. Using only first-order terms, build regression models using the following techniques:

(a) All possible regressions. Find the C_p and S values.

(b) Forward selection.

(c) Backward elimination.

(d) Comment on the models obtained. Which model would you prefer?

6-48. Consider the engine data of Exercise 6-38. Using only first-order terms, build regression models using the following techniques:

(a) All possible regressions. Find the C_p and S values.
(b) Forward selection.
(c) Backward elimination.
(d) Comment on the models obtained. Which model would you prefer?

6-49. An article in *Electronic Packaging and Production* (Vol. 42, 2002) reported the effect of X-ray inspection of integrated circuits. The rads (radiation dose) were studied as a function of current (in milliamps) and exposure time (in minutes).

Rads	mAmps	Exposure Time
7.4	10	0.25
14.8	10	0.5
29.6	10	1
59.2	10	2
88.8	10	3
296	10	10
444	10	15
592	10	20
11.1	15	0.25
22.2	15	0.5
44.4	15	1
88.8	15	2
133.2	15	3
444	15	10
666	15	15
888	15	20
14.8	20	0.25
29.6	20	0.5
59.2	20	1
118.4	20	2
177.6	20	3
592	20	10
888	20	15
1184	20	20
22.2	30	0.25
44.4	30	0.5
88.8	30	1
177.6	30	2
266.4	30	3
888	30	10
1332	30	15
1776	30	20

Rads	mAmps	Exposure Time
29.6	40	0.25
59.2	40	0.5
118.4	40	1
236.8	40	2
355.2	40	3
1184	40	10
1776	40	15
2368	40	20

(a) Fit a multiple linear regression model to these data with rads as the response.
(b) Estimate σ^2 and standard errors of the regression coefficients.
(c) Test for significance of β_1 and β_2. Use $\alpha = 0.05$.
(d) Use the model to predict rads when the current is 15 milliamps and the exposure time is 5 seconds. Construct a 90% PI.
(e) Use the model to compute the mean response rads when the current is 15 milliamps and the exposure time is 5 seconds. Construct a 95% PI.
(f) Interpret parts (d) and (e) and comment on the comparison between the 95% PI and 95% CI.

6-50. An article in *Cancer Epidemiology, Biomarkers and Prevention* (Vol. 5, 1996, pp. 849–852) describes a pilot study to assess the use of toenail arsenic concentrations as an indicator of ingestion of arsenic-containing water. Twenty-one participants were interviewed regarding use of their private (unregulated) wells for drinking and cooking, and each provided a sample of water and toenail clippings. The table below showed the data of age (years), sex of person (1 = male, 2 = female), proportion of times household well was used for drinking (1 ≤ 1/4, 2 = 1/4, 3 = 1/2, 4 = 3/4, 5 ≥ 3/4), proportion of times household well was used for cooking (1 ≤ 1/4, 2 = 1/4, 3 = 1/2, 4 = 3/4, 5 ≥ 3/4), arsenic in water (ppm), and arsenic in toenails (ppm) respectively.

Age	Sex	Drink Use	Cook Use	Arsenic Water	Arsenic Nails
44	2	5	5	0.00087	0.119
45	2	4	5	0.00021	0.118
44	1	5	5	0	0.099
66	2	3	5	0.00115	0.118
37	1	2	5	0	0.277
45	2	5	5	0	0.358
47	1	5	5	0.00013	0.08
38	2	4	5	0.00069	0.158
41	2	3	2	0.00039	0.31
49	2	4	5	0	0.105
72	2	5	5	0	0.073

Age	Sex	Drink Use	Cook Use	Arsenic Water	Arsenic Nails
45	2	1	5	0.046	0.832
53	1	5	5	0.0194	0.517
86	2	5	5	0.137	2.252
8	2	5	5	0.0214	0.851
32	2	5	5	0.0175	0.269
44	1	5	5	0.0764	0.433
63	2	5	5	0	0.141
42	1	5	5	0.0165	0.275
62	1	5	5	0.00012	0.135
36	1	5	5	0.0041	0.175

(a) Fit a multiple linear regression model using arsenic concentration in nails as the response and age, drink use, cook use, and arsenic in the water as the regressors.
(b) Estimate σ^2 and the standard errors of the regression coefficients.
(c) Use the model to predict the arsenic in nails when the age is 30, the drink use is category 5, the cook use is category 5, and arsenic in the water is 0.135 ppm.

TEAM EXERCISE

6-51.   Identify a situation in which two or more variables of interest may be related but the mechanistic model relating the variables is unknown. Collect a random sample of data for these variables and perform the following analyses.

(a) Build a simple or multiple linear regression model. Comment on your results.
(b) Test for significance of the regression model. Comment on your results.
(c) Test for significance of the individual regression coefficients. Comment on your results.
(d) Construct CIs on the individual regression coefficients. Comment on your results.
(e) Select a value for the regressor variables and construct a CI on the mean response. Comment on your results.
(f) Select two values for the regressor variables and use the model to make predictions. Construct prediction intervals for these values. Comment on your results.
(g) Perform a residual analysis and compute the coefficient of multiple determination. Comment on your results.

IMPORTANT TERMS AND CONCEPTS

Adjusted R^2
All possible regressions
Analysis of variance (ANOVA)
Backward elimination
Coefficient of determination, R^2
Confidence interval on mean response
Confidence interval on regression coefficients
Contour plot
Cook's distance measure, D_i

C_p statistic
Empirical model
Forward selection
Indicator variables
Influential observations
Interaction
Intercept
Least squares normal equations
Mechanistic model
Method of least squares
Model
Model adequacy
Multicollinearity
Multiple regression

Outliers
Polynomial regression
Population correlation coefficient, ρ
Prediction interval
Regression analysis
Regression coefficients
Regression model
Regression sum of squares
Regressor variable
Residual analysis
Residual sum of squares
Residuals
Response variable

Sample correlation coefficient, r
Significance of regression
Simple linear regression
Standard errors of model coefficients
Standardized residuals
Stepwise regression
Studentized residuals
t-tests on regression coefficients
Unbiased estimators
Variance inflation factor

7 Design of Engineering Experiments

CHAPTER OUTLINE

LEARNING OBJECTIVES

After careful study of this chapter, you should be able to do the following:

1. Design and conduct engineering experiments involving several factors using the factorial design approach.

2. Know how to analyze and interpret main effects and interactions.

3. Understand how the ANOVA is used to analyze the data from these experiments.

4. Assess model adequacy with residual plots.

5. Know how to use the two-level series of factorial designs.

6. Understand how two-level factorial designs can be run in blocks.

7. Design and conduct two-level fractional factorial designs.

7-1 THE STRATEGY OF EXPERIMENTATION

Recall from Chapter 1 that engineers conduct tests or **experiments** as a natural part of their work. Statistically based experimental design techniques are particularly useful in the engineering world for improving the performance of a manufacturing process. They also have extensive application in the development of new processes. Most processes can be described in terms of several **controllable variables,** such as temperature, pressure, and feed rate. By using designed experiments, engineers can determine which subset of the process variables have the most influence on process performance. The results of such an experiment can lead to

1. Improved process yield
2. Reduced variability in the process and closer conformance to nominal or target requirements
3. Reduced design and development time
4. Reduced cost of operation

Experimental design methods are also useful in **engineering design** activities, where new products are developed and existing ones improved. Some typical applications of statistically designed experiments in engineering design include

1. Evaluation and comparison of basic design configurations
2. Evaluation of different materials
3. Selection of design parameters so that the product will work well under a wide variety of field conditions (or so that the design will be robust)
4. Determination of key product design parameters that affect product performance

The use of experimental design in the engineering design process can result in products that are easier to manufacture, have better field performance and reliability than their competitors, and can be designed, developed, and produced in less time.

Designed experiments are usually employed **sequentially,** hence the term **sequential experimentation.** That is, the first experiment with a complex system (perhaps a manufacturing process) that has many controllable variables is often a **screening experiment** designed to determine which variables are most important. Subsequent experiments are then used to refine this information and determine which adjustments to these critical variables are required to improve the process. Finally, the objective of the experimenter is **optimization**—that is, to determine which levels of the critical variables result in the best process performance. This is the KISS principle: "keep it small and sequential." When small steps are completed, the knowledge gained can improve the subsequent experiments.

Every experiment involves a sequence of activities:

1. **Conjecture**—the original hypothesis that motivates the experiment.
2. **Experiment**—the test performed to investigate the conjecture.
3. **Analysis**—the statistical analysis of the data from the experiment.
4. **Conclusion**—what has been learned about the original conjecture from the experiment will lead to a revised conjecture, a new experiment, and so forth.

Statistical methods are essential to good experimentation. All experiments are designed experiments; some of them are poorly designed, and as a result, valuable resources are used ineffectively. Statistically designed experiments permit efficiency and economy in the experimental process, and the use of statistical methods in examining the data results in **scientific objectivity** when drawing conclusions.

In this chapter we focus on experiments that include two or more factors that the experimenter thinks may be important. The **factorial experimental design** will be introduced as a powerful technique for this type of problem. Generally, in a factorial experimental design, experimental trials (or runs) are performed at all combinations of factor levels. For example, if a chemical engineer is interested in investigating the effects of reaction time and reaction temperature on the yield of a process, and if two levels of time (1 and 1.5 hr) and two levels of temperature (125 and 150°F) are considered important, a factorial experiment would consist of making the experimental runs at each of the four possible combinations of these levels of reaction time and reaction temperature.

Most of the statistical concepts introduced previously can be extended to the factorial experiments of this chapter. We will also introduce several graphical methods that are useful in analyzing the data from designed experiments.

7-2 FACTORIAL EXPERIMENTS

When several factors are of interest in an experiment, a factorial experiment should be used. As noted previously, in these experiments factors are varied together.

> By a **factorial experiment** we mean that in each complete replicate of the experiment all possible combinations of the levels of the factors are investigated.

Thus, if there are two factors A and B with a levels of factor A and b levels of factor B, each replicate contains all ab treatment combinations.

The effect of a factor is defined as the change in response produced by a change in the level of the factor. It is called a **main effect** because it refers to the primary factors in the study. For example, consider the data in Table 7-1. This is a factorial experiment with two factors, A and B, each at two levels (A_{low}, A_{high}, and B_{low}, B_{high}). The main effect of factor A is the difference between the average response at the high level of A and the average response at the low level of A, or

$$A = \frac{30 + 40}{2} - \frac{10 + 20}{2} = 20$$

That is, changing factor A from the low level to the high level causes an average response increase of 20 units. Similarly, the main effect of B is

$$B = \frac{20 + 40}{2} - \frac{10 + 30}{2} = 10$$

In some experiments, the difference in response between the levels of one factor is not the same at all levels of the other factors. When this occurs, there is an **interaction** between the factors. For example, consider the data in Table 7-2. At the low level of factor B, the A effect is

$$A = 30 - 10 = 20$$

Table 7-1   A Factorial Experiment without Interaction

Factor A	Factor B	
	B_{low}	B_{high}
A_{low}	10	20
A_{high}	30	40

Table 7-2   A Factorial Experiment with Interaction

Factor A	Factor B	
	B_{low}	B_{high}
A_{low}	10	20
A_{high}	30	0

and at the high level of factor B, the A effect is

$$A = 0 - 20 = -20$$

Because the effect of A depends on the level chosen for factor B, there is interaction between A and B.

When an interaction is large, the corresponding main effects have very little practical meaning. For example, by using the data in Table 7-2, we find the main effect of A as

$$A = \frac{30 + 0}{2} - \frac{10 + 20}{2} = 0$$

and we would be tempted to conclude that there is no factor A effect. However, when we examined the effects of A at *different levels of factor B*, we saw that this was not the case. The effect of factor A depends on the levels of factor B. Thus, knowledge of the AB interaction is more useful than knowledge of the main effect. A significant interaction can mask the significance of main effects. Consequently, when interaction is present, the main effects of the factors involved in the interaction may not have much meaning.

It is easy to estimate the interaction effect in factorial experiments such as those illustrated in Tables 7-1 and 7-2. In this type of experiment, when both factors have two levels, the AB interaction effect is the difference in the diagonal averages. This represents one-half the difference between the A effects at the two levels of B. For example, in Table 7-1, we find the AB interaction effect to be

$$AB = \frac{20 + 30}{2} - \frac{10 + 40}{2} = 0$$

Thus, there is no interaction between A and B. In Table 7-2, the AB interaction effect is

$$AB = \frac{20 + 30}{2} - \frac{10 + 0}{2} = 20$$

As we noted before, the interaction effect in these data is very large.

The concept of interaction can be illustrated graphically in several ways. Figure 7-1 is a plot of the data in Table 7-1 against the levels of A for both levels of B. Note that the B_{low} and B_{high} lines are approximately parallel, indicating that factors A and B do not interact significantly. Figure 7-2 presents a similar plot for the data in Table 7-2. In this graph, the B_{low} and B_{high} lines are not parallel, indicating the interaction between factors A and B. Such graphical displays are called **two-factor interaction plots.** They are often useful in presenting the results of experiments, and many computer software programs used for analyzing data from designed experiments will construct these graphs automatically.

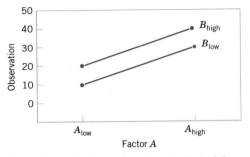

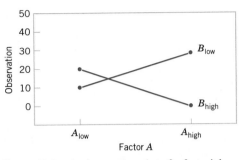

Figure 7-1 An interaction plot of a factorial experiment, no interaction.

Figure 7-2 An interaction plot of a factorial experiment, with interaction.

Figures 7-3 and 7-4 present another graphical illustration of the data from Tables 7-1 and 7-2. In Fig. 7-3 we have shown a **three-dimensional surface plot** of the data from Table 7-1, where the low and high levels are set at -1 and 1, respectively, for both A and B. The equations for these surfaces are discussed later in the chapter. These data contain no interaction, and the surface plot is a plane lying above the A–B space. The slope of the plane in the A and B directions is proportional to the main effects of factors A and B, respectively. Figure 7-4 is a surface plot for the data from Table 7-2. Note that the effect of the interaction in these data is to "twist" the plane so that there is curvature in the response function. **Factorial experiments are the only way to discover interactions between variables.**

An alternative to the factorial design that is (unfortunately) used in practice is to change the factors *one at a time* rather than to vary them simultaneously. To illustrate this one-factor-at-a-time procedure, suppose that we are interested in finding the values of temperature and pressure that maximize the yield of a chemical process. Suppose that we fix temperature at 155° F (the current operating level) and perform five runs at different levels of time—say, 0.5, 1.0, 1.5, 2.0, and 2.5 hours. The results of this series of runs are shown in Fig. 7-5. This figure indicates that maximum yield is achieved at about 1.7 hours of reaction time. To optimize temperature, the engineer then fixes time at 1.7 hours (the apparent optimum) and performs five runs at different temperatures—say, 140, 150, 160, 170, and 180°F. The results of this set of runs are plotted in Fig. 7-6. Maximum yield occurs at about 155°F. Therefore, we would conclude that running the process at 155°F and 1.7 hours is the best set of operating conditions, resulting in yields of around 75%.

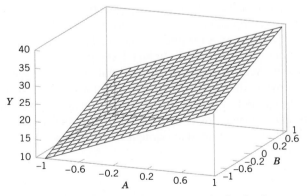

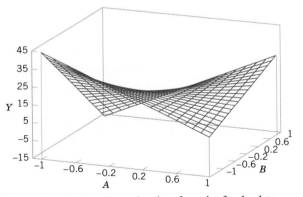

Figure 7-3 Three-dimensional surface plot for the data from Table 7-1, showing main effects of the two factors A and B.

Figure 7-4 Three-dimensional surface plot for the data from Table 7-2, showing the effect of the A and B interaction.

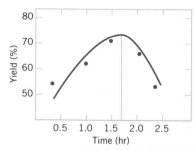

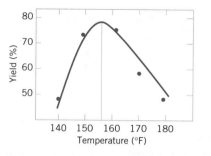

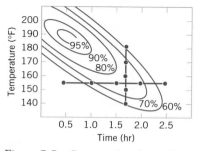

Figure 7-5   Yield versus reaction time with temperature constant at 155°F.

Figure 7-6   Yield versus temperature with reaction time constant at 1.7 hours.

Figure 7-7   Contour plot of a yield function and an optimization experiment using the one-factor-at-a-time method.

Figure 7-7 displays the **contour plot** of yield as a function of temperature and time with the one-factor-at-a-time experiments superimposed on the contours. Clearly, this one-factor-at-a-time approach has failed dramatically here, because the true optimum is at least 20 yield points higher and occurs at much lower reaction times and higher temperatures. The failure to discover the importance of the shorter reaction times is particularly important because this could have significant impact on production volume or capacity, production planning, manufacturing cost, and total productivity.

The one-factor-at-a-time approach has failed here because it cannot detect the interaction between temperature and time. Factorial experiments are the only way to detect interactions. Furthermore, the one-factor-at-a-time method is inefficient. It will require more experimentation than a factorial, and as we have just seen, there is no assurance that it will produce the correct results.

7-3 2^k FACTORIAL DESIGN

Factorial designs are frequently used in experiments involving several factors where it is necessary to study the joint effect of the factors on a response. However, several special cases of the general factorial design are important because they are widely employed in research work and because they form the basis of other designs of considerable practical value.

The most important of these special cases is that of k factors, each at only two levels. These levels may be quantitative, such as two values of temperature, pressure, or time; or they may be qualitative, such as two machines, two operators, the "high" and "low" levels of a factor, or perhaps the presence and absence of a factor. A complete replicate of such a design requires $2 \times 2 \times \cdots \times 2 = 2^k$ observations and is called a 2^k **factorial design.**

The 2^k design is particularly useful in the early stages of experimental work, when many factors are likely to be investigated. It provides the smallest number of runs for which k factors can be studied in a complete factorial design. Because there are only two levels for each factor, we must assume that the response is approximately linear over the range of the factor levels chosen. The 2^k design is a basic building block that is used to begin the study of a system.

7-3.1 2^2 Example

The simplest type of 2^k design is the 2^2—that is, two factors A and B, each at two levels. We usually think of these levels as the low and high levels of the factor. The 2^2 design is shown in

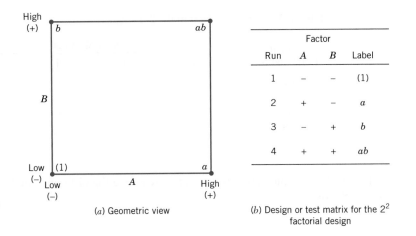

High (+) b ab

B

Low (−)   (1) a
Low (−) A High (+)

Factor			
Run	A	B	Label
1	−	−	(1)
2	+	−	a
3	−	+	b
4	+	+	ab

Figure 7-8 The 2^2 factorial design.

(a) Geometric view

(b) Design or test matrix for the 2^2 factorial design

Fig. 7-8. Note that the design can be represented geometrically as a square with the $2^2 = 4$ runs, or treatment combinations, forming the corners of the square (Fig. 7-8a). In the 2^2 design it is customary to denote the low and high levels of the factors A and B by the signs − and +, respectively. This is sometimes called the **geometric notation** for the design. Figure 7-8b shows the test, or design, matrix for the 2^2 design. Each row of the matrix is a run in the design and the −,+ signs in each row identify the factor settings for that run.

A special notation is used to label the treatment combinations. In general, a treatment combination is represented by a series of lowercase letters. If a letter is present, the corresponding factor is run at the high level in that treatment combination; if it is absent, the factor is run at its low level. For example, treatment combination a indicates that factor A is at the high level and factor B is at the low level. The treatment combination with both factors at the low level is represented by (1). This notation is used throughout the 2^k design series. For example, the treatment combination in a 2^4 with A and C at the high level and B and D at the low level is denoted by ac.

The effects of interest in the 2^2 design are the main effects A and B and the two-factor interaction AB. Let the letters (1), a, b, and ab also represent the totals of all n observations taken at each of these design points. It is easy to estimate the effects of these factors. To estimate the main effect of A, we would average the observations on the right side of the square in Fig. 7-8a where A is at the high level, and subtract from this the average of the observations on the left side of the square, where A is at the low level, or

Main Effect of A

$$A = \bar{y}_{A+} - \bar{y}_{A-} = \frac{a + ab}{2n} - \frac{b + (1)}{2n} = \frac{1}{2n}[a + ab - b - (1)] \qquad (7\text{-}1)$$

Similarly, the main effect of B is found by averaging the observations on the top of the square, where B is at the high level, and subtracting the average of the observations on the bottom of the square, where B is at the low level:

Main Effect of B

$$B = \bar{y}_{B+} - \bar{y}_{B-} = \frac{b + ab}{2n} - \frac{a + (1)}{2n} = \frac{1}{2n}[b + ab - a - (1)] \qquad (7\text{-}2)$$

Table 7-3   Signs for Effects in the 2^2 Design

Treatment Combination	Factorial Effect			
	I	*A*	*B*	*AB*
(1)	+	−	−	+
a	+	+	−	−
b	+	−	+	−
ab	+	+	+	+

Finally, the *AB* interaction is estimated by taking the difference in the diagonal averages in Fig. 7-8*a*, or

AB Interaction Effect

$$AB = \frac{ab + (1)}{2n} - \frac{a + b}{2n} = \frac{1}{2n}\left[ab + (1) - a - b\right] \tag{7-3}$$

The quantities in brackets in equations 7-1, 7-2, and 7-3 are called **contrasts.** For example, the *A* contrast is

$$\text{Contrast}_A = a + ab - b - (1)$$

In these equations, the contrast coefficients are always either $+1$ or -1. A table of plus and minus signs, such as Table 7-3, can be used to determine the sign on each treatment combination for a particular contrast. The column headings for Table 7-3 are the main effects *A* and *B*, the *AB* interaction, and *I*, which represents the total. The row headings are the treatment combinations. Note that the signs in the *AB* column are the product of signs from columns *A* and *B*. To generate a contrast from this table, multiply the signs in the appropriate column by the treatment combinations listed in the rows and add. For example, $\text{contrast}_{AB} = [(1)] + [-a] + [-b] + [ab] = ab + (1) - a - b$.

Contrasts are used in calculating both the effect estimates and the sums of squares for *A*, *B*, and the *AB* interaction. The sums of squares formulas are

$$SS_A = \frac{[a + ab - b - (1)]^2}{4n}$$

$$SS_B = \frac{[b + ab - a - (1)]^2}{4n} \tag{7-4}$$

$$SS_{AB} = \frac{[ab + (1) - a - b]^2}{4n}$$

The analysis of variance is completed by computing the total sum of squares SS_T (with $4n - 1$ degrees of freedom) as usual and obtaining the error sum of squares SS_E [with $4(n - 1)$ degrees of freedom] by subtraction.

Table 7-4   The 2^2 Design for the Epitaxial Process Experiment

Treatment Combination	Factorial Effect			Thickness (μm)				Thickness (mm)		
	A	B	AB					Total	Average	Variance
(1)	−	−	+	14.037	14.165	13.972	13.907	56.081	14.020	0.0121
a	+	−	−	14.821	14.757	14.843	14.878	59.299	14.825	0.0026
b	−	+	−	13.880	13.860	14.032	13.914	55.686	13.922	0.0059
ab	+	+	+	14.888	14.921	14.415	14.932	59.156	14.789	0.0625

EXAMPLE 7-1

Epitaxial Process

An article in the *AT&T Technical Journal* (Vol. 65, March/April 1986, pp. 39–50) describes the application of two-level factorial designs to integrated circuit manufacturing. A basic processing step in this industry is to grow an epitaxial layer on polished silicon wafers. The wafers are mounted on a susceptor and positioned inside a bell jar. Chemical vapors are introduced through nozzles near the top of the jar. The susceptor is rotated, and heat is applied. These conditions are maintained until the epitaxial layer is thick enough.

Table 7-4 presents the results of a 2^2 factorial design with $n = 4$ replicates using the factors A = deposition time and B = arsenic flow rate. The two levels of deposition time are − = short and + = long, and the two levels of arsenic flow rate are − = 55% and + = 59%. The response variable is epitaxial layer thickness (μm). Find the estimate of the effects and assess the importance of the effects.

Solution. We find the estimates of the effects using equations 7-1, 7-2, and 7-3 as follows:

$$A = \frac{1}{2n}[a + ab - b - (1)]$$

$$= \frac{1}{2(4)}[59.299 + 59.156 - 55.686 - 56.081] = 0.836$$

$$B = \frac{1}{2n}[b + ab - a - (1)]$$

$$= \frac{1}{2(4)}[55.686 + 59.156 - 59.299 - 56.081] = -0.067$$

Estimate the Effects

$$AB = \frac{1}{2n}[ab + (1) - a - b]$$

$$= \frac{1}{2(4)}[59.156 + 56.081 - 59.299 - 55.686] = 0.032$$

The numerical estimates of the effects indicate that the effect of deposition time is large and has a positive direction (increasing deposition time increases thickness), because changing deposition time from low to high changes the mean epitaxial layer thickness by 0.836 μm. The effects of arsenic flow rate (B) and the AB interaction appear small.

The importance of these effects may be confirmed with the analysis of variance. The sums of squares for A, B, and AB are computed as follows:

$$SS_A = \frac{[a + ab - b - (1)]^2}{16} = \frac{6.688^2}{16} = 2.7956$$

Compute the Sum of Squares

$$SS_B = \frac{[b + ab - a - (1)]^2}{16} = \frac{-0.538^2}{16} = 0.0181$$

$$SS_{AB} = \frac{[ab + (1) - a - b]^2}{16} = \frac{0.252^2}{16} = 0.0040$$

$$SS_T = 14.037^2 + \cdots + 14.932^2 - \frac{(56.081 + \cdots + 59.156)^2}{16}$$

$$= 3.0672$$

Table 7-5   Analysis of Variance for the Epitaxial Process Experiment

Source of Variation	Sum of Squares	Degree of Freedom	Mean Square	f_0	P-Value
A (deposition time)	2.7956	1	2.7956	134.40	7.07 E-8
B (arsenic flow)	0.0181	1	0.0181	0.87	0.38
AB	0.0040	1	0.0040	0.19	0.67
Error	0.2495	12	0.0208		
Total	3.0672	15			

The analysis of variance is summarized in Table 7-5 and confirms our conclusions obtained by examining the magnitude and direction of the effects. Deposition time is the only factor that significantly affects epitaxial layer thickness, and from the direction of the effect estimates we know that longer deposition times lead to thicker epitaxial layers.   ■

7-3.2   Statistical Analysis

We present two related methods for determining which effects are significantly different from zero. In the first method, the magnitude of an effect is compared to its estimated standard error. In the second method, a regression model is used in which each effect is associated with a regression coefficient. Then the regression results developed in Chapter 6 can be used to conduct the analysis. The two methods produce identical results for two-level designs. One might choose the method that is easiest to interpret or the one that is used by the available computer software. A third method that uses normal probability plots is discussed later in this chapter.

Standard Errors of the Effects

The magnitude of the effects in Example 7-1 can be judged by comparing each effect to its estimated standard error. In a 2^k design with n replicates, there are a total of $N = n2^k$ measurements. An effect estimate is the difference between two means, and each mean is calculated from half the measurements. Consequently, the variance of an effect estimate is

$$V(\text{Effect}) = \frac{\sigma^2}{N/2} + \frac{\sigma^2}{N/2} = \frac{2\sigma^2}{N/2} = \frac{\sigma^2}{n2^{k-2}} \tag{7-5}$$

To obtain the estimated standard error of an effect, replace σ^2 by an estimate $\hat{\sigma}^2$ and take the square root of equation 7-5.

If there are n replicates at each of the 2^k runs in the design, and if $y_{i1}, y_{i2}, \ldots, y_{in}$ are the observations at the ith run,

$$\hat{\sigma}_i^2 = \frac{\sum_{j=1}^{n} (y_{ij} - \bar{y}_{i.})^2}{(n-1)} \qquad i = 1, 2, \ldots, 2^k$$

is an estimate of the variance at the ith run. The 2^k variance estimates can be pooled (averaged) to give an overall variance estimate

$$\hat{\sigma}^2 = \sum_{i=1}^{2^k} \frac{\hat{\sigma}_i^2}{2^k} \tag{7-6}$$

Table 7-6 t-Tests of the Effects for Example 7-1 Epitaxial Process Experiment

Effect	Effect Estimate	Estimated Standard Error	t Ratio	P-Value	Effect $\pm$ Two Estimated Standard Errors
A	0.836	0.072	11.61	0.00	0.836 ± 0.144
B	-0.067	0.072	-0.93	0.38	-0.067 ± 0.144
AB	0.032	0.072	0.44	0.67	0.032 ± 0.144

Degrees of freedom $= 2^k(n-1) = 2^2(4-1) = 12$.

Each $\hat{\sigma}_i^2$ is associated with $n-1$ degrees of freedom, and so $\hat{\sigma}^2$ is associated with $2^k(n-1)$ degrees of freedom.

EXAMPLE 7-1 (continued)

Epitaxial Process

To illustrate this approach for the epitaxial process experiment, we find that

$$\hat{\sigma}^2 = \frac{0.0121 + 0.0026 + 0.0059 + 0.0625}{4} = 0.0208$$

Compute and Use the Estimated Standard Error of Each Effect

and the estimated standard error of each effect is

$$se(\text{Effect}) = \sqrt{[\hat{\sigma}^2/(n2^{k-2})]} = \sqrt{[0.0208/(4 \cdot 2^{2-2})]} = 0.072$$

In Table 7-6, each effect is divided by this estimated standard error and the resulting t ratio is compared to a t distribution with $2^2 \cdot 3 = 12$ degrees of freedom. Recall that the t ratio is used to judge whether the effect is significantly different from zero. The significant effects are the important ones in the experiment. Two standard error limits on the effect estimates are also shown in Table 7-6. These intervals are approximate 95% CIs.

The magnitude and direction of the effects were examined previously, and the analysis in Table 7-6 confirms those earlier, tentative conclusions. Deposition time is the only factor that significantly affects epitaxial layer thickness, and from the direction of the effect estimates we know that longer deposition times lead to thicker epitaxial layers.

Regression Analysis

In any designed experiment, it is important to examine a **model** for predicting responses. Furthermore, there is a close relationship between the analysis of a designed experiment and a **regression model** that can be used easily to obtain predictions from a 2^k experiment.

For the epitaxial process experiment, an initial regression model is

$$Y = \beta_0 + \beta_1 x_1 + \beta_2 x_2 + \beta_{12} x_1 x_2 + \epsilon$$

Build the Regression Model

The deposition time and arsenic flow are represented by coded variables x_1 and x_2, respectively. The low and high levels of deposition time are assigned values $x_1 = -1$ and $x_1 = +1$, respectively, and the low and high levels of arsenic flow are assigned values $x_2 = -1$ and $x_2 = +1$, respectively. The cross-product term $x_1 x_2$ represents the effect of the interaction between these variables.

The least squares fitted model is

$$\hat{y} = 14.389 + \left(\frac{0.836}{2}\right) x_1 + \left(\frac{-0.067}{2}\right) x_2 + \left(\frac{0.032}{2}\right) x_1 x_2$$

where the intercept $\hat{\beta}_0$ is the grand average of all 16 observations. The estimated coefficient of x_1 is one-half the effect estimate for deposition time. The estimated coefficient is one-half the effect estimate because regression coefficients measure the effect of a unit change in x_1 on the mean of Y, and the effect estimate is based on a two-unit change from -1 to $+1$. Similarly, the estimated coefficient of x_2 is one-half of the effect of arsenic flow, and the estimated coefficient of the cross-product term is one-half of the interaction effect.

Test for Significance of
Regression Coefficients

The regression analysis is shown in Table 7-7. Note that mean square error equals the estimate of σ^2 calculated previously. Because the P-value associated with the F-test for the model in the analysis of variance (or ANOVA) portion of the display is small (less than 0.05), we conclude that one or more of the effects are important. The t-test for the hypothesis H_0: $\beta_i = 0$ versus H_1: $\beta_i \neq 0$ (for each coefficient β_1, β_2, and β_3 in the regression analysis) is identical to the one computed from the standard error of the effects in Table 7-6. Consequently, the results in Table 7-7 can be interpreted as t-tests of regression coefficients. Because each estimated regression coefficient is one-half of the effect estimate, the standard errors in Table 7-7 are one-half of those in Table 7-6. The t-test from a regression analysis is identical to the t-test obtained from the standard error of an effect in a 2^k design whenever the estimate $\hat{\sigma}^2$ is the same in both analyses.

Similar to a regression analysis, a simpler model that uses only the important effects is the preferred choice to predict the response. Because the t-tests for the main effect of B and the AB interaction effect are not significant, these terms are removed from the model. The model then becomes

$$\hat{y} = 14.389 + \left(\frac{0.836}{2}\right) x_1$$

That is, the estimated regression coefficient for any effect is the same, regardless of the model considered. Although this is not true in general for a regression analysis, an estimated regression coefficient does not depend on the model in a factorial experiment. Consequently, it is easy to assess model changes when data are collected in one of these experiments. One may also revise the estimate of σ^2 by using the mean square error obtained from the ANOVA table for the simpler model.

These analysis methods are summarized as follows.

Formulas for Two-Level Factorial Experiments with k Factors Each at Two Levels and N Total Trials

$$\text{Coefficient} = \frac{\text{effect}}{2}$$

$$se(\text{Effect}) = \sqrt{\frac{2\hat{\sigma}^2}{N/2}} = \sqrt{\frac{\hat{\sigma}^2}{n2^{k-2}}}$$

$$se(\text{Coefficient}) = \frac{1}{2}\sqrt{\frac{2\hat{\sigma}^2}{N/2}} = \frac{1}{2}\sqrt{\frac{\hat{\sigma}^2}{n2^{k-2}}} \tag{7-7}$$

$$t\ \text{ratio} = \frac{\text{effect}}{se(\text{effect})} = \frac{\text{coefficient}}{se(\text{coefficient})}$$

$$\hat{\sigma}^2 = \text{mean square error}$$

$$2^k(n-1) = \text{residual degrees of freedom}$$

Table 7-7 Regression Analysis for Example 7-1. The regression equation is
Thickness $= 14.4 + 0.418x_1 - 0.0336x_2 + 0.0158x_1x_2$

		Analysis of Variance			
Source	Sum of Squares	Degrees of Freedom	Mean Square	f_0	P-Value
Model	2.81764	3	0.93921	45.18	0.000
Error	0.24948	12	0.02079		
Total	3.06712	15			

Independent Variable	Coefficient Estimate	Standard Error of Coefficient	t for H_0 Coefficient $= 0$	P-Value
Intercept	14.3889	0.0360	399.17	0.000
A or X_1	0.41800	0.03605	11.60	0.000
B or X_2	-0.03363	0.03605	-0.93	0.369
AB or X_1X_2	0.01575	0.03605	0.44	0.670

7-3.3 Residual Analysis and Model Checking

The analysis of a 2^k design assumes that the observations are normally and independently distributed with the same variance in each treatment or factor level. These assumptions should be checked by examining the residuals. Residuals are calculated the same as in regression analysis. A **residual** is the difference between an observation y and its estimated (or fitted) value from the statistical model being studied, denoted as $\hat{y}$. Each residual is

$$e = y - \hat{y}$$

The normality assumption can be checked by constructing a normal probability plot of the residuals. To check the assumption of equal variances at each factor level, plot the residuals against the factor levels and compare the spread in the residuals. It is also useful to plot the residuals against $\hat{y}$; the variability in the residuals should not depend in any way on the value of $\hat{y}$. When a pattern appears in these plots, it usually suggests the need for transformation—that is, analyzing the data in a different metric. For example, if the variability in the residuals increases with $\hat{y}$, a transformation such as log y or $\sqrt{y}$ should be considered. In some problems, the dependence of residual scatter on the fitted value $\hat{y}$ is very important information. It may be desirable to select the factor level that results in maximum response; however, this level may also cause more variation in response from run to run.

The independence assumption can be checked by plotting the residuals against the time or run order in which the experiment was performed. A pattern in this plot, such as sequences of positive and negative residuals, may indicate that the observations are not independent. This suggests that time or run order is important or that variables that change over time are important and have not been included in the experimental design. This phenomenon should be studied in a new experiment. It is easy to obtain residuals from a 2^k design by fitting a regression model to the data.

EXAMPLE 7-1
(continued)
Epitaxial Process

For the epitaxial process experiment in Example 7-1, the regression model is

$$\hat{y} = 14.389 + \left(\frac{0.836}{2}\right)x_1$$

because the only active variable is deposition time.

This model can be used to obtain the predicted values at the four points that form the corners of the square in the design. For example, consider the point with low deposition time ($x_1 = -1$) and low arsenic flow rate. The predicted value is

$$\hat{y} = 14.389 + \left(\frac{0.836}{2}\right)(-1) = 13.971 \ \mu m$$

Analyze the Residuals of the Regression Model

and the residuals are

$$e_1 = 14.037 - 13.971 = 0.066 \qquad e_3 = 13.972 - 13.971 = 0.001$$
$$e_2 = 14.165 - 13.971 = 0.194 \qquad e_4 = 13.907 - 13.971 = -0.064$$

It is easy to verify that the remaining predicted values and residuals are, for low deposition time ($x_1 = -1$) and high arsenic flow rate, $\hat{y} = 14.389 + (0.836/2)(-1) = 13.971 \ \mu m$

$$e_5 = 13.880 - 13.971 = -0.091 \qquad e_7 = 14.032 - 13.971 = 0.061$$
$$e_6 = 13.860 - 13.971 = -0.111 \qquad e_8 = 13.914 - 13.971 = -0.057$$

for high deposition time ($x_1 = +1$) and low arsenic flow rate, $\hat{y} = 14.389 + (0.836/2)(+1) = 14.807 \ \mu m$

$$e_9 = 14.821 - 14.807 = 0.014 \qquad e_{11} = 14.843 - 14.807 = 0.036$$
$$e_{10} = 14.757 - 14.807 = -0.050 \qquad e_{12} = 14.878 - 14.807 = 0.071$$

and for high deposition time ($x_1 = +1$) and high arsenic flow rate, $\hat{y} = 14.389 + (0.836/2)(+1) = 14.807 \ \mu m$

$$e_{13} = 14.888 - 14.807 = 0.081 \qquad e_{15} = 14.415 - 14.807 = -0.392$$
$$e_{14} = 14.921 - 14.807 = 0.114 \qquad e_{16} = 14.932 - 14.807 = 0.125$$

A normal probability plot of these residuals is shown in Fig. 7-9. This plot indicates that one residual $e_{15} = -0.392$ is an outlier. Examining the four runs with high deposition time and high arsenic flow rate reveals that observation $y_{15} = 14.415$ is considerably smaller than the other three observations at that

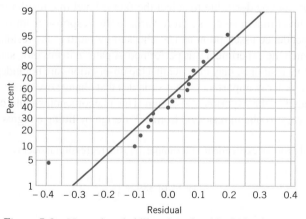

Figure 7-9   Normal probability plot of residuals for the epitaxial process experiment.

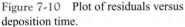

Figure 7-10   Plot of residuals versus deposition time.

Figure 7-11   Plot of residuals versus arsenic flow rate.

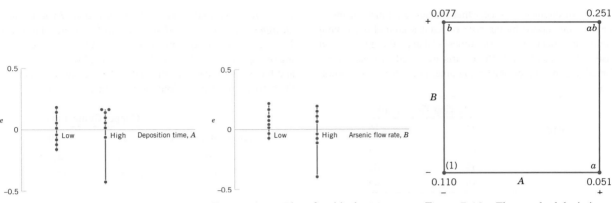

Figure 7-12   The standard deviation of epitaxial layer thickness at the four runs in the 2^2 design.

Interpret the Residual Plots

treatment combination. This adds some additional evidence to the tentative conclusion that observation 15 is an outlier. Another possibility is that some process variables affect the *variability* in epitaxial layer thickness. If we could discover which variables produce this effect, we could perhaps adjust these variables to levels that would minimize the variability in epitaxial layer thickness. This could have important implications in subsequent manufacturing stages. Figures 7-10 and 7-11 are plots of residuals versus deposition time and arsenic flow rate, respectively. Apart from that unusually large residual associated with y_{15}, there is no strong evidence that either deposition time or arsenic flow rate influences the variability in epitaxial layer thickness.

Figure 7-12 shows the standard deviation of epitaxial layer thickness at all four runs in the 2^2 design. These standard deviations were calculated using the data in Table 7-4. Note that the standard deviation of the four observations with A and B at the high level is considerably larger than the standard deviations at any of the other three design points. Most of this difference is attributable to the unusually low thickness measurement associated with y_{15}. The standard deviation of the four observations with A and B at the low level is also somewhat larger than the standard deviations at the remaining two runs. This could indicate that other process variables not included in this experiment may affect the variability in epitaxial layer thickness. Another experiment to study this possibility, involving other process variables, could be designed and conducted. (The original paper in the *AT&T Technical Journal* shows that two additional factors, not considered in this example, affect process variability.) ■

EXERCISES FOR SECTION 7-3

For each of the following designs in Exercises 7-1 through 7-8, answer the following questions.

(a) Compute the estimates of the effects and their standard errors for this design.

(b) Construct two-factor interaction plots and comment on the interaction of the factors.

(c) Use the *t* ratio to determine the significance of each effect with $\alpha = 0.05$. Comment on your findings.

(d) Compute an approximate 95% CI for each effect. Compare your results with those in part (c) and comment.

(e) Perform an analysis of variance of the appropriate regression model for this design. Include in your analysis hypothesis tests for each coefficient, as well as residual analysis. State your final conclusions about the adequacy of the model. Compare your results to part (c) and comment.

7-1. An experiment involves a storage battery used in the launching mechanism of a shoulder-fired ground-to-air missile. Two material types can be used to make the battery plates. The objective is to design a battery that is relatively unaffected by the ambient temperature. The output response from the battery is effective life in hours. Two temperature levels are selected, and a factorial experiment with four replicates is run. The data are as follows.

Material	Temperature (°F)			
	Low		High	
1	130	155	20	70
	74	180	82	58
2	138	110	96	104
	168	160	82	60

7-2. An engineer suspects that the surface finish of metal parts is influenced by the type of paint used and the drying time. She selects two drying times—20 and 30 minutes—and uses two types of paint. Three parts are tested with each combination of paint type and drying time. The data are as follows:

Paint	Drying Time (min)	
	20	30
1	74	78
	64	85
	50	92
2	92	66
	86	45
	68	85

7-3. An experiment was designed to identify a better ultra-filtration membrane for separating proteins and peptide drugs from fermentation broth. Two levels of an additive PVP (% wt) and time duration (hours) were investigated to determine the better membrane. The separation values (measured in %) resulting from these experimental runs are as follows:

PVP (% wt)	Time (hours)	
	1	3
2	69.6	80.0
	71.5	81.6
	70.0	83.0
	69.0	84.3
5	91.0	92.3
	93.2	93.4
	93.0	88.5
	87.2	95.6

7-4. An experiment was conducted to determine whether either firing temperature or furnace position affects the baked density of a carbon anode. The data are as follows:

Position	Temperature (°C)	
	800	825
1	570	1063
	565	1080
	583	1043
2	528	988
	547	1026
	521	1004

7-5. Johnson and Leone (*Statistics and Experimental Design in Engineering and the Physical Sciences*, John Wiley, 1977) describe an experiment conducted to investigate warping of copper plates. The two factors studied were temperature and the copper content of the plates. The response variable is the amount of warping. Some of the data are as follows:

Temperature (°C)	Copper Content (%)	
	40	80
50	17, 20	24, 22
100	16, 12	25, 23

7-6. An article in the *Journal of Testing and Evaluation* (Vol. 16, No. 6, 1988, pp. 508–515) investigated the effects of cyclic loading frequency and environment conditions on fatigue crack growth at a constant 22 MPa stress for a particular material. Some of the data from the experiment are shown here. The response variable is fatigue crack growth rate.

		Environment	
		H_2O	Salt H_2O
Frequency	10	2.06	1.90
		2.05	1.93
		2.23	1.75
		2.03	2.06
	1	3.20	3.10
		3.18	3.24
		3.96	3.98
		3.64	3.24

7-7. A article in the *IEEE Transactions on Electron Devices* (Vol. ED-33, 1986, p. 1754) describes a study on the effects of two variables—polysilicon doping and anneal conditions (time and temperature)—on the base current of a bipolar transistor. Some of the data from this experiment are as follows:

		Anneal (temperature/time)	
		900/180	1000/15
Polysilicon Doping	1×10^{20}	8.30	10.29
		8.90	10.30
	2×10^{20}	7.81	10.19
		7.75	10.10

7-8. An article in the *IEEE Transactions on Semiconductor Manufacturing* (Vol. 5, No. 3, 1992, pp. 214–222) describes an experiment to investigate the surface charge on a silicon wafer. The factors thought to influence induced surface charge are cleaning method (spin rinse dry, or SRD, and spin dry, or SD) and the position on the wafer where the charge was measured. The surface charge ($\times 10^{11}$ q/cm^3) response data are as shown.

	Test Position	
	L	**R**
SD	1.66	1.84
	1.90	1.84
	1.92	1.62
SRD	−4.21	−7.58
	−1.35	−2.20
	−2.08	−5.36

(left table labeled **Cleaning Method**)

7-9. Consider the analysis of a 2^2 designed experiment with 3 replicates. Use the partially complete Minitab output below to answer the following questions.
(a) Find all of the missing values in the analysis of variance table.
(b) Use this table to test for significance of the effects. Use $\alpha = 0.05$.
(c) Compute the estimated standard error of each effect.
(d) Find all of the missing values in the t-tests of the effects table below. Indicate which effects are significant. Use $\alpha = 0.05$.
(e) Write the least squares fitted model using only the significant terms.
(f) Use the model to predict the response when $x_1 = -1$ and $x_2 = 1$.

Source	DF	SS	MS	F	P
A	1	7.84083	7.84083	348.48	?
B	1	0.80083	?	35.59	?
Interaction	1	1.14083	1.14083	?	0.000
Error	8	0.18000	0.02250		
Total	11	9.96250			

Term	Effect	Coef	SE Coef	T	P
Constant		7.9750	0.04330	184.17	0.000
A	1.6167	0.8083	?	18.67	?
B	0.5167	?	0.04330	?	0.000
A*B	−0.6167	−0.3083	?	−7.12	?

7-10. Consider the analysis of a 2^2 designed experiment with 5 replicates. Use the partially complete Minitab output below to answer the following questions.
(a) Find all of the missing values in the analysis of variance table.
(b) Use this table to test for significance of the effects. Use $\alpha = 0.05$.
(c) Compute the estimated standard error of each effect.
(d) Find all of the missing values in the t-tests of the effects table below. Indicate which effects are significant. Use $\alpha = 0.05$.
(e) Write the least squares fitted model using only the significant terms.
(f) Use the model to predict the response when $x_1 = 1$ and $x_2 = 1$.

Analysis of Variance

Source	DF	SS	MS	F	P
Regression	?	3.3095	1.1032	183.86	?
Residual Error	16	0.0960	?		
Total	19	3.4055			

Predictor	Coef	SE Coef	T	P
Constant	10.3650	0.0173	598.42	0.000
A	0.40500	?	23.38	0.000
B	0.01500	0.01732	0.87	?
A*B	0.03500	0.01732	?	0.060

7-4 2^k DESIGN FOR $k \geq 3$ FACTORS

The methods presented in the previous section for factorial designs with $k = 2$ factors each at two levels can be easily extended to more than two factors. For example, consider $k = 3$ factors, each at two levels. This design is a 2^3 factorial design, and it has eight runs or treatment combinations. Geometrically, the design is a cube as shown in Fig. 7-13a, with the eight runs forming the corners of the cube. The test matrix or design matrix is shown in Fig. 7-13b. This design allows three main effects to be estimated (A, B, and C) along with three two-factor interactions (AB, AC, and BC) and a three-factor interaction (ABC).

The main effects can easily be estimated. Remember that the lowercase letters (1), a, b, ab, c, ac, bc, and abc represent the total of all n replicates at each of the eight runs in the design. As seen in Fig. 7-14a, note that the main effect of A can be estimated by averaging the four treatment combinations on the right-hand side of the cube, where A is at the high

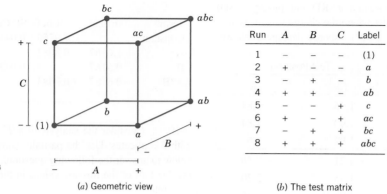

Run	A	B	C	Label
1	−	−	−	(1)
2	+	−	−	a
3	−	+	−	b
4	+	+	−	ab
5	−	−	+	c
6	+	−	+	ac
7	−	+	+	bc
8	+	+	+	abc

Figure 7-13 The 2^3 design.

(a) Geometric view   (b) The test matrix

level, and by subtracting from this quantity the average of the four treatment combinations on the left-hand side of the cube, where A is at the low level. This gives

$$A = \bar{y}_{A+} - \bar{y}_{A-}$$
$$= \frac{a + ab + ac + abc}{4n} - \frac{(1) + b + c + bc}{4n}$$

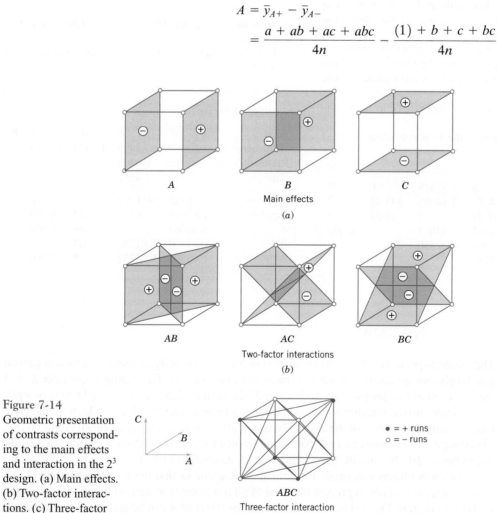

Main effects

(a)

Two-factor interactions

(b)

● = + runs
○ = − runs

ABC

Three-factor interaction

(c)

Figure 7-14
Geometric presentation of contrasts corresponding to the main effects and interaction in the 2^3 design. (a) Main effects. (b) Two-factor interactions. (c) Three-factor interaction.

This equation can be rearranged as

$$A = \bar{y}_{A+} - \bar{y}_{A-} = \frac{1}{4n}[a + ab + ac + abc - (1) - b - c - bc] \qquad (7\text{-}8)$$

In a similar manner, the effect of B is the difference in averages between the four treatment combinations in the back face of the cube (Fig. 7-14a) and the four in the front. This yields

$$B = \bar{y}_{B+} - \bar{y}_{B-} = \frac{1}{4n}[b + ab + bc + abc - (1) - a - c - ac] \qquad (7\text{-}9)$$

The effect of C is the difference in average response between the four treatment combinations in the top face of the cube (Fig. 7-14a) and the four in the bottom; that is,

$$C = \bar{y}_{C+} - \bar{y}_{C-} = \frac{1}{4n}[c + ac + bc + abc - (1) - a - b - ab] \qquad (7\text{-}10)$$

The two-factor interaction effects may be computed easily. A measure of the AB interaction is the difference between the average A effects at the two levels of B. By convention, one-half of this difference is called the AB interaction. Symbolically,

B	Average A Effect
High (+)	$\dfrac{[(abc - bc) + (ab - b)]}{2n}$
Low (−)	$\dfrac{\{(ac - c) + [a - (1)]\}}{2n}$
Difference	$\dfrac{[abc - bc + ab - b - ac + c - a + (1)]}{2n}$

Because the AB interaction is one-half of this difference,

$$AB = \frac{1}{4n}[abc - bc + ab - b - ac + c - a + (1)] \qquad (7\text{-}11)$$

We could write equation 7-11 as follows:

$$AB = \frac{abc + ab + c + (1)}{4n} - \frac{bc + b + ac + a}{4n}$$

In this form, the AB interaction is easily seen to be the difference in averages between runs on two diagonal planes in the cube in Fig. 7-14b. Using similar logic and referring to Fig. 7-14b, we find that the AC and BC interactions are

$$AC = \frac{1}{4n}\left[(1) - a + b - ab - c + ac - bc + abc\right] \qquad (7\text{-}12)$$

$$BC = \frac{1}{4n}\left[(1) + a - b - ab - c - ac + bc + abc\right] \qquad (7\text{-}13)$$

The ABC interaction is defined as the average difference between the AB interactions for the two different levels of C. Thus,

$$ABC = \frac{1}{4n}\left\{[abc - bc] - [ac - c] - [ab - b] + [a - (1)]\right\}$$

or

$$ABC = \frac{1}{4n}\left[abc - bc - ac + c - ab + b + a - (1)\right] \qquad (7\text{-}14)$$

As before, we can think of the ABC interaction as the difference in two averages. If the runs in the two averages are isolated, they define the vertices of the two tetrahedra that comprise the cube in Fig. 7-14c.

In equations 7-8 through 7-14, the quantities in brackets are contrasts in the treatment combinations. A table of plus and minus signs can be developed from the contrasts and is shown in Table 7-8. Signs for the main effects are determined by associating a plus with the high level and a minus with the low level. Once the signs for the main effects have been established, the signs for the remaining columns can be obtained by multiplying the appropriate preceding columns, row by row. For example, the signs in the AB column are the products of the A and B column signs in each row. The contrast for any effect can easily be obtained from this table.

Table 7-8 has several interesting properties:

1. Except for the identity column I, each column has an equal number of plus and minus signs.

2. The sum of products of signs in any two columns is zero; that is, the columns in the table are **orthogonal.**

3. Multiplying any column by column I leaves the column unchanged; that is, I is an **identity element.**

Table 7-8 Algebraic Signs for Calculating Effects in the 2^3 Design

Treatment Combination	Factorial Effect							
	I	A	B	AB	C	AC	BC	ABC
(1)	+	−	−	+	−	+	+	−
a	+	+	−	−	−	−	+	+
b	+	−	+	−	−	+	−	+
ab	+	+	+	+	−	−	−	−
c	+	−	−	+	+	−	−	+
ac	+	+	−	−	+	+	−	−
bc	+	−	+	−	+	−	+	−
abc	+	+	+	+	+	+	+	+

4. The product of any two columns yields a column in the table, for example $A \times B = AB$, and $AB \times ABC = A^2B^2C = C$, because any column multiplied by itself is the identity column.

The estimate of any main effect or interaction in a 2^k design is determined by multiplying the treatment combinations in the first column of the table by the signs in the corresponding main effect or interaction column, adding the result to produce a contrast, and then dividing the contrast by one-half the total number of runs in the experiment.

EXAMPLE 7-2
Surface
Roughness

A mechanical engineer is studying the surface roughness of a part produced in a metal cutting operation. A 2^3 factorial design in the factors feed rate (A), depth of cut (B), and tool angle (C), with $n = 2$ replicates, is run. The levels for the three factors are: low A = 20 in./min, high A = 30 in./min; low B = 0.025 in., high B = 0.040 in.; low C = 15°, high C = 25°. Table 7-9 presents the observed surface roughness data. Estimate the effects and construct an approximate 95% confidence interval on each.

Solution. The main effects may be estimated using equations 7-8 through 7-14. The effect of A, for example, is

$$A = \frac{1}{4n}[a + ab + ac + abc - (1) - b - c - bc]$$

Compute Effects

$$= \frac{1}{4(2)}[22 + 27 + 23 + 30 - 16 - 20 - 21 - 18]$$

$$= \frac{1}{8}[27] = 3.375$$

Table 7-9 Surface Roughness Data for Example 7-2

Treatment Combinations	Design Factors			Surface Roughness	Total	Average	Variance
	A	B	C				
(1)	−1	−1	−1	9, 7	16	8	2.0
a	1	−1	−1	10, 12	22	11	2.0
b	−1	1	−1	9, 11	20	10	2.0
ab	1	1	−1	12, 15	27	13.5	4.5
c	−1	−1	1	11, 10	21	10.5	0.5
ac	1	−1	1	10, 13	23	11.5	4.5
bc	−1	1	1	10, 8	18	9	2.0
abc	1	1	1.	16, 14	30	15	2.0
Average						11.0625	2.4375

It is easy to verify that the other effects are

$$
\begin{array}{ll}
B = 1.625 & C = 0.875 \\
AB = 1.375 & AC = 0.125 \\
BC = -0.625 & ABC = 1.125
\end{array}
$$

Examining the magnitude of the effects clearly shows that feed rate (factor A) is dominant, followed by depth of cut (B) and the AB interaction, although the interaction effect is relatively small.

For the surface roughness experiment, we find from pooling the variances at each of the eight treatments as in equation 7-6 that $\hat{\sigma}^2 = 2.4375$ and the estimated standard error of each effect is

$$
se(\text{effect}) = \sqrt{\frac{\hat{\sigma}^2}{n2^{k-2}}} = \sqrt{\frac{2.4375}{2 \cdot 2^{3-2}}} = 0.78
$$

Therefore, two standard error limits on the effect estimates are

<table>
<tr><td>*Compute and Interpret*</td><td>A:</td><td>3.375 ± 1.56</td><td>B:</td><td>1.625 ± 1.56</td></tr>
<tr><td>*Approximate*</td><td>C:</td><td>0.875 ± 1.56</td><td>AB:</td><td>1.375 ± 1.56</td></tr>
<tr><td>*Confidence Intervals on*</td><td>AC:</td><td>0.125 ± 1.56</td><td>BC:</td><td>-0.625 ± 1.56</td></tr>
<tr><td>*Effects*</td><td>ABC:</td><td>1.125 ± 1.56</td><td></td><td></td></tr>
</table>

These intervals are approximate 95% confidence intervals. They indicate that the two main effects A and B are important, but that the other effects are not, because the intervals for all effects except A and B include zero.

This CI approach is a nice method of analysis. With relatively simple modifications, it can be used in situations where only a few of the design points have been replicated. Normal probability plots can also be used to judge the significance of effects. We will illustrate that method in the next section.

Regression Model and Residual Analysis

We may obtain the residuals from a 2^k design by using the method demonstrated earlier for the 2^2 design. As an example, consider the surface roughness experiment. The three largest effects are A, B, and the AB interaction. The regression model used to obtain the predicted values is

$$
Y = \beta_0 + \beta_1 x_1 + \beta_2 x_2 + \beta_{12} x_1 x_2 + \epsilon
$$

where x_1 represents factor A, x_2 represents factor B, and $x_1 x_2$ represents the AB interaction.

EXAMPLE 7-2 (continued)

Surface Roughness

The regression coefficients β_1, β_2, and β_{12} are estimated by one-half the corresponding effect estimates, and β_0 is the grand average. Thus

$$
\hat{y} = 11.0625 + \left(\frac{3.375}{2}\right)x_1 + \left(\frac{1.625}{2}\right)x_2 + \left(\frac{1.375}{2}\right)x_1 x_2
$$

and the predicted values would be obtained by substituting the low and high levels of A and B into this equation. To illustrate this, at the treatment combination where A, B, and C are all at the low level, the predicted value is

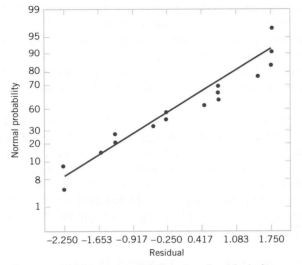

Figure 7-15   Normal probability plot of residuals from the surface roughness experiment.

Compute Predicted Values and Residuals

$$\hat{y} = 11.0625 + \left(\frac{3.375}{2}\right)(-1) + \left(\frac{1.625}{2}\right)(-1) + \left(\frac{1.375}{2}\right)(-1)(-1)$$

$$= 9.25$$

Because the observed values at this run are 9 and 7, the residuals are $9 - 9.25 = -0.25$ and $7 - 9.25 = -2.25$. Residuals for the other 14 runs are obtained similarly.

A normal probability plot of the residuals is shown in Fig. 7-15. Because the residuals lie approximately along a straight line, we do not suspect any problem with normality in the data. There are no indications of severe outliers. It would also be helpful to plot the residuals versus the predicted values and against each of the factors A, B, and C.

Projection of a 2^k Design

Any 2^k design will collapse or project into another 2^k design in fewer variables if one or more of the original factors is dropped. Sometimes this can provide additional insight into the remaining factors. For example, consider the surface roughness experiment. Because factor C and all its interactions are negligible, we could eliminate factor C from the design. The result is to collapse the cube in Fig. 7-13 into a square in the A–B plane; therefore, each of the four runs in the new design has four replicates. In general, if we delete h factors so that $r = k - h$ factors remain, the original 2^k design with n replicates will project into a 2^r design with $n2^h$ replicates.

7-5   SINGLE REPLICATE OF A 2^k DESIGN

As the number of factors in a factorial experiment grows, the number of effects that can be estimated also grows. For example, a 2^4 experiment has 4 main effects, 6 two-factor interactions, 4 three-factor interactions, and 1 four-factor interaction, and a 2^6 experiment has 6 main effects, 15 two-factor interactions, 20 three-factor interactions, 15 four-factor interactions, 6 five-factor interactions, and 1 six-factor interaction. In most situations the **sparsity of effects principle** applies; that is, the system is usually dominated by the main effects and low-order interactions. The three-factor and higher-order interactions are usually negligible. Therefore, when the number of factors is moderately large—say, $k \geq 4$ or 5—a common practice is to run only a single replicate of the 2^k design and then pool or combine the higher-order interactions as an estimate of error. Sometimes a single replicate of a 2^k design is called an **unreplicated** 2^k **factorial design**.

When analyzing data from unreplicated factorial designs, occasionally real high-order interactions occur. The use of an error mean square obtained by pooling high-order interactions is inappropriate in these cases. A simple method of analysis called a **normal probability plot of effects** can be used to overcome this problem. Construct a plot of the estimates of the effects on a normal probability scale. The effects that are negligible are normally distributed, with mean zero, and will tend to fall along a straight line on this plot, whereas significant effects will have nonzero means and will not lie along the straight line. We will illustrate this method in the next example.

EXAMPLE 7-3

Plasma Etch

An article in *Solid State Technology* ("Orthogonal Design for Process Optimization and Its Application in Plasma Etching," Vol. 30, May 1987, pp. 127–132) describes the application of factorial designs in developing a nitride etch process on a single-wafer plasma etcher. The process uses C_2F_6 as the reactant gas. It is possible to vary the gas flow, the power applied to the cathode, the pressure in the reactor chamber, and the spacing between the anode and the cathode (gap). Several response variables would usually be of interest in this process, but in this example we will concentrate on etch rate for silicon nitride.

We will use a single replicate of a 2^4 design to investigate this process. Because it is unlikely that the three- and four-factor interactions are significant, we will tentatively plan to combine them as an estimate of error. The factor levels used in the design are shown here:

	Design Factor			
Level	Gap (cm)	Pressure (mTorr)	C_2F_6 Flow (SCCM)	Power (w)
Low $(-)$	0.80	450	125	275
High $(+)$	1.20	550	200	325

Table 7-10 presents the test matrix and the data from the 16 runs of the 2^4 design. Table 7-11 is the table of plus and minus signs for the 2^4 design. Analyze this experiment.

Solution. The signs in the columns of this table can be used to estimate the factor effects. For example, the estimate of factor A is

$$A = \frac{1}{8}[a + ab + ac + abc + ad + abd + acd + abcd - (1) - b$$
$$- c - bc - d - bd - cd - bcd]$$
$$= \frac{1}{8}(669 + 650 + 642 + 635 + 749 + 868 + 860 + 729$$
$$- 550 - 604 - 633 - 601 - 1037 - 1052 - 1075 - 1063)$$
$$= -101.625$$

Thus the effect of increasing the gap between the anode and the cathode from 0.80 to 1.20 cm is to decrease the mean etch rate by 101.625 Å/min.

Table 7-10   The 2^4 Design for the Plasma Etch Experiment

A (gap)	B (pressure)	C (C_2F_6 flow)	D (power)	Etch Rate (Å/min)
−1	−1	−1	−1	550
1	−1	−1	−1	669
−1	1	−1	−1	604
1	1	−1	−1	650
−1	−1	1	−1	633
1	−1	1	−1	642
−1	1	1	−1	601
1	1	1	−1	635
−1	−1	−1	1	1037
1	−1	−1	1	749
−1	1	−1	1	1052
1	1	−1	1	868
−1	−1	1	1	1075
1	−1	1	1	860
−1	1	1	1	1063
1	1	1	1	729

Table 7-11   Contrast Constants for the 2^4 Design

							Factorial Effect								
	A	B	AB	C	AC	BC	ABC	D	AD	BD	ABD	CD	ACD	BCD	ABCD
(1)	−	−	+	−	+	+	−	−	+	+	−	+	−	−	+
a	+	−	−	−	−	+	+	−	−	+	+	+	+	−	−
b	−	+	−	−	+	−	+	−	+	−	+	+	−	+	−
ab	+	+	+	−	−	−	−	−	−	−	−	+	+	+	+
c	−	−	+	+	−	−	+	−	+	+	−	−	+	+	−
ac	+	−	−	+	+	−	−	−	−	+	+	−	−	+	+
bc	−	+	−	+	−	+	−	−	+	−	+	−	+	−	+
abc	+	+	+	+	+	+	+	−	−	−	−	−	−	−	−
d	−	−	+	−	+	+	−	+	−	−	+	+	+	+	−
ad	+	−	−	−	−	+	+	+	+	−	−	−	−	+	+
bd	−	+	−	−	+	−	+	+	−	+	−	−	+	−	+
abd	+	+	+	−	−	−	−	+	+	+	+	−	−	−	−
cd	−	−	+	+	−	−	+	+	−	−	+	+	−	−	+
acd	+	−	−	+	+	−	−	+	+	−	−	+	+	−	−
bcd	−	+	−	+	−	+	−	+	−	+	−	+	−	+	−
abcd	+	+	+	+	+	+	+	+	+	+	+	+	+	+	+

It is easy to verify that the complete set of effect estimates is

$$
\begin{aligned}
A &= -101.625 & B &= -1.625 \\
AB &= -7.875 & C &= 7.375 \\
AC &= -24.875 & BC &= -43.875 \\
ABC &= -15.625 & D &= 306.125
\end{aligned}
$$

$$AD = -153.625 \qquad BD = -0.625$$
$$ABD = 4.125 \qquad CD = -2.125$$
$$ACD = 5.625 \qquad BCD = -25.375$$
$$ABCD = -40.125$$

Interpret Normal Probability Plot of Effects

The normal probability plot of these effects from the plasma etch experiment is shown in Fig. 7-16. Clearly, the main effects of A and D and the AD interaction are significant because they fall far from the line passing through the other points. The analysis of variance summarized in Table 7-12 confirms these findings. Note that in the analysis of variance we have pooled the three- and four-factor interactions to form the error mean square. If the normal probability plot had indicated that any of these interactions were important, we would not have included them in the error term. Consequently

$$\hat{\sigma}^2 = 2037.4 \text{ and } se(\text{coefficient}) = \frac{1}{2}\sqrt{\frac{2(2037.4)}{16/2}} = 11.28$$

Because $A = -101.625$, the effect of increasing the gap between the cathode and anode is to decrease the etch rate. However, $D = 306.125$; thus, applying higher power levels will increase the etch rate. Figure 7-17 is a plot of the AD interaction. This plot indicates that the effect of changing the gap width at low power settings is small but that increasing the gap at high power settings dramatically reduces the etch rate. High etch rates are obtained at high power settings and narrow gap widths.

The residuals from the experiment can be obtained from the regression model

Compute Predicted Values and Residuals

$$\hat{y} = 776.0625 - \left(\frac{101.625}{2}\right)x_1 + \left(\frac{306.125}{2}\right)x_4 - \left(\frac{153.625}{2}\right)x_1 x_4$$

For example, when both A and D are at the low level, the predicted value is

$$\hat{y} = 776.0625 - \left(\frac{101.625}{2}\right)(-1) + \left(\frac{306.125}{2}\right)(-1) - \left(\frac{153.625}{2}\right)(-1)(-1)$$

$$= 597$$

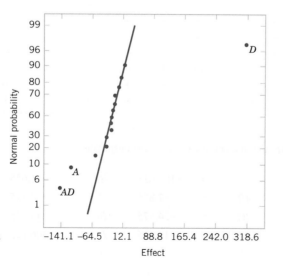

Figure 7-16 Normal probability plot of effects from the plasma etch experiment.

Table 7-12   Analysis for Example 7-3 Plasma Etch Experiment

Source	Sum of Squares	Degrees of Freedom	Mean Square	f_0	P-Value
		Analysis of Variance			
Model	521234	10	52123.40	25.58	0.000
Error	10187	5	2037.40		
Total	531421	15			

Independent Variable	Effect Estimate	Coefficient Estimate	Standard Error of Coefficient	t for H_0 Coefficient $= 0$	P-Value
Intercept		776.06	11.28	68.77	0.000
A	−101.63	−50.81	11.28	−4.50	0.006
B	−1.63	−0.81	11.28	−0.07	0.945
C	7.38	3.69	11.28	0.33	0.757
D	306.12	153.06	11.28	13.56	0.000
AB	−7.88	−3.94	11.28	−0.35	0.741
AC	−24.87	−12.44	11.28	−1.10	0.321
AD	−153.62	−76.81	11.28	−6.81	0.001
BC	−43.87	−21.94	11.28	−1.94	0.109
BD	−0.62	−0.31	11.28	−0.03	0.979
CD	−2.12	−1.06	11.28	−0.09	0.929

and the four residuals at this treatment combination are

$$e_1 = 550 - 597 = -47$$
$$e_2 = 604 - 597 = 7$$
$$e_3 = 633 - 597 = 36$$
$$e_4 = 601 - 597 = 4$$

The residuals at the other three treatment combinations (A high, D low), (A low, D high), and (A high, D high) are obtained similarly. A normal probability plot of the residuals is shown in Fig. 7-18. The plot is satisfactory. It would also be helpful to plot the residuals versus the predicted values and against each of the factors.

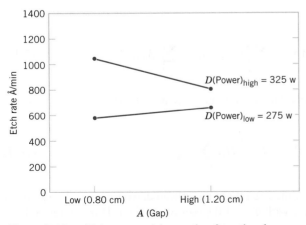

Figure 7-17   *AD* (gap-power) interaction from the plasma etch experiment.

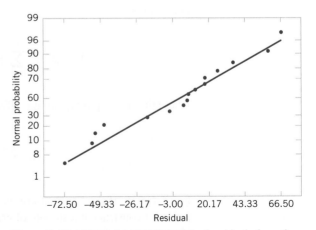

Figure 7-18   Normal probability plot of residuals from the plasma etch experiment.

7-6   CENTER POINTS AND BLOCKING IN 2^k DESIGNS

7-6.1   Addition of Center Points

A potential concern in the use of two-level factorial designs is the assumption of linearity in the factor effects. Of course, perfect linearity is unnecessary, and the 2^k system will work quite well even when the linearity assumption holds only approximately. However, there is a method of replicating certain points in the 2^k factorial that will provide protection against curvature as well as allow an independent estimate of error to be obtained. The method consists of adding **center points** to the 2^k design. These consist of n_C replicates run at the point $x_i = 0$, $i = 1, 2, \ldots, k$. One important reason for adding the replicate runs at the design center is that center points do not affect the usual effects estimates in a 2^k design. We assume that the k factors are quantitative. If some of the factors are categorical (such as Tool A and Tool B), the method can be modified.

To illustrate the approach, consider a 2^2 design with one observation at each of the factorial points $(-, -)$, $(+, -)$, $(-, +)$, and $(+, +)$ and n_C observations at the center point $(0, 0)$. Figure 7-19 illustrates the situation. Let $\bar{y}_F$ be the average of the four runs at the four factorial points, and let $\bar{y}_C$ be the average of the n_C run at the center point. If the difference $\bar{y}_F - \bar{y}_C$ is small, the center points lie on or near the plane passing through the factorial points, and there is no curvature. On the other hand, if $\bar{y}_F - \bar{y}_C$ is large, curvature is present.

A **t-test statistic** for curvature is given by

$$t_{\text{Curvature}} = \frac{\bar{y}_F - \bar{y}_C}{\sqrt{\hat{\sigma}^2 \left(\dfrac{1}{n_F} + \dfrac{1}{n_C} \right)}} \qquad (7\text{-}15)$$

where n_F is the number of factorial design points and n_C is the number of center points.

More specifically, when points are added to the center of the 2^k design, the model we may entertain is

$$Y = \beta_0 + \sum_{j=1}^{k} \beta_j x_j + \sum \sum_{i<j} \beta_{ij} x_i x_j + \sum_{j=1}^{k} \beta_{jj} x_j^2 + \epsilon \qquad (7\text{-}16)$$

where the β_{jj} are pure quadratic effects. The test for curvature actually tests the hypotheses

$$H_0: \sum_{j=1}^{k} \beta_{jj} = 0 \qquad H_1: \sum_{j=1}^{k} \beta_{jj} \neq 0 \qquad (7\text{-}17)$$

Furthermore, if the factorial points in the design are unreplicated, we may use the n_C center points to construct an estimate of error with $n_C - 1$ degrees of freedom. This is referred to as a **pure error** estimate.

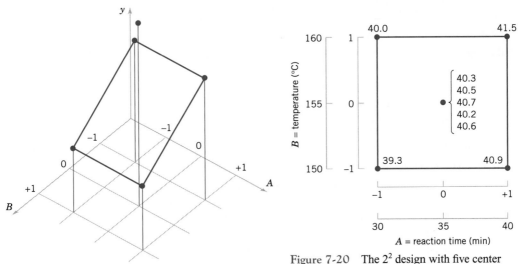

Figure 7-19   A 2^2 design with center points.

Figure 7-20   The 2^2 design with five center points for Example 7-4.

EXAMPLE 7-4

Process Yield

A chemical engineer is studying the percent conversion or yield of a process. There are two variables of interest, reaction time and reaction temperature. Because she is uncertain about the assumption of linearity over the region of exploration, the engineer decides to conduct a 2^2 design (with a single replicate of each factorial run) augmented with five center points. The design and the yield data are shown in Fig. 7-20.

Solution. Table 7-13 summarizes the analysis for this experiment. The estimate of pure error is calculated from the center points as follows:

$$\hat{\sigma}^2 = \frac{\sum_{\text{center points}} (y_i - \bar{y}_C)^2}{n_C - 1} = \frac{\sum_{i=1}^{5} (y_i - 40.46)^2}{4} = \frac{0.1720}{4} = 0.0430$$

The average of the points in the factorial portion of the design is $\bar{y}_F = 40.425$, and the average of the points at the center is $\bar{y}_C = 40.46$. The difference $\bar{y}_F - \bar{y}_C = 40.425 - 40.46 = -0.035$ appears to be small. The curvature t ratio is computed from equation 7-15 as follows:

$$t_{\text{Curvature}} = \frac{\bar{y}_F - \bar{y}_C}{\sqrt{\hat{\sigma}^2 \left(\frac{1}{n_F} + \frac{1}{n_C} \right)}} = \frac{-0.035}{\sqrt{0.0430 \left(\frac{1}{4} + \frac{1}{5} \right)}} = -0.252$$

The analysis indicates that there is no evidence of curvature in the response over the region of exploration; that is, the null hypothesis $H_0: \Sigma_{j=1}^2 \beta_{jj} = 0$ cannot be rejected.

Table 7-13 displays output from Minitab for this example. The effect of A is $(41.5 + 40.9 - 40.0 - 39.3)/2 = 1.55$, and the other effects are obtained similarly. The pure-error estimate (0.043) agrees with our previous result. Recall from regression modeling that the square of a t-ratio is an F-ratio. Consequently, Minitab uses $0.252^2 = 0.06$ as an F-ratio to obtain an identical test for curvature. The sum of squares for curvature is an intermediate step in the calculation of the F-ratio that equals the square of the t-ratio when the estimate of σ^2 is omitted. That is,

$$SS_{\text{Curvature}} = \frac{(\bar{y}_F - \bar{y}_C)^2}{\frac{1}{n_F} + \frac{1}{n_C}} \tag{7-18}$$

Table 7-13 Analysis for Example 7-4 Process Yield from Minitab

Factorial Design

Full Factorial Design

Factors:	2	Base Design:	2, 4
Runs:	9	Replicates:	1
Blocks:	none	Center pts (total):	5

All terms are free from aliasing

Fractional Factorial Fit

Estimated Effects and Coefficients for y

Term	Effect	Coef	StDev Coef	T	P
Constant		40.4444	0.06231	649.07	0.000
A	1.5500	0.7750	0.09347	8.29	0.000
B	0.6500	0.3250	0.09347	3.48	0.018
A*B	−0.0500	−0.0250	0.09347	−0.27	0.800

Analysis of Variance for y

Source	DF	Seq SS	Adj SS	Adj MS	F	P
Main Effects	2	2.82500	2.82500	1.41250	40.42	0.001
2-Way Interactions	1	0.00250	0.00250	0.00250	0.07	0.800
Residual Error	5	0.17472	0.17472	0.03494		
Curvature	1	0.00272	0.00272	0.00272	0.06	0.814
Pure Error	4	0.17200	0.17200	0.04300		
Total	8	3.00222				

Furthermore, Minitab adds the sum of squares for curvature and for pure error to obtain the residual sum of squares (0.17472) with 5 degrees of freedom. The residual mean square (0.03494) is a pooled estimate of σ^2, and it is used in the calculation of the t-ratio for the A, B and AB effects. The pooled estimate is close to the pure-error estimate in this example because curvature is negligible. If curvature were significant, the pooling would not be appropriate. The estimate of the intercept β_0 (40.444) is the mean of all nine measurements. ∎

7-6.2 Blocking and Confounding

It is often impossible to run all the observations in a 2^k factorial design under homogeneous conditions. Drawing on the notions originally introduced in Section 5-8.2, blocking is the design technique that is appropriate for this general situation. However, in many situations the block size is smaller than the number of runs in the complete replicate. In these cases, **confounding** is a useful procedure for running the 2^k design in 2^p blocks where the number of runs in a block is less than the number of treatment combinations in one complete replicate. The technique causes certain interaction effects to be indistinguishable from blocks, or **confounded with blocks.** We will illustrate confounding in the 2^k factorial design in 2^p blocks, where $p < k$.

Consider a 2^2 design. Suppose that each of the $2^2 = 4$ treatment combinations requires 4 hours of laboratory analysis. Thus, 2 days are required to perform the experiment. If days are considered as blocks, we must assign two of the four treatment combinations to each day.

This design is shown in Fig. 7-21. Note that block 1 contains the treatment combinations (1) and ab and that block 2 contains a and b. The contrasts for estimating the main effects of factors A and B are

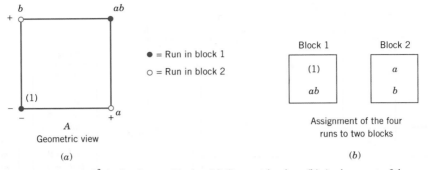

Figure 7-21   A 2^2 design in two blocks. (a) Geometric view. (b) Assignment of the four runs to two blocks.

$$\text{Contrast}_A = ab + a - b - (1)$$
$$\text{Contrast}_B = ab + b - a - (1)$$

Note that these contrasts are unaffected by blocking because in each contrast there is one plus and one minus treatment combination from each block. That is, any difference between block 1 and block 2 that increases the readings in one block by an additive constant cancels out. The contrast for the AB interaction is

$$\text{Contrast}_{AB} = ab + (1) - a - b$$

Because the two treatment combinations with the plus signs, ab and (1), are in block 1 and the two with the minus signs, a and b, are in block 2, the block effect and the AB interaction are identical. That is, the AB interaction is confounded with blocks.

The reason for this is apparent from the table of plus and minus signs for the 2^2 design shown in Table 7-3. From the table we see that all treatment combinations that have a plus on AB are assigned to block 1, whereas all treatment combinations that have a minus sign on AB are assigned to block 2.

This scheme can be used to confound any 2^k design in two blocks. As a second example, consider a 2^3 design run in two blocks. From the table of plus and minus signs, shown in Table 7-8, we assign the treatment combinations that are minus in the ABC column to block 1 and those that are plus in the ABC column to block 2. The resulting design is shown in Fig. 7-22.

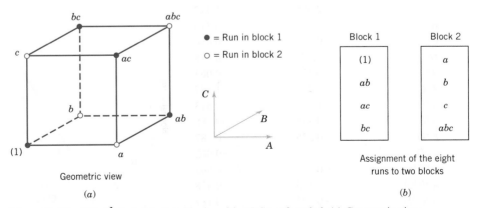

Figure 7-22   The 2^3 design in two blocks with ABC confounded. (a) Geometric view. (b) Assignment of the eight runs to two blocks.

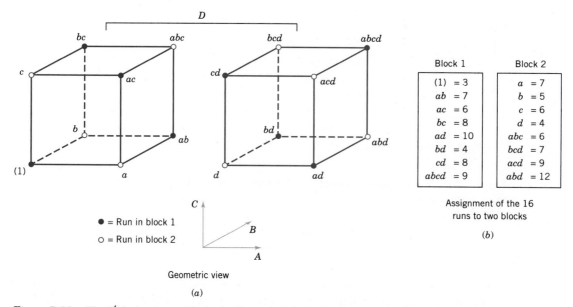

Block 1		Block 2	
(1)	= 3	a	= 7
ab	= 7	b	= 5
ac	= 6	c	= 6
bc	= 8	d	= 4
ad	= 10	abc	= 6
bd	= 4	bcd	= 7
cd	= 8	acd	= 9
abcd	= 9	abd	= 12

Assignment of the 16
runs to two blocks

(b)

● = Run in block 1
○ = Run in block 2

Geometric view

(a)

Figure 7-23 The 2^4 design in two blocks for Example 7-8. (a) Geometric view. (b) Assignment of the 16 runs to two blocks.

EXAMPLE 7-5
Missile Miss
Distance

An experiment is performed to investigate the effect of four factors on the terminal miss distance of a shoulder-fired ground-to-air-missile. The four factors are target type (A), seeker type (B), target altitude (C), and target range (D). Each factor may be conveniently run at two levels, and the optical tracking system will allow terminal miss distance to be measured to the nearest foot. Two different operators or gunners are used in the flight test and, because there may be differences between operators, the test engineers decided to conduct the 2^4 design in two blocks with $ABCD$ confounded.

The experimental design and the resulting data are shown in Fig. 7-23. The effect estimates obtained from Minitab are shown in Table 7-14. A normal probability plot of the effects in Fig. 7-24 reveals

Table 7-14 Minitab Effect Estimates for Example 7-5

Estimated Effects and Coefficients for Distance		
Term	Effect	Coef
Constant		6.938
Block		0.063
A	2.625	1.312
B	0.625	0.313
C	0.875	0.438
D	1.875	0.938
$A*B$	−0.125	−0.063
$A*C$	−2.375	−1.187
$A*D$	1.625	0.813
$B*C$	−0.375	−0.188
$B*D$	−0.375	−0.187
$C*D$	−0.125	−0.062
$A*B*C$	−0.125	−0.063
$A*B*D$	0.875	0.438
$A*C*D$	−0.375	−0.187
$B*C*D$	−0.375	−0.187

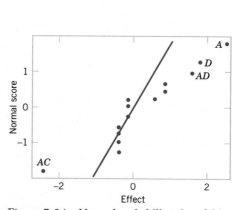

Figure 7-24 Normal probability plot of the effects from Minitab, Example 7-5.

Table 7-15   Analysis of Variance for Example 7-5

Source of Variation	Sum of Squares	Degrees of Freedom	Mean Square	f_0	P-Value
Blocks ($ABCD$)	0.0625	1	0.0625	0.06	—
A	27.5625	1	27.5625	25.94	0.0070
B	1.5625	1	1.5625	1.47	0.2920
C	3.0625	1	3.0625	2.88	0.1648
D	14.0625	1	14.0625	13.24	0.0220
AB	0.0625	1	0.0625	0.06	—
AC	22.5625	1	22.5625	21.24	0.0100
AD	10.5625	1	10.5625	9.94	0.0344
BC	0.5625	1	0.5625	0.53	—
BD	0.5625	1	0.5625	0.53	—
CD	0.0625	1	0.0625	0.06	—
Error ($ABC + ABD + ACD + BCD$)	4.2500	4	1.0625		
Total	84.9375	15			

that A (target type), D (target range), AD, and AC have large effects. A confirming analysis of variance, pooling the three-factor interactions as error, is shown in Table 7-15. Because the AC and AD interactions are significant, it is logical to conclude that A (target type), C (target altitude), and D (target range) all have important effects on the miss distance and that there are interactions between target type and altitude and target type and range. Note that the $ABCD$ effect is treated as blocks in this analysis. ■

It is possible to confound the 2^k design in four blocks of 2^{k-2} observations each. To construct the design, two effects are chosen to confound with blocks. A third effect, the **generalized interaction** of the two effects initially chosen, is also confounded with blocks. The generalized interaction of two effects is found by multiplying their respective letters and reducing the exponents modulus 2.

For example, consider the 2^4 design in four blocks. If AC and BD are confounded with blocks, their generalized interaction is $(AC)(BD) = ABCD$. The design is constructed by a partition of the treatments according to the signs of AC and BD. It is easy to verify that the four blocks are

Block 1	Block 2	Block 3	Block 4
$AC+, BD+$	$AC-, BD+$	$AC+, BD-$	$AC-, BD-$
(1)	a	b	ab
ac	c	abc	bc
bd	abd	d	ad
$abcd$	bcd	acd	cd

This general procedure can be extended to confounding the 2^k design in 2^p blocks, where $p < k$. Start by selecting p effects to be confounded, such that no effect chosen is a generalized interaction of the others. Then the blocks can be constructed from the p defining contrasts $L_1, L_2, \ldots, L_p$ that are associated with these effects. In addition to the p effects chosen to be confounded, exactly $2^p - p - 1$ additional effects are confounded with blocks; these are the generalized interactions of the original p effects chosen. Care should be taken so as not to confound effects of potential interest.

For more information on confounding in the 2^k factorial design, refer to Montgomery (2005a, Chapter 7). This book contains guidelines for selecting factors to confound with blocks so that main effects and low-order interactions are not confounded. In particular, the book contains a table of suggested confounding schemes for designs with up to seven factors and a range of block sizes, some of which are as small as two runs.

EXERCISES FOR SECTIONS 7-4, 7-5, AND 7-6

7-11. An engineer is interested in the effect of cutting speed (*A*), metal hardness (*B*), and cutting angle (*C*) on the life of a cutting tool. Two levels of each factor are chosen, and two replicates of a 2^3 factorial design are run. The tool life data (in hours) are shown in the following table.

Treatment Combination	Replicate I	Replicate II
(1)	221	311
a	325	435
b	354	348
ab	552	472
c	440	453
ac	406	377
bc	605	500
abc	392	419

(a) Analyze the data from this experiment using *t*-ratios with $\alpha = 0.05$.

(b) Find an appropriate regression model that explains tool life in terms of the variables used in the experiment.

(c) Analyze the residuals from this experiment.

7-12. Four factors are thought to influence the taste of a soft-drink beverage: type of sweetener (*A*), ratio of syrup to water (*B*), carbonation level (*C*), and temperature (*D*). Each factor can be run at two levels, producing a 2^4 design.

Treatment Combination	Replicate I	Replicate II
(1)	159	163
a	168	175
b	158	163
ab	166	168
c	175	178
ac	179	183
bc	173	168
abc	179	182
d	164	159
ad	187	189
bd	163	159
abd	185	191
cd	168	174

Treatment Combination	Replicate I	Replicate II
acd	197	199
bcd	170	174
abcd	194	198

At each run in the design, samples of the beverage are given to a test panel consisting of 20 people. Each tester assigns the beverage a point score from 1 to 10. Total score is the response variable, and the objective is to find a formulation that maximizes total score. Two replicates of this design are run, and the results are as shown. Analyze the data using *t*-ratios and draw conclusions. Use $\alpha = 0.05$ in the statistical tests.

7-13. Consider the experiment in Exercise 7-12. Determine an appropriate model and plot the residuals against the levels of factors *A*, *B*, *C*, and *D*. Also construct a normal probability plot of the residuals. Comment on these plots and the most important factors influencing taste.

7-14. The data shown here represent a single replicate of a 2^5 design that is used in an experiment to study the compressive strength of concrete. The factors are mix (*A*), time (*B*), laboratory (*C*), temperature (*D*), and drying time (*E*).

(1)	= 700	*e*	= 800
a	= 900	*ae*	= 1200
b	= 3400	*be*	= 3500
ab	= 5500	*abc*	= 6200
c	= 600	*ce*	= 600
ac	= 1000	*ace*	= 1200
bc	= 3000	*bce*	= 3000
abc	= 5300	*abce*	= 5500
d	= 1000	*de*	= 1900
ad	= 1100	*ade*	= 1500
bd	= 3000	*bde*	= 4000
abd	= 6100	*abde*	= 6500
cd	= 800	*cde*	= 1500
acd	= 1100	*acde*	= 2000
bcd	= 3300	*bcde*	= 3400
abcd	= 6000	*abcde*	= 6800

(a) Estimate the factor effects.

(b) Which effects appear important? Use a normal probability plot.

(c) Determine an appropriate model and analyze the residuals from this experiment. Comment on the adequacy of the model.

(d) If it is desirable to maximize the strength, in which direction would you adjust the process variables?

7-15. Consider a famous experiment reported by O. L. Davies (ed.), *The Design and Analysis of Industrial Experiments* (London: Oliver and Boyd, 1956). The following data were collected from an unreplicated experiment in which the investigator was interested in determining the effect of four factors on the yield of an isatin derivative used in a fabric-dying process. The four factors are each run at two levels as indicated: (A) acid strength at 87 and 93%, (B) reaction time at 15 and 30 min, (C) amount of acid 35 and 45 ml, and (D) temperature of reaction 60 and 70°C. The response is the yield of isatin in grams per 100 grams of base material. The data are as follows:

(1)	= 6.08	*d*	= 6.79
a	= 6.04	*ad*	= 6.68
b	= 6.53	*bd*	= 6.73
ab	= 6.43	*abd*	= 6.08
c	= 6.31	*cd*	= 6.77
ac	= 6.09	*acd*	= 6.38
bc	= 6.12	*bcd*	= 6.49
abc	= 6.36	*abcd*	= 6.23

(a) Estimate the effects and prepare a normal plot of the effects. Which interaction terms are negligible? Use *t*-ratios to confirm your findings.

(b) Based on your results in part (a), construct a model and analyze the residuals.

7-16. An experiment was run in a semiconductor fabrication plant in an effort to increase yield. Five factors, each at two levels, were studied. The factors (and levels) were A = aperture setting (small, large), B = exposure time (20% below nominal, 20% above nominal), C = development time (30 sec, 45 sec), D = mask dimension (small, large), and E = etch time (14.5 min, 15.5 min). The unreplicated 2^5 design shown here was run.

(1)	= 7	*e*	= 8
a	= 9	*ae*	= 12
b	= 34	*be*	= 35
ab	= 55	*abe*	= 52
c	= 16	*ce*	= 15
ac	= 20	*ace*	= 22
bc	= 40	*bce*	= 45
abc	= 60	*abce*	= 65
d	= 8	*de*	= 6
ad	= 10	*ade*	= 10
bd	= 32	*bde*	= 30
abd	= 50	*abde*	= 53
cd	= 18	*cde*	= 15
acd	= 21	*acde*	= 20
bcd	= 44	*bcde*	= 41
abcd	= 61	*abcde*	= 63

(a) Construct a normal probability plot of the effect estimates. Which effects appear to be large?

(b) Estimate σ^2 and use *t*-ratios to confirm your findings for part (a).

(c) Plot the residuals from an appropriate model on normal probability paper. Is the plot satisfactory?

(d) Plot the residuals versus the predicted yields and versus each of the five factors. Comment on the plots.

(e) Interpret any significant interactions.

(f) What are your recommendations regarding process operating conditions?

(g) Project the 2^5 design in this problem into a 2^r for $r < 5$ design in the important factors. Sketch the design and show the average and range of yields at each run. Does this sketch aid in data interpretation?

7-17. Consider the semiconductor experiment in Exercise 7-16. Suppose that a center point (replicated five times) could be added to this design and that the responses at the center are 45, 40, 41, 47, and 43.

(a) Estimate the error using the center points. How does this estimate compare to the estimate obtained in Exercise 7-16?

(b) Calculate the *t*-ratio for curvature and test at $\alpha = 0.05$.

7-18. Consider the data from Exercise 7-11, replicate I only. Suppose that a center point (with four replicates) is added to these eight runs. The tool life response at the center point is 425, 400, 437, and 418.

(a) Estimate the factor effects.

(b) Estimate pure error using the center points.

(c) Calculate the *t*-ratio for curvature and test at $\alpha = 0.05$.

(d) Test for main effects and interaction effects, using $\alpha = 0.05$.

(e) Give the regression model and analyze the residuals from this experiment.

7-19. Consider the data from the first replicate of Exercise 7-11. Suppose that these observations could not all be run under the same conditions. Set up a design to run these observations in two blocks of four observations each, with ABC confounded. Analyze the data.

7-20. Consider the data from the first replicate of Exercise 7-12. Construct a design with two blocks of eight observations each, with $ABCD$ confounded. Analyze the data.

7-21. Repeat Exercise 7-20 assuming that four blocks are required. Confound ABD and ABC (and consequently CD) with blocks.

7-22. Construct a 2^5 design in two blocks. Select the $ABCDE$ interaction to be confounded with blocks.

7-23. Construct a 2^5 design in four blocks. Select the appropriate effects to confound so that the highest possible interactions are confounded with blocks.

7-24. Consider the data from Exercise 7-15. Construct the design that would have been used to run this experiment in two blocks of eight runs each. Analyze the data and draw conclusions.

Term	Effect	Coef	SE Coef	T	P
Constant		6.0625	0.3903	15.53	0.000
A	3.3750	1.6875	0.3903	4.32	?
B	1.6250	0.8125	0.3903	2.08	?
C	0.8750	0.4375	0.3903	?	0.295
AB	1.3750	0.6875	?	?	?
AC	0.1250	0.0625	0.3903	0.16	?
BC	−0.6250	−0.3125	0.3903	?	0.446
ABC	1.1250	0.5625	0.3903	1.44	?

7-25. Consider the Minitab analysis results of a 2^3-designed experiment with two replicates.

(a) Find all of the missing values in the t-tests of the effects table above. Indicate which effects are significant using $\alpha = 0.1$.

(b) Write the least squares fitted model using only the significant terms.

(c). Use the model to predict the response when $x_1 = -1$, $x_2 = 1, x_3 = 1$.

7-26. Consider the Minitab analysis results of a 2^3-designed experiment with two replicates at the corner points and four replicates at a center point.

(a) Find all of the missing values for the t-tests and the F-tests in the two tables below. Indicate which effects are significant. Use $\alpha = 0.1$.

(b) Is there significant curvature? Indicate the lines in the Minitab output that provide the necessary information.

(c) Write the least squares fitted model using only the significant terms.

(d) Use the model to predict the response when $x_1 = -1$, $x_2 = -1, x_3 = -1$.

Term	Effect	Coef	SE Coef	T	P
Constant		14.97	0.6252	23.95	0.000
A	19.87	9.93	0.6252	15.89	?
B	−10.56	−5.28	0.6252	−8.45	?
C	0.34	0.17	0.6252	?	0.791
A*B	−29.96	−14.98	0.6252	?	?
A*C	0.54	0.27	0.6252	0.43	?
B*C	0.15	0.07	0.6252	?	?
A*B*C	0.68	0.34	0.6252	?	0.597
Ct Pt		−0.93	1.3980	-0.66	?

Analysis of Variance for Response

Source	DF	Seq SS	Adj SS	Adj MS	F	P
Main Effects	3	2025.69	2025.69	675.23	107.69	?
2-Way Interactions	3	3592.66	3592.66	1197.55	?	0.000
3-Way Interactions	1	1.85	1.85	1.85	0.30	?
Curvature	1	2.74	2.74	2.74	0.44	?
Residual Error	11	68.80	68.80	6.25		
Pure Error	11	68.80	68.80	6.25		
Total	19	5691.74				

7-7 FRACTIONAL REPLICATION OF A 2^k DESIGN

As the number of factors in a 2^k factorial design increases, the number of runs required increases rapidly. For example, a 2^5 requires 32 runs. In this design, only 5 degrees of freedom correspond to main effects, and 10 degrees of freedom correspond to two-factor interactions.

Sixteen of the 31 degrees of freedom are used to estimate high-order interactions—that is, three-factor and higher-order interactions. Often there is little interest in these high-order interactions, particularly when we first begin to study a process or system. If we can assume that certain high-order interactions are negligible, a **fractional factorial design** involving fewer than the complete set of 2^k runs can be used to obtain information on the main effects and low-order interactions. In this section, we will introduce fractional replications of the 2^k design.

A major use of fractional factorials is in screening experiments. These are experiments in which many factors are considered with the purpose of identifying those factors (if any) that have large effects. Screening experiments are usually performed in the early stages of a project when it is likely that many of the factors initially considered have little or no effect on the response. The factors that are identified as important are then investigated more thoroughly in subsequent experiments.

7-7.1 One-Half Fraction of a 2^k Design

A $\frac{1}{2}$ fraction of the 2^k design contains 2^{k-1} runs and is often called a 2^{k-1} fractional factorial design. As an example, consider the 2^{3-1} design—that is, a $\frac{1}{2}$ fraction of the 2^3. This design has only four runs, in contrast to the full factorial that would require eight runs. The table of plus and minus signs for the 2^3 design is shown in Table 7-16. Suppose we select the four treatment combinations a, b, c, and abc as our $\frac{1}{2}$ fraction. These treatment combinations are shown in the top half of Table 7-16 and in Fig. 7-25a. We will continue to use both the lowercase letter notation $(a, b, c, \ldots)$ and the geometric or plus and minus notation for the treatment combinations.

Table 7-16 Plus and Minus Signs for the 2^3 Factorial Design

Treatment Combination	Factorial Effect							
	I	A	B	C	AB	AC	BC	ABC
a	+	+	−	−	−	−	+	+
b	+	−	+	−	−	+	−	+
c	+	−	−	+	+	−	−	+
abc	+	+	+	+	+	+	+	+
ab	+	+	+	−	+	−	−	−
ac	+	+	−	+	−	+	−	−
bc	+	−	+	+	−	−	+	−
(1)	+	−	−	−	+	+	+	−

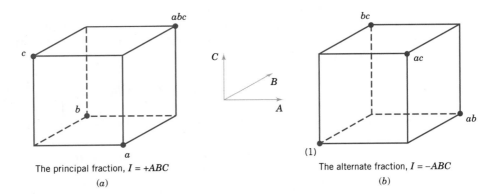

The principal fraction, $I = +ABC$

(a)

The alternate fraction, $I = -ABC$

(b)

Figure 7-25 The $\frac{1}{2}$ fractions of the 2^3 design. (a) The principal fraction, $I = +ABC$. (b) The alternate fraction, $I = -ABC$.

Note that the 2^{3-1} design is formed by selecting only those treatment combinations that yield a plus on the ABC effect. Thus, ABC is called the **generator** of this particular fraction. Furthermore, the identity element I is also plus for the four runs, so we call

$$I = ABC$$

the **defining relation** for the design.

The treatment combinations in the 2^{3-1} designs yield three degrees of freedom associated with the main effects. From the upper half of Table 7-16, we obtain the estimates of the main effects as linear combinations of the observations, say,

$$A = \frac{1}{2}[a - b - c + abc]$$

$$B = \frac{1}{2}[-a + b - c + abc]$$

$$C = \frac{1}{2}[-a - b + c + abc]$$

It is also easy to verify that the estimates of the two-factor interactions should be the following linear combinations of the observations:

$$BC = \frac{1}{2}[a - b - c + abc]$$

$$AC = \frac{1}{2}[-a + b - c + abc]$$

$$AB = \frac{1}{2}[-a - b + c + abc]$$

Thus, the linear combination of observations in column A estimates both the main effect of A and the BC interaction. That is, the linear combination estimates the sum of these two effects $A + BC$. Similarly, B estimates $B + AC$, and C estimates $C + AB$. Two or more effects that have this property are called **aliases.** In our 2^{3-1} design, A and BC are aliases, B and AC are aliases, and C and AB are aliases. Aliasing is the direct result of fractional replication. In many practical situations, it will be possible to select the fraction so that the main effects and low-order interactions of interest will be aliased only with high-order interactions (which are probably negligible).

The alias structure for this design is found by using the defining relation $I = ABC$. Multiplying any effect by the defining relation yields the aliases for that effect. In our example, the alias of A is

$$A = A \cdot ABC = A^2BC = BC$$

because $A \cdot I = A$ and $A^2 = I$. The aliases of B and C are

$$B = B \cdot ABC = AB^2C = AC$$

and

$$C = C \cdot ABC = ABC^2 = AB$$

Now suppose that we had chosen the other $\frac{1}{2}$ fraction—that is, the treatment combinations in Table 7-16 associated with minus on ABC. These four runs are shown in the lower half of Table 7-16 and in Fig. 7-25b. The defining relation for this design is $I = -ABC$. The aliases are $A = -BC$, $B = -AC$, and $C = -AB$. Thus, estimates of A, B, and C that result from this fraction really estimate $A - BC$, $B - AC$, and $C - AB$. In practice, it usually does not matter which $\frac{1}{2}$ fraction we select. The fraction with the plus sign in the defining relation is usually called the **principal fraction,** and the other fraction is usually called the **alternate fraction.**

Note that if we had chosen AB as the generator for the fractional factorial,

$$A = A \cdot AB = B$$

and the two main effects of A and B would be aliased. This typically loses important information.

Sometimes we use **sequences** of fractional factorial designs to estimate effects. For example, suppose we had run the principal fraction of the 2^{3-1} design with generator ABC. However, if after running the principal fraction important effects are aliased, it is possible to estimate them by running the *alternate* fraction. Then, the full factorial design is completed and the effects can be estimated by the usual calculation. Because the experiment has been split over two time periods, it has been confounded with blocks. One might be concerned that changes in the experimental conditions could bias the estimates of the effects. However, it can be shown that if the result of a change in the experimental conditions is to add a constant to all the responses, only the ABC interaction effect is biased as a result of confounding; the remaining effects are not affected. Thus, by combining a sequence of two fractional factorial designs, we can isolate both the main effects and the two-factor interactions. This property makes the fractional factorial design highly useful in experimental problems because we can run sequences of small, efficient experiments, combine information across *several* experiments, and take advantage of learning about the process we are experimenting with as we go along. This is an illustration of the concept of sequential experimentation.

A 2^{k-1} design may be constructed by writing down the treatment combinations for a full factorial with $k - 1$ factors, called the **basic design,** and then adding the kth factor by identifying its plus and minus levels with the plus and minus signs of the highest-order interaction. Therefore, a 2^{3-1} fractional factorial is constructed by writing down the basic design as a full 2^2 factorial and then equating factor C with the $\pm AB$ interaction. Thus, to construct the principal fraction, we would use $C = +AB$ as follows:

Basic Design		Fractional Design		
Full 2^2		$2^{3-1}, I = +ABC$		
A	B	A	B	$C = AB$
−	−	−	−	+
+	−	+	−	−
−	+	−	+	−
+	+	+	+	+

To obtain the alternate fraction we would equate the last column to $C = -AB$.

EXAMPLE 7-6
Plasma Etch

To illustrate the use of a $\frac{1}{2}$ fraction, consider the plasma etch experiment described in Example 7-3. Suppose that we decide to use a 2^{4-1} design with $I = ABCD$ to investigate the four factors gap (A), pressure (B), C_2F_6 flow rate (C), and power setting (D). This design would be constructed by writing down as the basic design a 2^3 in the factors A, B, and C and then setting the levels of the fourth factor $D = ABC$. The design and the etch rate for each trial are shown in Table 7-17. The design is shown graphically in Fig. 7-26. We are interested in how this reduced design affects our results.

Table 7-17   The 2^{4-1} Design with Defining Relation $I = ABCD$

A	B	C	$D = ABC$	Treatment Combination	Etch Rate
−	−	−	−	(1)	550
+	−	−	+	ad	749
−	+	−	+	bd	1052
+	+	−	−	ab	650
−	−	+	+	cd	1075
+	−	+	−	ac	642
−	+	+	−	bc	601
+	+	+	+	$abcd$	729

Solution. In this design, the main effects are aliased with the three-factor interactions; note that the alias of A is

$$A \cdot I = A \cdot ABCD$$
$$A = A^2BCD = BCD$$

and similarly

$$B = ACD \qquad C = ABD \qquad D = ABC$$

The two-factor interactions are aliased with each other. For example, the alias of AB is CD:

$$AB \cdot I = AB \cdot ABCD$$
$$AB = A^2B^2CD = CD$$

Examine the Aliases

The other aliases are

$$AC = BD \qquad \text{and} \qquad AD = BC$$

The estimates of the main effects and their aliases are found using the four columns of signs in Table 7-17. For example, from column A we obtain the estimated effect as the difference between the averages of the four + runs and the four − runs.

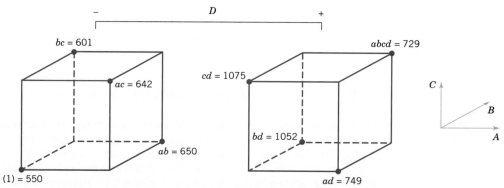

Figure 7-26   The 2^{4-1} design for the experiment of Example 7-6.

$$A + BCD = \frac{1}{4}(-550 + 749 - 1052 + 650 - 1075 + 642 - 601 + 729)$$
$$= -127.00$$

The other columns produce

$$B + ACD = 4.00$$
$$C + ABD = 11.50$$

and

$$D + ABC = 290.50$$

Clearly, $A + BCD$ and $D + ABC$ are large, and if we believe that the three-factor interactions are negligible, the main effects A (gap) and D (power setting) significantly affect etch rate.

The interactions are estimated by forming the AB, AC, and AD columns and adding them to the table. For example, the signs in the AB column are $+, -, -, +, +, -, -, +$, and this column produces the estimate

Interpret the Interaction Effects

$$AB + CD = \frac{1}{4}(550 - 749 - 1052 + 650 + 1075 - 642 - 601 + 729)$$
$$= -10.00$$

From the AC and AD columns we find

$$AC + BD = -25.50$$
$$AD + BC = -197.50$$

The $AD + BC$ estimate is large; the most straightforward interpretation of the results is that because the main effects A and D are large, this is the AD interaction. Thus, the results obtained from the 2^{4-1} design agree with the full factorial results in Example 7-3. ∎

Normality Probability Plots and Residuals
The normal probability plot is useful in assessing the significance of effects from a fractional factorial design, particularly when many effects are to be estimated. We strongly recommend this approach. Residuals can be obtained from a fractional factorial by the regression model method shown previously. These residuals should be graphically analyzed as we have discussed before, both to assess the validity of the underlying model assumptions and to gain additional insight into the experimental situation.

Projection of a 2^{k-1} Design
If one or more factors from a one-half fraction of a 2^k can be dropped, the design will project into a full factorial design. For example, Fig. 7-27 presents a 2^{3-1} design. Note that this design will project into a full factorial in any two of the three original factors. Thus, if we think that at most two of the three factors are important, the 2^{3-1} design is an excellent design for identifying the significant factors. This **projection property** is highly useful in factor screening, because it allows negligible factors to be eliminated, resulting in a stronger experiment in the active factors that remain.

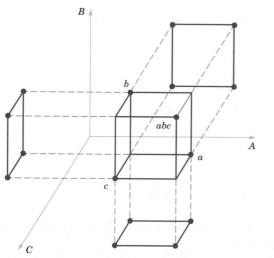

Figure 7-27   Projection of a 2^{3-1} design into three 2^2 designs.

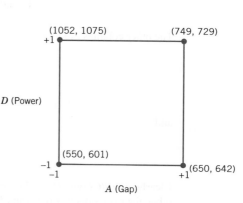

Figure 7-28   The 2^2 design obtained by dropping factors B and C from the plasma etch experiment in Example 7-6.

In the 2^{4-1} design used in the plasma etch experiment in Example 7-6, we found that two of the four factors (B and C) could be dropped. If we eliminate these two factors, the remaining columns in Table 7-16 form a 2^2 design in the factors A and D, with two replicates. This design is shown in Fig. 7-28. The main effects of A and D and the strong two-factor AD interaction are clearly evident from this graph.

Design Resolution

The concept of design resolution is a useful way to catalog fractional factorial designs according to the alias patterns they produce. Designs of resolution III, IV, and V are particularly important. The definitions of these terms and an example of each follow.

1. **Resolution III designs.**   These are designs in which no main effects are aliased with any other main effect, but main effects are aliased with two-factor interactions and some two-factor interactions may be aliased with each other. The 2^{3-1} design with $I = ABC$ is a resolution III design. We usually employ a Roman numeral subscript to indicate design resolution; thus, this $\frac{1}{2}$ fraction is a 2^{3-1}_{III} design.

2. **Resolution IV designs.**   These are designs in which no main effect is aliased with any other main effect or two-factor interactions, but two-factor interactions are aliased with each other. The 2^{4-1} design with $I = ABCD$ used in Example 7-9 is a resolution IV design (2^{4-1}_{IV}).

3. **Resolution V designs.**   These are designs in which no main effect or two-factor interaction is aliased with any other main effect or two-factor interaction, but two-factor interactions are aliased with three-factor interactions. A 2^{5-1} design with $I = ABCDE$ is a resolution V design (2^{5-1}_{V}).

Resolution III and IV designs are particularly useful in factor screening experiments. A resolution IV design provides good information about main effects and will provide some information about all two-factor interactions.

7-7.2   Smaller Fractions: 2^{k-p} Fractional Factorial Designs

Although the 2^{k-1} design is valuable in reducing the number of runs required for an experiment, we frequently find that smaller fractions will provide almost as much useful information at even greater economy. In general, a 2^k design may be run in a $1/2^p$ fraction called a 2^{k-p} fractional factorial design. Thus, a $1/4$ fraction is called a 2^{k-2} design, a $1/8$ fraction is called a 2^{k-3} design, a $1/16$ fraction a 2^{k-4} design, and so on.

To illustrate the $1/4$ fraction, consider an experiment with six factors and suppose that the engineer is primarily interested in main effects but would also like to get some information about the two-factor interactions. A 2^{6-1} design would require 32 runs and would have 31 degrees of freedom for estimating effects. Because there are only 6 main effects and 15 two-factor interactions, the $\frac{1}{2}$ fraction is inefficient—it requires too many runs. Suppose we consider a $1/4$ fraction, or a 2^{6-2} design. This design contains 16 runs and, with 15 degrees of freedom, will allow all 6 main effects to be estimated, with some capability for examining the two-factor interactions.

To generate this design, we would write down a 2^4 design in the factors A, B, C, and D as the basic design, and then add two columns for E and F. To find the new columns we could select the two **design generators** $I = ABCE$ and $I = BCDF$. Thus, column E would be found from $E = ABC$, and column F would be $F = BCD$. That is, columns $ABCE$ and $BCDF$ are equal to the identity column. However, we know that the product of any two columns in the table of plus and minus signs for a 2^k design is simply another column in the table; therefore, the product of $ABCE$ and $BCDF$ that equals $ABCE\,(BCDF) = AB^2C^2DEF = ADEF$ is also an identity column. Consequently, the **complete defining relation** for the 2^{6-2} design is

$$I = ABCE = BCDF = ADEF$$

We refer to each term in a defining relation (such as $ABCE$ above) as a **word.** To find the alias of any effect, simply multiply the effect by each word in the foregoing defining relation. For example, the alias of A is

$$A = BCE = ABCDF = DEF$$

The complete alias relationships for this design are shown in Table 7-18. In general, the resolution of a 2^{k-p} design is equal to the number of letters in the shortest word in the complete defining

Table 7-18   Alias Structure for the 2_{IV}^{6-2} Design with $I = ABCE = BCDF = ADEF$

$A = BCE = DEF = ABCDF$	$AB = CE = ACDF = BDEF$
$B = ACE = CDF = ABDEF$	$AC = BE = ABDF = CDEF$
$C = ABE = BDF = ACDEF$	$AD = EF = BCDE = ABCF$
$D = BCF = AEF = ABCDE$	$AE = BC = DF = ABCDEF$
$E = ABC = ADF = BCDEF$	$AF = DE = BCEF = ABCD$
$F = BCD = ADE = ABCEF$	$BD = CF = ACDE = ABEF$
$ABD = CDE = ACF = BEF$	$BF = CD = ACEF = ABDE$
$ACD = BDE = ABF = CEF$	

relation. Therefore, this is a resolution IV design; main effects are aliased with three-factor and higher interactions, and two-factor interactions are aliased with each other. This design provides good information on the main effects and gives some idea about the strength of the two-factor interactions. The construction and analysis of the design are illustrated in Example 7-7.

EXAMPLE 7-7

Injection-
Molding
Experiment

Parts manufactured in an injection-molding process are showing excessive shrinkage, which is causing problems in assembly operations upstream from the injection-molding area. In an effort to reduce the shrinkage, a quality-improvement team has decided to use a designed experiment to study the injection-molding process. The team investigates six factors—mold temperature (A), screw speed (B), holding time (C), cycle time (D), gate size (E), and holding pressure (F)—each at two levels, with the objective of learning how each factor affects shrinkage and obtaining preliminary information about how the factors interact.

The team decides to use a 16-run two-level fractional factorial design for these six factors. The design is constructed by writing down a 2^4 as the basic design in the factors A, B, C, and D and then setting $E = ABC$ and $F = BCD$ as discussed previously. Table 7-19 shows the design, along with the observed shrinkage ($\times 10$) for the test part produced at each of the 16 runs in the design. Analyze this fractional factorial design and identify the factor levels that reduce average part shrinkage with low part-to-part variability.

Solution. Minitab is useful for analyzing fractional factorial designs. Table 7-20 shows the Minitab output for the 2^{6-2} fractional factorial design in this example. The design generators and aliases are displayed when the design is initially created. The effect estimates and ANOVA table are displayed when the design is analyzed. The F-tests are not shown in the ANOVA output because nonsignificant effects have not yet been pooled into an estimate of error.

A normal probability plot of the effect estimates from this experiment is shown in Fig. 7-29. The only large effects are A (mold temperature), B (screw speed), and the AB interaction. In light of the alias relationship in Table 7-18, it seems reasonable to tentatively adopt these conclusions. The plot of the AB interaction in Fig. 7-30 shows that the process is insensitive to temperature if the screw speed is at the

Table 7-19   A 2_{IV}^{6-2} Design for the Injection-Molding Experiment in Example 7-7

Run	A	B	C	D	$E = ABC$	$F = BCD$	Observed Shrinkage ($\times 10$)
1	−	−	−	−	−	−	6
2	+	−	−	−	+	−	10
3	−	+	−	−	+	+	32
4	+	+	−	−	−	+	60
5	−	−	+	−	+	+	4
6	+	−	+	−	−	+	15
7	−	+	+	−	−	−	26
8	+	+	+	−	+	−	60
9	−	−	−	+	−	+	8
10	+	−	−	+	+	+	12
11	−	+	−	+	+	−	34
12	+	+	−	+	−	−	60
13	−	−	+	+	+	−	16
14	+	−	+	+	−	−	5
15	−	+	+	+	−	+	37
16	+	+	+	+	+	+	52

Table 7-20   Analysis of the 2^{6-2} Fractional Factorial Design for Example 7-7 from Minitab

Factorial Design

Fractional Factorial Design

Factors:	6	Base Design:	6, 16	Resolution: IV
Runs:	16	Replicates:	1	Fraction 1/4
Blocks:	none	Center pts (total):	0	

Design Generators: E = ABC   F = BCD

Alias Structure

1 + ABCE + ADEF + BCDF
A + BCE + DEF + ABCDF
B + ACE + CDF + ABDEF
C + ABE + BDF + ACDEF
D + AEF + BCF + ABCDE
E + ABC + ADF + BCDEF
F + ADE + BCD + ABCEF
AB + CE + ACDF + BDEF
AC + BE + ABDF + CDEF
AD + EF + ABCF + BCDE
AE + BC + DF + ABCDEF
AF + DE + ABCD + BCEF
BD + CF + ABEF + ACDE
BF + CD + ABDE + ACEF
ABD + ACF + BEF + CDE
ABF + ACD + BDE + CEF

Fractional Factorial Fit

Estimated Effects and Coefficients for y

Term	Effect	Coef
Constant		27.313
A	13.875	6.938
B	35.625	17.812
C	−0.875	−0.438
D	1.375	0.687
E	0.375	0.187
F	0.375	0.188
A*B	11.875	5.938
A*C	−1.625	−0.813
A*D	−5.375	−2.688
A*E	−1.875	−0.937
A*F	0.625	0.313
B*D	−0.125	−0.062
B*F	−0.125	−0.062
A*B*D	0.125	0.062
A*B*F	−4.875	−2.437

Analysis of Variance for y

Source	DF	Seq SS	Adj SS	Adj MS	F	P
Main Effects	6	5858.37	5858.37	976.40	**	
2-Way Interactions	7	705.94	705.94	100.85	**	
3-Way Interactions	2	95.12	95.12	47.56	**	
Residual Error	0	0.00	0.00	0.00		
Total	15	6659.44				

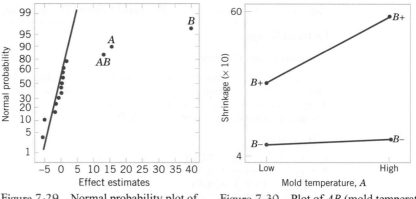

Figure 7-29   Normal probability plot of effects for Example 7-7.

Figure 7-30   Plot of AB (mold temperature–screw speed) interaction for Example 7-7.

low level but sensitive to temperature if the screw speed is at the high level. With the screw speed at a low level, the process should produce an average shrinkage of around 10% regardless of the temperature level chosen.

Interpret the Results   Based on this initial analysis, the team decides to set both the mold temperature and the screw speed at the low level. This set of conditions should reduce the mean shrinkage of parts to around 10%. However, the variability in shrinkage from part to part is still a potential problem. In effect, the mean shrinkage can be adequately reduced by the above modifications; however, the part-to-part variability in shrinkage over a production run could still cause problems in assembly. One way to address this issue is to see if any of the process factors affect the variability in parts shrinkage.

Figure 7-31 presents the normal probability plot of the residuals. This plot appears satisfactory. The plots of residuals versus each factor were then constructed. One of these plots, that for residuals versus factor C (holding time), is shown in Fig. 7-32. The plot reveals much less scatter in the residuals at the low holding time than at the high holding time. These residuals were obtained in the usual way from a model for predicted shrinkage

$$\hat{y} = \hat{\beta}_0 + \hat{\beta}_1 x_1 + \hat{\beta}_2 x_2 + \hat{\beta}_{12} x_1 x_2$$
$$= 27.3125 + 6.9375 x_1 + 17.8125 x_2 + 5.9375 x_1 x_2$$

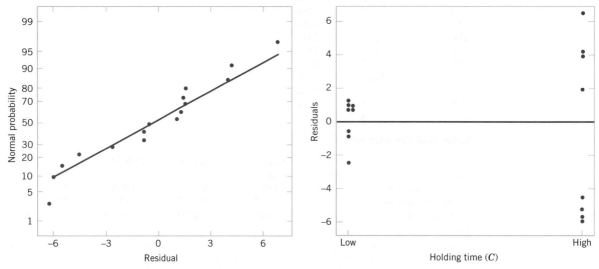

Figure 7-31   Normal probability plot of residuals for Example 7-7.

Figure 7-32   Residuals versus holding time (C) for Example 7-7.

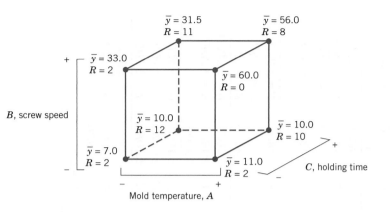

Figure 7-33
Average shrinkage and range of shrinkage in factors A, B, and C for Example 7-7.

where x_1, x_2, and x_1x_2 are coded variables that correspond to the factors A and B and the AB interaction. The residuals are then

$$e = y - \hat{y}$$

The regression model used to produce the residuals essentially removes the location effects of A, B, and AB from the data; the residuals therefore contain information about unexplained variability. Figure 7-32 indicates that there is a pattern in the variability and that the variability in the shrinkage of parts may be smaller when the holding time is at the low level.

Select the Preferred Process Settings

Figure 7-33 shows the data from this experiment projected onto a cube in the factors A, B, and C. The average observed shrinkage and the range of observed shrinkage are shown at each corner of the cube. From inspection of this figure, we see that running the process with the screw speed (B) at the low level is the key to reducing average parts shrinkage. If B is low, virtually any combination of temperature (A) and holding time (C) will result in low values of average parts shrinkage. However, from examining the ranges of the shrinkage values at each corner of the cube, it is immediately clear that setting the holding time (C) at the low level is the most appropriate choice if we wish to keep the part-to-part variability in shrinkage low during a production run. ∎

The concepts used in constructing the 2^{6-2} fractional factorial design in Example 7-7 can be extended to the construction of any 2^{k-p} fractional factorial design. In general, a 2^k fractional factorial design containing 2^{k-p} runs is called a $1/2^p$ fraction of the 2^k design or, more simply, a 2^{k-p} fractional factorial design. These designs require the selection of p independent generators. The defining relation for the design consists of the p generators initially chosen and their $2^p - p - 1$ generalized interactions.

The alias structure may be found by multiplying each effect column by the defining relation. Care should be exercised in choosing the generators so that effects of potential interest are not aliased with each other. Each effect has $2^p - 1$ aliases. For moderately large values of k, we usually assume higher-order interactions (say, third- or fourth-order or higher) to be negligible, and this greatly simplifies the alias structure.

It is important to select the p generators for the 2^{k-p} fractional factorial design in such a way that we obtain the best possible alias relationships. A reasonable criterion is to select the generators so that the resulting 2^{k-p} design has the highest possible design resolution. Montgomery (2005a) presents a table of recommended generators for 2^{k-p} fractional factorial designs for $k \le 11$ factors and up to $n \le 128$ runs. A portion of this table is reproduced here as Table 7-21. In this table, the generators are shown with either + or − choices; selection of all generators as + will give a principal fraction, whereas if any generators are − choices, the design will be one of the alternate fractions for the same family. The suggested generators in this table will result in a design of the highest possible resolution. Montgomery (2005a) also gives a table of alias relationships for these designs.

Table 7-21 Selected 2^{k-p} Fractional Factorial Designs

Number of Factors k	Fraction	Number of Runs	Design Generators
3	2^{3-1}_{III}	4	$C = \pm AB$
4	2^{4-1}_{IV}	8	$D = \pm ABC$
5	2^{5-1}_{V}	16	$E = \pm ABCD$
	2^{5-2}_{III}	8	$D = \pm AB$, $E = \pm AC$
6	2^{6-1}_{VI}	32	$F = \pm ABCDE$
	2^{6-2}_{IV}	16	$E = \pm ABC$, $F = \pm BCD$
	2^{6-3}_{III}	8	$D = \pm AB$, $E = \pm AC$, $F = \pm BC$
7	2^{7-1}_{VII}	64	$G = \pm ABCDEF$
	2^{7-2}_{IV}	32	$F = \pm ABCD$, $G = \pm ABDE$
	2^{7-3}_{IV}	16	$E = \pm ABC$, $F = \pm BCD$, $G = \pm ACD$
	2^{7-4}_{III}	8	$D = \pm AB$, $E = \pm AC$, $F = \pm BC$, $G = \pm ABC$
8	2^{8-2}_{V}	64	$G = \pm ABCD$, $H = \pm ABEF$
	2^{8-3}_{IV}	32	$F = \pm ABC$, $G = \pm ABD$, $H = \pm BCDE$
	2^{8-4}_{IV}	16	$E = \pm BCD$, $F = \pm ACD$, $G = \pm ABC$, $H = \pm ABD$
9	2^{9-2}_{VI}	128	$H = \pm ACDFG$, $J = \pm BCEFG$
	2^{9-3}_{IV}	64	$G = \pm ABCD$, $H = \pm ACEF$, $J = \pm CDEF$
	2^{9-4}_{IV}	32	$F = \pm BCDE$, $G = \pm ACDE$, $H = \pm ABDE$, $J = \pm ABCE$
	2^{9-5}_{III}	16	$E = \pm ABC$, $F = \pm BCD$, $G = \pm ACD$, $H = \pm ABD$, $J = \pm ABCD$
10	2^{10-3}_{V}	128	$H = \pm ABCG$, $J = \pm ACDE$, $K = \pm ACDF$
	2^{10-4}_{IV}	64	$G = \pm BCDF$, $H = \pm ACDF$, $J = \pm ABDE$, $K = \pm ABCE$
	2^{10-5}_{IV}	32	$F = \pm ABCD$, $G = \pm ABCE$, $H = \pm ABDE$, $J = \pm ACDE$, $K = \pm BCDE$
	2^{10-6}_{III}	16	$E = \pm ABC$, $F = \pm BCD$, $G = \pm ACD$, $H = \pm ABD$, $J = \pm ABCD$, $K = \pm AB$
11	2^{11-5}_{IV}	64	$G = \pm CDE$, $H = \pm ABCD$, $J = \pm ABF$, $K = \pm BDEF$, $L = \pm ADEF$
	2^{11-6}_{IV}	32	$F = \pm ABC$, $G = \pm BCD$, $H = \pm CDE$, $J = \pm ACD$, $K = \pm ADE$, $L = \pm BDE$
	2^{11-7}_{III}	16	$E = \pm ABC$, $F = \pm BCD$, $G = \pm ACD$, $H = \pm ABD$, $J = \pm ABCD$, $K = \pm AB$, $L = \pm AC$

Source: Montgomery (2005a).

Table 7-22 Generators, Defining Relation, and Aliases for the 2_{IV}^{7-3} Fractional Factorial Design Discussed in Example 7-8

Generators and Defining Relation

$$E = ABC \qquad F = BCD \qquad G = ACD$$

$$I = ABCE = BCDF = ADEF = ACDG = BDEG = ABFG = CEFG$$

Aliases

$$A = BCE = DEF = CDG = BFG \qquad AB = CE = FG$$
$$B = ACE = CDF = DEG = AFG \qquad AC = BE = DG$$
$$C = ABE = BDF = ADG = EFG \qquad AD = EF = CG$$
$$D = BCF = AEF = ACG = BEG \qquad AE = BC = DF$$
$$E = ABC = ADF = BDG = CFG \qquad AF = DE = BG$$
$$F = BCD = ADE = ABG = CEG \qquad AG = CD = BF$$
$$G = ACD = BDE = ABF = CEF \qquad BD = CF = EG$$

$$ABD = CDE = ACF = BEF = BCG = AEG = DFG$$

EXAMPLE 7-8

Aliases with Seven Factors

To illustrate the use of Table 7-21, suppose that we have seven factors and that we are interested in estimating the seven main effects and obtaining some insight regarding the two-factor interactions. We are willing to assume that three-factor and higher interactions are negligible. This information suggests that a resolution IV design would be appropriate.

Solution. Table 7-21 shows that two resolution IV fractions are available: the 2_{IV}^{7-2} with 32 runs and the 2_{IV}^{7-3} with 16 runs. The aliases involving main effects and two- and three-factor interactions for the 16-run design are presented in Table 7-22. Note that all seven main effects are aliased with three-factor interactions. All the two-factor interactions are aliased in groups of three. Therefore, this design will satisfy our objectives; that is, it will allow the estimation of the main effects, and it will give some insight regarding two-factor interactions. It is not necessary to run the 2_{IV}^{7-2} design, which would require 32 runs. The construction of the 2_{IV}^{7-3}

Table 7-23 A 2_{IV}^{7-3} Fractional Factorial Design Discussed in Example 7-8

	Basic Design						
Run	A	B	C	D	E = ABC	F = BCD	G = ACD
1	−	−	−	−	−	−	−
2	+	−	−	−	+	−	+
3	−	+	−	−	+	+	−
4	+	+	−	−	−	+	+
5	−	−	+	−	+	+	+
6	+	−	+	−	−	+	−
7	−	+	+	−	−	−	+
8	+	+	+	−	+	−	−
9	−	−	−	+	−	+	+
10	+	−	−	+	+	+	−
11	−	+	−	+	+	−	+
12	+	+	−	+	−	−	−
13	−	−	+	+	+	−	−
14	+	−	+	+	−	−	+
15	−	+	+	+	−	+	−
16	+	+	+	+	+	+	+

Table 7-24   A 2_{III}^{7-4} Fractional Factorial Design Discussed in Example 7-9

A	B	C	$D = AB$	$E = AC$	$F = BC$	$G = ABC$
$-$	$-$	$-$	$+$	$+$	$+$	$-$
$+$	$-$	$-$	$-$	$-$	$+$	$+$
$-$	$+$	$-$	$-$	$+$	$-$	$+$
$+$	$+$	$-$	$+$	$-$	$-$	$-$
$-$	$-$	$+$	$+$	$-$	$-$	$+$
$+$	$-$	$+$	$-$	$+$	$-$	$-$
$-$	$+$	$+$	$-$	$-$	$+$	$-$
$+$	$+$	$+$	$+$	$+$	$+$	$+$

design is shown in Table 7-23. Note that it was constructed by starting with the 16-run 2^4 design in A, B, C, and D as the basic design and then adding the three columns $E = ABC$, $F = BCD$, and $G = ACD$ as suggested in Table 7-21. Thus, the generators for this design are $I = ABCE$, $I = BCDF$, and $I = ACDG$. The complete defining relation is $I = ABCE = BCDF = ADEF = ACDG = BDEG = CEFG = ABFG$. This defining relation was used to produce the aliases in Table 7-22. For example, the alias relationship of A is

$$A = BCE = ABCDF = DEF = CDG = ABDEG = ACEFG = BFG$$

which, if we ignore interactions higher than three factors, agrees with those in Table 7-22.   ■

EXAMPLE 7-9

Saturated
Fractional
Factorial
Design

For seven factors, reduce the number of runs even further.

Solution. The 2^{7-4} design is an eight-run experiment accommodating seven variables. This is a 1/16th fraction and is obtained by first writing down a 2^3 design as the basic design in the factors A, B, and C, and then forming the four new columns from $I = ABD$, $I = ACE$, $I = BCF$, and $I = ABCG$, as suggested in Table 7-20. The design is shown in Table 7-24.

The complete defining relation is found by multiplying the generators together two, three, and finally four at a time, producing

$$I = ABD = ACE = BCF = ABCG = BCDE = ACDF = CDG = ABEF$$
$$= BEG = AFG = DEF = ADEG = CEFG = BDFG = ABCDEFG$$

The alias of any main effect is found by multiplying that effect through each term in the defining relation. For example, the alias of A is

$$A = BD = CE = ABCF = BCG = ABCDE = CDF = ACDG$$
$$= BEF = ABEG = FG = ADEF = DEG = ACEFG = ABDFG$$
$$= BCDEFG$$

This design is of resolution III, because the main effect is aliased with two-factor interactions. If we assume that all three-factor and higher interactions are negligible, the aliases of the seven main effects are

$$A = BD = CE = FG$$
$$B = AD = CF = EG$$
$$C = AE = BF = DG$$
$$D = AB = CG = EF$$

$$E = AC = BG = DF$$
$$F = BC = AG = DE$$
$$G = CD = BE = AF$$

This 2_{III}^{7-4} design is called a **saturated fractional factorial** because all the available degrees of freedom are used to estimate main effects. It is possible to combine sequences of these resolution III fractional factorials to separate the main effects from the two-factor interactions. The procedure is illustrated in Montgomery (2005a). ∎

EXERCISES FOR SECTION 7-7

7-27. R. D. Snee ("Experimenting with a Large Number of Variables," in *Experiments in Industry: Design, Analysis and Interpretation of Results,* by R. D. Snee, L. D. Hare, and J. B. Trout, eds., ASQC, 1985) describes an experiment in which a 2^{5-1} design with $I = ABCDE$ was used to investigate the effects of five factors on the color of a chemical product. The factors are A = solvent/reactant, B = catalyst/reactant, C = temperature, D = reactant purity, and E = reactant pH. The results obtained are as follows:

e	$= -0.63$	d	$= 6.79$
a	$= 2.51$	ade	$= 6.47$
b	$= -2.68$	bde	$= 3.45$
abe	$= 1.66$	abd	$= 5.68$
c	$= 2.06$	cde	$= 5.22$
ace	$= 1.22$	acd	$= 4.38$
bce	$= -2.09$	bcd	$= 4.30$
abc	$= 1.93$	$abcde$	$= 4.05$

(a) Write down the complete defining relation and the aliases from the design.
(b) Estimate the effects and prepare a normal probability plot of the effects. Which effects are active?
(c) Interpret the effects and develop an appropriate model for the response.
(d) Plot the residuals from your model against the fitted values. Also construct a normal probability plot of the residuals. Comment on the results.

7-28. Montgomery (2005a) describes a 2^{4-1} fractional factorial design used to study four factors in a chemical process. The factors are A = temperature, B = pressure, C = concentration, and D = stirring rate, and the response is filtration rate. The design and the data are shown in Table 7-25 below.
(a) Write down the complete defining relation and the aliases from the design.
(b) Estimate the effects and prepare a normal probability plot of the effects. Which effects are active?
(c) Interpret the effects and develop an appropriate model for the response.
(d) Plot the residuals from your model against the fitted values. Also construct a normal probability plot of the residuals. Comment on the results.

7-29. An article in *Industrial and Engineering Chemistry* ("More on Planning Experiments to Increase Research Efficiency," Vol. 62, 1970, pp. 60–65) uses a 2^{5-2} design to investigate the effect on process yield of A = condensation temperature, B = amount of material 1, C = solvent volume, D = condensation time, and E = amount of material 2. The results obtained are as follows:

e	$= 23.2$	cd	$= 23.8$
ab	$= 15.5$	ace	$= 23.4$
ad	$= 16.9$	bde	$= 16.8$
bc	$= 16.2$	$abcde$	$= 18.1$

Table 7-25 Data for Exercise 7-28

Run	A	B	C	$D = ABC$	Treatment Combination	Filtration Rate
1	−	−	−	−	(1)	45
2	+	−	−	+	ad	100
3	−	+	−	+	bd	45
4	+	+	−	−	ab	65
5	−	−	+	+	cd	75
6	+	−	+	−	ac	60
7	−	+	+	−	bc	80
8	+	+	+	+	$abcd$	96

(a) Write down the complete defining relation and the aliases from the design. Verify that the design generators used were $I = ACE$ and $I = BDE$.
(b) Estimate the effects and prepare a normal probability plot of the effects. Which effects are active? Verify that the AB and AD interactions are available to use as error.
(c) Interpret the effects and develop an appropriate model for the response.
(d) Plot the residuals from your model against the fitted values. Also construct a normal probability plot of the residuals. Comment on the results.

7-30. Suppose that in Exercise 7-12 it was possible to run only a $\frac{1}{2}$ fraction of replicate I for the 2^4 design. Construct the design and use only the data from the eight runs you have generated to perform the analysis.

7-31. Construct the table of the treatment combinations tested for the 2_{IV}^{6-2} recommended in Table 7-21.

7-32. Suppose that in Exercise 7-14 only a $\frac{1}{4}$ fraction of the 2^5 design could be run. Construct the design and analyze the data that are obtained by selecting only the response for the eight runs in your design.

7-33. Construct a 2_{III}^{6-3} fractional factorial design. Write down the aliases, assuming that only main effects and two-factor interactions are of interest.

7-34. Consider the 2_{IV}^{7-3} design generated by Minitab and shown below. Confirm that the design generators are E = ABC, F = BCD, G = ACD.

A	B	C	D	E	F	G
1	-1	-1	1	1	1	-1
1	-1	1	1	-1	-1	1
-1	-1	1	-1	1	1	1
-1	1	-1	-1	1	1	-1
1	1	1	1	1	1	1
-1	-1	1	1	1	-1	-1
1	-1	1	-1	-1	1	-1
1	1	-1	1	-1	-1	-1
-1	1	-1	1	1	-1	1
-1	1	1	1	-1	-1	1
1	1	-1	-1	-1	1	1
-1	-1	-1	-1	-1	-1	-1
-1	1	1	-1	-1	-1	1
1	-1	-1	-1	1	-1	1
-1	-1	-1	1	-1	1	1
1	1	1	-1	1	-1	-1

7-35. Consider the 2_{IV}^{8-4} design generated by Minitab and shown below. Confirm that the design generators are E = BCD, F = ACD, G = ABC, H = ABD.

A	B	C	D	E	F	G	H
-1	1	1	-1	-1	1	-1	1
-1	-1	-1	-1	-1	-1	-1	-1
-1	1	-1	-1	1	-1	1	1
-1	-1	1	1	-1	-1	1	1
1	-1	-1	-1	-1	1	1	1
1	-1	1	-1	1	-1	-1	1
1	1	-1	-1	1	1	-1	-1
1	1	1	-1	-1	-1	1	-1
-1	1	1	1	1	-1	-1	-1
-1	1	-1	1	-1	1	1	-1
1	-1	1	1	-1	1	-1	-1
1	-1	-1	1	1	-1	1	-1
-1	-1	-1	1	1	1	-1	1
-1	-1	1	-1	1	1	1	-1
1	1	1	1	1	1	1	1
1	1	-1	1	-1	-1	-1	1

7-8 RESPONSE SURFACE METHODS AND DESIGNS

Response surface methodology, or RSM, is a collection of mathematical and statistical techniques that are useful for modeling and analysis in applications where a response of interest is influenced by several variables and the objective is to **optimize** this response. For example, suppose that a chemical engineer wishes to find the levels of temperature (x_1) and feed concentration (x_2) that maximize the yield (y) of a process. The process yield is a function of the levels of temperature and feed concentration—say,

$$Y = f(x_1, x_2) + \epsilon \tag{7-19}$$

where ϵ represents the noise or error observed in the response Y. If we denote the expected response by $E(Y) = f(x_1, x_2)$, the surface represented by $f(x_1, x_2)$ is called a **response surface.**

We may represent the response surface graphically as shown in Fig. 7-34, where $f(x_1, x_2)$ is plotted versus the levels of x_1 and x_2. Note that the response is represented as a surface plot in a three-dimensional space. To help visualize the shape of a response surface, we often plot the contours of the response surface as shown in Fig. 7-35. In the contour plot, lines of constant response are drawn in the x_1, x_2 plane. Each contour corresponds to a particular height of the response surface. The contour plot is helpful in studying the levels of x_1 and x_2 that result in changes in the shape or height of the response surface.

In most RSM problems, the form of the relationship between the response and the independent variables is unknown. Thus, the first step in RSM is to find a suitable approximation for the true relationship between Y and the independent variables. Usually, a low-order polynomial in some region of the independent variables is employed. If the response is well modeled by a linear function of the independent variables, the approximating function is the **first-order model**

$$Y = \beta_0 + \beta_1 x_1 + \beta_2 x_2 + \cdots + \beta_k x_k + \epsilon$$

If there is curvature in the system, a polynomial of higher degree must be used, such as the **second-order model**

$$Y = \beta_0 + \sum_{i=1}^{k} \beta_i x_i + \sum_{i=1}^{k} \beta_{ii} x_i^2 + \sum\sum_{i<j} \beta_{ij} x_i x_j + \epsilon \tag{7-20}$$

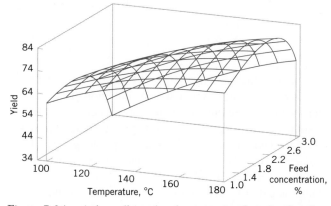

Figure 7-34   A three-dimensional response surface showing the expected yield as a function of temperature and feed concentration.

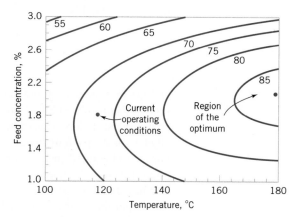

Figure 7-35   A contour plot of the yield response surface in Figure 7-34.

Many RSM problems use one or both of these approximating polynomials. Of course, it is unlikely that a polynomial model will be a reasonable approximation of the true functional relationship over the entire space of the independent variables, but for a relatively small region they usually work quite well.

The method of least squares, discussed in Chapter 6, is used to estimate the parameters in the approximating polynomials. The response surface analysis is then done in terms of the fitted surface. If the fitted surface is an adequate approximation of the true response function, analysis of the fitted surface will be approximately equivalent to analysis of the actual system.

RSM is a **sequential** procedure. Often, when we are at a point on the response surface that is remote from the optimum, such as the current operating conditions in Fig. 7-35, there is little curvature in the system and the first-order model will be appropriate. Our objective here is to lead the experimenter rapidly and efficiently to the general vicinity of the optimum. Once the region of the optimum has been found, a more elaborate model such as the second-order model may be employed, and an analysis may be performed to locate the optimum. From Fig. 7-35, we see that the analysis of a response surface can be thought of as "climbing a hill," where the top of the hill represents the point of maximum response. If the true optimum is a point of minimum response, we may think of "descending into a valley."

The eventual objective of RSM is to determine the optimum operating conditions for the system or to determine a region of the factor space in which operating specifications are satisfied. Also, note that the word "optimum" in RSM is used in a special sense. The hill-climbing procedures of RSM guarantee convergence to a local optimum only. Nevertheless, experiments carried out using this approach are called **optimization experiments.**

7-8.1 Method of Steepest Ascent

Frequently, the initial estimate of the optimum operating conditions for the system will be far from the actual optimum. In such circumstances, the objective of the experimenter is to move rapidly to the general vicinity of the optimum. We wish to use a simple and economically efficient experimental procedure. When we are remote from the optimum, we usually assume that a first-order model is an adequate approximation to the true surface in a small region of the x's.

The **method of steepest ascent** is a procedure for moving sequentially along the path of steepest ascent—that is, in the direction of the maximum increase in the response. Of course, if **minimization** is desired, we are talking about the **method of steepest descent.** The fitted first-order model is

$$\hat{y} = \hat{\beta}_0 + \sum_{i=1}^{k} \hat{\beta}_i x_i \qquad (7\text{-}21)$$

and the first-order response surface—that is, the contours of $\hat{y}$—is a series of parallel lines such as that shown in Fig. 7-36. The direction of steepest ascent is the direction in which $\hat{y}$ increases most rapidly. This direction is normal to the fitted response surface contours. We usually take as the **path of steepest ascent** the line through the center of the region of interest and normal to the fitted surface contours. As shown in Example 7-10, the steps along the path are proportional to the regression coefficients $\{\hat{\beta}_i\}$. The experimenter determines the actual step size based on process knowledge or other practical considerations.

Experiments are conducted along the path of steepest ascent until no further increase in response is observed. Then a new first-order model may be fit, a new direction of steepest ascent determined, and further experiments conducted in that direction until the experimenter feels that the process is near the optimum.

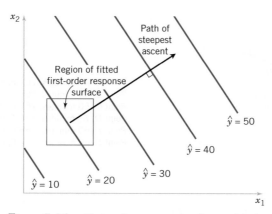

Figure 7-36   First-order response surface and path of steepest ascent.

EXAMPLE 7-10

Process Yield

In Example 7-4 we described an experiment on a chemical process in which two factors, reaction time (x_1) and reaction temperature (x_2), affect the percent conversion or yield (Y). Figure 7-20 shows the 2^2 design plus five center points used in this study. The engineer found that both factors were important, there was no interaction, and there was no curvature in the response surface. Therefore, the first-order model

$$Y = \beta_0 + \beta_1 x_1 + \beta_2 x_2 + \epsilon$$

should be appropriate. Now, the effect estimate of time is 1.55 and the effect estimate of temperature is 0.65, and because the regression coefficients $\hat{\beta}_1$ and $\hat{\beta}_2$ are one-half of the corresponding effect estimates, the fitted first-order model is

$$\hat{y} = 40.44 + 0.775x_1 + 0.325x_2$$

Figure 7-37 shows the contour plot and three-dimensional surface plot of this model. Figure 7-37 also shows the relationship between the **coded variables** x_1 and x_2 (which defined the high and low levels of the

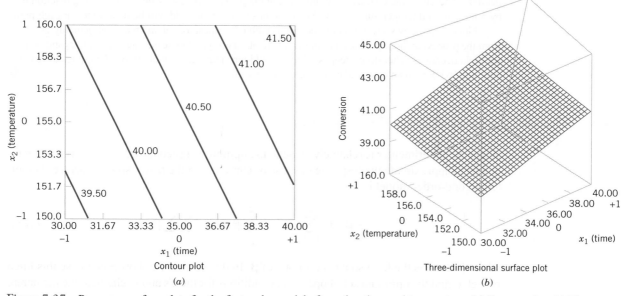

Figure 7-37   Response surface plots for the first-order model of reaction time and temperature. (a) Contour plot. (b) Three-dimensional surface plot.

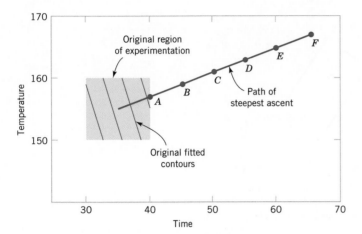

Point A: 40 minutes, 157°F, $y = 40.5$
Point B: 45 minutes, 159°F, $y = 51.3$
Point C: 50 minutes, 161°F, $y = 59.6$
Point D: 55 minutes, 163°F, $y = 67.1$
Point E: 60 minutes, 165°F, $y = 63.6$
Point F: 65 minutes, 167°F, $y = 60.7$

Figure 7-38   Steepest ascent experiment for the first-order model of reaction time and temperature.

factors) and the original variables time (in minutes) and temperature (in °F). Use this fitted first-order model to identify the direction of maximum increase in process yield.

Solution. From examining these plots (or the fitted model), we see that to move away from the design center—the point ($x_1 = 0$, $x_2 = 0$)—along the path of steepest ascent, we would move 0.775 unit in the x_1 direction for every 0.325 unit in the x_2 direction. Thus, the path of steepest ascent passes through the point ($x_1 = 0$, $x_2 = 0$) and has a slope 0.325/0.775. The engineer decides to use 5 minutes of reaction time as the basic step size. Now, 5 minutes of reaction time is equivalent to a step in the *coded* variable x_1 of $\Delta x_1 = 1$. Therefore, the steps along the path of steepest ascent are

$$\Delta x_1 = 1.000$$
$$\Delta x_2 = (\hat{\beta}_2/\hat{\beta}_1)\Delta x_1 = (0.325/0.775)\Delta x_1 = 0.42 \tag{7-22}$$

A change of $\Delta x_2 = 0.42$ in the coded variable x_2 is equivalent to about 2°F in the original variable temperature. Therefore, the engineer will move along the path of steepest ascent by increasing reaction time by 5 minutes and temperature by 2°F. An actual observation on yield will be determined at each point.

Figure 7-38 shows several points along this path of steepest ascent and the yields actually observed from the process at those points. At points A–D the observed yield increases steadily, but beyond point D the yield decreases. Therefore, steepest ascent would terminate in the vicinity of 55 minutes of reaction time and 163°F with an observed percent conversion of 67%. ∎

7-8.2 Analysis of a Second-Order Response Surface

When the experimenter is relatively close to the optimum, a second-order model is usually required to approximate the response because of curvature in the true response surface. The fitted second-order model is

$$\hat{y} = \hat{\beta}_0 + \sum_{i=1}^{k} \hat{\beta}_i x_i + \sum_{i=1}^{k} \hat{\beta}_{ii} x_i^2 + \sum\sum_{i<j} \hat{\beta}_{ij} x_i x_j \tag{7-23}$$

where $\hat{\beta}$ denotes the least squares estimate of β. In this section we show how to use this fitted model to find the optimum set of operating conditions for the x's and to characterize the nature of the response surface.

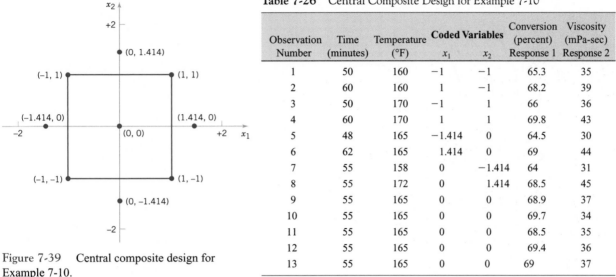

Figure 7-39   Central composite design for Example 7-10.

Table 7-26   Central Composite Design for Example 7-10

Observation Number	Time (minutes)	Temperature (°F)	Coded Variables x_1	x_2	Conversion (percent) Response 1	Viscosity (mPa-sec) Response 2
1	50	160	−1	−1	65.3	35
2	60	160	1	−1	68.2	39
3	50	170	−1	1	66	36
4	60	170	1	1	69.8	43
5	48	165	−1.414	0	64.5	30
6	62	165	1.414	0	69	44
7	55	158	0	−1.414	64	31
8	55	172	0	1.414	68.5	45
9	55	165	0	0	68.9	37
10	55	165	0	0	69.7	34
11	55	165	0	0	68.5	35
12	55	165	0	0	69.4	36
13	55	165	0	0	69	37

EXAMPLE 7-10 (continued)

Process Yield

The method of steepest ascent terminated at a reaction time of 55 minutes and a temperature of 163°F. The experimenter decides to fit a second-order model in this region. Table 7-26 and Fig. 7-39 show the experimental design, which consists of a 2^2 design centered at 55 minutes and 165°F, five center points, and four runs along the coordinate axes called axial runs. This type of design is called a **central composite design** (CCD), and it is a very popular design for fitting second-order response surfaces.

Central Composite Design

Solution. Two response variables were measured during this phase of the experiment: percent conversion (yield) and viscosity. The least squares quadratic model for the yield response is

$$\hat{y}_1 = 69.1 + 1.633x_1 + 1.083x_2 - 0.969x_1^2 - 1.219x_2^2 + 0.225x_1x_2$$

The analysis of variance for this model is shown in Table 7-27. Because the coefficient of the x_1x_2 term is not significant, one might choose to remove this term from the model.

Table 7-27   Analysis of Variance for the Quadratic Model, Yield Response for Example 7-10

Source of Variation	Sum of Squares	Degrees of Freedom	Mean Square	f_0	P-value
Model	45.89	5	9.178	14.93	0.0013
Residual	4.30	7	0.615		
Total	50.19	12			

Independent Variable	Coefficient Estimate	Standard Error	t for H_0 Coefficient $= 0$	P-value
Intercept	69.100	0.351	197.1	0.0000
x_1	1.633	0.277	5.891	0.0006
x_2	1.083	0.277	3.907	0.0058
x_1^2	−0.969	0.297	−3.259	0.0139
x_2^2	−1.219	0.297	−4.100	0.0046
x_1x_2	0.225	0.392	0.5740	0.5839

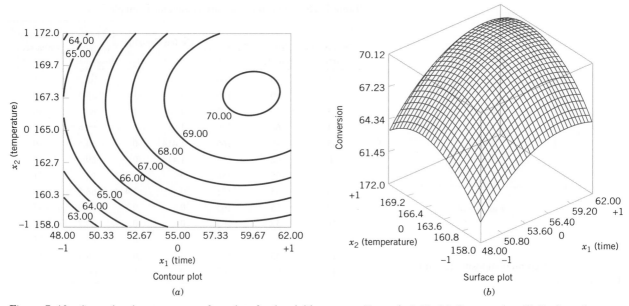

Figure 7-40   Second-order response surface plots for the yield response, Example 7-10. (*a*) Contour plot. (*b*) Surface plot.

Reconciling Two
Responses

Figure 7-40 shows the response surface contour plot and the three-dimensional surface plot for this model. From examination of these plots, the maximum yield is about 70%, obtained at approximately 60 minutes of reaction time and 167°F.

The viscosity response is adequately described by the first-order model

$$\hat{y}_2 = 37.08 + 3.85x_1 + 3.10x_2$$

Table 7-28 summarizes the analysis of variance for this model. The response surface is shown graphically in Fig. 7-41. Note that viscosity increases as both time and temperature increase.

As in most response surface problems, the experimenter in this example had conflicting objectives regarding the two responses. The objective was to maximize yield, but the acceptable range for viscosity was $38 \le y_2 \le 42$. When there are only a few independent variables, an easy way to solve this problem is to overlay the response surfaces to find the optimum. Figure 7-42 shows the overlay plot of both responses, with the contours $y_1 = 69\%$ conversion, $y_2 = 38$, and $y_2 = 42$ highlighted. The shaded areas

Table 7-28   Analysis of Variance for the First-Order Model, Viscosity Response for Example 7-10

Source	Sum of Squares	Degrees of Freedom	Mean Square	f_0	*P*-value
Model	195.4	2	97.72	15.89	0.0008
Residual	61.5	10	6.15		
Total	256.9	12			

Independent Variable	Coefficient Estimate	Degrees of Freedom	Standard Error of Coefficient	*t* for H_0 Coefficient $= 0$	*P*-value
Intercept	37.08	1	0.69	53.91	
x_1	3.85	1	0.88	4.391	0.0014
x_2	3.10	1	0.88	3.536	0.0054

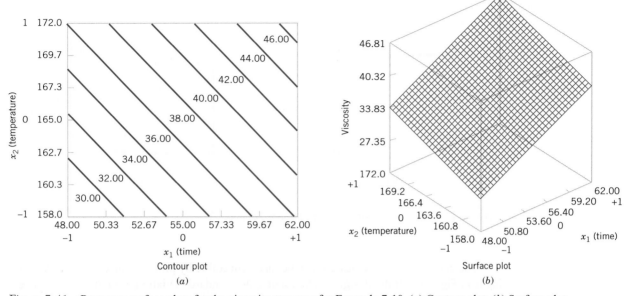

Figure 7-41   Response surface plots for the viscosity response for Example 7-10. (*a*) Contour plot. (*b*) Surface plot.

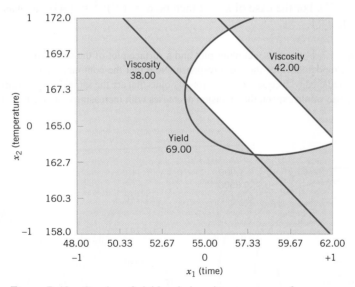

Figure 7-42   Overlay of yield and viscosity response surfaces, Example 7-10.

on this plot identify infeasible combinations of time and temperature. This graph shows that several combinations of time and temperature will be satisfactory. ∎

Example 7-10 illustrates the use of a central composite design for fitting a second-order response surface model. These designs are widely used in practice because they are relatively efficient with respect to the number of runs required. In general, a CCD in k factors requires 2^k factorial runs, $2k$ axial runs, and at least one center point (three to five center points are typically used). Designs for $k = 2$ and $k = 3$ factors are shown in Fig. 7-43.

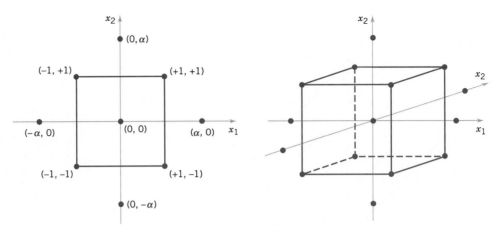

Figure 7-43   Central composite designs for $k = 2$ and $k = 3$.

The central composite design may be made **rotatable** by proper choice of the axial spacing α in Fig. 7-43. If the design can be rotated, the standard deviation of predicted response $\hat{y}$ is constant at all points that are the same distance from the center of the design. For rotatability, choose $\alpha = (F)^{1/4}$, where F is the number of points in the factorial part of the design (usually $F = 2^k$). For the case of $k = 2$ factors, $\alpha = (2^2)^{1/4} = 1.414$, as was used in the design in Example 7-11.

Figure 7-44 presents a contour plot and a surface plot of the standard deviation of prediction for the quadratic model used for the yield response. Note that the contours are concentric circles, implying that yield is predicted with equal precision for all points that are the same distance from the center of the design. Also, as one would expect, the precision decreases with increasing distance from the design center.

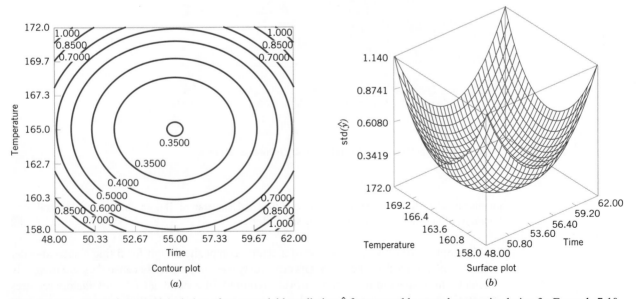

Figure 7-44   Plots of standard deviation of process yield prediction $\hat{y}$ for a rotatable central composite design for Example 7-10. (a) Contour plot. (b) Surface plot.

EXERCISES FOR SECTION 7-8

7-36.   An article in *Rubber Age* (Vol. 89, 1961, pp. 453–458) describes an experiment on the manufacture of a product in which two factors were varied: reaction time (hr) and temperature (°C). These factors are coded as $x_1 = (\text{time} - 12)/8$ and $x_2 = (\text{temperature} - 250)/30$. The following data were observed where y is the yield (in percent).

Run Number	x_1	x_2	y
1	−1	0	83.8
2	1	0	81.7
3	0	0	82.4
4	0	0	82.9
5	0	−1	84.7
6	0	1	75.9
7	0	0	81.2
8	−1.414	−1.414	81.3
9	−1.414	1.414	83.1
10	1.414	−1.414	85.3
11	1.414	1.414	72.7
12	0	0	82.0

(a) Plot the points at which the experimental runs were made.
(b) Fit a second-order model to the data. Is the second-order model adequate?
(c) Plot the yield response surface. What recommendations would you make about the operating conditions for this process?

7-37.   Consider the experimental design in the table below. This experiment was run in a chemical process.

x_1	x_2	x_3	y_1	y_2
−1	−1	−1	480	68
0	−1	−1	530	95
1	−1	−1	590	86
−1	0	−1	490	184
0	0	−1	580	220
1	0	−1	660	230
−1	1	−1	490	220
0	1	−1	600	280
1	1	−1	720	310
−1	−1	0	410	134
0	−1	0	450	189
1	−1	0	530	210
−1	0	0	400	230

x_1	x_2	x_3	y_1	y_2
0	0	0	510	300
1	0	0	590	330
−1	1	0	420	270
0	1	0	540	340
1	1	0	640	380
−1	−1	1	340	164
0	−1	1	390	250
1	−1	1	450	300
−1	0	1	340	250
0	0	1	420	340
1	0	1	520	400
−1	1	1	360	250
0	1	1	470	370
1	1	1	560	440

(a) The response y_1 is the viscosity of the product. Fit an appropriate response surface model.
(b) The response y_2 is the conversion, in grams. Fit an appropriate response surface model.
(c) Where would you recommend that we set x_1, x_2, and x_3 if the objective is to maximize conversion while keeping viscosity in the range $450 < y_1 < 500$?

7-38.   A manufacturer of cutting tools has developed two empirical equations for tool life (y_1) and tool cost (y_2). Both models are functions of tool hardness (x_1) and manufacturing time (x_2). The equations are

$$\hat{y}_1 = 10 + 5x_1 + 2x_2$$
$$\hat{y}_2 = 23 + 3x_1 + 4x_2$$

and both equations are valid over the range $-1.5 \le x_i \le 1.5$. Suppose that tool life must exceed 12 hours and cost must be below \$27.50.

(a) Is there a feasible set of operating conditions?
(b) Where would you run this process?

7-39.   An article in *Tappi* (Vol. 43, 1960, pp. 38–44) describes an experiment that investigated the ash value of paper pulp (a measure of inorganic impurities). Two variables, temperature T in degrees Celsius and time t in hours, were studied, and some of the results are shown in the following table. The coded predictor variables shown are

$$x_1 = \frac{(T - 775)}{115} \qquad x_2 = \frac{(t - 3)}{1.5}$$

and the response y is (dry ash value in %) $\times 10^3$.

x_1	x_2	y
−1	−1	211
1	−1	92
−1	1	216
1	1	99
−1.5	0	222
1.5	0	48
0	−1.5	168
0	1.5	179
0	0	122
0	0	175
0	0	157
0	0	146

x_1	x_2	x_3	y_1	y_2
1	0	−1	340.67	16.17
−1	1	−1	112.33	27.57
0	1	−1	256.33	4.62
1	1	−1	271.67	23.63
−1	−1	0	81.00	0.00
0	−1	0	101.67	17.67
1	−1	0	357.00	32.91
−1	0	0	171.33	15.01
0	0	0	372.00	0.00
1	0	0	501.67	92.50
−1	1	0	264.00	63.50
0	1	0	427.00	88.61
1	1	0	730.67	21.08
−1	−1	1	220.67	133.82
0	−1	1	239.67	23.46
1	−1	1	422.00	18.52
−1	0	1	199.00	29.44
0	0	1	485.33	44.67
1	0	1	673.67	158.21
−1	1	1	176.67	55.51
0	1	1	501.00	138.94
1	1	1	1010.00	142.45

(a) What type of design has been used in this study? Can the design be rotated?

(b) Fit a quadratic model to the data. Is this model satisfactory?

(c) If it is important to minimize the ash value, where would you run the process?

7-40. In their book, *Empirical Model Building and Response Surfaces* (Hoboken, NJ: John Wiley & Sons, 1987), G. E. P. Box and N. R. Draper describe an experiment with three factors. The data shown in the following table are a variation of the original experiment on page 247 of their book. Suppose that these data were collected in a semiconductor manufacturing process.

x_1	x_2	x_3	y_1	y_2
−1	−1	−1	24.00	12.49
0	−1	−1	120.33	8.39
1	−1	−1	213.67	42.83
−1	0	−1	86.00	3.46
0	0	−1	136.63	80.41

(a) The response y_1 is the average of three readings on resistivity for a single wafer. Fit a quadratic model to this response.

(b) The response y_2 is the standard deviation of the three resistivity measurements. Fit a linear model to this response.

(c) Where would you recommend that we set x_1, x_2, and x_3 if the objective is to hold mean resistivity at 500 and minimize the standard deviation?

7-9 FACTORIAL EXPERIMENTS WITH MORE THAN TWO LEVELS

The 2^k full and fractional factorial designs are usually used in the initial stages of experimentation. After the most important effects have been identified, one might run a factorial experiment with more than two levels to obtain details of the relationship between the response and the factors. The basic analysis of variance (ANOVA) can be modified to analyze the results from this type of experiment.

The ANOVA decomposes the total variability in the data into component parts and then compares the various elements in this decomposition. For an experiment with two factors

The **sum of squares identity for a two-factor analysis of variance** is

$$\sum_{i=1}^{a} \sum_{j=1}^{b} \sum_{k=1}^{n} (y_{ijk} - \bar{y}...)^2 = bn \sum_{i=1}^{a} (\bar{y}_{i..} - \bar{y}_{...})^2$$

$$+ an \sum_{j=1}^{b} (\bar{y}_{.j.} - \bar{y}_{...})^2$$

$$+ n \sum_{i=1}^{a} \sum_{j=1}^{b} (\bar{y}_{ij.} - \bar{y}_{i..} - \bar{y}_{.j.} + \bar{y}_{...})^2$$

$$+ \sum_{i=1}^{a} \sum_{j=1}^{b} \sum_{k=1}^{n} (y_{ijk} - \bar{y}_{ij.})^2 \qquad (7\text{-}25)$$

(with a levels for factor A and b levels for factor B), the total variability is measured by the total sum of squares of the observations

$$SS_T = \sum_{i=1}^{a} \sum_{j=1}^{b} \sum_{k=1}^{n} (y_{ijk} - \bar{y}...)^2 \qquad (7\text{-}24)$$

and the sum of squares decomposition follows. The notation is defined in Table 7-29.

The sum of squares identity may be written symbolically as

$$SS_T = SS_A + SS_B + SS_{AB} + SS_E$$

corresponding to each of the terms in equation 7-25. There are $abn - 1$ total degrees of freedom. The main effects A and B have $a - 1$ and $b - 1$ degrees of freedom, whereas the interaction effect AB has $(a - 1)(b - 1)$ degrees of freedom. Within each of the ab cells in Table 7-29, there are $n - 1$ degrees of freedom for the n replicates, and observations in the same cell can differ only because of random error. Therefore, there are $ab(n - 1)$ degrees of freedom for error. Therefore, the degrees of freedom are partitioned according to

$$abn - 1 = (a - 1) + (b - 1) + (a - 1)(b - 1) + ab(n - 1)$$

Table 7-29   Data Arrangement for a Two-Factor Factorial Design

		Factor B				Totals	Averages
		1	2	$\cdots$	b		
Factor A	1	$y_{111}, y_{112},$ $\ldots, y_{11n}$	$y_{121}, y_{122},$ $\ldots, y_{12n}$		$y_{1b1}, y_{1b2},$ $\ldots, y_{1bn}$	$y_{1..}$	$\bar{y}_{1..}$
	2	$y_{211}, y_{212},$ $\ldots, y_{21n}$	$y_{221}, y_{222},$ $\ldots, y_{22n}$		$y_{2b1}, y_{2b2},$ $\ldots, y_{2bn}$	$y_{2..}$	$\bar{y}_{2..}$
	$\vdots$						
	a	$y_{a11}, y_{a12},$ $\ldots, y_{a1n}$	$y_{a21}, y_{a22},$ $\ldots, y_{a2n}$		$y_{ab1}, y_{ab2},$ $\ldots, y_{abn}$	$y_{a..}$	$\bar{y}_{a..}$
Totals		$y_{.1.}$	$y_{.2.}$		$y_{.b.}$	$y_{...}$	
Averages		$\bar{y}_{.1.}$	$\bar{y}_{.2.}$		$\bar{y}_{.b.}$		$\bar{y}_{...}$

Table 7-30   Analysis of Variance Table for a Two-Factor Factorial

Source of Variation	Sum of Squares	Degrees of Freedom	Mean Square	F_0
A treatments	SS_A	$a-1$	$MS_A = \dfrac{SS_A}{a-1}$	$\dfrac{MS_A}{MS_E}$
B treatments	SS_B	$b-1$	$MS_B = \dfrac{SS_B}{b-1}$	$\dfrac{MS_B}{MS_E}$
Interaction	SS_{AB}	$(a-1)(b-1)$	$MS_{AB} = \dfrac{SS_{AB}}{(a-1)(b-1)}$	$\dfrac{MS_{AB}}{MS_E}$
Error	SS_E	$ab(n-1)$	$MS_E = \dfrac{SS_E}{ab(n-1)}$	
Total	SS_T	$abn-1$		

If we divide each of the sum of squares by the corresponding number of degrees of freedom, we obtain the mean squares for A, B, the interaction, and error:

$$MS_A = \frac{SS_A}{a-1} \qquad MS_B = \frac{SS_B}{b-1}$$
$$MS_{AB} = \frac{SS_{AB}}{(a-1)(b-1)} \qquad MS_E = \frac{SS_E}{ab(n-1)} \tag{7-26}$$

To test that the row, column, and interaction effects are zero, we would use the ratios

$$F_0 = \frac{MS_A}{MS_E} \quad F_0 = \frac{MS_B}{MS_E} \text{ and } F_0 = \frac{MS_{AB}}{MS_E} \tag{7-27}$$

respectively. Each test statistic is compared to an F distribution with $a-1$, $b-1$, and $(a-1) \times (b-1)$ degrees of freedom in the numerator and $ab(n-1)$ degrees of freedom in the denominator. This analysis is summarized in Table 7-30.

It is usually best to conduct the test for interaction first and then to evaluate the main effects. If interaction is not significant, interpretation of the tests on the main effects is straightforward. However, when interaction is significant, the main effects of the factors involved in the interaction may not have much practical interpretative value. Knowledge of the interaction is usually more important than knowledge about the main effects.

EXAMPLE 7-11

Aircraft Primer Paint

Aircraft primer paints are applied to aluminum surfaces by two methods, dipping and spraying. The purpose of the primer is to improve paint adhesion, and some parts can be primed using either application method. The process engineering group responsible for this operation is interested in learning whether three different primers differ in their adhesion properties. A factorial experiment was performed to investigate the effect of paint primer type and application method on paint adhesion. Three specimens were painted with each primer using each application method, a finish paint was applied, and the adhesion force was measured. The data from the experiment are shown in Table 7-31. Perform an analysis to determine the best choices for the application methods.

Solution. The sums of squares required to perform the ANOVA are computed from Minitab and summarized in Table 7-32. The experimenter has decided to use $\alpha = 0.05$. Because $f_{0.05,2,12} = 3.89$ and

Table 7-31   Adhesion Force Data for Example 7-11 for Primer Type ($i = 1, 2, 3$) and Application Method ($j = 1, 2$) with $n = 3$ Replicates

	Primer Type	Dipping	Spraying	Totals $y_{i..}$
	1	4.0, 4.5, 4.3	5.4, 4.9, 5.6	28.7
	2	5.6, 4.9, 5.4	5.8, 6.1, 6.3	34.1
	3	3.8, 3.7, 4.0	5.5, 5.0, 5.0	27.0
Totals	$y_{.j.}$	40.2	49.6	$y_{...} = 89.8$

Table 7-32   Analysis of Variance for Aircraft Primer Paint, Example 7-11

Source of Variation	Sum of Squares	Degrees of Freedom	Mean Square	f_0	P-value
Primer types	4.58	2	2.29	28.63	$2.7 \times$ E-5
Application methods	4.91	1	4.91	61.38	$5.0 \times$ E-7
Interaction	0.24	2	0.12	1.50	0.2621
Error	0.99	12	0.08		
Total	10.72	17			

$f_{0.05,1,12} = 4.75$, we conclude that the main effects of primer type and application method affect adhesion force. Furthermore, because $1.5 < f_{0.05,2,12}$, there is no indication of interaction between these factors. The last column of Table 7-32 shows the P-value for each F-ratio. Note that the P-values for the two test statistics for the main effects are considerably less than 0.05, whereas the P-value for the test statistic for the interaction is greater than 0.05.

A graph of the cell adhesion force averages $\{\bar{y}_{ij.}\}$ versus levels of primer type for each application method is shown in Fig. 7-45. The averages are available in the Minitab computer output in Table 7-34. The no-interaction conclusion is obvious in this graph because the two curves are nearly parallel. Furthermore, because a large response indicates greater adhesion force, we conclude that spraying is the best application method and that primer type 2 is most effective. ■

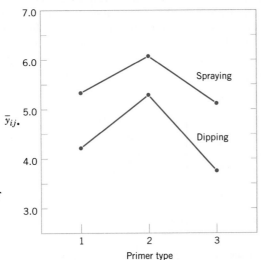

Figure 7-45   Graph of average adhesion force versus primer types for both application methods in Example 7-11.

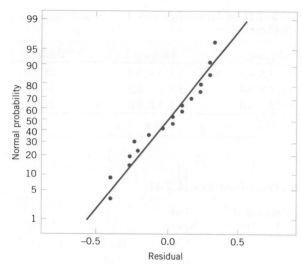

Figure 7-46   Normal probability plot of the residuals from Example 7-12.

Table 7-33   Residuals for the Aircraft Primer Experiment in Example 7-11

	Application Method	
Primer Type	Dipping	Spraying
1	$-0.27, \quad 0.23, 0.03$	$0.10, -0.40, \quad 0.30$
2	$0.30, -0.40, 0.10$	$-0.27, \quad 0.03, \quad 0.23$
3	$-0.03, -0.13, 0.17$	$0.33, -0.17, -0.17$

Model Adequacy Checking

Just as in the other experiments discussed in this chapter, the residuals from a factorial experiment play an important role in assessing model adequacy. In general, the residuals from a two-factor factorial are

$$e_{ijk} = y_{ijk} - \bar{y}_{ij}.$$

That is, the residuals are just the difference between the observations and the corresponding cell averages. If interaction is negligible, then the cell averages could be replaced by a better predictor, but we only consider the simpler case.

Analyze Residuals

Table 7-33 presents the residuals for the aircraft primer paint data in Example 7-11. The normal probability plot of these residuals is shown in Fig. 7-46. This plot has tails that do not fall exactly along a straight line passing through the center of the plot, indicating some potential problems with the normality assumption, but the deviation from normality does not appear severe. Figures 7-47 and 7-48 plot the residuals versus the levels of primer types and application methods, respectively. There is some indication that primer type 3 results in slightly lower variability in adhesion force than the other two primers. The graph of residuals versus fitted values in Fig. 7-49 does not reveal any unusual or diagnostic pattern.

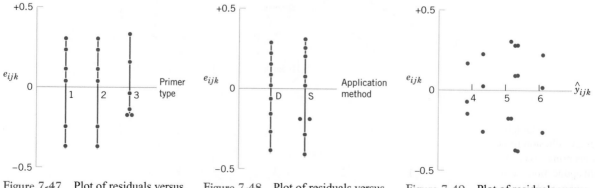

Figure 7-47   Plot of residuals versus primer type.

Figure 7-48   Plot of residuals versus application method.

Figure 7-49   Plot of residuals versus predicted values $\hat{y}_{ijk}$.

Table 7-34   Analysis of Variance from Minitab for Example 7-12 Aircraft Primer Paint

Analysis of Variance (Balanced Designs)

Factor	Type	Levels	Values		
primer	fixed	3	1	2	3
method	fixed	2	1	2	

Analysis of Variance for y

Source	DF	SS	MS	F	P
primer	2	4.5811	2.2906	27.86	0.000
method	1	4.9089	4.9089	59.70	0.000
primer*method	2	0.2411	0.1206	1.47	0.269
Error	12	0.9867	0.0822		
Total	17	10.7178			

Means

primer	N	y
1	6	4.7833
2	6	5.6833
3	6	4.5000

method	N	y
1	9	4.4667
2	9	5.5111

primer	method	N	y
1	1	3	4.2667
1	2	3	5.3000
2	1	3	5.3000
2	2	3	6.0667
3	1	3	3.8333
3	2	3	5.1667

**Interpret Minitab
Output**

Table 7-34 shows some of the output from the analysis of variance procedure in Minitab for the aircraft primer paint experiment in Example 7-11. The means table presents the sample means by primer type, by application method, and by cell (AB). The standard error for each mean is computed as $\sqrt{MS_E/m}$, where m is the number of observations in each sample mean. For example, each cell has $m = 3$ observations, so the standard error of a cell mean is $\sqrt{MS_E/3} = \sqrt{0.0822/3} = 0.1655$. A 95% CI can be determined from the mean plus or minus the standard error times the multiplier $t_{0.025,12} = 2.179$. Minitab (and many other programs) will also produce the residual plots and interaction plot shown previously.

EXERCISES FOR SECTION 7-9

7-41.   Consider the experiment in Exercise 7-2. Suppose that the experiment was actually carried out on three types of drying times and two types of paint. The data are:

Paint	Drying Time (min)		
	20	25	30
1	74	73	78
	64	61	85
	50	44	92

Paint	Drying Time (min)		
	20	25	30
2	92	98	66
	86	73	45
	68	88	85

(a) Perform the analysis of variance with $\alpha = 0.05$. What is your conclusion about the significance of the interaction effect?

(b) Assess the adequacy of the model by analyzing the residuals. What is your conclusion?

(c) If the smaller values are desirable, what levels of the factors do you recommend to obtain the necessary surface finish?

7-42. Consider the experiment in Exercise 7-4. Suppose that the experiment was actually carried out on three levels of temperature and two positions. The data are:

Position	Temperature (°C)		
	800	825	850
1	570	1063	565
	565	1080	510
	583	1043	590
2	528	988	526
	547	1026	538
	521	1004	532

(a) Perform the analysis of variance with $\alpha = 0.05$. What is your conclusion about the significance of the interaction effect?

(b) Assess the adequacy of the model by analyzing the residuals. What is your conclusion?

(c) If higher density values are desirable, what levels of the factors do you recommend?

7-43. The percentage of hardwood concentration in raw pulp, the freeness, and the cooking time of the pulp are being investigated for their effects on the strength of paper. The data from a three-factor factorial experiment are shown in Table 7-35 below.

(a) Use a statistical software package to perform the analysis of variance. Use $\alpha = 0.05$.

(b) Find P-values for the F-ratios in part (a) and interpret your results.

(c) The residuals are found by $e_{ijkl} = y_{ijkl} - \bar{y}_{ijk.}$. Graphically analyze the residuals from this experiment.

7-44. The quality control department of a fabric finishing plant is studying the effect of several factors on dyeing for a blended cotton/synthetic cloth used to manufacture shirts. Three operators, three cycle times, and two temperatures were selected, and three small specimens of cloth were dyed under

Table 7-35   Data for Exercise 7-43

Percentage of Hardwood Concentration	Cooking Time 1.5 Hours			Cooking Time 2.0 Hours		
	Freeness			Freeness		
	350	500	650	350	500	650
10	96.6	97.7	99.4	98.4	99.6	100.6
	96.0	96.0	99.8	98.6	100.4	100.9
15	98.5	96.0	98.4	97.5	98.7	99.6
	97.2	96.9	97.6	98.1	96.0	99.0
20	97.5	95.6	97.4	97.6	97.0	98.5
	96.6	96.2	98.1	98.4	97.8	99.8

Table 7-36   Data for Exercise 7-44

Cycle Time	Temperature					
	300°			350°		
	Operator			Operator		
	1	2	3	1	2	3
40	23	27	31	24	38	34
	24	28	32	23	36	36
	25	26	28	28	35	39
50	36	34	33	37	34	34
	35	38	34	39	38	36
	36	39	35	35	36	31
60	28	35	26	26	36	28
	24	35	27	29	37	26
	27	34	25	25	34	34

each set of conditions. The finished cloth was compared to a standard, and a numerical score was assigned. The results are shown in Table 7-36 on the previous page.

(a) Perform the analysis of variance with $\alpha = 0.05$. Interpret your results.

(b) The residuals may be obtained from $e_{ijkl} = y_{ijkl} - \bar{y}_{ijk\cdot\cdot}$. Graphically analyze the residuals from this experiment.

7-45. Consider the Minitab analysis results of a two-factor experiment, A and B. Factor A was run with two levels and factor B with three, with two replicates. Find all of the missing values in the ANOVA table below and summarize your findings.

Source	DF	SS	MS	F	P
A	1	61.675	61.6748	?	0.005
B	2	82.644	?	12.84	?
A*B	2	7.959	3.9795	?	?
Error	6	19.305	3.2174		
Total	11	?			

7-46. Consider the Minitab analysis results of a three-factor experiment, A, B, and C as described on the right. Find

all of the missing values in the ANOVA table and summarize your findings.

Factor	Type	Levels	Values
A	fixed	2	$-1, 1$
B	fixed	2	$-1, 1$
C	fixed	3	$-1, 0, 1$

Analysis of Variance for y, using Adjusted SS for Tests

Source	DF	Seq SS	Adj SS	Adj MS	F	P
A	1	362.551	362.551	362.551	90.84	0.000
B	1	15.415	15.415	15.415	3.86	?
C	2	240.613	240.613	120.306	?	0.000
A*B	1	0.522	0.522	0.522	0.13	?
A*C	2	62.322	62.322	31.161	?	?
B*C	2	3.553	3.553	1.777	0.45	?
A*B*C	2	15.724	15.724	7.862	?	?
Error	12	47.891	47.891	3.991		
Total	23	748.591				

SUPPLEMENTAL EXERCISES

7-47. An article in *Process Engineering* (No. 71, 1992, pp. 46–47) presents a two-factor factorial experiment used to investigate the effect of pH and catalyst concentration on product viscosity (cSt). The data are as follows.

		Catalyst Concentration	
		2.5	2.7
pH	5.6	192, 199, 189, 198	178, 186, 179, 188
	5.9	185, 193, 185, 192	197, 196, 204, 204

(a) Test for main effects and interactions using $\alpha = 0.05$. What are your conclusions?

(b) Graph the interaction and discuss the information provided by this plot.

(c) Analyze the residuals from this experiment.

7-48. Heat treating of metal parts is a widely used manufacturing process. An article in the *Journal of Metals* (Vol. 41, 1989) describes an experiment to investigate flatness distortion from heat treating for three types of gears and two heat-treating times. Some of the data are as follows.

	Time (minutes)	
Gear Type	90	120
20-tooth	0.0265	0.0560
	0.0340	0.0650
24-tooth	0.0430	0.0720
	0.0510	0.0880

(a) Is there any evidence that flatness distortion is different for the different gear types? Is there any indication that heat treating time affects the flatness distortion? Do these factors interact? Use $\alpha = 0.05$.

(b) Construct graphs of the factor effects that aid in drawing conclusions from this experiment.

(c) Analyze the residuals from this experiment. Comment on the validity of the underlying assumptions.

7-49. An article in the *Textile Research Institute Journal* (Vol. 54, 1984, pp. 171–179) reported the results of an experiment that studied the effects of treating fabric with selected inorganic salts on the flammability of the material. Two

Table 7-37   Data for Exercise 7-49

Level	Salt					
	Untreated	MgCl$_2$	NaCl	CaCO$_3$	CaCl$_2$	Na$_2$CO$_3$
1	812	752	739	733	725	751
	827	728	731	728	727	761
	876	764	726	720	719	755
2	945	794	741	786	756	910
	881	760	744	771	781	854
	919	757	727	779	814	848

application levels of each salt were used, and a vertical burn test was used on each sample. (This finds the temperature at which each sample ignites.) The burn test data are shown in Table 7-37 above.

(a) Test for differences between salts, application levels, and interactions. Use $\alpha = 0.01$.

(b) Draw a graph of the interaction between salt and application level. What conclusions can you draw from this graph?

(c) Analyze the residuals from this experiment.

7-50.   An article in the *IEEE Transactions on Components, Hybrids, and Manufacturing Technology* (Vol. 15, 1992, pp. 225–230) describes an experiment for investigating a method for aligning optical chips onto circuit boards. The method involves placing solder bumps onto the bottom of the chip. The experiment used two solder bump sizes and two alignment methods. The response variable is alignment accuracy (μm). The data are as follows.

Solder Bump Size (diameter in μm)	Alignment Method	
	1	2
75	4.60	1.05
	4.53	1.00
130	2.33	0.82
	2.44	0.95

(a) Is there any indication that either solder bump size or alignment method affects the alignment accuracy? Is there any evidence of interaction between these factors? Use $\alpha = 0.05$.

(b) What recommendations would you make about this process?

(c) Analyze the residuals from this experiment. Comment on model adequacy.

7-51.   An article in *Solid State Technology* (Vol. 29, 1984, pp. 281–284) describes the use of factorial experiments in photolithography, an important step in the process of manufacturing integrated circuits. The variables in this experiment (all at two levels) are prebake temperature (A), prebake time

(B), and exposure energy (C), and the response variable is delta-line width, the difference between the line on the mask and the printed line on the device. The data are as follows: $(1) = -2.30$, $a = -9.87$, $b = -18.20$, $ab = -30.20$, $c = -23.80$, $ac = -4.30$, $bc = -3.80$, and $abc = -14.70$.

(a) Estimate the factor effects.

(b) Suppose that a center point is added to this design and four replicates are obtained: -10.50, -5.30, -11.60, and -7.30. Calculate an estimate of experimental error.

(c) Test the significance of main effects, interactions, and curvature. At $\alpha = 0.05$, what conclusions can you draw?

(d) What model would you recommend for predicting the delta-line width response, based on the results of this experiment?

(e) Analyze the residuals from this experiment, and comment on model adequacy.

7-52.   An article in the *Journal of Coatings Technology* (Vol. 60, 1988, pp. 27–32) describes a 2^4 factorial design used for studying a silver automobile basecoat. The response variable is distinctness of image (DOI). The variables used in the experiment are

A = Percent of polyester by weight of polyester/melamine (low value = 50%, high value = 70%)

B = Percent cellulose acetate butyrate carboxylate (low value = 15%, high value = 30%)

C = Percent aluminum stearate (low value = 1%, high value = 3%)

D = Percent acid catalyst (low value = 0.25%, high value = 0.50%)

The responses are $(1) = 63.8$, $a = 77.6$, $b = 68.8$, $ab = 76.5$, $c = 72.5$, $ac = 77.2$, $bc = 77.7$, $abc = 84.5$, $d = 60.6$, $ad = 64.9$, $bd = 72.7$, $abd = 73.3$, $cd = 68.0$, $acd = 76.3$, $bcd = 76.0$, and $abcd = 75.9$.

(a) Estimate the factor effects.

(b) From a normal probability plot of the effects, identify a tentative model for the data from this experiment.

(c) Using the apparently negligible factors as an estimate of error, test for significance of the factors identified in part (b). Use $\alpha = 0.05$.

Table 7-38   Data for Exercise 7-53

| | Factors | | | | Surface Roughness |
| | V | F | P | G | |
Run	(in/min)	(lb/min)	(kpsi)	(mesh no.)	(μm)
1	6	2.0	38	80	104
2	2	2.0	38	80	98
3	6	2.0	30	80	103
4	2	2.0	30	80	96
5	6	1.0	38	80	137
6	2	1.0	38	80	112
7	6	1.0	30	80	143
8	2	1.0	30	80	129
9	6	2.0	38	170	88
10	2	2.0	38	170	70
11	6	2.0	30	170	110
12	2	2.0	30	170	110
13	6	1.0	38	170	102
14	2	1.0	38	170	76
15	6	1.0	30	170	98
16	2	1.0	30	170	68
17	4	1.5	34	115	95
18	4	1.5	34	115	98
19	4	1.5	34	115	100
20	4	1.5	34	115	97
21	4	1.5	34	115	94
22	4	1.5	34	115	93
23	4	1.5	34	115	91

(d) What model would you use to describe the process, based on this experiment? Interpret the model.

(e) Analyze the residuals from the model in part (d) and comment on your findings.

7-53.   An article in the *Journal of Manufacturing Systems* (Vol. 10, 1991, pp. 32–40) describes an experiment to investigate the effect of four factors P = waterjet pressure, F = abrasive flow rate, G = abrasive grain size, and V = jet traverse speed on the surface roughness of a waterjet cutter. A 2^4 design with seven center points is shown in Table 7-38 above.

(a) Estimate the factor effects.

(b) Form a tentative model by examining a normal probability plot of the effects.

(c) Is the model in part (b) a reasonable description of the process? Use $\alpha = 0.05$.

(d) Interpret the results of this experiment.

(e) Analyze the residuals from this experiment.

7-54.   Construct a 2_{IV}^{4-1} design for the problem in Exercise 7-52. Select the data for the eight runs that would have been required

for this design. Analyze these runs and compare your conclusions to those obtained in Exercise 7-52 for the full factorial.

7-55.   Construct a 2_{IV}^{4-1} design for the problem in Exercise 7-53. Select the data for the eight runs that would have been required for this design, plus the center points. Analyze these data and compare your conclusions to those obtained in Exercise 7-53 for the full factorial.

7-56.   Construct a 2_{IV}^{8-4} design in 16 runs. What are the alias relationships in this design?

7-57.   Construct a 2_{III}^{5-2} design in eight runs. What are the alias relationships in this design?

7-58.   In a process development study on yield, four factors were studied, each at two levels: time (A), concentration (B), pressure (C), and temperature (D). A single replicate of a 2^4 design was run, and the resulting data are shown in Table 7-39 on the next page.

(a) Plot the effect estimates on a normal probability scale. Which factors appear to have large effects?

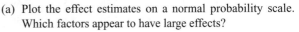

Table 7-39 Data for Exercise 7-58

Run Number	Actual Run Order	A	B	C	D	Yield (lbs)	Factor Levels	Low (−)	High (+)
1	5	−	−	−	−	12	A (h)	2.5	3
2	9	+	−	−	−	18	B (%)	14	18
3	8	−	+	−	−	13	C (psi)	60	80
4	13	+	+	−	−	16	D (°C)	225	250
5	3	−	−	+	−	17			
6	7	+	−	+	−	15			
7	14	−	+	+	−	20			
8	1	+	+	+	−	15			
9	6	−	−	−	+	10			
10	11	+	−	−	+	25			
11	2	−	+	−	+	13			
12	15	+	+	−	+	24			
13	4	−	−	+	+	19			
14	16	+	−	+	+	21			
15	10	−	+	+	+	17			
16	12	+	+	+	+	23			

(b) Conduct an analysis of variance using the normal probability plot in part (a) for guidance in forming an error term. What are your conclusions?

(c) Analyze the residuals from this experiment. Does your analysis indicate any potential problems?

(d) Can this design be collapsed into a 2^3 design with two replicates? If so, sketch the design with the average and range of yield shown at each point in the cube. Interpret the results.

7-59. An article in the *Journal of Quality Technology* (Vol. 17, 1985, pp. 198–206) describes the use of a replicated fractional factorial to investigate the effect of five factors on the free height of leaf springs used in an automotive application. The factors are A = furnace temperature, B = heating time, C = transfer time, D = hold down time, and E = quench oil temperature. The data are shown in the following table.

A	B	C	D	E		Free Height	
−	−	−	−	−	7.78	7.78	7.81
+	−	−	+	−	8.15	8.18	7.88
−	+	−	+	−	7.50	7.56	7.50
+	+	−	−	−	7.59	7.56	7.75
−	−	+	+	−	7.54	8.00	7.88
+	−	+	−	−	7.69	8.09	8.06
−	+	+	−	−	7.56	7.52	7.44
+	+	+	+	−	7.56	7.81	7.69
−	−	−	−	+	7.50	7.56	7.50
+	−	−	+	+	7.88	7.88	7.44

A	B	C	D	E		Free Height	
−	+	−	+	+	7.50	7.56	7.50
+	+	−	−	+	7.63	7.75	7.56
−	−	+	+	+	7.32	7.44	7.44
+	−	+	−	+	7.56	7.69	7.62
−	+	+	−	+	7.18	7.18	7.25
+	+	+	+	+	7.81	7.50	7.59

(a) What is the generator for this fraction? Write out the alias structure.

(b) Analyze the data. What factors influence mean free height?

(c) Calculate the range of free height for each run. Is there any indication that any of these factors affects variability in free height?

(d) Analyze the residuals from this experiment and comment on your findings.

7-60. Consider the experiment described in Exercise 7-58. Find 95% CIs on the factor effects that appear important. Use the normal probability plot to provide guidance concerning the effects that can be combined to provide an estimate of error.

7-61. An article in *Rubber Chemistry and Technology* (Vol. 47, 1974, pp. 825–836) describes an experiment that studies the Mooney viscosity of rubber to several variables, including silica filler (parts per hundred) and oil filler (parts per hundred). Some of the data from this experiment are shown here, where

$$x_1 = \frac{silica - 60}{15} \qquad x_2 = \frac{oil - 21}{15}$$

Coded Levels		
x_1	x_2	y
-1	-1	13.71
1	-1	14.15
-1	1	12.87
1	1	13.53
-1.4	0	12.99
1.4	0	13.89
0	-1.4	14.16
0	1.4	12.90
0	0	13.75
0	0	13.66
0	0	13.86
0	0	13.63
0	0	13.74

Fit a quadratic model to these data. What values of x_1 and x_2 will maximize the Mooney viscosity?

7-62. An article in *Oikos: A Journal of Ecology*, "Regulation of Root Vole Population Dynamics by Food Supply and Predation: A Two-Factor Experiment," (Vol. 109, 2005, pp.

387–395), investigated how food supply interacts with predation in the regulation of root vole (*Microtus oeconomus* Pallas) population dynamics. A replicated two-factor field experiment manipulating both food supply and predation condition for root voles was conducted. Four treatments were applied: $-P$, $+F$ (no-predator, food-supplemented); $+P$, $+F$ (predator-access, food-supplemented); $-P$, $-F$ (no-predator, non-supplemented); $+P$, $-F$ (predator-access, food-supplemented). The population density of root voles (voles ha^{-1}) for each treatment combination in each is shown below.

Food Supply (F)	Predation (P)	Replicates		
$+1$	-1	88.589	114.059	200.979
$+1$	$+1$	56.949	97.079	78.759
-1	-1	65.439	89.089	172.339
-1	$+1$	40.799	47.959	74.439

(a) What is an appropriate statistical model for this experiment?
(b) Analyze the data and draw conclusions.
(c) Analyze the residuals from this experiment. Are there any problems with model adequacy?

TEAM EXERCISE

7-63. The project consists of planning, designing, conducting, and analyzing an experiment, using appropriate experimental design principles. The context of the project experiment is limited only by your imagination. Students have conducted experiments directly connected to their own research interests, a project that they are involved with at work (something for the industrial participants or the part-timers in industry to think about), or if all else fails, you could conduct a "household" experiment (such as how do varying factors such as type of cooking oil, amount of oil, cooking temperature, pan type, brand of popcorn, etc., affect the yield and taste of popcorn).

The major requirement is that the experiment must involve at least three factors. Each of the interim steps requires information about the problem, the factors, the responses that will be observed, and the specific details of the design that will be used. Your final report should include a clear statement of objectives, the procedures and techniques used, appropriate analyses, and specific conclusions that state what you learned from the experiment.

IMPORTANT TERMS AND CONCEPTS

2^k factorial design

2^{k-p} fractional factorial design

Aliases

Blocking and confounding

Center points in a 2^k factorial design

Central composite design

Contour plot

Factorial design

First-order model

Interaction

Interaction plot

Main effect of a factor

Method of steepest ascent

Normal probability plot of effects

Optimization experiments

Regression model

Residual analysis

Residuals

Response surface

Screening experiment

Second-order model

Sequential experimentation

8 Statistical Process Control

CHAPTER OUTLINE

LEARNING OBJECTIVES

After careful study of this chapter, you should be able to do the following:

1. Understand the role of statistical tools in quality improvement.
2. Understand the different types of variability and rational subgroups, and how a control chart is used to detect assignable causes.
3. Understand the general form of a Shewhart control chart and how to apply zone rules (such as the Western Electric rules) and pattern analysis to detect assignable causes.
4. Construct and interpret control charts for variables such as $\overline{X}$, R, S, and individual charts.
5. Construct and interpret control charts for attributes such as P and U charts.
6. Calculate and interpret process capability ratios.
7. Calculate the ARL performance for a Shewhart control chart.
8. Use ANOVA to study the performance of a measurement system.

8-1 QUALITY IMPROVEMENT AND STATISTICS

The quality of products and services has become a major decision factor in most businesses today. Regardless of whether the consumer is an individual, a corporation, a military defense program, or a retail store, when the consumer is making purchase decisions, he or she is likely to consider quality to be equal in importance to cost and schedule. Consequently, **quality improvement** has become a major concern to many U.S. corporations. This chapter is about **statistical process control** (SPC), a collection of tools that are essential in quality-improvement activities.

Quality means **fitness for use**. For example, you or I may purchase automobiles that we expect to be free of manufacturing defects and that should provide reliable and economical transportation, a retailer buys finished goods with the expectation that they are properly packaged and arranged for easy storage and display, and a manufacturer buys raw material and expects to process it with no rework or scrap. In other words, all consumers expect that the products and services they buy will meet their requirements. Those requirements define fitness for use.

Quality or fitness for use is determined through the interaction of **quality of design** and **quality of conformance**. By quality of design, we mean the different grades or levels of performance, reliability, serviceability, and function that are the result of deliberate engineering and management decisions. By quality of conformance, we mean the systematic **reduction of variability** and **elimination of defects** until every unit produced is identical and defect-free.

Some confusion exists in our society about quality improvement; some people still think that it means gold-plating a product or spending more money to develop a product or process. This thinking is wrong. Quality improvement means the systematic **elimination of waste**. Examples of waste include scrap and rework in manufacturing, inspection and test, errors on documents (such as engineering drawings, checks, purchase orders, and plans), customer complaint hotlines, warranty costs, and the time required to do things over again that could have been done right the first time. A successful quality-improvement effort can eliminate much of this waste and lead to lower costs, higher productivity, increased customer satisfaction, increased business reputation, higher market share, and ultimately higher profits for the company.

Statistical methods play a vital role in quality improvement. Some applications are outlined next:

1. In product design and development, statistical methods, including designed experiments, can be used to compare different materials, components, or ingredients and to help determine both system and component tolerances. This application can significantly lower development costs and reduce development time.

2. Statistical methods can be used to determine the capability of a manufacturing process. Statistical process control can be used to systematically improve a process by reducing variability.

3. Experimental design methods can be used to investigate improvements in the process. These improvements can lead to higher yields and lower manufacturing costs.

4. Life testing provides reliability and other performance data about the product. This can lead to new and improved designs and products that have longer useful lives and lower operating and maintenance costs.

Some of these applications have been illustrated in earlier chapters of this book. It is essential that engineers, scientists, and managers have an in-depth understanding of these statistical tools in any industry or business that wants to be a high-quality, low-cost producer. In this chapter we provide an introduction to the basic methods of statistical quality control that, along with experimental design, form the basis of a successful quality-improvement effort.

8-2 STATISTICAL PROCESS CONTROL

SPC has its origins in the 1920s. Dr. Walter A. Shewhart of the Bell Telephone Laboratories was one of the early pioneers of the field. In 1924 he wrote a memorandum illustrating a control chart, one of the basic SPC tools. World War II saw the widespread dissemination of these methods to U.S. industry. Dr. W. Edwards Deming and Dr. Joseph M. Juran were instrumental in spreading the methodology after World War II.

SPC is a set of problem-solving tools that may be applied to any process. The major tools of SPC[1] are

1. Histogram
2. Pareto chart
3. Cause-and-effect diagram
4. Defect-concentration diagram
5. Control chart
6. Scatter diagram
7. Check sheet

Although these tools are an important part of SPC, they comprise only the technical aspect of the subject. An equally important element of SPC is attitude—a desire of all individuals in the organization for continuous improvement in quality and productivity through the systematic reduction of variability. The control chart is the most powerful of the SPC tools. For complete discussion of these methods, see Montgomery (2005b).

8-3 INTRODUCTION TO CONTROL CHARTS

8-3.1 Basic Principles

In any production process, regardless of how well designed or carefully maintained it is, a certain amount of inherent or natural variability will always exist. This natural variability, or "background noise," is the cumulative effect of many small, essentially unavoidable causes. When the background noise in a process is relatively small, we usually consider it an acceptable level of process performance. In the framework of statistical process control, this natural variability is often called a "stable system of chance causes." A process that is operating with only **chance causes** of variation present is said to be in statistical control. In other words, the chance causes are an inherent part of the process.

Other kinds of variability may occasionally be present in the output of a process. This variability in key quality characteristics usually arises from three sources: improperly adjusted machines, operator errors, or defective raw materials. Such variability is generally large when compared to the background noise, and it usually represents an unacceptable level of process performance. We refer to these sources of variability that are not part of the chance cause pattern as **assignable causes**. A process that is operating in the presence of assignable causes is said to be out of control.[2]

---

[1]Some prefer to include the experimental design methods discussed in Chapter 7 as part of the SPC toolkit. We did not do so because we think of SPC as an online approach to quality improvement using techniques founded on passive observation of the process, whereas design of experiments is an active approach in which deliberate changes are made to the process variables. As such, designed experiments are often referred to as offline quality control.

[2]The terminology *chance* and *assignable* causes was developed by Dr. Walter A. Shewhart. Today, some writers use *common* cause instead of *chance* cause and *special* cause instead of *assignable* cause.

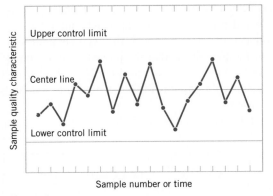

Figure 8-1   A typical control chart.

Production processes will often operate in the in-control state, producing acceptable product for relatively long periods of time. Occasionally, however, assignable causes will occur, seemingly at random, resulting in a "shift" to an out-of-control state where a large proportion of the process output does not conform to requirements. A major objective of statistical process control is to quickly detect the occurrence of assignable causes or process shifts so that investigation of the process and corrective action may be undertaken before many nonconforming units are manufactured. The control chart is an online process-monitoring technique widely used for this purpose.

Recall the following from Chapter 1: Fig. 1-17 illustrates that adjustments to common causes of variation increase the variation of a process, whereas Fig. 1-18 illustrates that actions should be taken in response to assignable causes of variation. Control charts may also be used to estimate the parameters of a production process and, through this information, to determine the capability of a process to meet specifications. The control chart can also provide information that is useful in improving the process. Finally, remember that the eventual goal of SPC is the *elimination of variability in the process*. Although it may not be possible to eliminate variability completely, the control chart helps reduce it as much as possible.

A typical control chart is shown in Fig. 8-1, which is a graphical display of a quality characteristic that has been measured or computed from a sample versus the sample number or time. Often, the samples are selected at periodic intervals, such as every hour. The chart contains a center line (CL) that represents the average value of the quality characteristic corresponding to the in-control state. (That is, only chance causes are present.) Two other horizontal lines, called the upper control limit (UCL) and the lower control limit (LCL), are also shown on the chart. These control limits are chosen so that if the process is in control, nearly all of the sample points will fall between them. In general, as long as the points plot within the control limits, the process is assumed to be in control, and no action is necessary. However, a point that plots outside of the control limits is interpreted as evidence that the process is out of control, and investigation and corrective action are required to find and eliminate the assignable cause or causes responsible for this behavior. The sample points on the control chart are usually connected with straight-line segments, so that it is easier to visualize how the sequence of points has evolved over time.

Even if all the points plot inside the control limits, if they behave in a systematic or nonrandom manner, this is an indication that the process is out of control. For example, if 18 of the last 20 points plotted above the center line but below the upper control limit and only two of these points plotted below the center line but above the lower control limit, we would be very suspicious that something was wrong. If the process is in control, all the plotted points should have an essentially random pattern. Methods designed to find sequences or nonrandom patterns can be applied to control charts as an aid in detecting out-of-control conditions. A particular nonrandom pattern usually appears on a control chart for a reason, and if that reason can be found and eliminated, process performance can be improved.

There is a close connection between control charts and hypothesis testing. Essentially, the control chart is a test of the hypothesis that the process is in a state of statistical control. A point plotting within the control limits is equivalent to failing to reject the hypothesis of statistical control, and a point plotting outside the control limits is equivalent to rejecting the hypothesis of statistical control.

We may give a general *model* for a control chart.

General Model for a Control Chart

Let W be a sample statistic that measures some quality characteristic of interest, and suppose that the mean of W is μ_W and the standard deviation of W is σ_W.[3] Then the **center line** (CL), the **upper control limit** (UCL), and the **lower control limit** (LCL) become

$$UCL = \mu_W + k\sigma_W$$
$$CL = \mu_W$$
$$LCL = \mu_W - k\sigma_W \tag{8-1}$$

where k is the "distance" of the control limits from the center line, expressed in standard deviation units.

A common choice is $k = 3$. This general theory of control charts was first proposed by Dr. Walter A. Shewhart, and control charts developed according to these principles are often called **Shewhart control charts**.

The control chart is a device for describing exactly what is meant by statistical control; as such, it may be used in a variety of ways. In many applications, it is used for online process monitoring. That is, sample data are collected and used to construct the control chart, and if the sample values of $\bar{x}$ (say) fall within the control limits and do not exhibit any systematic pattern, we say the process is in control at the level indicated by the chart. Note that we may be interested here in determining *both* whether the past data came from a process that was in control and whether future samples from this process indicate statistical control.

The most important use of a control chart is to *improve* the process. We have found that, generally,

1. Most processes do not operate in a state of statistical control.

2. Consequently, the routine and attentive use of control charts will identify assignable causes. If these causes can be eliminated from the process, variability will be reduced and the process will be improved.

This process-improvement activity using the control chart is illustrated in Fig. 8-2. Note that:

3. The control chart will only *detect* assignable causes. Management, operator, and engineering *action* will usually be necessary to eliminate the assignable cause. An action plan for responding to control chart signals is vital.

---

[3]Note that sigma refers to the standard deviation of the statistic plotted on the chart (i.e., σ_W), not the standard deviation of the quality characteristic.

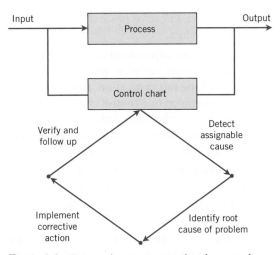

Figure 8-2   Process improvement using the control chart.

In identifying and eliminating assignable causes, it is important to find the underlying **root cause** of the problem and to attack it. A cosmetic solution will not result in any real, long-term process improvement. Developing an effective system for corrective action is an essential component of an effective SPC implementation.

We may also use the control chart as an *estimating device*. That is, from a control chart that exhibits statistical control, we may estimate certain process parameters, such as the mean, standard deviation, and fraction nonconforming or fallout. These estimates may then be used to determine the *capability* of the process to produce acceptable products. Such **process capability studies** have considerable impact on many management decision problems that occur over the product cycle, including make-or-buy decisions, plant and process improvements that reduce process variability, and contractual agreements with customers or suppliers regarding product quality.

Control charts may be classified into two general types. Many quality characteristics can be measured and expressed as numbers on some continuous scale of measurement. In such cases, it is convenient to describe the quality characteristic with a measure of central tendency and a measure of variability. Control charts for central tendency and variability are collectively called **variables control charts.** The $\overline{X}$ chart is the most widely used chart for monitoring central tendency, whereas charts based on either the sample range or the sample standard deviation are used to control process variability. Many quality characteristics are not measured on a continuous scale or even a quantitative scale. In these cases, we may judge each unit of product as either conforming or nonconforming on the basis of whether or not it possesses certain attributes, or we may count the number of nonconformities (defects) appearing on a unit of product. Control charts for such quality characteristics are called **attributes control charts.**

Control charts have had a long history of use in industry. There are at least five reasons for their popularity; control charts

1. **Are a proven technique for improving productivity.** A successful control chart program will reduce scrap and rework, which are the primary productivity killers in *any* operation. If you reduce scrap and rework, productivity increases, cost decreases, and production capacity (measured in the number of *good* parts per hour) increases.

2. **Are effective in defect prevention.** The control chart helps keep the process in control, which is consistent with the "do it right the first time" philosophy. It is never cheaper to sort out the "good" units from the "bad" later on than it is to build it right

initially. If you do not have effective process control, you are paying someone to make a nonconforming product.

3. **Prevent unnecessary process adjustments.** A control chart can distinguish between background noise and abnormal variation; no other device, including a human operator, is as effective in making this distinction. If process operators adjust the process based on periodic tests unrelated to a control chart program, they will often overreact to the background noise and make unneeded adjustments. These unnecessary adjustments can usually result in a deterioration of process performance. In other words, the control chart is consistent with the "if it isn't broken, don't fix it" philosophy.

4. **Provide diagnostic information.** Frequently, the pattern of points on the control chart will contain information that is of diagnostic value to an experienced operator or engineer. This information allows the operator to implement a change in the process that will improve its performance.

5. **Provide information about process capability.** The control chart provides information about the value of important process parameters and their stability over time, which allows an estimate of process capability to be made. This information is of tremendous use to product and process designers.

Control charts are among the most effective management control tools, and they are as important as cost controls and material controls. Modern computer technology has made it easy to implement control charts in any type of process because data collection and analysis can be performed on a microcomputer or a local area network terminal in real time, online at the work center.

8-3.2 Design of a Control Chart

To illustrate these ideas, we give a simplified example of a control chart. In manufacturing automobile engine piston rings, the inside diameter of the rings is a critical quality characteristic. The process mean inside ring diameter is 74 mm, and it is known that the standard deviation of ring diameter is 0.01 mm. A control chart for average ring diameter is shown in Fig. 8-3. Every hour a random sample of five rings is taken, the average ring diameter of the sample (say, $\bar{x}$) is computed, and $\bar{x}$ is plotted on the chart. Because this control chart utilizes the sample mean $\bar{X}$ to monitor the process mean, it is usually called an $\bar{X}$ control chart. Note that all

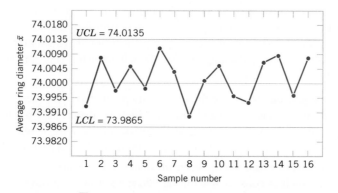

Figure 8-3 $\bar{X}$ control chart for piston ring diameter.

the points fall within the control limits, so the chart indicates that the process is in statistical control.

Consider how the control limits were determined. The process average is 74 mm, and the process standard deviation is $\sigma = 0.01$ mm. Now, if samples of size $n = 5$ are taken, the standard deviation of the sample average $\overline{X}$ is

$$\sigma_{\overline{X}} = \frac{\sigma}{\sqrt{n}} = \frac{0.01}{\sqrt{5}} = 0.0045$$

Therefore, if the process is in control with a mean diameter of 74 mm, by using the central limit theorem to assume that $\overline{X}$ is approximately normally distributed, we would expect approximately $100(1 - \alpha)$ % of the sample mean diameters $\overline{X}$ to fall between $74 + z_{\alpha/2} (0.0045)$ and $74 - z_{\alpha/2} (0.0045)$. As discussed above, we customarily choose the constant $z_{\alpha/2}$ to be 3, so the upper and lower control limits become

$$UCL = 74 + 3(0.0045) = 74.0135$$

and

$$LCL = 74 - 3(0.0045) = 73.9865$$

as shown on the control chart. These are the three-sigma control limits referred to above. Note that the use of three-sigma limits implies that $\alpha = 0.0027$; that is, the probability that the point plots outside the control limits when the process is in control is 0.0027. The width of the control limits is inversely related to the sample size n for a given multiple of sigma. Choosing the control limits is equivalent to setting up the critical region for testing the hypothesis

$$H_0: \mu = 74$$
$$H_1: \mu \neq 74$$

where $\sigma = 0.01$ is known. Essentially, the control chart tests this hypothesis repeatedly at different points in time.

In designing a control chart, we must specify both the sample size to use and the frequency of sampling. In general, larger samples will make it easier to detect small shifts in the process. When choosing the sample size, we must keep in mind the size of the shift that we are trying to detect. If we are interested in detecting a relatively large process shift, we use smaller sample sizes than those that would be employed if the shift of interest were relatively small.

We must also determine the frequency of sampling. The most desirable situation from the point of view of detecting shifts would be to take large samples very frequently; however, this is usually not economically feasible. The general problem is one of *allocating sampling effort*. That is, either we take small samples at short intervals or larger samples at longer intervals. Current industry practice tends to favor smaller, more frequent samples, particularly in high-volume manufacturing processes, or where a great many types of assignable causes can occur. Furthermore, as automatic sensing and measurement technology develops, it is becoming possible to greatly increase frequencies. Ultimately, every unit can be tested as it is manufactured. This capability will not eliminate the need for control charts because the test system does not prevent defects. The increased data will increase the effectiveness of process control and improve quality.

8-3.3   Rational Subgroups

A fundamental idea in the use of control charts is to collect sample data according to what Shewhart called the **rational subgroup** concept. Generally, this means that subgroups or samples should be selected so that to the extent possible, the variability of the observations within a subgroup include all the chance or natural variability and exclude the assignable variability. Then, the control limits will represent bounds for all the chance variability and not the assignable variability. Consequently, assignable causes will tend to generate points that are outside of the control limits, whereas chance variability will tend to generate points that are within the control limits.

When control charts are applied to production processes, the time order of production is a logical basis for rational subgrouping. Even though time order is preserved, it is still possible to form subgroups erroneously. If some of the observations in the subgroup are taken at the end of one 8-hour shift and the remaining observations are taken at the start of the next 8-hour shift, any differences between shifts might not be detected. Time order is frequently a good basis for forming subgroups because it allows us to detect assignable causes that occur over time.

Two general approaches to constructing rational subgroups are used. In the first approach, each subgroup consists of units that were produced at the same time (or as close together as possible). This approach is used when the primary purpose of the control chart is to detect process shifts. It minimizes variability due to assignable causes *within* a sample, and it maximizes variability *between* samples if assignable causes are present. It also provides better estimates of the standard deviation of the process in the case of variables control charts. This approach to rational subgrouping essentially gives a "snapshot" of the process at each point in time where a sample is collected.

In the second approach, each sample consists of units of product that are representative of *all* units that have been produced since the last sample was taken. Essentially, each subgroup is a *random sample* of *all* process output over the sampling interval. This method of rational subgrouping is often used when the control chart is employed to make decisions about the acceptance of all units of product that have been produced since the last sample. In fact, if the process shifts to an out-of-control state and then back in control again *between* samples, it is sometimes argued that the first method of rational subgrouping defined above will be ineffective against these types of shifts, and so the second method must be used.

When the rational subgroup is a random sample of all units produced over the sampling interval, considerable care must be taken in interpreting the control charts. If the process mean drifts between several levels during the interval between samples, the range of observations within the sample may consequently be relatively large. It is the within-sample variability that determines the width of the control limits on an $\overline{X}$ chart, so this practice will result in wider limits on the $\overline{X}$ chart. This makes it harder to detect shifts in the mean. In fact, we can often make *any* process appear to be in statistical control just by stretching out the interval between observations in the sample. It is also possible for shifts in the process average to cause points on a control chart for the range or standard deviation to plot out of control, even though no shift in process variability has taken place.

There are other bases for forming rational subgroups. For example, suppose a process consists of several machines that pool their output into a common stream. If we sample from this common stream of output, it will be very difficult to detect whether or not some of the machines are out of control. A logical approach to rational subgrouping here is to apply control chart techniques to the output for each individual machine. Sometimes this concept needs to be applied to different heads on the same machine, different workstations, different operators, and so forth.

The rational subgroup concept is very important. The proper selection of samples requires careful consideration of the process, with the objective of obtaining as much useful information as possible from the control chart analysis.

8-3.4   Analysis of Patterns on Control Charts

A control chart may indicate an out-of-control condition either when one or more points fall beyond the control limits, or when the plotted points exhibit some nonrandom pattern of behavior. For example, consider the $\overline{X}$ chart shown in Fig. 8-4. Although all 25 points fall within the control limits, the points do not indicate statistical control because their pattern is very nonrandom in appearance. Specifically, we note that 19 of the 25 points plot below the center line, whereas only 6 of them plot above. If the points are truly random, we should expect a more even distribution of them above and below the center line. We also observe that following the fourth point, five points in a row increase in magnitude. This arrangement of points is called a **run.** Because the observations are increasing, we could call it a run up; similarly, a sequence of decreasing points is called a run down. This control chart has an unusually long run up (beginning with the fourth point) and an unusually long run down (beginning with the eighteenth point).

In general, we define a run as a sequence of observations of the same type. In addition to runs up and runs down, we could define the types of observations as those above and below the center line, respectively, so that two points in a row above the center line would be a run of length 2.

A run of length 8 or more points has a very low probability of occurrence in a random sample of points. Consequently, any type of run of length 8 or more is often taken as a signal of an out-of-control condition. For example, 8 consecutive points on one side of the center line will indicate that the process is out of control.

Although runs are an important measure of nonrandom behavior on a control chart, other types of patterns may also indicate an out-of-control condition. For example, consider the $\overline{X}$ chart in Fig. 8-5. Note that the plotted sample averages exhibit a cyclic behavior, yet they all fall within the control limits. Such a pattern may indicate a problem with the process, such as operator fatigue, raw material deliveries, and heat or stress buildup. The yield may be improved by eliminating or reducing the sources of variability causing this cyclic behavior (see Fig. 8-6). In Fig. 8-6, *LSL* and *USL* denote the lower and upper specification limits of the process. These limits represent bounds within which acceptable product must fall, and they are often based on customer requirements.

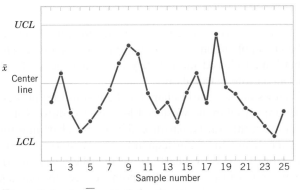

Figure 8-4   An $\overline{X}$ control chart.

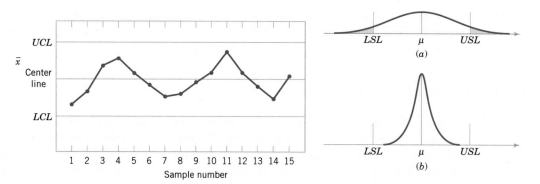

Figure 8-5   An $\overline{X}$ chart with a cyclic pattern.

Figure 8-6   (a) Variability with the cyclic pattern. (b) Variability with the cyclic pattern eliminated.

The problem is one of **pattern recognition**—that is, recognizing systematic or nonrandom **patterns on the control chart** and identifying the reason for this behavior. The ability to interpret a particular pattern in terms of assignable causes requires experience and knowledge of the process. That is, we must not only know the statistical principles of control charts, but we must also have a good understanding of the process.

The Western Electric Handbook (1956) suggests a set of decision rules for detecting nonrandom patterns on control charts. These rules are as follows:

The **Western Electric rules** would signal that the process is out of control if either

1. One point plots outside three-sigma control limits.
2. Two out of three consecutive points plot beyond a two-sigma limit.
3. Four out of five consecutive points plot at a distance of one sigma or beyond from the center line.
4. Eight consecutive points plot on one side of the center line.

We have found these rules very effective in practice for enhancing the sensitivity of control charts. Rules 2 and 3 apply to one side of the center line at a time. That is, a point above the *upper* two-sigma limit followed immediately by a point below the *lower* two-sigma limit would not signal an out-of-control alarm.

Figure 8-7 shows an $\overline{X}$ control chart for the piston ring process with the one-, two-, and three-sigma limits used in the Western Electric procedure. Note that these inner limits, sometimes called **warning limits,** partition the control chart into three zones A, B, and C on each side of the center line. Consequently, the Western Electric rules are sometimes called the **zone rules** for control charts. Note that the last four points fall in zone B or beyond. Thus, since four of five consecutive points exceed the one-sigma limit, the Western Electric procedure will conclude that the pattern is nonrandom and the process is out of control.

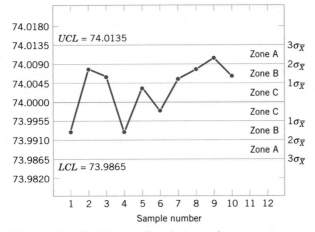

Figure 8-7   The Western Electric zone rules.

8-4 $\overline{X}$ AND R CONTROL CHARTS

When dealing with a quality characteristic that can be expressed as a measurement, it is customary to monitor both the mean value of the quality characteristic and its variability. Control over the average quality is exercised by the control chart for averages, usually called the $\overline{X}$ chart. Process variability can be controlled by either a range chart (R chart) or a standard deviation chart (S chart), depending on how the population standard deviation is estimated. We will discuss only the R chart.

Suppose that the process mean and standard deviation μ and σ are known and that we can assume that the quality characteristic has a normal distribution. Consider the $\overline{X}$ chart. As discussed previously, we can use μ as the center line for the control chart, and we can place the upper and lower three-sigma limits at $UCL = \mu + 3\sigma/\sqrt{n}$ and $LCL = \mu - 3\sigma/\sqrt{n}$, respectively.

When the parameters μ and σ are unknown, we usually estimate them on the basis of preliminary samples, taken when the process is thought to be in control. We recommend the use of at least 20 to 25 preliminary samples. Suppose m preliminary samples are available, each of size n. Typically, n will be 4, 5, or 6; these relatively small sample sizes are widely used and often arise from the construction of rational subgroups.

***

Grand Mean

Let the sample mean for the ith sample be $\overline{X}_i$. Then we estimate the mean of the population, μ, by the *grand mean*

$$\overline{\overline{X}} = \frac{1}{m} \sum_{i=1}^{m} \overline{X}_i \tag{8-2}$$

***

Thus, we may take $\overline{\overline{X}}$ as the center line on the $\overline{X}$ control chart.

We may estimate σ from either the standard deviation or the range of the observations within each sample. Because it is more frequently used in practice, we confine our discussion to the range method. The sample size is relatively small, so there is little loss in efficiency in estimating σ from the sample ranges.

The relationship between the range R of a sample from a normal population with known parameters and the standard deviation of that population is needed. Because R is a random variable, the quantity $W = R/\sigma$, called the relative range, is also a random variable. The parameters of the distribution of W have been determined for any sample size n.

The mean of the distribution of W is called d_2, and a table of d_2 for various n is given in Table VII of Appendix A. The standard deviation of W is called d_3. Because $R = \sigma W$,

$$\mu_R = d_2\sigma$$
$$\sigma_R = d_3\sigma$$

Average Range and Estimator of σ

Let R_i be the range of the ith sample, and let

$$\bar{R} = \frac{1}{m}\sum_{i=1}^{m} R_i \tag{8-3}$$

be the average range. Then $\bar{R}$ is an estimator of μ_R and an estimator of σ is

$$\hat{\sigma} = \frac{\bar{R}}{d_2} \tag{8-4}$$

Therefore, we may use as our upper and lower control limits for the $\bar{X}$ chart

$$UCL = \bar{\bar{X}} + \frac{3}{d_2\sqrt{n}}\,\bar{R}$$

$$LCL = \bar{\bar{X}} - \frac{3}{d_2\sqrt{n}}\,\bar{R} \tag{8-5}$$

Define the constant

$$A_2 = \frac{3}{d_2\sqrt{n}} \tag{8-6}$$

Now, once we have computed the sample values $\bar{\bar{x}}$ and $\bar{r}$, the parameters of the $\bar{X}$ control chart may be defined as follows.

The $\bar{X}$ Control Chart

The center line and upper and lower control limits for an $\bar{X}$ control chart are

$$UCL = \bar{\bar{x}} + A_2\bar{r}$$

$$CL = \bar{\bar{x}}$$

$$LCL = \bar{\bar{x}} - A_2\bar{r} \tag{8-7}$$

where the constant A_2 is tabulated for various sample sizes in Appendix A Table VII.

The parameters of the R chart may also be easily determined. The center line will obviously be $\overline{R}$. To determine the control limits, we need an estimate of σ_R, the standard deviation of R. Once again, assuming that the process is in control, the distribution of the relative range, W, will be useful. Because σ is unknown, we may estimate σ_R as

$$\hat{\sigma}_R = d_3 \frac{\overline{R}}{d_2}$$

and we would use as the upper and lower control limits on the R chart

$$UCL = \overline{R} + \frac{3d_3}{d_2} \overline{R} = \left(1 + \frac{3d_3}{d_2} \right) \overline{R}$$

$$LCL = \overline{R} - \frac{3d_3}{d_2} \overline{R} = \left(1 - \frac{3d_3}{d_2} \right) \overline{R}$$

Setting $D_3 = 1 - 3d_3/d_2$ and $D_4 = 1 + 3d_3/d_2$ leads to the following definition.

The R Control Chart

The center line and upper and lower control limits for an R chart are

$$UCL = D_4 \overline{r}$$
$$CL = \overline{r} \qquad\qquad (8\text{-}8)$$
$$LCL = D_3 \overline{r}$$

where $\overline{r}$ is the sample average range, and the constants D_3 and D_4 are tabulated for various sample sizes in Appendix A Table VII.

When preliminary samples are used to construct limits for control charts, these limits are customarily treated as trial values. Therefore, the m sample means and ranges should be plotted on the appropriate charts, and any points that exceed the control limits should be investigated. If assignable causes for these points are discovered, they should be eliminated and new limits for the control charts determined. In this way, the process may be eventually brought into statistical control and its inherent capabilities assessed. Other changes in process centering and dispersion may then be contemplated. Also, we often study the R chart first because if the process variability is not constant over time, the control limits calculated for the $\overline{X}$ chart can be misleading.

EXAMPLE 8-1
Vane Opening

A component part for a jet aircraft engine is manufactured by an investment casting process. The vane opening on this casting is an important functional parameter of the part. We will illustrate the use of $\overline{X}$ and R control charts to assess the statistical stability of this process. Table 8-1 presents 20 samples of five parts each. The values given in the table have been coded by using the last three digits of the dimension; that is, 31.6 should be 0.50316 inch. Construct a control chart.

Solution. The quantities $\overline{\overline{x}} = 33.32$ and $\overline{r} = 5.8$ are shown at the foot of Table 8-1. The value of A_2 for samples of size 5 is $A_2 = 0.577$. Then the trial control limits for the $\overline{X}$ chart are

$$\overline{\overline{x}} \pm A_2 \overline{r} = 33.32 \pm (0.577)(5.8) = 33.32 \pm 3.35$$

Table 8-1   Vane Opening Measurements for Example 8-1

Sample Number	x_1	x_2	x_3	x_4	x_5	$\bar{x}$	r
1	33	29	31	32	33	31.6	4
2	33	31	35	37	31	33.4	6
3	35	37	33	34	36	35.0	4
4	30	31	33	34	33	32.2	4
5	33	34	35	33	34	33.8	2
6	38	37	39	40	38	38.4	3
7	30	31	32	34	31	31.6	4
8	29	39	38	39	39	36.8	10
9	28	33	35	36	43	35.0	15
10	38	33	32	35	32	34.0	6
11	28	30	28	32	31	29.8	4
12	31	35	35	35	34	34.0	4
13	27	32	34	35	37	33.0	10
14	33	33	35	37	36	34.8	4
15	35	37	32	35	39	35.6	7
16	33	33	27	31	30	30.8	6
17	35	34	34	30	32	33.0	5
18	32	33	30	30	33	31.6	3
19	25	27	34	27	28	28.2	9
20	35	35	36	33	30	33.8	6
						$\bar{\bar{x}} = 33.32$	$\bar{r} = 5.8$

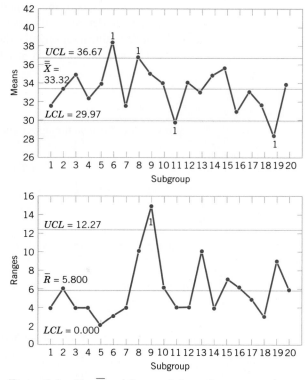

Figure 8-8   The $\bar{X}$ and R control charts for vane opening for Example 8-1.

or

$$UCL = 36.67$$
$$LCL = 29.97$$

For the R chart, the trial control limits are

$$UCL = D_4 \bar{r} = (2.115)(5.8) = 12.27$$
$$LCL = D_3 \bar{r} = (0)(5.8) = 0$$

The $\bar{X}$ and R control charts with these trial control limits are shown in Fig. 8-8. Note that samples 6, 8, 11, and 19 are out of control on the $\bar{X}$ chart and that sample 9 is out of control on the R chart. (These points are labeled with a 1 because they violate the first Western Electric rule.) Suppose that all of these assignable causes can be traced to a defective tool in the wax-molding area. We should discard these five samples and recompute the limits for the $\bar{X}$ and R charts. These new revised limits are, for the $\bar{X}$ chart,

$$UCL = \bar{\bar{x}} + A_2 \bar{r} = 33.21 + (0.577)(5.0) = 36.10$$
$$LCL = \bar{\bar{x}} - A_2 \bar{r} = 33.21 - (0.577)(5.0) = 30.33$$

and for the R chart,

$$UCL = D_4 \bar{r} = (2.115)(5.0) = 10.57$$
$$LCL = D_3 \bar{r} = (0)(5.0) = 0$$

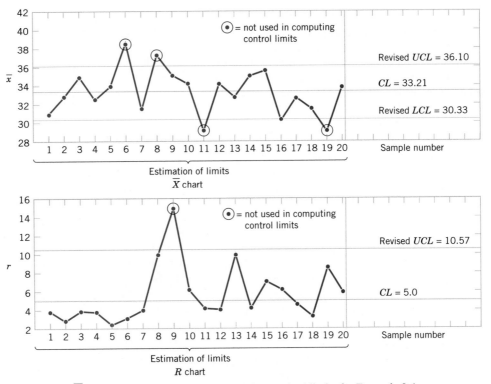

Figure 8-9 $\overline{X}$ and R control charts for vane opening, revised limits for Example 8-1.

The revised control charts are shown in Fig. 8-9. Note that we have treated the first 20 preliminary samples as **estimation data** with which to establish control limits. These limits can now be used to judge the statistical control of future production. As each new sample becomes available, the values of $\overline{x}$ and r should be computed and plotted on the control charts. It may be desirable to revise the limits periodically, even if the process remains stable. The limits should always be revised when process improvements are made.   ■

Computer Construction of $\overline{X}$ and R Control Charts

Many computer programs construct $\overline{X}$ and R control charts. Figures 8-8 and 8-9 show charts similar to those produced by Minitab for the vane-opening data in Example 8-1. This program will allow the user to select any multiple of sigma as the width of the control limits and utilize the Western Electric rules to detect out-of-control points. The program will also prepare a summary report as in Table 8-2 and exclude subgroups from the calculation of the control limits.

Table 8-2   Summary Report from Minitab for the Vane Opening
            Data in Example 8-1

Test Results for Xbar Chart
TEST 1. One point more than 3.00 sigmas from center line.
Test Failed at points: 6 8 11 19

Test Results for R Chart
TEST 1. One point more than 3.00 sigmas from center line.
Test Failed at points: 9

EXERCISES FOR SECTION 8-4

8-1. An extrusion die is used to produce aluminum rods. The diameter of the rods is a critical quality characteristic. The following table shows values for 20 samples of three rods each. Specifications on the rods are 0.4030 ± 0.0010 inch. The values given are the last three digits of the measurement; that is, 36 is read as 0.4036.

	Observation				Observation		
Sample	1	2	3	Sample	1	2	3
1	36	33	34	11	20	30	33
2	30	34	31	12	30	32	38
3	33	32	29	13	34	35	30
4	35	30	34	14	36	39	37
5	33	31	33	15	38	33	34
6	32	34	33	16	33	43	35
7	27	36	35	17	36	39	37
8	32	36	41	18	35	34	31
9	32	33	39	19	36	33	37
10	36	40	37	20	34	33	31

(a) Using all the data, find trial control limits for $\overline{X}$ and R charts, construct the chart, and plot the data.
(b) Use the trial control limits from part (a) to identify out-of-control points. If necessary, revise your control limits, assuming that any samples that plot outside the control limits can be eliminated.

8-2. Twenty samples of size 4 are drawn from a process at 1-hour intervals, and the following data are obtained:

$$\sum_{i=1}^{20} \bar{x}_i = 378.50 \qquad \sum_{i=1}^{20} r_i = 7.80$$

(a) Find trial control limits for $\overline{X}$ and R charts.
(b) Assuming that the process is in control, estimate the process mean and standard deviation.

8-3. The overall length of a skew used in a knee replacement device is monitored using $\overline{X}$ and R charts. The following table gives the length for 20 samples of size 4. (Measurements are coded from 2.00 mm; that is, 15 is 2.15 mm.)

	Observation					Observation			
Sample	1	2	3	4	Sample	1	2	3	4
1	16	18	15	13	8	17	13	17	16
2	16	15	17	16	9	15	11	13	16
3	15	16	20	16	10	15	18	14	13
4	14	16	14	12	11	14	14	15	13
5	14	15	13	16	12	15	13	15	16
6	16	14	16	15	13	13	17	16	15
7	16	16	14	15	14	11	14	14	21

	Observation					Observation			
Sample	1	2	3	4	Sample	1	2	3	4
15	14	15	14	13	18	16	14	13	19
16	18	15	16	14	19	17	19	17	13
17	14	16	19	16	20	12	15	12	17

(a) Using all the data, find trial control limits for $\overline{X}$ and R charts, construct the chart, and plot the data.
(b) Use the trial control limits from part (a) to identify out-of-control points. If necessary, revise your control limits, assuming that any samples that plot outside the control limits can be eliminated.

8-4. Samples of size $n = 6$ are collected from a process every hour. After 20 samples have been collected, we calculate $\bar{\bar{x}} = 20.0$ and $\bar{r}/d_2 = 1.4$. Find trial control limits for $\overline{X}$ and R charts.

8-5. Control charts for $\overline{X}$ and R are to be set up for an important quality characteristic. The sample size is $n = 4$, and $\bar{x}$ and r are computed for each of 25 preliminary samples. The summary data are

$$\sum_{i=1}^{25} \bar{x}_i = 7657 \qquad \sum_{i=1}^{25} r_i = 1180$$

(a) Find trial control limits for $\overline{X}$ and R charts.
(b) Assuming that the process is in control, estimate the process mean and standard deviation.

8-6. The thickness of a metal part is an important quality parameter. Data on thickness (in inches) are given here, for 25 samples of five parts each.

Sample Number	x_1	x_2	x_3	x_4	x_5
1	0.0629	0.0636	0.0640	0.0635	0.0640
2	0.0630	0.0631	0.0622	0.0625	0.0627
3	0.0628	0.0631	0.0633	0.0633	0.0630
4	0.0634	0.0630	0.0631	0.0632	0.0633
5	0.0619	0.0628	0.0630	0.0619	0.0625
6	0.0613	0.0629	0.0634	0.0625	0.0628
7	0.0630	0.0639	0.0625	0.0629	0.0627
8	0.0628	0.0627	0.0622	0.0625	0.0627
9	0.0623	0.0626	0.0633	0.0630	0.0624
10	0.0631	0.0631	0.0633	0.0631	0.0630
11	0.0635	0.0630	0.0638	0.0635	0.0633
12	0.0623	0.0630	0.0630	0.0627	0.0629
13	0.0635	0.0631	0.0630	0.0630	0.0630
14	0.0645	0.0640	0.0631	0.0640	0.0642
15	0.0619	0.0644	0.0632	0.0622	0.0635

Sample Number	x_1	x_2	x_3	x_4	x_5
16	0.0631	0.0627	0.0630	0.0628	0.0629
17	0.0616	0.0623	0.0631	0.0620	0.0625
18	0.0630	0.0630	0.0626	0.0629	0.0628
19	0.0636	0.0631	0.0629	0.0635	0.0634
20	0.0640	0.0635	0.0629	0.0635	0.0634
21	0.0628	0.0625	0.0616	0.0620	0.0623
22	0.0615	0.0625	0.0619	0.0619	0.0622
23	0.0630	0.0632	0.0630	0.0631	0.0630
24	0.0635	0.0629	0.0635	0.0631	0.0633
25	0.0623	0.0629	0.0630	0.0626	0.0628

(a) Using all the data, find trial control limits for $\overline{X}$ and R charts, construct the chart, and plot the data. Is the process in statistical control?

(b) Use the trial control limits from part (a) to identify out-of-control points. List the sample numbers of the out-of-control points. Continue eliminating points and revising control limits until the charts are based only on in-control observations.

8-7.   The copper content of a plating bath is measured three times per day, and the results are reported in ppm. The values for 25 days are shown in the table on the right.

	Observation				Observation		
Sample	1	2	3	Sample	1	2	3
1	5.10	6.10	5.50	14	7.59	7.93	6.90
2	5.70	5.59	5.29	15	6.72	6.79	5.23
3	6.31	5.00	6.07	16	6.30	5.37	7.08
4	6.83	8.10	7.96	17	6.33	6.33	5.80
5	5.42	5.29	6.71	18	6.91	6.05	6.03
6	7.03	7.29	7.54	19	8.05	6.52	8.51
7	6.57	5.89	7.08	20	6.39	5.07	6.86
8	5.96	7.52	7.29	21	5.63	6.42	5.39
9	8.15	6.69	6.06	22	6.51	6.90	7.40
10	6.11	5.14	6.68	23	6.91	6.87	6.83
11	6.49	5.68	5.51	24	6.28	6.09	6.71
12	5.12	4.26	4.49	25	5.07	7.17	6.11
13	5.59	5.21	4.94				

(a) Using all the data, find trial control limits for $\overline{X}$ and R charts, construct the chart, and plot the data. Is the process in statistical control?

(b) If necessary, revise the control limits computed in part (a), assuming any samples that plot outside the control limits can be eliminated. Continue to eliminate points outside the control limits and revise, until all points plot between control limits.

8-5   CONTROL CHARTS FOR INDIVIDUAL MEASUREMENTS

In many situations, the sample size used for process control is $n = 1$; that is, the sample consists of an individual unit. Some examples of these situations are as follows.

1.   Automated inspection and measurement technology is used, and every unit manufactured is analyzed.

2.   The production rate is very slow, and it is inconvenient to allow sample sizes of $n > 1$ to accumulate before being analyzed.

3.   Repeat measurements on the process differ only because of laboratory or analysis error, as in many chemical processes.

4.   In process plants, such as papermaking, measurements on some parameters such as coating thickness *across* the roll will differ very little and produce a standard deviation that is much too small if the objective is to control coating thickness *along* the roll.

In such situations, the **control chart for individuals** is useful. The control chart for individuals uses the **moving range** of two successive observations to estimate the process variability. The moving range is defined as $MR_i = |X_i - X_{i-1}|$. It is also possible to establish a control chart on the moving range. The parameters for these charts are defined as follows.

The procedure is illustrated in Example 8-2.

Control Chart for Individuals

The center line and upper and lower control limits for a control chart for individuals are

$$UCL = \bar{x} + 3\,\frac{\overline{mr}}{d_2}$$

$$CL = \bar{x} \tag{8-9}$$

$$LCL = \bar{x} - 3\,\frac{\overline{mr}}{d_2}$$

and for a control chart for moving ranges

$$UCL = D_4\,\overline{mr}$$

$$CL = \overline{mr}$$

$$LCL = D_3\,\overline{mr}$$

The factor d_2 is given in Appendix A Table VII.

EXAMPLE 8-2

Chemical
Process
Concentration

Table 8-3 shows 20 observations on concentration for the output of a chemical process. The observations are taken at 1-hour intervals. If several observations are taken at the same time, the observed concentration reading will differ only because of measurement error. Because the measurement error is small, only one observation is taken each hour. Construct a control chart.

Solution. To set up the control chart for individuals, note that the sample average of the 20 concentration readings is $\bar{x} = 99.1$ and that the moving ranges of two observations are shown in the last column of Table 8-3. The average of the 19 moving ranges is $\overline{mr} = 2.59$. To set up the moving-range chart, we note that $D_3 = 0$ and $D_4 = 3.267$ for $n = 2$. Therefore, the moving-range chart has center line $\overline{mr} = 2.59$, $LCL = 0$, and $UCL = D_4\overline{mr} = (3.267)(2.59) = 8.46$. The control chart is shown as the lower control chart in Fig. 8-10. This control chart was constructed by Minitab. Because no points exceed the upper control limit, we may now set up the control chart for individual concentration measurements. If a moving range of $n = 2$ observations is used, $d_2 = 1.128$. For the data in Table 8-3 we have

$$UCL = \bar{x} + 3\,\frac{\overline{mr}}{d_2} = 99.1 + 3\,\frac{2.59}{1.128} = 105.99$$

$$CL = \bar{x} = 99.1$$

$$LCL = \bar{x} - 3\,\frac{\overline{mr}}{d_2} = 99.1 - 3\,\frac{2.59}{1.128} = 92.21$$

The control chart for individual concentration measurements is shown as the upper control chart in Fig. 8-10. There is no indication of an out-of-control condition. ∎

Table 8-3   Chemical Process Concentration Measurements in Example 8-2

Observation	Concentration x	Moving Range mr
1	102.0	
2	94.8	7.2
3	98.3	3.5
4	98.4	0.1
5	102.0	3.6
6	98.5	3.5
7	99.0	0.5
8	97.7	1.3
9	100.0	2.3
10	98.1	1.9
11	101.3	3.2
12	98.7	2.6
13	101.1	2.4
14	98.4	2.7
15	97.0	1.4
16	96.7	0.3
17	100.3	3.6
18	101.4	1.1
19	97.2	4.2
20	101.0	3.8
	$\bar{x} = 99.1$	$\overline{mr} = 2.59$

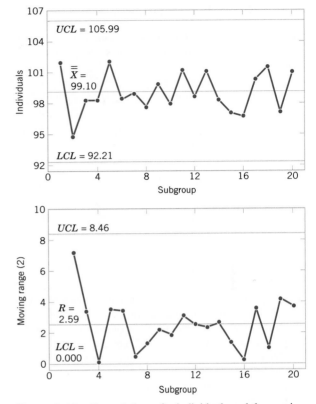

Figure 8-10   Control charts for individuals and the moving range (from Minitab) for the chemical process concentration data in Example 8-2.

The chart for individuals can be interpreted much like an ordinary $\overline{X}$ control chart. A shift in the process average will result in either a point (or points) outside the control limits, or a pattern consisting of a run on one side of the center line.

Some care should be exercised in interpreting patterns on the moving-range chart. The moving ranges are correlated, and this correlation may often induce a pattern of runs or cycles on the chart. The individual measurements are assumed to be uncorrelated, however, and any apparent pattern on the individuals' control chart should be carefully investigated.

The control chart for individuals is very insensitive to small shifts in the process mean. For example, if the size of the shift in the mean is one standard deviation, the average number of points to detect this shift is 43.9. This result is shown later in the chapter. Although the performance of the control chart for individuals is much better for large shifts, in many situations the shift of interest is not large and more rapid shift detection is desirable. In these cases, we recommend the *cumulative sum control chart* or an *exponentially weighted moving-average chart* (Montgomery, 2005b).

Some individuals have suggested that limits narrower than three-sigma be used on the chart for individuals to enhance its ability to detect small process shifts. This is a dangerous suggestion because narrower limits will dramatically increase false alarms such that the charts may be ignored and become useless. If you are interested in detecting small shifts, use the cumulative sum or exponentially weighted moving-average control chart referred to above.

EXERCISES FOR SECTION 8-5

8-8. Twenty successive hardness measurements are made on a metal alloy, and the data are shown in the following table.

Observation	Hardness	Observation	Hardness
1	54	11	49
2	52	12	53
3	54	13	55
4	52	14	54
5	51	15	56
6	55	16	52
7	54	17	55
8	62	18	51
9	49	19	55
10	54	20	52

(a) Using all the data, compute trial control limits for individual observations and moving-range $n = 2$ charts. Construct the chart and plot the data. Determine whether the process is in statistical control. If not, assume that assignable causes can be found to eliminate these samples and revise the control limits.
(b) Estimate the process mean and standard deviation for the in-control process.

8-9. In a semiconductor manufacturing process CVD metal thickness was measured on 30 wafers obtained over approximately 2 weeks. Data are shown in the following table.

Wafer	x	Wafer	x
1	16.8	16	15.4
2	14.9	17	14.3
3	18.3	18	16.1
4	16.5	19	15.8
5	17.1	20	15.9
6	17.4	21	15.2
7	15.9	22	16.7
8	14.4	23	15.2
9	15.0	24	14.7
10	15.7	25	17.9
11	17.1	26	14.8
12	15.9	27	17.0
13	16.4	28	16.2
14	15.8	29	15.6
15	15.4	30	16.3

(a) Using all the data, compute trial control limits for individual observations and moving-range charts. Construct the chart and plot the data. Determine whether the process is in statistical control. If not, assume assignable causes can be found to eliminate these samples and revise the control limits.
(b) Estimate the process mean and standard deviation for the in-control process.

8-10. The diameter of individual holes is measured in consecutive order by an automatic sensor. The results of measuring 25 holes are as follows.

Sample	Diameter	Sample	Diameter
1	14.06	14	20.68
2	23.70	15	16.33
3	15.10	16	16.29
4	22.46	17	9.59
5	35.26	18	15.83
6	22.74	19	15.65
7	20.14	20	19.80
8	11.62	21	21.64
9	10.21	22	28.38
10	8.29	23	21.58
11	16.49	24	8.38
12	15.34	25	17.00
13	14.08		

(a) Using all the data, compute trial control limits for individual observations and moving-range $n = 2$ charts. Construct the control chart and plot the data. Determine whether the process is in statistical control. If not, assume that assignable causes can be found to eliminate these samples and revise the control limits.
(b) Estimate the process mean and standard deviation for the in-control process.

8-11. The viscosity of a chemical intermediate is measured every hour. Twenty samples consisting of a single observation are as follows.

Sample	Viscosity	Sample	Viscosity
1	378	11	462
2	438	12	502
3	487	13	449
4	515	14	470
5	485	15	501
6	474	16	470
7	486	17	512
8	548	18	530
9	502	19	462
10	440	20	491

(a) Using all the data, compute trial control limits for individual observations and moving-range $n = 2$ charts. Determine whether the process is in statistical control. If not, assume that assignable causes can be found to eliminate these samples and revise the control limits.

(b) Estimate the process mean and standard deviation for the in-control process.

8-6   PROCESS CAPABILITY

It is usually necessary to obtain some information about the **capability** of the process—that is, the performance of the process when it is operating in control. Two graphical tools, the **tolerance chart** (or tier chart) and the **histogram,** are helpful in assessing process capability. The tolerance chart for all 20 samples from the vane-manufacturing process is shown in Fig. 8-11. The specifications on vane opening are 0.5030 ± 0.0010 in. In terms of the coded data, the upper specification limit is $USL = 40$ and the lower specification limit is $LSL = 20$, and these limits are shown on the chart in Fig. 8-11. Each measurement is plotted on the tolerance chart. Measurements from the same subgroup are connected with lines. The tolerance chart is useful in revealing patterns over time in the individual measurements, or it may show that a particular value of $\bar{x}$ or r was produced by one or two unusual observations in the sample. For example, note the two unusual observations in sample 9 and the single unusual observation in sample 8. Note also that it is appropriate to plot the specification limits on the tolerance chart because it is a chart of individual measurements. **It is never appropriate to plot specification limits on a control chart or to use the specifications in determining the control limits.** Specification limits and control limits are unrelated. Finally, note from Fig. 8-11 that the process is running off-center from the nominal dimension of 30 (or 0.5030 inch).

The histogram for the vane-opening measurements is shown in Fig. 8-12. The observations from samples 6, 8, 9, 11, and 19 (corresponding to out-of-control points on either the $\bar{X}$ or R chart) have been deleted from this histogram. The general impression from examining this histogram is that the process is capable of meeting the specification but that it is running off-center.

Another way to express process capability is in terms of an index that is defined as follows.

The **process capability ratio** C_p is

$$C_p = \frac{USL - LSL}{6\sigma} \qquad (8\text{-}10)$$

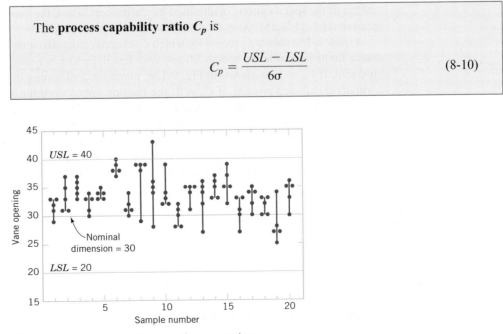

Figure 8-11   Tolerance diagram of vane openings.

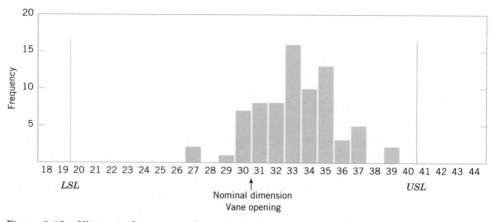

Figure 8-12   Histogram for vane opening.

The numerator of C_p is the width of the specifications. The three sigma limits on either side of the process mean are sometimes called **natural tolerance limits** because these represent limits that an in-control process should meet with most of the units produced. Consequently, six sigma is often referred to as the width of the process. For the vane opening, where our sample size is 5, we could estimate σ as

$$\hat{\sigma} = \frac{\bar{r}}{d_2} = \frac{5.0}{2.326} = 2.15$$

Therefore, C_p is estimated to be

$$\hat{C}_p = \frac{USL - LSL}{6\hat{\sigma}} = \frac{40 - 20}{6(2.15)} = 1.55$$

The process capability ratio (C_p) has a natural interpretation: $(1/C_p)\,100$ is simply the percentage of the specifications' width used by the process. Thus, the vane-opening process uses approximately $(1/1.55)100 = 64.5\%$ of the specifications' width.

Figure 8-13*a* shows a process for which C_p exceeds unity. Because the process natural tolerance limits lie inside the specifications, very few defective or nonconforming units will be produced. If $C_p = 1$, as shown in Fig. 8-13*b*, more nonconforming units result. In fact, for a normally distributed process, if $C_p = 1$, the fraction nonconforming is 0.27%, or 2700 parts per million. Finally, when the C_p is less than unity, as in Fig. 8-13*c*, the process is very yield sensitive and a large number of nonconforming units will be produced.

The definition of the C_p given in equation 8-10 implicitly assumes that the process is centered at the nominal dimension. If the process is running off-center, its **actual capability** will be less than indicated by the C_p. It is convenient to think of C_p as a measure of **potential**

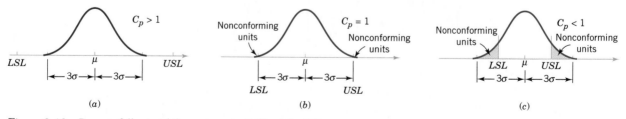

Figure 8-13   Process fallout and the process capability ratio (C_p).

capability—that is, capability with a centered process. If the process is not centered, then a measure of actual capability is often used. This ratio, called C_{pk}, is defined next.

The **process capability ratio C_{pk}** is

$$C_{pk} = \min\left[\frac{USL - \mu}{3\sigma}, \frac{\mu - LSL}{3\sigma}\right] \qquad (8\text{-}11)$$

In effect, C_{pk} is a **one-sided** process capability ratio that is calculated relative to the specification limit nearest to the process mean. For the vane-opening process, we find that the estimate of the process capability ratio C_{pk} is

$$\hat{C}_{pk} = \min\left[\frac{USL - \bar{\bar{x}}}{3\hat{\sigma}}, \frac{\bar{\bar{x}} - LSL}{3\hat{\sigma}}\right]$$

$$= \min\left[\frac{40 - 3321}{3(2.15)} = 1.05, \frac{3321 - 20}{3(2.15)} = 2.05\right] = 1.05$$

Note that if $C_p = C_{pk}$, the process is centered at the nominal dimension. Because $\hat{C}_{pk} = 1.05$ for the vane-opening process and $\hat{C}_p = 1.55$, the process is obviously running off-center, as was first noted in Figs. 8-11 and 8-12. This off-center operation was ultimately traced to an oversized wax tool. Changing the tooling resulted in a substantial improvement in the process (Montgomery, 2005b).

The fractions of nonconforming output (or fallout) below the lower specification limit and above the upper specification limit are often of interest. Suppose that the output from a normally distributed process in statistical control is denoted as X. The fractions are determined from

$$P(X < LSL) = P(Z < (LSL - \mu)/\sigma)$$
$$P(X > USL) = P(Z > (USL - \mu)/\sigma)$$

EXAMPLE 8-3
Electrical
Current

For an electronic manufacturing process, a current has specifications of 100 ± 10 milliamperes. The process mean μ and standard deviation σ are 107.0 and 1.5, respectively. Estimate C_p, C_{pk}, and the probability of not meeting specification.

Solution. $\hat{C}_p = (110 - 90)/(6 \cdot 1.5) = 2.22$ and $\hat{C}_{pk} = (110 - 107)/(3 \cdot 1.5) = 0.67$

The small C_{pk} indicates that the process is likely to produce currents outside of the specification limits. From the normal distribution in Appendix A Table I,

$$P(X < LSL) = P(Z < (90 - 107)/1.5) = P(Z < -11.33) = 0$$
$$P(X > USL) = P(Z > (110 - 107)/1.5) = P(Z > 2) = 0.023$$

For this example, the relatively large probability of exceeding the USL is a warning of potential problems with this criterion even if none of the measured observations in a preliminary sample exceeds this limit. We emphasize that the fraction-nonconforming calculation assumes that the observations are normally distributed and the process is in control. Departures from

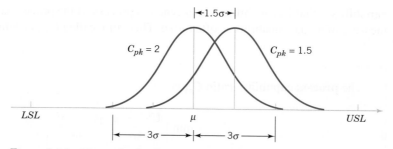

Figure 8-14   Mean of a six-sigma process shifts by 1.5 standard deviations.

normality can seriously affect the results. The calculation should be interpreted as an approximate guideline for process performance. To make matters worse, μ and σ need to be estimated from the data available, and a small sample size can result in poor estimates that further degrade the calculation.

Montgomery (2005b) provides guidelines on appropriate values of the C_p and a table relating fallout for a normally distributed process in statistical control to the value of C_p. Many U.S. companies use $C_p = 1.33$ as a minimum acceptable target and $C_p = 1.66$ as a minimum target for strength, safety, or critical characteristics. Some companies require that internal processes and those at suppliers achieve a $C_{pk} = 2.0$. Figure 8-14 illustrates a process with $C_p = C_{pk} = 2.0$. Assuming a normal distribution, the calculated fallout for this process is 0.0018 parts per million. A process with $C_{pk} = 2.0$ is referred to as a **six-sigma process** because the distance from the process mean to the nearest specification is six standard deviations. The reason that such a large process capability is often required is that it is difficult to maintain a process mean at the center of the specifications for long periods of time. A common model that is used to justify the importance of a six-sigma process is illustrated in Fig. 8-14. If the process mean shifts off-center by 1.5 standard deviations, the C_{pk} decreases to $4.5\sigma/3\sigma = 1.5$. Assuming a normally distributed process, the fallout of the shifted process is **3.4 parts per million.** Consequently, the mean of a six-sigma process can shift 1.5 standard deviations from the center of the specifications and still maintain a fallout of 3.4 parts per million.

We repeat that process capability calculations are meaningful only for stable processes; that is, processes that are in control. A process capability ratio indicates whether or not the natural or chance variability in a process is acceptable relative to the specifications.

EXERCISES FOR SECTION 8-6

8-12. Six standard deviations of a normally distributed process use 66.7% of the specification band. It is centered at the nominal dimension, located halfway between the upper and lower specification limits.

(a) Estimate C_p and C_{pk}. Interpret these ratios.
(b) What fallout level (fraction defective) is produced?

8-13. Reconsider Exercise 8-1. Use the revised control limits and process estimates.

(a) Estimate C_p and C_{pk}. Interpret these ratios.
(b) What percentage of defectives is being produced by this process?

8-14. Reconsider Exercise 8-2, where the specification limits are 18.50 ± 0.50.

(a) What conclusions can you draw about the ability of the process to operate within these limits? Estimate the percentage of defective items that will be produced.
(b) Estimate C_p and C_{pk}. Interpret these ratios.

8-15. Reconsider Exercise 8-3. Using the process estimates, what is the fallout level if the coded specifications are 15 ± 3 mm? Estimate C_p and interpret this ratio.

8-16. Six standard deviations of a normally distributed process use 85% of the specification band. It is centered at the nominal dimension, located halfway between the upper and lower specification limits.

(a) Estimate C_p and C_{pk}. Interpret these ratios.
(b) What fallout level (fraction defective) is produced?

8-17. Reconsider Exercise 8-5. Suppose that the quality characteristic is normally distributed with specification at 300 ± 40. What is the fallout level? Estimate $\mathbf{C}_p$ and $\mathbf{C}_{pk}$ and interpret these ratios.

8-18. Reconsider Exercise 8-4. Assuming that both charts exhibit statistical control and that the process specifications are at 20 ± 5, estimate $\mathbf{C}_p$ and $\mathbf{C}_{pk}$ and interpret these ratios.

8-19. Reconsider Exercise 8-7. Given that the specifications are at 6.0 ± 0.5, estimate $\mathbf{C}_p$ and $\mathbf{C}_{pk}$ for the in-control process and interpret these ratios.

8-20. Reconsider Exercise 8-6. What are the natural tolerance limits of this process?

8-21. Reconsider Exercise 8-11. What are the natural tolerance limits of this process?

8-7 ATTRIBUTE CONTROL CHARTS

8-7.1 P Chart (Control Chart for Proportions) and nP Chart

Often it is desirable to classify a product as either defective or nondefective on the basis of comparison with a standard. This classification is usually done to achieve economy and simplicity in the inspection operation. For example, the diameter of a ball bearing may be checked by determining whether it will pass through a gauge consisting of circular holes cut in a template. This kind of measurement would be much simpler than directly measuring the diameter with a device such as a micrometer. Control charts for attributes are used in these situations. Attribute control charts often require a considerably larger sample size than do their measurements counterparts. In this section, we will discuss the **fraction-defective control chart,** or **P chart.** Sometimes the P chart is called the **control chart for fraction nonconforming.**

Suppose D is the number of defective units in a random sample of size n. We assume that D is a binomial random variable with unknown parameter p. The fraction defective

$$\hat{P} = \frac{D}{n}$$

of each sample is plotted on the chart. Furthermore, the variance of the statistic $\hat{P}$ is

$$\sigma_{\hat{P}}^2 = \frac{p(1-p)}{n}$$

Therefore, a P chart for fraction defective could be constructed using p as the center line and control limits at

$$\text{UCL} = p + 3\sqrt{\frac{p(1-p)}{n}}$$

$$\text{LCL} = p - 3\sqrt{\frac{p(1-p)}{n}} \tag{8-12}$$

However, the true process fraction defective is almost always unknown and must be estimated using the data from preliminary samples.

Suppose that m preliminary samples each of size n are available, and let D_i be the number of defectives in the ith sample. The $\hat{P}_i = D_i/n$ is the sample fraction defective in the ith sample. The average fraction defective is

$$\overline{P} = \frac{1}{m} \sum_{i=1}^{m} \hat{P}_i = \frac{1}{mn} \sum_{i=1}^{m} D_i \tag{8-13}$$

Now $\bar{P}$ may be used as an estimator of p in the center line and control limit calculations.

The *P* Chart

The center line and upper and lower control limits for the *P* chart are

$$UCL = \bar{p} + 3\sqrt{\frac{\bar{p}(1 - \bar{p})}{n}}$$

$$CL = \bar{p}$$

$$LCL = \bar{p} - 3\sqrt{\frac{\bar{p}(1 - \bar{p})}{n}} \qquad (8\text{-}14)$$

where $\bar{p}$ is the observed value of the average fraction defective.

These control limits are based on the normal approximation to the binomial distribution. When p is small, the normal approximation may not always be adequate. In such cases, we may use control limits obtained directly from a table of binomial probabilities. If $\bar{p}$ is small, the lower control limit may be a negative number. If this should occur, it is customary to consider zero as the lower control limit.

EXAMPLE 8-4

Ceramic
Substrate

We have 20 preliminary samples, each of size 100; the number of defectives in each sample is shown in Table 8-4. Construct a fraction-defective control chart for this ceramic substrate production line.

Solution. Assume that the samples are numbered in the sequence of production. Note that $\bar{p} = (800/2000) = 0.40$; therefore, the trial parameters for the control chart are

$$UCL = 0.40 + 3\sqrt{\frac{(0.40)(0.60)}{100}} = 0.55$$

$$CL = 0.40$$

$$LCL = 0.40 - 3\sqrt{\frac{(0.40)(0.60)}{100}} = 0.25$$

Table 8-4 Number of Defectives in Samples of 100 Ceramic Substrates in Example 8-4

Sample	No. of Defectives	Sample	No. of Defectives
1	44	11	36
2	48	12	52
3	32	13	35
4	50	14	41
5	29	15	42
6	31	16	30
7	46	17	46
8	52	18	38
9	44	19	26
10	48	20	30

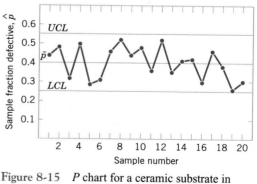

Figure 8-15 *P* chart for a ceramic substrate in Example 8-4.

Interpret the Results

The control chart displaying the fraction-defective for each sample is shown in Fig. 8-15 (page 440). All samples are in control. If they were not, we would search for assignable causes of variation and revise the limits accordingly. This chart can be used for controlling future production.

Although this process exhibits statistical control, its defective rate ($\bar{p} = 0.40$) is very poor. We should take appropriate steps to investigate the process to determine why such a large number of defective units are being produced. Defective units should be analyzed to determine the specific types of defects present. Once the defect types are known, process changes should be investigated to determine their impact on defect levels. Designed experiments may be useful in this regard. ∎

Computer software also produces an ***nP* chart.** This is simply a control chart of $n\hat{P} = D$, the number of defectives in a sample. The points, center line, and control limits for this chart are multiples (times n) of the corresponding elements of a P chart. The use of an nP chart avoids the fractions in a P chart.

The *nP* Chart

The center line and upper and lower control limits for the nP chart are

$$UCL = n\bar{p} + 3\sqrt{n\bar{p}(1 - \bar{p})}$$
$$CL = n\bar{p}$$
$$LCL = n\bar{p} - 3\sqrt{n\bar{p}(1 - \bar{p})}$$

where $\bar{p}$ is the observed value of the average fraction defective.

For the data in Example 8-4, the center line is $n\bar{p} = 100(0.4) = 40$ and the upper and lower control limits for the nP chart are $UCL = 100(0.4) + 3\sqrt{100(0.4)(0.6)} = 54.70$ and $LCL = 100(0.4) - 3\sqrt{100(0.4)(0.6)} = 25.30$. The number of defectives in Table 8-4 would be plotted on such a chart.

8-7.2 U Chart (Control Chart for Average Number of Defects per Unit) and C Chart

It is sometimes necessary to monitor the number of defects in a unit of product rather than the fraction defective. Suppose that in the production of cloth it is necessary to control the number of defects per yard or that in assembling an aircraft wing the number of missing rivets must be controlled. In these situations, we may use the control chart for defects per unit, or the ***U*** **chart.** Many defects-per-unit situations can be modeled by the Poisson distribution.

If each sample consists of n units and there are C total defects in the sample,

$$U = \frac{C}{n}$$

is the average number of defects per unit. A U chart may be constructed for such data. If there are m samples, and the number of defects in these samples is $C_1, C_2, \ldots, C_m$, the estimator of the average number of defects per unit is

$$\bar{U} = \frac{1}{m}\sum_{i=1}^{m} U_i = \frac{1}{mn}\sum_{i=1}^{m} C_i \qquad (8\text{-}15)$$

The parameters of the U chart are defined as follows.

The U Chart

The center line and upper and lower control limits on the U chart are

$$UCL = \bar{u} + 3\sqrt{\frac{\bar{u}}{n}}$$

$$CL = \bar{u}$$

$$LCL = \bar{u} - 3\sqrt{\frac{\bar{u}}{n}} \qquad (8\text{-}16)$$

where $\bar{u}$ is the average number of defects per unit.

If the number of defects in a unit is a Poisson random variable with parameter λ, the mean and variance of this distribution are both λ. Each point on the chart is U, the average number of defects per unit. Therefore, the mean of U is λ and the variance of U is λ/n. Usually λ is unknown and $\overline{U}$ is the estimator of λ that is used to set the control limits.

These control limits are based on the normal approximation to the Poisson distribution. When λ is small, the normal approximation may not always be adequate. In such cases, we may use control limits obtained directly from a table of Poisson probabilities. If $\bar{u}$ is small, the lower control limit may be a negative number. If this should occur, it is customary to consider zero as the lower control limit.

EXAMPLE 8-5

Printed Circuit Boards

Printed circuit boards are assembled by a combination of manual assembly and automation. A flow solder machine is used to make the mechanical and electrical connections of the leaded components to the board. The boards are run through the flow solder process almost continuously, and every hour five boards are selected and inspected for process-control purposes. The number of defects in each sample of five boards is noted. Results for 20 samples are shown in Table 8-5. Construct a U chart.

Solution. The center line for the U chart is

$$\bar{u} = \frac{1}{20} \sum_{i=1}^{20} u_i = \frac{32}{20} = 1.6$$

Table 8-5 Number of Defects in Samples of Five Printed Circuit Boards for Example 8-5

Sample	Number of Defects c_i	Defects per Unit u_i	Sample	Number of Defects c_i	Defects per Unit u_i
1	6	1.2	11	9	1.8
2	4	0.8	12	15	3.0
3	8	1.6	13	8	1.6
4	10	2.0	14	10	2.0
5	9	1.8	15	8	1.6
6	12	2.4	16	2	0.4
7	16	3.2	17	7	1.4
8	2	0.4	18	1	0.2
9	3	0.6	19	7	1.4
10	10	2.0	20	13	2.6

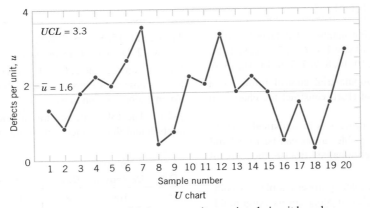

Figure 8-16 U chart of defects per unit on printed circuit boards.

and the upper and lower control limits are

$$UCL = \bar{u} + 3\sqrt{\frac{\bar{u}}{n}} = 1.6 + 3\sqrt{\frac{1.6}{5}} = 3.3$$

$$LCL = \bar{u} - 3\sqrt{\frac{\bar{u}}{n}} = 1.6 - 3\sqrt{\frac{1.6}{5}} = 0$$

Interpret the Results

The control chart is plotted in Fig. 8-16. Because LCL is negative, it is set to zero. From the control chart in Fig. 8-16, we see that the process is in control. However, eight defects per group of five circuit boards is too many (about $8/5 = 1.6$ defects/board), and the process needs improvement. An investigation needs to be made of the specific types of defects found on the printed circuit boards. This will usually suggest potential avenues for process improvement.

Computer software also produces a **C chart.** This is simply a control chart of C, the total number of defects in a sample. The use of a C chart avoids the fractions that can occur in a U chart.

The C Chart

The center line and upper and lower control limits for the C chart are

$$UCL = \bar{c} + 3\sqrt{\bar{c}}$$
$$CL = \bar{c}$$
$$LCL = \bar{c} - 3\sqrt{\bar{c}}$$

(8-17)

where $\bar{c}$ is the average number of defects in a sample.

For the data in Example 8-5

$$\bar{c} = \frac{1}{20}\sum_{i=1}^{20} c_i = \left(\frac{1}{20}\right)160 = 8$$

and the upper and lower control limits for the C chart are $UCL = 8 + 3\sqrt{8} = 16.5$ and $LCL = 8 - 3\sqrt{8} = -0.5$, which is set to zero. The number of defects in Table 8-5 would be plotted on such a chart.

EXERCISES FOR SECTION 8-7

8-22. Suppose the following number of defects has been found in successive samples of size 100: 6, 7, 3, 9, 6, 9, 4, 14, 3, 5, 6, 9, 6, 10, 9, 2, 8, 4, 8, 10, 10, 8, 7, 7, 7, 6, 14, 18, 13, 6.

(a) Using all the data, compute trial control limits for a fraction-defective control chart, construct the chart, and plot the data.

(b) Determine whether the process is in statistical control. If not, assume that assignable causes can be found and out-of-control points eliminated. Revise the control limits.

8-23. Using an injection molding process, a plastics company produces interchangeable cell phone covers. After molding, the covers are sent through an intricate painting process. Quality control engineers inspect the covers and record the paint blemishes. The number of blemishes found in 20 samples of 5 covers are as follows: 2, 1, 5, 5, 3, 3, 1, 3, 4, 5, 4, 4, 1, 5, 2, 2, 3, 1, 4, 4.

(a) Using all the data, compute trial control limits for a U control chart, construct the chart, and plot the data.

(b) Can we conclude that the process is in control using a U chart? If not, assume that assignable causes can be found, list points, and revise the control limits.

8-24. The following represent the number of defects per 1000 feet in rubber-covered wire: 1, 1, 3, 7, 8, 10, 5, 13, 0, 19, 24, 6, 9, 11, 15, 8, 3, 6, 7, 4, 9, 20, 11, 7, 18, 10, 6, 4, 0, 9, 7, 3, 1, 8, 12. Do the data come from a controlled process?

8-25. Consider the data in Exercise 8-23. Set up a C chart for this process. Compare it to the U chart in Exercise 8-23. Comment on your findings.

8-26. The following are the numbers of defective solder joints found during successive samples of 500 solder joints.

Day	No. of Defectives	Day	No. of Defectives	Day	No. of Defectives
1	106	8	36	15	101
2	116	9	69	16	64
3	164	10	74	17	51
4	89	11	42	18	74
5	99	12	37	19	71
6	40	13	25	20	43
7	112	14	88	21	80

(a) Using all the data, compute trial control limits for both a P chart and an nP chart, construct the charts, and plot the data.

(b) Determine whether the process is in statistical control. If not, assume that assignable causes can be found and out-of-control points eliminated. Revise the control limits.

8-8 CONTROL CHART PERFORMANCE

Specifying the control limits is one of the critical decisions that must be made in designing a control chart. By moving the control limits farther from the center line, we decrease the risk of a type I error—that is, the risk of a point falling beyond the control limits, indicating an out-of-control condition when no assignable cause is present. However, widening the control limits will also increase the risk of a type II error—that is, the risk of a point falling between the control limits when the process is really out of control. If we move the control limits closer to the center line, the opposite effect is obtained: The risk of type I error is increased, whereas the risk of type II error is decreased.

The control limits on a Shewhart control chart are customarily located a distance of plus or minus three standard deviations of the variable plotted on the chart from the center line; that is, the constant k in equation 8-1 should be set equal to 3. These limits are called **three-sigma control limits.**

A way to evaluate decisions regarding sample size and sampling frequency is through the **average run length (ARL)** of the control chart. Essentially, the ARL is the average number of points that must be plotted before a point indicates an out-of-control condition. For any Shewhart control chart, the ARL can be calculated from the mean of a geometric random variable (Montgomery, 2005b) as

$$ARL = \frac{1}{p} \tag{8-18}$$

where p is the probability that any point exceeds the control limits. Thus, for an $\overline{X}$ chart with three-sigma limits, $p = 0.0027$ is the probability that a single point falls outside the limits when the process is in control, so

$$\text{ARL} = \frac{1}{p} = \frac{1}{0.0027} \cong 370$$

is the average run length of the $\overline{X}$ chart when the process is in control. That is, even if the process remains in control, an out-of-control signal will be generated every 370 points, on the average.

Consider the piston ring process discussed earlier, and suppose we are sampling every hour. Thus, we will have a **false alarm** about every 370 hours on the average. Suppose that we are using a sample size of $n = 5$ and that when the process goes out of control the mean shifts to 74.0135 mm. Then, the probability that $\overline{X}$ falls between the control limits of Fig. 8-3 is equal to

$$P[73.9865 \le \overline{X} \le 74.0135 \text{ when } \mu = 74.0135]$$

$$= P\left[\frac{73.9865 - 74.0135}{0.0045} \le Z \le \frac{74.0135 - 74.0135}{0.0045} \right]$$

$$= P[-6 \le Z \le 0] = 0.5$$

Therefore, p in equation 8-18 is 0.50, and the out-of-control ARL is

$$\text{ARL} = \frac{1}{p} = \frac{1}{0.5} = 2$$

That is, the control chart will require two samples to detect the process shift, on the average, so 2 hours will elapse between the shift and its detection (*again on the average*). Suppose this approach is unacceptable because production of piston rings with a mean diameter of 74.0135 mm results in excessive scrap costs and delays final engine assembly. How can we reduce the time needed to detect the out-of-control condition? One method is to sample more frequently. For example, if we sample every half hour, only 1 hour will elapse (on the average) between the shift and its detection. The second possibility is to increase the sample size. For example, if we use $n = 10$, the control limits in Fig. 8-3 narrow to 73.9905 and 74.0095. The probability of $\overline{X}$ falling between the control limits when the process mean is 74.0135 mm is approximately 0.1, so $p = 0.9$, and the out-of-control ARL is

$$\text{ARL} = \frac{1}{p} = \frac{1}{0.9} = 1.11$$

Thus, the larger sample size would allow the shift to be detected about twice as quickly as the old one. If it became important to detect the shift in the first hour after it occurred, two control chart designs would work:

Design 1	Design 2
Sample size: $n = 5$	Sample size: $n = 10$
Sampling frequency: every half hour	Sampling frequency: every hour

Table 8-6 Average Run Length (ARL) for an $\overline{X}$ Chart with Three-Sigma Control Limits

Magnitude of Process Shift	ARL $n = 1$	ARL $n = 4$
0	370.4	370.4
0.5σ	155.2	43.9
1.0σ	43.9	6.3
1.5σ	15.0	2.0
2.0σ	6.3	1.2
3.0σ	2.0	1.0

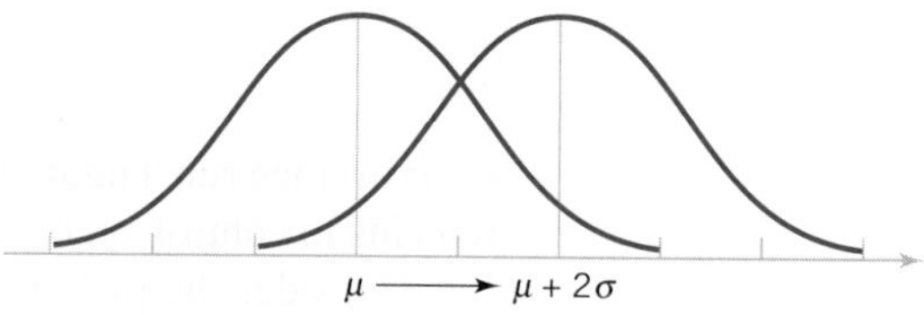

Figure 8-17 Process mean shift of two sigmas.

Table 8-6 provides average run lengths for an $\overline{X}$ chart with three-sigma control limits. The average run lengths are calculated for shifts in the process mean from 0 to 3.0σ and for sample sizes of $n = 1$ and $n = 4$ by using $1/p$, where p is the probability that a point plots outside of the control limits. Figure 8-17 illustrates a shift in the process mean of 2σ.

EXERCISES FOR SECTION 8-8

8-27. Consider the $\overline{X}$ control chart in Fig. 8-3. Suppose that the mean shifts to 74.010 mm.

(a) What is the probability that this shift will be detected on the next sample?
(b) What is the ARL after the shift?

8-28. An $\overline{X}$ chart uses samples of size 6. The center line is at 100, and the upper and lower three-sigma control limits are at 106 and 94, respectively.

(a) What is the process σ?
(b) Suppose the process mean shifts to 105. Find the probability that this shift will be detected on the next sample.
(c) Find the ARL to detect the shift in part (b).

8-29. Consider the revised $\overline{X}$ control chart in Exercise 8-1 with $\hat{\sigma} = 2.922$, $UCL = 39.34$, $LCL = 29.22$, and $n = 3$. Suppose that the mean shifts to 39.0.

(a) What is the probability that this shift will be detected on the next sample?
(b) What is the ARL after the shift?

8-30. Consider the $\overline{X}$ control chart in Exercise 8-2 with $\bar{r} = 0.39$, $UCL = 19.209$, $LCL = 18.641$, and $n = 4$. Suppose that the mean shifts to 19.1.

(a) What is the probability that this shift will be detected on the next sample?
(b) What is the ARL after the shift?

8-31. Consider the revised $\overline{X}$ control chart in Exercise 8-3 with $\bar{r} = 3.895$, $UCL = 17.98$, $LCL = 12.31$, and $n = 4$. Suppose that the mean shifts to 12.8.

(a) What is the probability that this shift will be detected on the next sample?
(b) What is the ARL after the shift?

8-32. Consider the $\overline{X}$ control chart in Exercise 8-4 with $\hat{\sigma} = 1.40$, $UCL = 21.71$, $LCL = 18.29$, and $n = 6$. Suppose that the mean shifts to 18.5.

(a) What is the probability that this shift will be detected on the next sample?
(b) What is the ARL after the shift?

8-33. Consider the $\overline{X}$ control chart in Exercise 8-5 with $\bar{r} = 47.2$, $UCL = 340.69$, $LCL = 271.87$, and $n = 4$. Suppose that the mean shifts to 310.

(a) What is the probability that this shift will be detected on the next sample?
(b) What is the ARL after the shift?

8-34. Consider the revised $\overline{X}$ control chart in Exercise 8-6 with $\hat{\sigma} = 0.00024$, $UCL = 0.06331$, $LCL = 0.06266$, and $n = 5$. Suppose that the mean shifts to 0.0630.

(a) What is the probability that this shift will be detected on the next sample?
(b) What is the ARL after the shift?

8-35. Consider the revised $\overline{X}$ control chart in Exercise 8-7 with $\hat{\sigma} = 0.671$, $UCL = 7.385$, $LCL = 5.061$, and $n = 3$. Suppose that the mean shifts to 6.80.

(a) What is the probability that this shift will be detected on the next sample?
(b) What is the ARL after the shift?

8-9   MEASUREMENT SYSTEMS CAPABILITY

An important component of many engineering studies is the performance of the gauge or test instrument used to produce measurements on the system of interest. In any problem involving measurements, some of the observed variability will arise from the experimental units that are being measured and some will be due to measurement error. There are two types of error associated with a gauge or measurement device, **precision** and **accuracy.** These two components of measurement error are illustrated in Fig. 8-18. In this figure, the bull's eye of the target is considered to be the true value of the measured characteristic. Accuracy refers to the ability to measure the true value of the characteristic correctly on average, and precision reflects the inherent variability in the measurements. In this section we describe some methods for evaluating the **precision** of a measurement device or system. Determining accuracy often requires the use of a standard, for which the true value of the measured characteristic is known. Often the accuracy feature of a measurement system or device can be modified by making adjustments to the device or by using a properly constructed calibration curve. The regression methods of Chapter 6 can be used to construct calibration curves.

Some data from a measurement system study in the semiconductor industry are shown in Table 8-7. An electronic tool was used to measure the resistivity of 20 randomly selected silicon wafers following a process step during which a layer was deposited on the wafer surface. The technician who was responsible for the setup and operation of the measurement tool measured each wafer twice. Each measurement on all 20 wafers was made in random order.

A very simple model can be used to describe the measurements in Table 8-7:

$$\sigma^2_{\text{Total}} = \sigma^2_{\text{Wafer}} + \sigma^2_{\text{Gauge}} \tag{8-19}$$

where σ^2_{Total} is the variance of the observed measurements, σ^2_{Wafer} is the component of the total variance that is due to the wafers, and σ^2_{Gauge} is the component of the total variance due to the

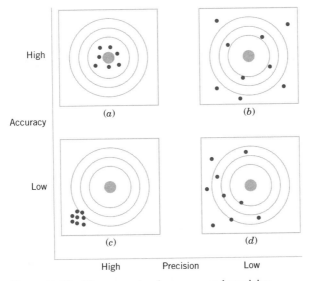

Figure 8-18   The concepts of accuracy and precision.
(*a*) The gauge is accurate and precise. (*b*) The gauge is accurate but not precise. (*c*) The gauge is not accurate but it is precise. (*d*) The gauge is neither accurate nor precise.

Table 8-7 Resistivity Measurements on 20 Silicon Wafers (ohms/cm^2)

Wafer	Meas. 1	Meas. 2	Wafer	Meas. 1	Meas. 2
1	3712	3710	11	3711	3709
2	3680	3683	12	3712	3715
3	3697	3700	13	3728	3721
4	3671	3668	14	3694	3698
5	3692	3689	15	3704	3702
6	3688	3690	16	3686	3685
7	3691	3694	17	3705	3706
8	3696	3701	18	3678	3680
9	3705	3709	19	3723	3724
10	3678	3681	20	3672	3669

gauge or measurement tool. Figure 8-19 shows $\overline{X}$ and R charts (from Minitab) for the data in Table 8-7. The $\overline{X}$ chart indicates that there are many out-of-control points because the control chart is showing the discriminating capability of the measuring instrument—literally the ability of the device to distinguish between different units of product. Notice that this is a somewhat different interpretation for an $\overline{X}$ control chart. The R chart directly reflects the magnitude of measurement error because the range values are the difference between measurements made on the same wafer using the same measurement tool. The R chart is in control, indicating that the operator is not experiencing any difficulty making consistent measurements. Nor is there any indication that measurement variability is increasing with time. Out-of-control points or patterns on the R chart can indicate that the operator/measuring tool combination is experiencing some difficulty in making consistent measurements.

We may also consider the measurement system study in Table 8-7 as a **single-factor completely randomized experiment** with parts as treatments. Recall from Section 5-8 that the analysis of variance (ANOVA) can be used to analyze data from this type of experiment. The

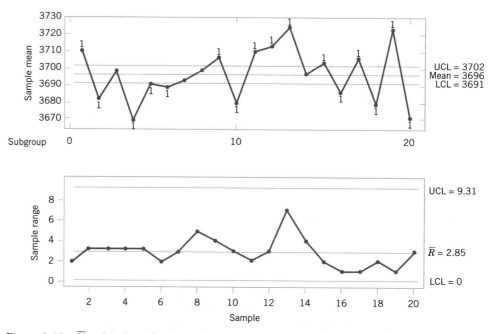

Figure 8-19 $\overline{X}$ and R charts for the resistivity measurements in Table 8-7.

ANOVA model for the single-factor experiment is

$$Y_{ij} = \mu + \tau_i + \epsilon_{ij} \begin{cases} i = 1, 2, \ldots, a \\ j = 1, 2, \ldots, n \end{cases} \tag{8-20}$$

where we have $a = 20$ wafers and $n = 2$ replicates. In the model of equation 8-20, τ_i is the treatment effect; in this case, the effect of the ith part, and the random error component ϵ_{ij} represents the variability of the measurement tool. Therefore, the variance of ϵ_{ij} is σ^2_{Gauge}. Because the wafers used in the study were selected at random, the treatment effects τ_i are random variables and the variance of these treatment effects is σ^2_{Wafer}. The Minitab ANOVA for the data in Table 8-7 is summarized in Table 8-8. Notice that the F-statistic for wafers is significant, implying that there are differences in the parts used in the study.

Because the treatment effects in this experiment are random, we can use the ANOVA results to estimate σ^2_{Gauge} and σ^2_{Wafer}. It turns out that (see Montgomery [2005b] for details)

$$E(MS_{Wafer}) = \sigma^2_{Gauge} + n\sigma^2_{Wafer}$$
$$E(MS_{Error}) = \sigma^2_{Gauge} \tag{8-21}$$

We can substitute the calculated values of the mean squares for their expected values in equation 8-21 and solve the resulting equations for the estimates of the two variance components, σ^2_{Gauge} and σ^2_{Wafer}. This yields

$$514.80 = \hat{\sigma}^2_{Gauge} + 2\hat{\sigma}^2_{Wafer}$$
$$5.02 = \hat{\sigma}^2_{Gauge}$$

with resulting solution

$$\hat{\sigma}^2_{Gauge} = 5.02$$
$$\hat{\sigma}^2_{Wafer} = \frac{514.80 - \hat{\sigma}^2_{Gauge}}{2} = \frac{514.80 - 5.02}{2} = 254.89$$

Notice that the variability of the measurement system, σ^2_{Gauge}, is considerably smaller than the variability in the wafers. This is a desirable situation.

The ANOVA approach is a very useful way to evaluate measurement systems. It can be extended to more complex types of experiments. Table 8-9 presents an expanded study of the tool for measuring resistivity of silicon wafers. In the original study, the 20 wafers were measured on the first shift, and an operator from that shift was responsible for setup and operation of the measurement tool. In the expanded study, the 20 wafers were measured on two additional shifts, and operators from those shifts did the setup and ran the measurement tool.

Table 8-8 One-Way ANOVA: Resistivity versus Wafers

Analysis of Variance

Source	DF	SS	MS	F	P
Wafers	19	9781.28	514.80	102.45	0.000
Error	20	100.50	5.02		
Total	39	9881.77			

Table 8-9   Resistivity Measurements on 20 Silicon Wafers (ohms/cm^2), Expanded Study

Wafer	Shift 1 Meas. 1	Shift 1 Meas. 2	Shift 2 Meas. 1	Shift 2 Meas. 2	Shift 3 Meas. 1	Shift 3 Meas. 2
1	3712	3710	3710	3708	3713	3710
2	3680	3683	3775	3679	3681	3682
3	3697	3700	3692	3695	3697	3700
4	3671	3668	3666	3664	3671	3669
5	3692	3689	3691	3683	3694	3687
6	3688	3690	3683	3687	3687	3689
7	3691	3694	3685	3690	3692	3694
8	3696	3701	3688	3695	3695	3701
9	3705	3709	3697	3704	3704	3709
10	3678	3681	3677	3676	3679	3680
11	3711	3709	3700	3704	3710	3709
12	3712	3715	3702	3709	3712	3714
13	3728	3721	3722	3716	3729	3722
14	3694	3698	3689	3695	3694	3698
15	3704	3702	3696	3697	3704	3703
16	3686	3685	3681	3683	3687	3684
17	3705	3706	3699	3701	3704	3707
18	3678	3680	3676	3676	3677	3679
19	3723	3724	3721	3720	3723	3724
20	3672	3669	3666	3665	3672	3668

This new study is a two-factor factorial experiment, with parts and shifts as the two design factors. Both parts and shifts are considered as random factors. The ANOVA model for this experiment is

$$Y_{ijk} = \mu + \tau_i + \beta_j + (\tau\beta)_{ij} + \epsilon_{ijk} \quad \begin{cases} i = 1, 2, \ldots, 20 \\ j = 1, 2, 3 \\ k = 1, 2 \end{cases} \tag{8-22}$$

where τ_i is the wafer effect with variance component σ^2_{Wafer}, β_j is the shift effect with variance component σ^2_{Shift}, $(\tau\beta)_{ij}$ is the interaction between wafers and shifts with variance component $\sigma^2_{S \times W}$, and ϵ_{ijk} is a term that reflects the observed variability when the measuring tool is applied to the same part on the same shift. The variance component associated with ϵ_{ijk} is denoted $\sigma^2_{\text{Repeatability}}$, and **repeatability** is a component of the total variability associated with the measurement tool. The other component of measurement variability is called **reproducibility,** and it reflects the variability associated with the shifts (which arise from the setup procedure for the tool, drift in the tool over time, and different operators). The reproducibility variance component is

$$\sigma^2_{\text{Reproducibility}} = \sigma^2_{\text{Shift}} + \sigma^2_{S \times W}$$

Notice that the interaction variance component from the ANOVA model (equation 8-22) is included in the reproducibility variance, and it also contains some of the shift-to-shift

Table 8-10 ANOVA: Resistivity versus Shift, Wafer

Factor	Type	Levels	Values						
Shift	random	3	1	2	3				
Wafer	random	20	1	2	3	4	5	6	7
			8	9	10	11	12	13	14
			15	16	17	18	19	20	

Analysis of Variance for Resistivity

Source	DF	SS	MS	F	P
Shift	2	142.72	71.36	0.83	0.446
Wafer	19	27510.03	1447.90	16.75	0.000
Shift*Wafer	38	3284.62	86.44	1.03	0.447
Error	60	5019.00	83.65		
Total	119	35956.37			

Source	Variance component	Error term	Expected Mean Square for Each Term (using unrestricted model)
1 Shift	−0.377	1	(4) + 2(3) + 40(1)
2 Wafer	226.910	2	(4) + 2(3) + 6(2)
3 Shift*Wafer	1.394	3	(4) + 2(3)
4 Error	83.650	4	(4)

variability. In this way, ANOVA is used to estimate the repeatability (R) and reproducibility (R) of a measurement system, which is sometimes called a **gauge R & R study.**

The two-factor ANOVA procedure from Section 7-9 can be applied to the data in Table 8-9. The Minitab output (from Balanced ANOVA) is shown in Table 8-10.

In the Minitab analysis, we have specified that both factors, wafers and shifts, are random factors. Minitab has provided estimates of the variance components associated with the ANOVA model in equation 8-22, σ^2_{Shift} (denoted 1 in Table 8-10), σ^2_{Wafer} (denoted 2 in the table), $\sigma^2_{S\times W}$ (denoted 3 in the table), and $\sigma^2_{Repeatability}$ (denoted 4 in the table). The estimates of the variance components were obtained by solving the equations derived from the expected mean squares essentially as we did with the single-factor experiment. The Minitab output also contains the expected mean squares using the 1, 2, 3, 4, notation for the variance components.

We can now estimate the two components of the gauge variability,

$$\sigma^2_{Gauge} = \sigma^2_{Repeatability} + \sigma^2_{Reproducibility}$$

The results are $\sigma^2_{Repeatability} = 83.650$, and the two components of $\sigma^2_{Reproducibility}$ are $\sigma^2_{Shift} = -0.377$ and $\sigma^2_{S\times W} = 1.394$. Notice that the estimate of one of the variance components is negative. The ANOVA method of variance component estimation sometimes produces negative estimates of one or more variance components. This is usually taken as evidence that the variance component is really zero. Furthermore, the F-tests on these two factors indicate that they are not significant, so this is evidence that both of the variance components associated with $\sigma^2_{Reproducibility}$ are really zero. Therefore, the only significant component of overall gauge variability is the component due to repeatability, and the operators are very consistent about how the tool is set up and run over the different shifts.

EXERCISES FOR SECTION 8-9

8-36. Consider the wafers measured in example data given in Table 8-7 for two measurements on 20 different wafers. Assume that a third measurement is recorded on these 20 wafers, respectively. The measurements are as follows: 3703, 3694, 3691, 3689, 3696, 3692, 3693, 3698, 3694, 3697, 3693, 3698, 3694, 3694, 3700, 3693, 3690, 3699, 3695, 3686.

(a) Perform an analysis of variance to determine if there is a significant difference in the parts used in this study, based on these three measurements.
(b) Find $\hat{\sigma}^2_{Total}$, $\hat{\sigma}^2_{Gauge}$, and $\hat{\sigma}^2_{Wafer}$.
(c) What percentage of the total variability is due to the gauge? Do we have a desirable situation, with respect to the variability of the gauge?

8-37. The process engineer is concerned with the device used to measure purity level of a steel alloy. To assess the device variability he measures 10 specimens with the same instrument twice. The resultant measurements follow:

Measurement 1	Measurement 2
2.0	2.3
2.1	2.4
2.0	2.2
2.2	2.2
1.6	2.4
2.0	2.2
1.9	2.0
2.1	1.6
1.9	2.6
2.0	2.0

(a) Perform an analysis of variance to determine if there is a significant difference in the specimens used in this study, based on these two measurements.
(b) Find $\hat{\sigma}^2_{Total}$, $\hat{\sigma}^2_{Gauge}$, and $\hat{\sigma}^2_{Specimen}$.
(c) What percentage of the total variability is due to the gauge? Do we have a desirable situation, with respect to the variability of the gauge?

8-38. An implantable defibrillator is a small unit that senses erratic heart signals from a patient. If the signal does not normalize within a few seconds, the unit charges itself to 650 V and delivers a powerful shock to the patient's heart, restoring the normal beat. The quality control department of the manufacturer of these defibrillators is responsible for checking the output voltage of these assembled devices. To check the variability of the volt meters, the department performed a designed experiment, measuring each of eight units twice. The data collected are as follows:

Measurement 1	Measurement 2
640.9	650.6
644.2	647.1
659.9	654.4
656.2	648.3
646.6	652.1
656.3	647.0
659.6	655.8
657.8	651.8

(a) Perform an analysis of variance to determine if there is a significant difference in the units used in this study.
(b) Find $\hat{\sigma}^2_{Total}$, $\hat{\sigma}^2_{Gauge}$, and $\hat{\sigma}^2_{Unit}$.
(c) What percentage of the total variability is due to the volt meter? Comment on your results.

8-39. Consider the following Minitab ANOVA table used to analyze multiple measurements on each of 10 parts.

Source	DF	SS	MS	F	P
Part	9	143186	15910	27.72	0.000
Error	10	5740	574		
Total	19	148925			

(a) How many replicate measurements were made of these 10 parts?
(b) Estimate σ^2_{Total}, σ^2_{Gauge}, and σ^2_{Part}.
(c) What percentage of the total variability is due to the gauge? Do we have a desirable situation, with respect to the variability of the gauge?

8-40. Consider the following Minitab ANOVA table used to analyze the multiple measurements on each of 10 parts.

Source	DF	SS	MS	F	P
Part	9	708642	78738	15.71	0.000
Error	20	100263	5013		
Total	29	808905			

(a) How many replicate measurements were made of these 10 parts?
(b) Estimate σ^2_{Total}, σ^2_{Gauge}, and σ^2_{Part}.
(c) What percentage of the total variability is due to the gauge? Do we have a desirable situation, with respect to the variability of the gauge?

8-41. Asphalt compressive strength is measured in units of psi. To test the repeatability and reproducibility of strength measurements, two engineers in charge of the quality system make two measurements on six specimens. The data are as follows:

Operator 1		Operator 2	
Meas. 1	Meas. 2	Meas. 1	Meas. 2
1501.22	1510.00	1505.66	1504.74
1498.06	1512.40	1504.64	1501.82
1506.44	1513.54	1499.84	1492.95
1496.35	1541.54	1502.19	1507.04
1502.03	1499.46	1503.08	1498.43
1499.90	1521.83	1515.57	1512.84

(a) Perform an ANOVA on these data.
(b) Find $\sigma^2_{Repeatability}$ and $\sigma^2_{Reproducibility}$.
(c) Comment on the significance of the overall gauge capability.

8-42.   A handheld caliper is used to measure the diameter of fuse pins in an aircraft engine. A repeatability and reproducibility study is carried out to determine the variability of the gauge and the operators. Two operators measure five pistons three times. The data are as follows (1-1 equals operator 1, measurement 1; 1-2 equals operator 1, measurement 2, and so forth):

1-1	1-2	1-3	2-1	2-2	2-3	3-1	3-2	3-3
65.03	64.42	59.89	63.59	63.41	59.10	63.45	59.99	65.13
64.09	60.60	64.98	64.70	63.11	64.33	63.73	61.69	64.43
64.53	59.65	67.57	63.87	63.99	69.94	64.27	63.98	66.00
65.57	61.68	67.03	65.89	62.41	58.51	63.71	62.74	64.11
63.98	63.84	64.08	63.90	67.38	65.21	63.41	60.59	63.35

(a) Perform an ANOVA on these data.
(b) Find $\sigma^2_{Repeatability}$ and $\sigma^2_{Reproducibility}$.
(c) Comment on the significance of the overall gauge capability.

8-43.   Consider a gauge study in which two operators measure 15 parts six times. The result of their analysis is given in the following Minitab output.

Factor	Type	Levels
Part	random	15
Operator	random	2

Factor	Values
Part	1, 2, 3, 4, 5, 6, 7, 8, 9, 10, 11, 12, 13, 14, 15
Operator	1, 2

Analysis of Variance for Response

Source	DF	SS	MS	F	P
Part	14	2025337	144667	1.61	0.193
Operator	1	661715	661715	7.35	0.017
Part* Operator	14	1260346	90025	25.98	0.000
Error	150	519690	3465		
Total	179	4467088			

$S = 58.8609$ R-Sq = 88.37%   R-Sq(adj) = 86.12%

Source	Variance component	Error term	Expected Mean Square for Each Term (using unrestricted model)
1 Part	4554	3	(4) + 6 (3) + 12 (1)
2 Operator	6352	3	(4) + 6 (3) + 90 (2)
3 Part* Operator	14427	4	(4) + 6 (3)
4 Error	3465		(4)

(a) Find $\sigma^2_{Repeatablity}$ and $\sigma^2_{Reproducibility}$.
(b) Comment on the significance of the overall gauge capability.

8-44.   Consider a gauge study in which two operators measure 15 parts five times. The result of their analysis is given in the following Minitab output.

Factor	Type	Levels
Part	random	15
Operator	random	2

Factor	Values
Part	1, 2, 3, 4, 5, 6, 7, 8, 9, 10, 11, 12, 13, 14, 15
Operator	1, 2

Analysis of Variance for Response

Source	DF	SS	MS	F	P
Part	14	2773567	198112	2.45	0.053
Operator	1	518930	518930	6.41	0.024
Part* Operator	14	1133722	80980	7.80	0.000
Error	120	1245396	10378		
Total	149	5671615			

Source	Variance component	Error term	Expected Mean Square for Each Term (using unrestricted model)
1 Part	11713	3	(4) + 5 (3) + 10 (1)
2 Operator	5839	3	(4) + 5 (3) + 75 (2)
3 Part* Operator	14120	4	(4) + 5 (3)
4 Error	10378		(4)

(a) Find $\sigma^2_{Repeatablity}$ and $\sigma^2_{Reproducibility}$.
(b) Comment on the significance of the overall gauge capability.

SUPPLEMENTAL EXERCISES

8-45. The diameter of fuse pins used in an aircraft engine application is an important quality characteristic. Twenty-five samples of three pins each are as follows (in mm).

Sample Number	Diameter		
1	64.030	64.002	64.019
2	63.995	63.992	64.001
3	63.988	64.024	64.021
4	64.002	63.996	63.993
5	63.992	64.007	64.015
6	64.009	63.994	63.997
7	63.995	64.006	63.994
8	63.985	64.003	63.993
9	64.008	63.995	64.009
10	63.998	74.000	63.990
11	63.994	63.998	63.994
12	64.004	64.000	64.007
13	63.983	64.002	63.998
14	64.006	63.967	63.994
15	64.012	64.014	63.998
16	64.000	63.984	64.005
17	63.994	64.012	63.986
18	64.006	64.010	64.018
19	63.984	64.002	64.003
20	64.000	64.010	64.013
21	63.988	64.001	64.009
22	64.004	63.999	63.990
23	64.010	63.989	63.990
24	64.015	64.008	63.993
25	63.982	63.984	63.995

(a) Set up $\overline{X}$ and R charts for this process. If necessary, revise limits so that no observations are out of control.
(b) Estimate the process mean and standard deviation.
(c) Suppose the process specifications are at 64 ± 0.02. Calculate an estimate of C_p. Does the process meet a minimum capability level of $C_p \geq 1.33$?
(d) Calculate an estimate of C_{pk}. Use this ratio to draw conclusions about process capability.
(e) To make this process a six-sigma process, the variance σ^2 would have to be decreased such that $C_{pk} = 2.0$. What should this new variance value be?
(f) Suppose the mean shifts to 64.005. What is the probability that this shift will be detected on the next sample? What is the ARL after the shift?

8-46. Plastic bottles for liquid laundry detergent are formed by blow molding. Twenty samples of $n = 100$ bottles are inspected in time order of production, and the number defective

in each sample is reported. The data are as follows: 9, 11, 10, 8, 3, 8, 8, 10, 3, 5, 9, 8, 8, 8, 6, 10, 17, 11, 9, 10.

(a) Set up a P chart for this process. Is the process in statistical control?
(b) Suppose that instead of $n = 100$, $n = 200$. Use the data given to set up a P chart for this process. Is the process in statistical control?
(c) Compare your control limits for the P charts in parts (a) and (b). Explain why they differ. Also, explain why your assessment about statistical control differs for the two sizes of n.

8-47. Cover cases for a personal computer are manufactured by injection molding. Samples of five cases are taken from the process periodically, and the number of defects is noted. The results for twenty-five samples follow: 3, 2, 0, 1, 4, 3, 2, 4, 1, 0, 2, 3, 2, 8, 0, 2, 4, 3, 5, 0, 2, 1, 9, 3, 2.

(a) Using all the data, find trial control limits for the U chart for this process.
(b) Use the trial control limits from part (a) to identify out-of-control points. If necessary, revise your control limits.
(c) Suppose that instead of samples of 5 cases, the sample size was 10. Repeat parts (a) and (b). Explain how this change alters your answers to parts (a) and (b).

8-48. Consider the data in Exercise 8-47.

(a) Using all the data, find trial control limits for a C chart for this process.
(b) Use the trial control limits of part (a) to identify out-of-control points. If necessary, revise your control limits.
(c) Suppose that instead of samples of 5 cases, the sample was 10 cases. Repeat parts (a) and (b). Explain how this alters your answers to parts (a) and (b).

8-49. Suppose that a process is in control and an $\overline{X}$ chart is used with a sample size of 4 to monitor the process. Suddenly there is a mean shift of 1.75.

(a) If three-sigma control limits are in use on the $\overline{X}$ chart, what is the probability that this shift will remain undetected for three consecutive samples?
(b) If two-sigma control limits are in use on the $\overline{X}$ chart, what is the probability that this shift will remain undetected for three consecutive samples?
(c) Compare your answers to parts (a) and (b) and explain why they differ. Also, which limits would you recommend using and why?

8-50. Consider the control chart for individuals with three-sigma limits.

(a) Suppose that a shift in the process mean of magnitude σ occurs. Verify that the ARL for detecting the shift is ARL = 43.9.
(b) Find the ARL for detecting a shift of magnitude two sigma in the process mean.
(c) Find the ARL for detecting a shift of magnitude three sigma in the process mean.
(d) Compare your answers to parts (a), (b), and (c) and explain why the ARL for detection is decreasing as the magnitude of the shift increases.

8-51. Consider a control chart for individuals, applied to a continuous 24-hour chemical process with observations taken every hour.

(a) If the chart has three-sigma limits, verify that the in-control ARL is ARL = 370. How many false alarms would occur each 30-day month, on the average, with this chart?

(b) Suppose that the chart has two-sigma limits. Does this reduce the ARL for detecting a shift in the mean of magnitude σ? (Recall that the ARL for detecting this shift with three-sigma limits is 43.9.)

(c) Find the in-control ARL if two-sigma limits are used on the chart. How many false alarms would occur each month with this chart? Is this in-control ARL performance satisfactory? Explain your answer.

8-52. The depth of a keyway is an important part quality characteristic. Samples of size $n = 5$ are taken every 4 hours from the process and 20 samples are given as follows.

Sample	Observation				
	1	2	3	4	5
1	139.9	138.8	139.85	141.1	139.8
2	140.7	139.3	140.55	141.6	140.1
3	140.8	139.8	140.15	141.9	139.9
4	140.6	141.1	141.05	141.2	139.6
5	139.8	138.9	140.55	141.7	139.6
6	139.8	139.2	140.55	141.2	139.4
7	140.1	138.8	139.75	141.2	138.8
8	140.3	140.6	140.65	142.5	139.9
9	140.1	139.1	139.05	140.5	139.1
10	140.3	141.1	141.25	142.6	140.9
11	138.4	138.1	139.25	140.2	138.6
12	139.4	139.1	139.15	140.3	137.8
13	138.0	137.5	138.25	141.0	140.0
14	138.0	138.1	138.65	139.5	137.8
15	141.2	140.5	141.45	142.5	141.0
16	141.2	141.0	141.95	141.9	140.1
17	140.2	140.3	141.45	142.3	139.6
18	139.6	140.3	139.55	141.7	139.4
19	136.2	137.2	137.75	138.3	137.7
20	138.8	137.7	140.05	140.8	138.9

(a) Using all the data, find trial control limits for $\overline{X}$ and R charts. Is the process in control?

(b) Use the trial control limits from part (a) to identify out-of-control points. If necessary, revise your control limits. Then, estimate the process standard deviation.

(c) Suppose that the specifications are at 140 ± 2. Using the results from part (b), what statements can you make about process capability? Compute estimates of the appropriate process capability ratios.

(d) To make this process a six-sigma process, the variance σ^2 would have to be decreased such that $C_{pk} = 2.0$. What should this new variance value be?

(e) Suppose the mean shifts to 139.7. What is the probability that this shift will be detected on the next sample? What is the ARL after the shift?

8-53. A process is controlled by a P chart using samples of size 100. The center line on the chart is 0.05.

(a) What is the probability that the control chart detects a shift to 0.06 on the first sample following the shift?

(b) What is the probability that the control chart does not detect a shift to 0.06 on the first sample following the shift but does detect it on the second sample?

(c) Suppose that instead of a shift in the mean to 0.06, the mean shifts to 0.08. Repeat parts (a) and (b).

(d) Compare your answers for a shift to 0.06 and for a shift to 0.08. Explain why they differ. Also, explain why a shift to 0.08 is easier to detect.

8-54. Suppose the average number of defects in a unit is known to be 8. If the mean number of defects in a unit shifts to 16, what is the probability that it will be detected by the U chart on the first sample following the shift

(a) if the sample size is $n = 5$?

(b) if the sample size is $n = 8$?

Use a normal approximation for U.

8-55. Suppose the average number of defects in a unit is known to be 10. If the mean number of defects in a unit shifts to 14, what is the probability that it will be detected by the U chart on the first sample following the shift

(a) if the sample size is $n = 3$?

(b) if the sample size is $n = 6$?

Use a normal approximation for U.

8-56. Suppose that an $\overline{X}$ control chart with two-sigma limits is used to control a process. Find the probability that a false out-of-control signal will be produced on the next sample. Compare this with the corresponding probability for the chart with three-sigma limits and discuss. Comment on when you would prefer to use two-sigma limits instead of three-sigma limits.

8-57. Consider an $\overline{X}$ control chart with k-sigma control limits. Develop a general expression for the probability that a point will plot outside the control limits when the process mean has shifted by δ units from the center line.

8-58. Consider the $\overline{X}$ control chart with two-sigma limits in Exercise 8-56.

(a) Find the probability of no signal on the first sample but a signal on the second.

(b) What is the probability that there will not be a signal in three samples?

8-59. Suppose a process has a $C_p = 2$, but the mean is exactly three standard deviations above the upper specification limit. What is the probability of making a product outside the specification limits?

TEAM EXERCISE

8-60. Obtain time-ordered data from a process of interest. Use the data to construct appropriate control charts and comment on the control of the process. Can you make any recommendations to improve the process? If appropriate, calculate appropriate measures of process capability.

IMPORTANT TERMS AND CONCEPTS

Assignable cause
Attributes control charts
Average run length
C chart
Chance cause
Control chart for individuals
Control limits
Fraction defective

Gauge R & R study
Measurement systems capability
Moving range control chart
nP chart
P chart
Patterns on control charts
Process capability

Process capability ratio C_p
Process capability ratio C_{pk}
R control chart
Rational subgroups
Repeatability
Reproducibility
Shewhart control charts

SPC
Three-sigma control limits
U chart
Variables control charts
Western Electric rules
$\overline{X}$ control chart

APPENDICES

Appendix A
Statistical
Tables and
Charts

Table I   Cumulative Standard Normal Distribution

$$\Phi(z) = P(Z \le z) = \int_{-\infty}^{z} \frac{1}{\sqrt{2\pi}} e^{\frac{-u^2}{2}} \, du$$

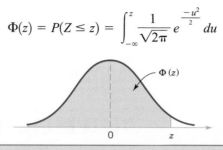

z	−0.09	−0.08	−0.07	−0.06	−0.05	−0.04	−0.03	−0.02	−0.01	−0.00	z
−3.9	0.000033	0.000034	0.000036	0.000037	0.000039	0.000041	0.000042	0.000044	0.000046	0.000048	−3.9
−3.8	0.000050	0.000052	0.000054	0.000057	0.000059	0.000062	0.000064	0.000067	0.000069	0.000072	−3.8
−3.7	0.000075	0.000078	0.000082	0.000085	0.000088	0.000092	0.000096	0.000100	0.000104	0.000108	−3.7
−3.6	0.000112	0.000117	0.000121	0.000126	0.000131	0.000136	0.000142	0.000147	0.000153	0.000159	−3.6
−3.5	0.000165	0.000172	0.000179	0.000185	0.000193	0.000200	0.000208	0.000216	0.000224	0.000233	−3.5
−3.4	0.000242	0.000251	0.000260	0.000270	0.000280	0.000291	0.000302	0.000313	0.000325	0.000337	−3.4
−3.3	0.000350	0.000362	0.000376	0.000390	0.000404	0.000419	0.000434	0.000450	0.000467	0.000483	−3.3
−3.2	0.000501	0.000519	0.000538	0.000557	0.000577	0.000598	0.000619	0.000641	0.000664	0.000687	−3.2
−3.1	0.000711	0.000736	0.000762	0.000789	0.000816	0.000845	0.000874	0.000904	0.000935	0.000968	−3.1
−3.0	0.001001	0.001035	0.001070	0.001107	0.001144	0.001183	0.001223	0.001264	0.001306	0.001350	−3.0
−2.9	0.001395	0.001441	0.001489	0.001538	0.001589	0.001641	0.001695	0.001750	0.001807	0.001866	−2.9
−2.8	0.001926	0.001988	0.002052	0.002118	0.002186	0.002256	0.002327	0.002401	0.002477	0.002555	−2.8
−2.7	0.002635	0.002718	0.002803	0.002890	0.002980	0.003072	0.003167	0.003264	0.003364	0.003467	−2.7
−2.6	0.003573	0.003681	0.003793	0.003907	0.004025	0.004145	0.004269	0.004396	0.004527	0.004661	−2.6
−2.5	0.004799	0.004940	0.005085	0.005234	0.005386	0.005543	0.005703	0.005868	0.006037	0.006210	−2.5
−2.4	0.006387	0.006569	0.006756	0.006947	0.007143	0.007344	0.007549	0.007760	0.007976	0.008198	−2.4
−2.3	0.008424	0.008656	0.008894	0.009137	0.009387	0.009642	0.009903	0.010170	0.010444	0.010724	−2.3
−2.2	0.011011	0.011304	0.011604	0.011911	0.012224	0.012545	0.012874	0.013209	0.013553	0.013903	−2.2
−2.1	0.014262	0.014629	0.015003	0.015386	0.015778	0.016177	0.016586	0.017003	0.017429	0.017864	−2.1
−2.0	0.018309	0.018763	0.019226	0.019699	0.020182	0.020675	0.021178	0.021692	0.022216	0.022750	−2.0
−1.9	0.023295	0.023852	0.024419	0.024998	0.025588	0.026190	0.026803	0.027429	0.028067	0.028717	−1.9
−1.8	0.029379	0.030054	0.030742	0.031443	0.032157	0.032884	0.033625	0.034379	0.035148	0.035930	−1.8
−1.7	0.036727	0.037538	0.038364	0.039204	0.040059	0.040929	0.041815	0.042716	0.043633	0.044565	−1.7
−1.6	0.045514	0.046479	0.047460	0.048457	0.049471	0.050503	0.051551	0.052616	0.053699	0.054799	−1.6
−1.5	0.055917	0.057053	0.058208	0.059380	0.060571	0.061780	0.063008	0.064256	0.065522	0.066807	−1.5
−1.4	0.068112	0.069437	0.070781	0.072145	0.073529	0.074934	0.076359	0.077804	0.079270	0.080757	−1.4
−1.3	0.082264	0.083793	0.085343	0.086915	0.088508	0.090123	0.091759	0.093418	0.095098	0.096801	−1.3
−1.2	0.098525	0.100273	0.102042	0.103835	0.105650	0.107488	0.109349	0.111233	0.113140	0.115070	−1.2
−1.1	0.117023	0.119000	0.121001	0.123024	0.125072	0.127143	0.129238	0.131357	0.133500	0.135666	−1.1
−1.0	0.137857	0.140071	0.142310	0.144572	0.146859	0.149170	0.151505	0.153864	0.156248	0.158655	−1.0
−0.9	0.161087	0.163543	0.166023	0.168528	0.171056	0.173609	0.176185	0.178786	0.181411	0.184060	−0.9
−0.8	0.186733	0.189430	0.192150	0.194894	0.197662	0.200454	0.203269	0.206108	0.208970	0.211855	−0.8
−0.7	0.214764	0.217695	0.220650	0.223627	0.226627	0.229650	0.232695	0.235762	0.238852	0.241964	−0.7
−0.6	0.245097	0.248252	0.251429	0.254627	0.257846	0.261086	0.264347	0.267629	0.270931	0.274253	−0.6
−0.5	0.277595	0.280957	0.284339	0.287740	0.291160	0.294599	0.298056	0.301532	0.305026	0.308538	−0.5
−0.4	0.312067	0.315614	0.319178	0.322758	0.326355	0.329969	0.333598	0.337243	0.340903	0.344578	−0.4
−0.3	0.348268	0.351973	0.355691	0.359424	0.363169	0.366928	0.370700	0.374484	0.378281	0.382089	−0.3
−0.2	0.385908	0.389739	0.393580	0.397432	0.401294	0.405165	0.409046	0.412936	0.416834	0.420740	−0.2
−0.1	0.424655	0.428576	0.432505	0.436441	0.440382	0.444330	0.448283	0.452242	0.456205	0.460172	−0.1
0.0	0.464144	0.468119	0.472097	0.476078	0.480061	0.484047	0.488033	0.492022	0.496011	0.500000	0.0

Table I   Cumulative Standard Normal Distribution (*continued*)

$$\Phi(z) = P(Z \le z) = \int_{-\infty}^{z} \frac{1}{\sqrt{2\pi}} e^{\frac{-u^2}{2}} du$$

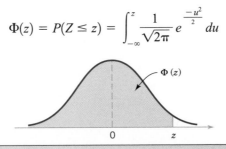

z	0.00	0.01	0.02	0.03	0.04	0.05	0.06	0.07	0.08	0.09	z
0.0	0.500000	0.503989	0.507978	0.511967	0.515953	0.519939	0.523922	0.527903	0.531881	0.535856	0.0
0.1	0.539828	0.543795	0.547758	0.551717	0.555760	0.559618	0.563559	0.567495	0.571424	0.575345	0.1
0.2	0.579260	0.583166	0.587064	0.590954	0.594835	0.598706	0.602568	0.606420	0.610261	0.614092	0.2
0.3	0.617911	0.621719	0.625516	0.629300	0.633072	0.636831	0.640576	0.644309	0.648027	0.651732	0.3
0.4	0.655422	0.659097	0.662757	0.666402	0.670031	0.673645	0.677242	0.680822	0.684386	0.687933	0.4
0.5	0.691462	0.694974	0.698468	0.701944	0.705401	0.708840	0.712260	0.715661	0.719043	0.722405	0.5
0.6	0.725747	0.729069	0.732371	0.735653	0.738914	0.742154	0.745373	0.748571	0.751748	0.754903	0.6
0.7	0.758036	0.761148	0.764238	0.767305	0.770350	0.773373	0.776373	0.779350	0.782305	0.785236	0.7
0.8	0.788145	0.791030	0.793892	0.796731	0.799546	0.802338	0.805106	0.807850	0.810570	0.813267	0.8
0.9	0.815940	0.818589	0.821214	0.823815	0.826391	0.828944	0.831472	0.833977	0.836457	0.838913	0.9
1.0	0.841345	0.843752	0.846136	0.848495	0.850830	0.853141	0.855428	0.857690	0.859929	0.862143	1.0
1.1	0.864334	0.866500	0.868643	0.870762	0.872857	0.874928	0.876976	0.878999	0.881000	0.882977	1.1
1.2	0.884930	0.886860	0.888767	0.890651	0.892512	0.894350	0.896165	0.897958	0.899727	0.901475	1.2
1.3	0.903199	0.904902	0.906582	0.908241	0.909877	0.911492	0.913085	0.914657	0.916207	0.917736	1.3
1.4	0.919243	0.920730	0.922196	0.923641	0.925066	0.926471	0.927855	0.929219	0.930563	0.931888	1.4
1.5	0.933193	0.934478	0.935744	0.936992	0.938220	0.939429	0.940620	0.941792	0.942947	0.944083	1.5
1.6	0.945201	0.946301	0.947384	0.948449	0.949497	0.950529	0.951543	0.952540	0.953521	0.954486	1.6
1.7	0.955435	0.956367	0.957284	0.958185	0.959071	0.959941	0.960796	0.961636	0.962462	0.963273	1.7
1.8	0.964070	0.964852	0.965621	0.966375	0.967116	0.967843	0.968557	0.969258	0.969946	0.970621	1.8
1.9	0.971283	0.971933	0.972571	0.973197	0.973810	0.974412	0.975002	0.975581	0.976148	0.976705	1.9
2.0	0.977250	0.977784	0.978308	0.978822	0.979325	0.979818	0.980301	0.980774	0.981237	0.981691	2.0
2.1	0.982136	0.982571	0.982997	0.983414	0.983823	0.984222	0.984614	0.984997	0.985371	0.985738	2.1
2.2	0.986097	0.986447	0.986791	0.987126	0.987455	0.987776	0.988089	0.988396	0.988696	0.988989	2.2
2.3	0.989276	0.989556	0.989830	0.990097	0.990358	0.990613	0.990863	0.991106	0.991344	0.991576	2.3
2.4	0.991802	0.992024	0.992240	0.992451	0.992656	0.992857	0.993053	0.993244	0.993431	0.993613	2.4
2.5	0.993790	0.993963	0.994132	0.994297	0.994457	0.994614	0.994766	0.994915	0.995060	0.995201	2.5
2.6	0.995339	0.995473	0.995604	0.995731	0.995855	0.995975	0.996093	0.996207	0.996319	0.996427	2.6
2.7	0.996533	0.996636	0.996736	0.996833	0.996928	0.997020	0.997110	0.997197	0.997282	0.997365	2.7
2.8	0.997445	0.997523	0.997599	0.997673	0.997744	0.997814	0.997882	0.997948	0.998012	0.998074	2.8
2.9	0.998134	0.998193	0.998250	0.998305	0.998359	0.998411	0.998462	0.998511	0.998559	0.998605	2.9
3.0	0.998650	0.998694	0.998736	0.998777	0.998817	0.998856	0.998893	0.998930	0.998965	0.998999	3.0
3.1	0.999032	0.999065	0.999096	0.999126	0.999155	0.999184	0.999211	0.999238	0.999264	0.999289	3.1
3.2	0.999313	0.999336	0.999359	0.999381	0.999402	0.999423	0.999443	0.999462	0.999481	0.999499	3.2
3.3	0.999517	0.999533	0.999550	0.999566	0.999581	0.999596	0.999610	0.999624	0.999638	0.999650	3.3
3.4	0.999663	0.999675	0.999687	0.999698	0.999709	0.999720	0.999730	0.999740	0.999749	0.999758	3.4
3.5	0.999767	0.999776	0.999784	0.999792	0.999800	0.999807	0.999815	0.999821	0.999828	0.999835	3.5
3.6	0.999841	0.999847	0.999853	0.999858	0.999864	0.999869	0.999874	0.999879	0.999883	0.999888	3.6
3.7	0.999892	0.999896	0.999900	0.999904	0.999908	0.999912	0.999915	0.999918	0.999922	0.999925	3.7
3.8	0.999928	0.999931	0.999933	0.999936	0.999938	0.999941	0.999943	0.999946	0.999948	0.999950	3.8
3.9	0.999952	0.999954	0.999956	0.999958	0.999959	0.999961	0.999963	0.999964	0.999966	0.999967	3.9

Table II   Percentage Points $t_{\alpha,v}$ of the t Distribution

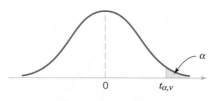

v \ α	.40	.25	.10	.05	.025	.01	.005	.0025	.001	.0005
1	.325	1.000	3.078	6.314	12.706	31.821	63.657	127.32	318.31	636.62
2	.289	.816	1.886	2.920	4.303	6.965	9.925	14.089	23.326	31.598
3	.277	.765	1.638	2.353	3.182	4.541	5.841	7.453	10.213	12.924
4	.271	.741	1.533	2.132	2.776	3.747	4.604	5.598	7.173	8.610
5	.267	.727	1.476	2.015	2.571	3.365	4.032	4.773	5.893	6.869
6	.265	.718	1.440	1.943	2.447	3.143	3.707	4.317	5.208	5.959
7	.263	.711	1.415	1.895	2.365	2.998	3.499	4.029	4.785	5.408
8	.262	.706	1.397	1.860	2.306	2.896	3.355	3.833	4.501	5.041
9	.261	.703	1.383	1.833	2.262	2.821	3.250	3.690	4.297	4.781
10	.260	.700	1.372	1.812	2.228	2.764	3.169	3.581	4.144	4.587
11	.260	.697	1.363	1.796	2.201	2.718	3.106	3.497	4.025	4.437
12	.259	.695	1.356	1.782	2.179	2.681	3.055	3.428	3.930	4.318
13	.259	.694	1.350	1.771	2.160	2.650	3.012	3.372	3.852	4.221
14	.258	.692	1.345	1.761	2.145	2.624	2.977	3.326	3.787	4.140
15	.258	.691	1.341	1.753	2.131	2.602	2.947	3.286	3.733	4.073
16	.258	.690	1.337	1.746	2.120	2.583	2.921	3.252	3.686	4.015
17	.257	.689	1.333	1.740	2.110	2.567	2.898	3.222	3.646	3.965
18	.257	.688	1.330	1.734	2.101	2.552	2.878	3.197	3.610	3.922
19	.257	.688	1.328	1.729	2.093	2.539	2.861	3.174	3.579	3.883
20	.257	.687	1.325	1.725	2.086	2.528	2.845	3.153	3.552	3.850
21	.257	.686	1.323	1.721	2.080	2.518	2.831	3.135	3.527	3.819
22	.256	.686	1.321	1.717	2.074	2.508	2.819	3.119	3.505	3.792
23	.256	.685	1.319	1.714	2.069	2.500	2.807	3.104	3.485	3.767
24	.256	.685	1.318	1.711	2.064	2.492	2.797	3.091	3.467	3.745
25	.256	.684	1.316	1.708	2.060	2.485	2.787	3.078	3.450	3.725
26	.256	.684	1.315	1.706	2.056	2.479	2.779	3.067	3.435	3.707
27	.256	.684	1.314	1.703	2.052	2.473	2.771	3.057	3.421	3.690
28	.256	.683	1.313	1.701	2.048	2.467	2.763	3.047	3.408	3.674
29	.256	.683	1.311	1.699	2.045	2.462	2.756	3.038	3.396	3.659
30	.256	.683	1.310	1.697	2.042	2.457	2.750	3.030	3.385	3.646
40	.255	.681	1.303	1.684	2.021	2.423	2.704	2.971	3.307	3.551
60	.254	.679	1.296	1.671	2.000	2.390	2.660	2.915	3.232	3.460
120	.254	.677	1.289	1.658	1.980	2.358	2.617	2.860	3.160	3.373
∞	.253	.674	1.282	1.645	1.960	2.326	2.576	2.807	3.090	3.291

v = degrees of freedom.

Table III   Percentage Points $\chi^2_{\alpha,\nu}$ of the Chi-Square Distribution

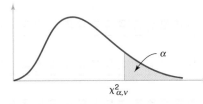

$\chi^2_{\alpha,\nu}$

ν \ α	.995	.990	.975	.950	.900	.500	.100	.050	.025	.010	.005
1	.00+	.00+	.00+	.00+	.02	.45	2.71	3.84	5.02	6.63	7.88
2	.01	.02	.05	.10	.21	1.39	4.61	5.99	7.38	9.21	10.60
3	.07	.11	.22	.35	.58	2.37	6.25	7.81	9.35	11.34	12.84
4	.21	.30	.48	.71	1.06	3.36	7.78	9.49	11.14	13.28	14.86
5	.41	.55	.83	1.15	1.61	4.35	9.24	11.07	12.83	15.09	16.75
6	.68	.87	1.24	1.64	2.20	5.35	10.65	12.59	14.45	16.81	18.55
7	.99	1.24	1.69	2.17	2.83	6.35	12.02	14.07	16.01	18.48	20.28
8	1.34	1.65	2.18	2.73	3.49	7.34	13.36	15.51	17.53	20.09	21.96
9	1.73	2.09	2.70	3.33	4.17	8.34	14.68	16.92	19.02	21.67	23.59
10	2.16	2.56	3.25	3.94	4.87	9.34	15.99	18.31	20.48	23.21	25.19
11	2.60	3.05	3.82	4.57	5.58	10.34	17.28	19.68	21.92	24.72	26.76
12	3.07	3.57	4.40	5.23	6.30	11.34	18.55	21.03	23.34	26.22	28.30
13	3.57	4.11	5.01	5.89	7.04	12.34	19.81	22.36	24.74	27.69	29.82
14	4.07	4.66	5.63	6.57	7.79	13.34	21.06	23.68	26.12	29.14	31.32
15	4.60	5.23	6.27	7.26	8.55	14.34	22.31	25.00	27.49	30.58	32.80
16	5.14	5.81	6.91	7.96	9.31	15.34	23.54	26.30	28.85	32.00	34.27
17	5.70	6.41	7.56	8.67	10.09	16.34	24.77	27.59	30.19	33.41	35.72
18	6.26	7.01	8.23	9.39	10.87	17.34	25.99	28.87	31.53	34.81	37.16
19	6.84	7.63	8.91	10.12	11.65	18.34	27.20	30.14	32.85	36.19	38.58
20	7.43	8.26	9.59	10.85	12.44	19.34	28.41	31.41	34.17	37.57	40.00
21	8.03	8.90	10.28	11.59	13.24	20.34	29.62	32.67	35.48	38.93	41.40
22	8.64	9.54	10.98	12.34	14.04	21.34	30.81	33.92	36.78	40.29	42.80
23	9.26	10.20	11.69	13.09	14.85	22.34	32.01	35.17	38.08	41.64	44.18
24	9.89	10.86	12.40	13.85	15.66	23.34	33.20	36.42	39.36	42.98	45.56
25	10.52	11.52	13.12	14.61	16.47	24.34	34.28	37.65	40.65	44.31	46.93
26	11.16	12.20	13.84	15.38	17.29	25.34	35.56	38.89	41.92	45.64	48.29
27	11.81	12.88	14.57	16.15	18.11	26.34	36.74	40.11	43.19	46.96	49.65
28	12.46	13.57	15.31	16.93	18.94	27.34	37.92	41.34	44.46	48.28	50.99
29	13.12	14.26	16.05	17.71	19.77	28.34	39.09	42.56	45.72	49.59	52.34
30	13.79	14.95	16.79	18.49	20.60	29.34	40.26	43.77	46.98	50.89	53.67
40	20.71	22.16	24.43	26.51	29.05	39.34	51.81	55.76	59.34	63.69	66.77
50	27.99	29.71	32.36	34.76	37.69	49.33	63.17	67.50	71.42	76.15	79.49
60	35.53	37.48	40.48	43.19	46.46	59.33	74.40	79.08	83.30	88.38	91.95
70	43.28	45.44	48.76	51.74	55.33	69.33	85.53	90.53	95.02	100.42	104.22
80	51.17	53.54	57.15	60.39	64.28	79.33	96.58	101.88	106.63	112.33	116.32
90	59.20	61.75	65.65	69.13	73.29	89.33	107.57	113.14	118.14	124.12	128.30
100	67.33	70.06	74.22	77.93	82.36	99.33	118.50	124.34	129.56	135.81	140.17

ν = degrees of freedom.

Table IV Percentage Points $f_{\alpha,u,v}$ of the F Distribution

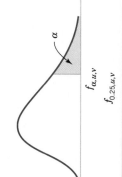

$f_{0.25,u,v}$

v \ u	1	2	3	4	5	6	7	8	9	10	12	15	20	24	30	40	60	120	∞
1	5.83	7.50	8.20	8.58	8.82	8.98	9.10	9.19	9.26	9.32	9.41	9.49	9.58	9.63	9.67	9.71	9.76	9.80	9.85
2	2.57	3.00	3.15	3.23	3.28	3.31	3.34	3.35	3.37	3.38	3.39	3.41	3.43	3.43	3.44	3.45	3.46	3.47	3.48
3	2.02	2.28	2.36	2.39	2.41	2.42	2.43	2.44	2.44	2.44	2.45	2.46	2.46	2.46	2.47	2.47	2.47	2.47	2.47
4	1.81	2.00	2.05	2.06	2.07	2.08	2.08	2.08	2.08	2.08	2.08	2.08	2.08	2.08	2.08	2.08	2.08	2.08	2.08
5	1.69	1.85	1.88	1.89	1.89	1.89	1.89	1.89	1.89	1.89	1.89	1.89	1.88	1.88	1.88	1.88	1.87	1.87	1.87
6	1.62	1.76	1.78	1.79	1.79	1.78	1.78	1.78	1.77	1.77	1.77	1.76	1.76	1.75	1.75	1.75	1.74	1.74	1.74
7	1.57	1.70	1.72	1.72	1.71	1.71	1.70	1.70	1.70	1.69	1.68	1.68	1.67	1.67	1.66	1.66	1.65	1.65	1.65
8	1.54	1.66	1.67	1.66	1.66	1.65	1.64	1.64	1.63	1.63	1.62	1.62	1.61	1.60	1.60	1.59	1.59	1.58	1.58
9	1.51	1.62	1.63	1.63	1.62	1.61	1.60	1.60	1.59	1.59	1.58	1.57	1.56	1.56	1.55	1.54	1.54	1.53	1.53
10	1.49	1.60	1.60	1.59	1.59	1.58	1.57	1.56	1.56	1.55	1.54	1.53	1.52	1.52	1.51	1.51	1.50	1.49	1.48
11	1.47	1.58	1.58	1.57	1.56	1.55	1.54	1.53	1.53	1.52	1.51	1.50	1.49	1.49	1.48	1.47	1.47	1.46	1.45
12	1.46	1.56	1.56	1.55	1.54	1.53	1.52	1.51	1.51	1.50	1.49	1.48	1.47	1.46	1.45	1.45	1.44	1.43	1.42
13	1.45	1.55	1.55	1.53	1.52	1.51	1.50	1.49	1.49	1.48	1.47	1.46	1.45	1.44	1.43	1.42	1.42	1.41	1.40
14	1.44	1.53	1.53	1.52	1.51	1.50	1.49	1.48	1.47	1.46	1.45	1.44	1.43	1.42	1.41	1.41	1.40	1.39	1.38
15	1.43	1.52	1.52	1.51	1.49	1.48	1.47	1.46	1.46	1.45	1.44	1.43	1.41	1.41	1.40	1.39	1.38	1.37	1.36
16	1.42	1.51	1.51	1.50	1.48	1.47	1.46	1.45	1.44	1.44	1.43	1.41	1.40	1.39	1.38	1.37	1.36	1.35	1.34
17	1.42	1.51	1.50	1.49	1.47	1.46	1.45	1.44	1.43	1.43	1.41	1.40	1.39	1.38	1.37	1.36	1.35	1.34	1.33
18	1.41	1.50	1.49	1.48	1.46	1.45	1.44	1.43	1.42	1.42	1.40	1.39	1.38	1.37	1.36	1.35	1.34	1.33	1.32
19	1.41	1.49	1.49	1.47	1.46	1.44	1.43	1.42	1.41	1.41	1.40	1.38	1.37	1.36	1.35	1.34	1.33	1.32	1.30
20	1.40	1.49	1.48	1.47	1.45	1.44	1.43	1.42	1.41	1.40	1.39	1.37	1.36	1.35	1.34	1.33	1.32	1.31	1.29
21	1.40	1.48	1.48	1.46	1.44	1.43	1.42	1.41	1.40	1.39	1.38	1.37	1.35	1.34	1.33	1.32	1.31	1.30	1.28
22	1.40	1.48	1.47	1.45	1.44	1.42	1.41	1.40	1.39	1.39	1.37	1.36	1.34	1.33	1.32	1.31	1.30	1.29	1.28
23	1.39	1.47	1.47	1.45	1.43	1.42	1.41	1.40	1.39	1.38	1.37	1.35	1.34	1.33	1.32	1.31	1.30	1.28	1.27
24	1.39	1.47	1.46	1.44	1.43	1.41	1.40	1.39	1.38	1.38	1.36	1.35	1.33	1.32	1.31	1.30	1.29	1.28	1.26
25	1.39	1.47	1.46	1.44	1.42	1.41	1.40	1.39	1.38	1.37	1.36	1.34	1.33	1.32	1.31	1.29	1.28	1.27	1.25
26	1.38	1.46	1.45	1.44	1.42	1.41	1.39	1.38	1.37	1.37	1.35	1.34	1.32	1.31	1.30	1.29	1.28	1.26	1.25
27	1.38	1.46	1.45	1.43	1.42	1.40	1.39	1.38	1.37	1.36	1.35	1.33	1.32	1.31	1.30	1.28	1.27	1.26	1.24
28	1.38	1.46	1.45	1.43	1.41	1.40	1.39	1.38	1.37	1.36	1.34	1.33	1.31	1.30	1.29	1.28	1.27	1.25	1.24
29	1.38	1.45	1.45	1.43	1.41	1.40	1.38	1.37	1.36	1.35	1.34	1.32	1.31	1.30	1.29	1.27	1.26	1.25	1.23
30	1.38	1.45	1.44	1.42	1.41	1.39	1.38	1.37	1.36	1.35	1.34	1.32	1.30	1.29	1.28	1.27	1.26	1.24	1.23
40	1.36	1.44	1.42	1.40	1.39	1.37	1.36	1.35	1.34	1.33	1.31	1.30	1.28	1.26	1.25	1.24	1.22	1.21	1.19
60	1.35	1.42	1.41	1.38	1.37	1.35	1.33	1.32	1.31	1.30	1.29	1.27	1.25	1.24	1.22	1.21	1.19	1.17	1.15
120	1.34	1.40	1.39	1.37	1.35	1.33	1.31	1.30	1.29	1.28	1.26	1.24	1.22	1.21	1.19	1.18	1.16	1.13	1.10
∞	1.32	1.39	1.37	1.35	1.33	1.31	1.29	1.28	1.27	1.25	1.24	1.22	1.19	1.18	1.16	1.14	1.12	1.08	1.00

Degrees of freedom for the numerator (u)

Degrees of freedom for the denominator (v)

Table IV Percentage Points $f_{\alpha,u,v}$ of the F Distribution (continued)

$$f_{0.10,u,v}$$

v \ u	1	2	3	4	5	6	7	8	9	10	12	15	20	24	30	40	60	120	∞
1	39.86	49.50	53.59	55.83	57.24	58.20	58.91	59.44	59.86	60.19	60.71	61.22	61.74	62.00	62.26	62.53	62.79	63.06	63.33
2	8.53	9.00	9.16	9.24	9.29	9.33	9.35	9.37	9.38	9.39	9.41	9.42	9.44	9.45	9.46	9.47	9.47	9.48	9.49
3	5.54	5.46	5.39	5.34	5.31	5.28	5.27	5.25	5.24	5.23	5.22	5.20	5.18	5.18	5.17	5.16	5.15	5.14	5.13
4	4.54	4.32	4.19	4.11	4.05	4.01	3.98	3.95	3.94	3.92	3.90	3.87	3.84	3.83	3.82	3.80	3.79	3.78	3.76
5	4.06	3.78	3.62	3.52	3.45	3.40	3.37	3.34	3.32	3.30	3.27	3.24	3.21	3.19	3.17	3.16	3.14	3.12	3.10
6	3.78	3.46	3.29	3.18	3.11	3.05	3.01	2.98	2.96	2.94	2.90	2.87	2.84	2.82	2.80	2.78	2.76	2.74	2.72
7	3.59	3.26	3.07	2.96	2.88	2.83	2.78	2.75	2.72	2.70	2.67	2.63	2.59	2.58	2.56	2.54	2.51	2.49	2.47
8	3.46	3.11	2.92	2.81	2.73	2.67	2.62	2.59	2.56	2.54	2.50	2.46	2.42	2.40	2.38	2.36	2.34	2.32	2.29
9	3.36	3.01	2.81	2.69	2.61	2.55	2.51	2.47	2.44	2.42	2.38	2.34	2.30	2.28	2.25	2.23	2.21	2.18	2.16
10	3.29	2.92	2.73	2.61	2.52	2.46	2.41	2.38	2.35	2.32	2.28	2.24	2.20	2.18	2.16	2.13	2.11	2.08	2.06
11	3.23	2.86	2.66	2.54	2.45	2.39	2.34	2.30	2.27	2.25	2.21	2.17	2.12	2.10	2.08	2.05	2.03	2.00	1.97
12	3.18	2.81	2.61	2.48	2.39	2.33	2.28	2.24	2.21	2.19	2.15	2.10	2.06	2.04	2.01	1.99	1.96	1.93	1.90
13	3.14	2.76	2.56	2.43	2.35	2.28	2.23	2.20	2.16	2.14	2.10	2.05	2.01	1.98	1.96	1.93	1.90	1.88	1.85
14	3.10	2.73	2.52	2.39	2.31	2.24	2.19	2.15	2.12	2.10	2.05	2.01	1.96	1.94	1.91	1.89	1.86	1.83	1.80
15	3.07	2.70	2.49	2.36	2.27	2.21	2.16	2.12	2.09	2.06	2.02	1.97	1.92	1.90	1.87	1.85	1.82	1.79	1.76
16	3.05	2.67	2.46	2.33	2.24	2.18	2.13	2.09	2.06	2.03	1.99	1.94	1.89	1.87	1.84	1.81	1.78	1.75	1.72
17	3.03	2.64	2.44	2.31	2.22	2.15	2.10	2.06	2.03	2.00	1.96	1.91	1.86	1.84	1.81	1.78	1.75	1.72	1.69
18	3.01	2.62	2.42	2.29	2.20	2.13	2.08	2.04	2.00	1.98	1.93	1.89	1.84	1.81	1.78	1.75	1.72	1.69	1.66
19	2.99	2.61	2.40	2.27	2.18	2.11	2.06	2.02	1.98	1.96	1.91	1.86	1.81	1.79	1.76	1.73	1.70	1.67	1.63
20	2.97	2.59	2.38	2.25	2.16	2.09	2.04	2.00	1.96	1.94	1.89	1.84	1.79	1.77	1.74	1.71	1.68	1.64	1.61
21	2.96	2.57	2.36	2.23	2.14	2.08	2.02	1.98	1.95	1.92	1.87	1.83	1.78	1.75	1.72	1.69	1.66	1.62	1.59
22	2.95	2.56	2.35	2.22	2.13	2.06	2.01	1.97	1.93	1.90	1.86	1.81	1.76	1.73	1.70	1.67	1.64	1.60	1.57
23	2.94	2.55	2.34	2.21	2.11	2.05	1.99	1.95	1.92	1.89	1.84	1.80	1.74	1.72	1.69	1.66	1.62	1.59	1.55
24	2.93	2.54	2.33	2.19	2.10	2.04	1.98	1.94	1.91	1.88	1.83	1.78	1.73	1.70	1.67	1.64	1.61	1.57	1.53
25	2.92	2.53	2.32	2.18	2.09	2.02	1.97	1.93	1.89	1.87	1.82	1.77	1.72	1.69	1.66	1.63	1.59	1.56	1.52
26	2.91	2.52	2.31	2.17	2.08	2.01	1.96	1.92	1.88	1.86	1.81	1.76	1.71	1.68	1.65	1.61	1.58	1.54	1.50
27	2.90	2.51	2.30	2.17	2.07	2.00	1.95	1.91	1.87	1.85	1.80	1.75	1.70	1.67	1.64	1.60	1.57	1.53	1.49
28	2.89	2.50	2.29	2.16	2.06	2.00	1.94	1.90	1.87	1.84	1.79	1.74	1.69	1.66	1.63	1.59	1.56	1.52	1.48
29	2.89	2.50	2.28	2.15	2.06	1.99	1.93	1.89	1.86	1.83	1.78	1.73	1.68	1.65	1.62	1.58	1.55	1.51	1.47
30	2.88	2.49	2.28	2.14	2.03	1.98	1.93	1.88	1.85	1.82	1.77	1.72	1.67	1.64	1.61	1.57	1.54	1.50	1.46
40	2.84	2.44	2.23	2.09	2.00	1.93	1.87	1.83	1.79	1.76	1.71	1.66	1.61	1.57	1.54	1.51	1.47	1.42	1.38
60	2.79	2.39	2.18	2.04	1.95	1.87	1.82	1.77	1.74	1.71	1.66	1.60	1.54	1.51	1.48	1.44	1.40	1.35	1.29
120	2.75	2.35	2.13	1.99	1.90	1.82	1.77	1.72	1.68	1.65	1.60	1.55	1.48	1.45	1.41	1.37	1.32	1.26	1.19
∞	2.71	2.30	2.08	1.94	1.85	1.77	1.72	1.67	1.63	1.60	1.55	1.49	1.42	1.38	1.34	1.30	1.24	1.17	1.00

Degrees of freedom for the numerator (u)

Degrees of freedom for the denominator (v)

Table IV Percentage Points $f_{\alpha,u,v}$ of the F Distribution (continued)

$f_{0.05,u,v}$

v \\ u	1	2	3	4	5	6	7	8	9	10	12	15	20	24	30	40	60	120	∞
1	161.4	199.5	215.7	224.6	230.2	234.0	236.8	238.9	240.5	241.9	243.9	245.9	248.0	249.1	250.1	251.1	252.2	253.3	254.3
2	18.51	19.00	19.16	19.25	19.30	19.33	19.35	19.37	19.38	19.40	19.41	19.43	19.45	19.45	19.46	19.47	19.48	19.49	19.50
3	10.13	9.55	9.28	9.12	9.01	8.94	8.89	8.85	8.81	8.79	8.74	8.70	8.66	8.64	8.62	8.59	8.57	8.55	8.53
4	7.71	6.94	6.59	6.39	6.26	6.16	6.09	6.04	6.00	5.96	5.91	5.86	5.80	5.77	5.75	5.72	5.69	5.66	5.63
5	6.61	5.79	5.41	5.19	5.05	4.95	4.88	4.82	4.77	4.74	4.68	4.62	4.56	4.53	4.50	4.46	4.43	4.40	4.36
6	5.99	5.14	4.76	4.53	4.39	4.28	4.21	4.15	4.10	4.06	4.00	3.94	3.87	3.84	3.81	3.77	3.74	3.70	3.67
7	5.59	4.74	4.35	4.12	3.97	3.87	3.79	3.73	3.68	3.64	3.57	3.51	3.44	3.41	3.38	3.34	3.30	3.27	3.23
8	5.32	4.46	4.07	3.84	3.69	3.58	3.50	3.44	3.39	3.35	3.28	3.22	3.15	3.12	3.08	3.04	3.01	2.97	2.93
9	5.12	4.26	3.86	3.63	3.48	3.37	3.29	3.23	3.18	3.14	3.07	3.01	2.94	2.90	2.86	2.83	2.79	2.75	2.71
10	4.96	4.10	3.71	3.48	3.33	3.22	3.14	3.07	3.02	2.98	2.91	2.85	2.77	2.74	2.70	2.66	2.62	2.58	2.54
11	4.84	3.98	3.59	3.36	3.20	3.09	3.01	2.95	2.90	2.85	2.79	2.72	2.65	2.61	2.57	2.53	2.49	2.45	2.40
12	4.75	3.89	3.49	3.26	3.11	3.00	2.91	2.85	2.80	2.75	2.69	2.62	2.54	2.51	2.47	2.43	2.38	2.34	2.30
13	4.67	3.81	3.41	3.18	3.03	2.92	2.83	2.77	2.71	2.67	2.60	2.53	2.46	2.42	2.38	2.34	2.30	2.25	2.21
14	4.60	3.74	3.34	3.11	2.96	2.85	2.76	2.70	2.65	2.60	2.53	2.46	2.39	2.35	2.31	2.27	2.22	2.18	2.13
15	4.54	3.68	3.29	3.06	2.90	2.79	2.71	2.64	2.59	2.54	2.48	2.40	2.33	2.29	2.25	2.20	2.16	2.11	2.07
16	4.49	3.63	3.24	3.01	2.85	2.74	2.66	2.59	2.54	2.49	2.42	2.35	2.28	2.24	2.19	2.15	2.11	2.06	2.01
17	4.45	3.59	3.20	2.96	2.81	2.70	2.61	2.55	2.49	2.45	2.38	2.31	2.23	2.19	2.15	2.10	2.06	2.01	1.96
18	4.41	3.55	3.16	2.93	2.77	2.66	2.58	2.51	2.46	2.41	2.34	2.27	2.19	2.15	2.11	2.06	2.02	1.97	1.92
19	4.38	3.52	3.13	2.90	2.74	2.63	2.54	2.48	2.42	2.38	2.31	2.23	2.16	2.11	2.07	2.03	1.98	1.93	1.88
20	4.35	3.49	3.10	2.87	2.71	2.60	2.51	2.45	2.39	2.35	2.28	2.20	2.12	2.08	2.04	1.99	1.95	1.90	1.84
21	4.32	3.47	3.07	2.84	2.68	2.57	2.49	2.42	2.37	2.32	2.25	2.18	2.10	2.05	2.01	1.96	1.92	1.87	1.81
22	4.30	3.44	3.05	2.82	2.66	2.55	2.46	2.40	2.34	2.30	2.23	2.15	2.07	2.03	1.98	1.94	1.89	1.84	1.78
23	4.28	3.42	3.03	2.80	2.64	2.53	2.44	2.37	2.32	2.27	2.20	2.13	2.05	2.01	1.96	1.91	1.86	1.81	1.76
24	4.26	3.40	3.01	2.78	2.62	2.51	2.42	2.36	2.30	2.25	2.18	2.11	2.03	1.98	1.94	1.89	1.84	1.79	1.73
25	4.24	3.39	2.99	2.76	2.60	2.49	2.40	2.34	2.28	2.24	2.16	2.09	2.01	1.96	1.92	1.87	1.82	1.77	1.71
26	4.23	3.37	2.98	2.74	2.59	2.47	2.39	2.32	2.27	2.22	2.15	2.07	1.99	1.95	1.90	1.85	1.80	1.75	1.69
27	4.21	3.35	2.96	2.73	2.57	2.46	2.37	2.31	2.25	2.20	2.13	2.06	1.97	1.93	1.88	1.84	1.79	1.73	1.67
28	4.20	3.34	2.95	2.71	2.56	2.45	2.36	2.29	2.24	2.19	2.12	2.04	1.96	1.91	1.87	1.82	1.77	1.71	1.65
29	4.18	3.33	2.93	2.70	2.55	2.43	2.35	2.28	2.22	2.18	2.10	2.03	1.94	1.90	1.85	1.81	1.75	1.70	1.64
30	4.17	3.32	2.92	2.69	2.53	2.42	2.33	2.27	2.21	2.16	2.09	2.01	1.93	1.89	1.84	1.79	1.74	1.68	1.62
40	4.08	3.23	2.84	2.61	2.45	2.34	2.25	2.18	2.12	2.08	2.00	1.92	1.84	1.79	1.74	1.69	1.64	1.58	1.51
60	4.00	3.15	2.76	2.53	2.37	2.25	2.17	2.10	2.04	1.99	1.92	1.84	1.75	1.70	1.65	1.59	1.53	1.47	1.39
120	3.92	3.07	2.68	2.45	2.29	2.17	2.09	2.02	1.96	1.91	1.83	1.75	1.66	1.61	1.55	1.50	1.43	1.35	1.25
∞	3.84	3.00	2.60	2.37	2.21	2.10	2.01	1.94	1.88	1.83	1.75	1.67	1.57	1.52	1.46	1.39	1.32	1.22	1.00

Degrees of freedom for the numerator (u)

Degrees of freedom for the denominator (v)

466

Table IV Percentage Points $f_{\alpha,u,v}$ of the F Distribution (*continued*)

$$f_{0.025,u,v}$$

v \\ u	1	2	3	4	5	6	7	8	9	10	12	15	20	24	30	40	60	120	∞
1	647.8	799.5	864.2	899.6	921.8	937.1	948.2	956.7	963.3	968.6	976.7	984.9	993.1	997.2	1001	1006	1010	1014	1018
2	38.51	39.00	39.17	39.25	39.30	39.33	39.36	39.37	39.39	39.40	39.41	39.43	39.45	39.46	39.46	39.47	39.48	39.49	39.50
3	17.44	16.04	15.44	15.10	14.88	14.73	14.62	14.54	14.47	14.42	14.34	14.25	14.17	14.12	14.08	14.04	13.99	13.95	13.90
4	12.22	10.65	9.98	9.60	9.36	9.20	9.07	8.98	8.90	8.84	8.75	8.66	8.56	8.51	8.46	8.41	8.36	8.31	8.26
5	10.01	8.43	7.76	7.39	7.15	6.98	6.85	6.76	6.68	6.62	6.52	6.43	6.33	6.28	6.23	6.18	6.12	6.07	6.02
6	8.81	7.26	6.60	6.23	5.99	5.82	5.70	5.60	5.52	5.46	5.37	5.27	5.17	5.12	5.07	5.01	4.96	4.90	4.85
7	8.07	6.54	5.89	5.52	5.29	5.12	4.99	4.90	4.82	4.76	4.67	4.57	4.47	4.42	4.36	4.31	4.25	4.20	4.14
8	7.57	6.06	5.42	5.05	4.82	4.65	4.53	4.43	4.36	4.30	4.20	4.10	4.00	3.95	3.89	3.84	3.78	3.73	3.67
9	7.21	5.71	5.08	4.72	4.48	4.32	4.20	4.10	4.03	3.96	3.87	3.77	3.67	3.61	3.56	3.51	3.45	3.39	3.33
10	6.94	5.46	4.83	4.47	4.24	4.07	3.95	3.85	3.78	3.72	3.62	3.52	3.42	3.37	3.31	3.26	3.20	3.14	3.08
11	6.72	5.26	4.63	4.28	4.04	3.88	3.76	3.66	3.59	3.53	3.43	3.33	3.23	3.17	3.12	3.06	3.00	2.94	2.88
12	6.55	5.10	4.47	4.12	3.89	3.73	3.61	3.51	3.44	3.37	3.28	3.18	3.07	3.02	2.96	2.91	2.85	2.79	2.72
13	6.41	4.97	4.35	4.00	3.77	3.60	3.48	3.39	3.31	3.25	3.15	3.05	2.95	2.89	2.84	2.78	2.72	2.66	2.60
14	6.30	4.86	4.24	3.89	3.66	3.50	3.38	3.29	3.21	3.15	3.05	2.95	2.84	2.79	2.73	2.67	2.61	2.55	2.49
15	6.20	4.77	4.15	3.80	3.58	3.41	3.29	3.20	3.12	3.06	2.96	2.86	2.76	2.70	2.64	2.59	2.52	2.46	2.40
16	6.12	4.69	4.08	3.73	3.50	3.34	3.22	3.12	3.05	2.99	2.89	2.79	2.68	2.63	2.57	2.51	2.45	2.38	2.32
17	6.04	4.62	4.01	3.66	3.44	3.28	3.16	3.06	2.98	2.92	2.82	2.72	2.62	2.56	2.50	2.44	2.38	2.32	2.25
18	5.98	4.56	3.95	3.61	3.38	3.22	3.10	3.01	2.93	2.87	2.77	2.67	2.56	2.50	2.44	2.38	2.32	2.26	2.19
19	5.92	4.51	3.90	3.56	3.33	3.17	3.05	2.96	2.88	2.82	2.72	2.62	2.51	2.45	2.39	2.33	2.27	2.20	2.13
20	5.87	4.46	3.86	3.51	3.29	3.13	3.01	2.91	2.84	2.77	2.68	2.57	2.46	2.41	2.35	2.29	2.22	2.16	2.09
21	5.83	4.42	3.82	3.48	3.25	3.09	2.97	2.87	2.80	2.73	2.64	2.53	2.42	2.37	2.31	2.25	2.18	2.11	2.04
22	5.79	4.38	3.78	3.44	3.22	3.05	2.93	2.84	2.76	2.70	2.60	2.50	2.39	2.33	2.27	2.21	2.14	2.08	2.00
23	5.75	4.35	3.75	3.41	3.18	3.02	2.90	2.81	2.73	2.67	2.57	2.47	2.36	2.30	2.24	2.18	2.11	2.04	1.97
24	5.72	4.32	3.72	3.38	3.15	2.99	2.87	2.78	2.70	2.64	2.54	2.44	2.33	2.27	2.21	2.15	2.08	2.01	1.94
25	5.69	4.29	3.69	3.35	3.13	2.97	2.85	2.75	2.68	2.61	2.51	2.41	2.30	2.24	2.18	2.12	2.05	1.98	1.91
26	5.66	4.27	3.67	3.33	3.10	2.94	2.82	2.73	2.65	2.59	2.49	2.39	2.28	2.22	2.16	2.09	2.03	1.95	1.88
27	5.63	4.24	3.65	3.31	3.08	2.92	2.80	2.71	2.63	2.57	2.47	2.36	2.25	2.19	2.13	2.07	2.00	1.93	1.85
28	5.61	4.22	3.63	3.29	3.06	2.90	2.78	2.69	2.61	2.55	2.45	2.34	2.23	2.17	2.11	2.05	1.98	1.91	1.83
29	5.59	4.20	3.61	3.27	3.04	2.88	2.76	2.67	2.59	2.53	2.43	2.32	2.21	2.15	2.09	2.03	1.96	1.89	1.81
30	5.57	4.18	3.59	3.25	3.03	2.87	2.75	2.65	2.57	2.51	2.41	2.31	2.20	2.14	2.07	2.01	1.94	1.87	1.79
40	5.42	4.05	3.46	3.13	2.90	2.74	2.62	2.53	2.45	2.39	2.29	2.18	2.07	2.01	1.94	1.88	1.80	1.72	1.64
60	5.29	3.93	3.34	3.01	2.79	2.63	2.51	2.41	2.33	2.27	2.17	2.06	1.94	1.88	1.82	1.74	1.67	1.58	1.48
120	5.15	3.80	3.23	2.89	2.67	2.52	2.39	2.30	2.22	2.16	2.05	1.94	1.82	1.76	1.69	1.61	1.53	1.43	1.31
∞	5.02	3.69	3.12	2.79	2.57	2.41	2.29	2.19	2.11	2.05	1.94	1.83	1.71	1.64	1.57	1.48	1.39	1.27	1.00

Degrees of freedom for the numerator (u)

Degrees of freedom for the denominator (v)

Table IV   Percentage Points $f_{\alpha,u,v}$ of the F Distribution (*continued*)

$$f_{0.01,u,v}$$

		Degrees of freedom for the numerator (u)																	
v	1	2	3	4	5	6	7	8	9	10	12	15	20	24	30	40	60	120	∞
1	4052	4999.5	5403	5625	5764	5859	5928	5982	6022	6056	6106	6157	6209	6235	6261	6287	6313	6339	6366
2	98.50	99.00	99.17	99.25	99.30	99.33	99.36	99.37	99.39	99.40	99.42	99.43	99.45	99.46	99.47	99.47	99.48	99.49	99.50
3	34.12	30.82	29.46	28.71	28.24	27.91	27.67	27.49	27.35	27.23	27.05	26.87	26.69	26.60	26.50	26.41	26.32	26.22	26.13
4	21.20	18.00	16.69	15.98	15.52	15.21	14.98	14.80	14.66	14.55	14.37	14.20	14.02	13.93	13.84	13.75	13.65	13.56	13.46
5	16.26	13.27	12.06	11.39	10.97	10.67	10.46	10.29	10.16	10.05	9.89	9.72	9.55	9.47	9.38	9.29	9.20	9.11	9.02
6	13.75	10.92	9.78	9.15	8.75	8.47	8.26	8.10	7.98	7.87	7.72	7.56	7.40	7.31	7.23	7.14	7.06	6.97	6.88
7	12.25	9.55	8.45	7.85	7.46	7.19	6.99	6.84	6.72	6.62	6.47	6.31	6.16	6.07	5.99	5.91	5.82	5.74	5.65
8	11.26	8.65	7.59	7.01	6.63	6.37	6.18	6.03	5.91	5.81	5.67	5.52	5.36	5.28	5.20	5.12	5.03	4.95	4.46
9	10.56	8.02	6.99	6.42	6.06	5.80	5.61	5.47	5.35	5.26	5.11	4.96	4.81	4.73	4.65	4.57	4.48	4.40	4.31
10	10.04	7.56	6.55	5.99	5.64	5.39	5.20	5.06	4.94	4.85	4.71	4.56	4.41	4.33	4.25	4.17	4.08	4.00	3.91
11	9.65	7.21	6.22	5.67	5.32	5.07	4.89	4.74	4.63	4.54	4.40	4.25	4.10	4.02	3.94	3.86	3.78	3.69	3.60
12	9.33	6.93	5.95	5.41	5.06	4.82	4.64	4.50	4.39	4.30	4.16	4.01	3.86	3.78	3.70	3.62	3.54	3.45	3.36
13	9.07	6.70	5.74	5.21	4.86	4.62	4.44	4.30	4.19	4.10	3.96	3.82	3.66	3.59	3.51	3.43	3.34	3.25	3.17
14	8.86	6.51	5.56	5.04	4.69	4.46	4.28	4.14	4.03	3.94	3.80	3.66	3.51	3.43	3.35	3.27	3.18	3.09	3.00
15	8.68	6.36	5.42	4.89	4.56	4.32	4.14	4.00	3.89	3.80	3.67	3.52	3.37	3.29	3.21	3.13	3.05	2.96	2.87
16	8.53	6.23	5.29	4.77	4.44	4.20	4.03	3.89	3.78	3.69	3.55	3.41	3.26	3.18	3.10	3.02	2.93	2.84	2.75
17	8.40	6.11	5.18	4.67	4.34	4.10	3.93	3.79	3.68	3.59	3.46	3.31	3.16	3.08	3.00	2.92	2.83	2.75	2.65
18	8.29	6.01	5.09	4.58	4.25	4.01	3.84	3.71	3.60	3.51	3.37	3.23	3.08	3.00	2.92	2.84	2.75	2.66	2.57
19	8.18	5.93	5.01	4.50	4.17	3.94	3.77	3.63	3.52	3.43	3.30	3.15	3.00	2.92	2.84	2.76	2.67	2.58	2.49
20	8.10	5.85	4.94	4.43	4.10	3.87	3.70	3.56	3.46	3.37	3.23	3.09	2.94	2.86	2.78	2.69	2.61	2.52	2.42
21	8.02	5.78	4.87	4.37	4.04	3.81	3.64	3.51	3.40	3.31	3.17	3.03	2.88	2.80	2.72	2.64	2.55	2.46	2.36
22	7.95	5.72	4.82	4.31	3.99	3.76	3.59	3.45	3.35	3.26	3.12	2.98	2.83	2.75	2.67	2.58	2.50	2.40	2.31
23	7.88	5.66	4.76	4.26	3.94	3.71	3.54	3.41	3.30	3.21	3.07	2.93	2.78	2.70	2.62	2.54	2.45	2.35	2.26
24	7.82	5.61	4.72	4.22	3.90	3.67	3.50	3.36	3.26	3.17	3.03	2.89	2.74	2.66	2.58	2.49	2.40	2.31	2.21
25	7.77	5.57	4.68	4.18	3.85	3.63	3.46	3.32	3.22	3.13	2.99	2.85	2.70	2.62	2.54	2.45	2.36	2.27	2.17
26	7.72	5.53	4.64	4.14	3.82	3.59	3.42	3.29	3.18	3.09	2.96	2.81	2.66	2.58	2.50	2.42	2.33	2.23	2.13
27	7.68	5.49	4.60	4.11	3.78	3.56	3.39	3.26	3.15	3.06	2.93	2.78	2.63	2.55	2.47	2.38	2.29	2.20	2.10
28	7.64	5.45	4.57	4.07	3.75	3.53	3.36	3.23	3.12	3.03	2.90	2.75	2.60	2.52	2.44	2.35	2.26	2.17	2.06
29	7.60	5.42	4.54	4.04	3.73	3.50	3.33	3.20	3.09	3.00	2.87	2.73	2.57	2.49	2.41	2.33	2.23	2.14	2.03
30	7.56	5.39	4.51	4.02	3.70	3.47	3.30	3.17	3.07	2.98	2.84	2.70	2.55	2.47	2.39	2.30	2.21	2.11	2.01
40	7.31	5.18	4.31	3.83	3.51	3.29	3.12	2.99	2.89	2.80	2.66	2.52	2.37	2.29	2.20	2.11	2.02	1.92	1.80
60	7.08	4.98	4.13	3.65	3.34	3.12	2.95	2.82	2.72	2.63	2.50	2.35	2.20	2.12	2.03	1.94	1.84	1.73	1.60
120	6.85	4.79	3.95	3.48	3.17	2.96	2.79	2.66	2.56	2.47	2.34	2.19	2.03	1.95	1.86	1.76	1.66	1.53	1.38
∞	6.63	4.61	3.78	3.32	3.02	2.80	2.64	2.51	2.41	2.32	2.18	2.04	1.88	1.79	1.70	1.59	1.47	1.32	1.00

Degrees of freedom for the denominator (v)

Chart V   Operating Characteristic Curves for the t-Test

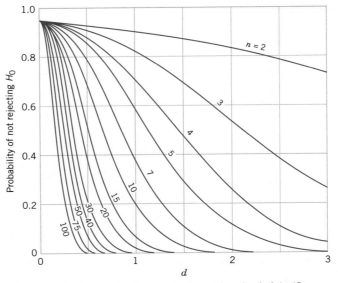

(a) OC curves for different values of n for the two-sided t-test for a level of significance $\alpha = 0.05$.

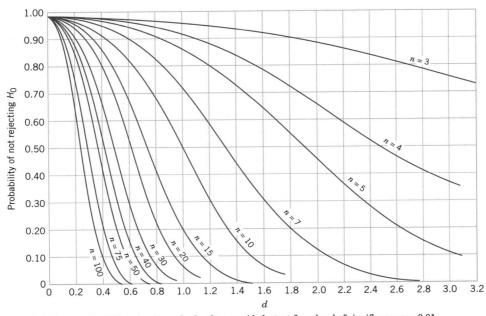

(b) OC curves for different values of n for the two-sided t-test for a level of significance $\alpha = 0.01$.

Source: These charts are reproduced with permission from "Operating Characteristics for the Common Statistical Tests of Significance," by C. L. Ferris, F. E. Grubbs, and C. L. Weaver, *Annals of Mathematical Statistics,* June 1946, and from *Engineering Statistics,* 2nd Edition, by A. H. Bowker and G. J. Lieberman, Prentice-Hall, 1972.

Chart V Operating Characteristic Curves for the *t*-Test (*continued*)

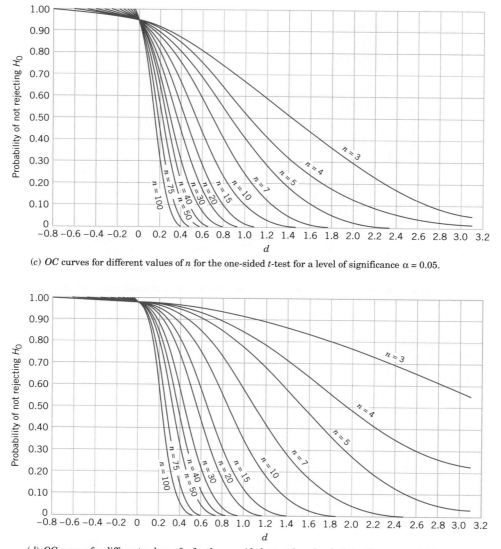

(c) *OC* curves for different values of *n* for the one-sided *t*-test for a level of significance $\alpha = 0.05$.

(d) *OC* curves for different values of *n* for the one-sided *t*-test for a level of significance $\alpha = 0.01$.

Table VI Factors for Normal Distribution Tolerance Intervals

	Values of k for Two-Sided Intervals								
Confidence Level	0.90			0.95			0.99		
Percent Contained γ	90	95	99	90	95	99	90	95	99
Sample Size									
2	15.978	18.800	24.167	32.019	37.674	48.430	160.193	188.491	242.300
3	5.847	6.919	8.974	8.380	9.916	12.861	18.930	22.401	29.055
4	4.166	4.943	6.440	5.369	6.370	8.299	9.398	11.150	14.527
5	3.949	4.152	5.423	4.275	5.079	6.634	6.612	7.855	10.260
6	3.131	3.723	4.870	3.712	4.414	5.775	5.337	6.345	8.301
7	2.902	3.452	4.521	3.369	4.007	5.248	4.613	5.488	7.187
8	2.743	3.264	4.278	3.136	3.732	4.891	4.147	4.936	6.468
9	2.626	3.125	4.098	2.967	3.532	4.631	3.822	4.550	5.966
10	2.535	3.018	3.959	2.839	3.379	4.433	3.582	4.265	5.594
11	2.463	2.933	3.849	2.737	3.259	4.277	3.397	4.045	5.308
12	2.404	2.863	3.758	2.655	3.162	4.150	3.250	3.870	5.079
13	2.355	2.805	3.682	2.587	3.081	4.044	3.130	3.727	4.893
14	2.314	2.756	3.618	2.529	3.012	3.955	3.029	3.608	4.737
15	2.278	2.713	3.562	2.480	2.954	3.878	2.945	3.507	4.605
16	2.246	2.676	3.514	2.437	2.903	3.812	2.872	3.421	4.492
17	2.219	2.643	3.471	2.400	2.858	3.754	2.808	3.345	4.393
18	2.194	2.614	3.433	2.366	2.819	3.702	2.753	3.279	4.307
19	2.172	2.588	3.399	2.337	2.784	3.656	2.703	3.221	4.230
20	2.152	2.564	3.368	2.310	2.752	3.615	2.659	3.168	4.161
21	2.135	2.543	3.340	2.286	2.723	3.577	2.620	3.121	4.100
22	2.118	2.524	3.315	2.264	2.697	3.543	2.584	3.078	4.044
23	2.103	2.506	3.292	2.244	2.673	3.512	2.551	3.040	3.993
24	2.089	2.489	3.270	2.225	2.651	3.483	2.522	3.004	3.947
25	2.077	2.474	3.251	2.208	2.631	3.457	2.494	2.972	3.904
30	2.025	2.413	3.170	2.140	2.529	3.350	2.385	2.841	3.733
40	1.959	2.334	3.066	2.052	2.445	3.213	2.247	2.677	3.518
50	1.916	2.284	3.001	1.996	2.379	3.126	2.162	2.576	3.385
60	1.887	2.248	2.955	1.958	2.333	3.066	2.103	2.506	3.293
70	1.865	2.222	2.920	1.929	2.299	3.021	2.060	2.454	3.225
80	1.848	2.202	2.894	1.907	2.272	2.986	2.026	2.414	3.173
90	1.834	2.185	2.872	1.889	2.251	2.958	1.999	2.382	3.130
100	1.822	2.172	2.854	1.874	2.233	2.934	1.977	2.355	3.096
∞	1.645	1.960	2.576	1.645	1.960	2.576	1.645	1.960	2.576

Table VI   Factors for Normal Distribution Tolerance Intervals (*continued*)

	Values of k for One-Sided Intervals								
Confidence Level	0.90			0.95			0.99		
Percent Coverage	90	95	99	90	95	99	90	95	99
Sample Size									
2	10.253	13.090	18.500	20.581	26.260	37.094	103.029	131.426	185.617
3	4.258	5.311	7.340	6.155	7.656	10.553	13.995	17.370	23.896
4	3.188	3.957	5.438	4.162	5.144	7.042	7.380	9.083	12.387
5	2.742	3.400	4.666	3.407	4.203	5.741	5.362	6.578	8.939
6	2.494	3.092	4.243	3.006	3.708	5.062	4.411	5.406	7.335
7	2.333	2.894	3.972	2.755	3.399	4.642	3.859	4.728	6.412
8	2.219	2.754	3.783	2.582	3.187	4.354	3.497	4.285	5.812
9	2.133	2.650	3.641	2.454	3.031	4.143	3.240	3.972	5.389
10	2.066	2.568	3.532	2.355	2.911	3.981	3.048	3.738	5.074
11	2.011	2.503	3.443	2.275	2.815	3.852	2.898	3.556	4.829
12	1.966	2.448	3.371	2.210	2.736	3.747	2.777	3.410	4.633
13	1.928	2.402	3.309	2.155	2.671	3.659	2.677	3.290	4.472
14	1.895	2.363	3.257	2.109	2.614	3.585	2.593	3.189	4.337
15	1.867	2.329	3.212	2.068	2.566	3.520	2.521	3.102	4.222
16	1.842	2.299	3.172	2.033	2.524	3.464	2.459	3.028	4.123
17	1.819	2.272	3.137	2.002	2.486	3.414	2.405	2.963	4.037
18	1.800	2.249	3.105	1.974	2.453	3.370	2.357	2.905	3.960
19	1.782	2.227	3.077	1.949	2.423	3.331	2.314	2.854	3.892
20	1.765	2.028	3.052	1.926	2.396	3.295	2.276	2.808	3.832
21	1.750	2.190	3.028	1.905	2.371	3.263	2.241	2.766	3.777
22	1.737	2.174	3.007	1.886	2.349	3.233	2.209	2.729	3.727
23	1.724	2.159	2.987	1.869	2.328	3.206	2.180	2.694	3.681
24	1.712	2.145	2.969	1.853	2.309	3.181	2.154	2.662	3.640
25	1.702	2.132	2.952	1.838	2.292	3.158	2.129	2.633	3.601
30	1.657	2.080	2.884	1.777	2.220	3.064	2.030	2.515	3.447
40	1.598	2.010	2.793	1.697	2.125	2.941	1.902	2.364	3.249
50	1.559	1.965	2.735	1.646	2.065	2.862	1.821	2.269	3.125
60	1.532	1.933	2.694	1.609	2.022	2.807	1.764	2.202	3.038
70	1.511	1.909	2.662	1.581	1.990	2.765	1.722	2.153	2.974
80	1.495	1.890	2.638	1.559	1.964	2.733	1.688	2.114	2.924
90	1.481	1.874	2.618	1.542	1.944	2.706	1.661	2.082	2.883
100	1.470	1.861	2.601	1.527	1.927	2.684	1.639	2.056	2.850
∞	1.28	1.645	1.960	1.28	1.645	1.960	1.28	1.645	1.960

Table VII   Factors for Constructing Variables Control Charts

	$\overline{X}$ Chart			R Chart		
	Factors for Control Limits			Factors for Control Limits		
n^a	A_1	A_2	d_2	D_3	D_4	n
2	3.760	1.880	1.128	0	3.267	2
3	2.394	1.023	1.693	0	2.575	3
4	1.880	.729	2.059	0	2.282	4
5	1.596	.577	2.326	0	2.115	5
6	1.410	.483	2.534	0	2.004	6
7	1.277	.419	2.704	.076	1.924	7
8	1.175	.373	2.847	.136	1.864	8
9	1.094	.337	2.970	.184	1.816	9
10	1.028	.308	3.078	.223	1.777	10
11	.973	.285	3.173	.256	1.744	11
12	.925	.266	3.258	.284	1.716	12
13	.884	.249	3.336	.308	1.692	13
14	.848	.235	3.407	.329	1.671	14
15	.816	.223	3.472	.348	1.652	15
16	.788	.212	3.532	.364	1.636	16
17	.762	.203	3.588	.379	1.621	17
18	.738	.194	3.640	.392	1.608	18
19	.717	.187	3.689	.404	1.596	19
20	.697	.180	3.735	.414	1.586	20
21	.679	.173	3.778	.425	1.575	21
22	.662	.167	3.819	.434	1.566	22
23	.647	.162	3.858	.443	1.557	23
24	.632	.157	3.895	.452	1.548	24
25	.619	.153	3.931	.459	1.541	25

$^a n > 25$: $A_1 = 3/\sqrt{n}$ where n = number of observations in sample.

Appendix B
Bibliography

INTRODUCTORY WORKS AND GRAPHICAL METHODS

Chambers, J., Cleveland, W., Kleiner, B., and Tukey, P. (1983), *Graphical Methods for Data Analysis,* Wadsworth & Brooks/Cole, Pacific Grove, CA. A very well-written presentation of graphical methods in statistics.

Freedman, D., Pisani, R., Purves R., and Adbikari, A. (1991), *Statistics,* 2nd ed., Norton, New York. An excellent introduction to statistical thinking, requiring minimal mathematical background.

Hoaglin, D., Mosteller, F., and Tukey, J. (1983), *Understanding Robust and Exploratory Data Analysis,* John Wiley & Sons, New York. Good discussion and illustration of techniques such as stem-and-leaf displays and box plots.

Tanur, J., et al. (eds.) (1989), *Statistics: A Guide to the Unknown,* 3rd ed., Wadsworth & Brooks/Cole, Pacific Grove, CA. Contains a collection of short nonmathematical articles describing different applications of statistics.

Tukey, J. (1977), *Exploratory Data Analysis,* Addison-Wesley, Reading, MA. Introduces many new descriptive and analytical methods. Not extremely easy to read.

PROBABILITY

Derman, C., Olkin, I., and Gleser, L. (1980), *Probability Models and Applications,* 2nd ed., Macmillan, New York. A comprehensive treatment of probability at a higher mathematical level than this book.

Hoel, P. G., Port, S. C., and Stone, C. J. (1971), *Introduction to Probability Theory,* Houghton Mifflin, Boston. A well-written and comprehensive treatment of probability theory and the standard discrete and continuous distributions.

Mosteller, F., Rourke, R., and Thomas, G. (1970), *Probability with Statistical Applications,* 2nd ed., Addison-Wesley, Reading, MA. A precalculus introduction to probability with many excellent examples.

Ross, S. (1998), *A First Course in Probability,* 5th ed., Macmillan, New York. More mathematically sophisticated than this book, but has many excellent examples and exercises.

ENGINEERING STATISTICS

Hines, W. W., Montgomery, D. C., Goldsman, D. M., and Borror, C. M. (2003), *Probability and Statistics in Engineering,* 4th ed., John Wiley & Sons, Hoboken, NJ. Covers many of the same topics as this book. More emphasis on probability and a higher mathematical level.

Montgomery, D. C., and Runger, G. C. (2006), *Applied Statistics and Probability for Engineers,* 4th ed., John Wiley & Sons, Hoboken, NJ. A more comprehensive book on engineering statistics at about the same level as this one.

Ross, S. (1987), *Introduction to Probability and Statistics for Engineers and Scientists,* John Wiley & Sons, New York. More tightly written and mathematically oriented than this book, but contains some good examples.

EMPIRICAL MODEL BUILDING

Daniel, C., and Wood, F. (1980), *Fitting Equations to Data,* 2nd ed., John Wiley & Sons, New York. An excellent reference containing many insights on data analysis.

Draper, N., and Smith, H. (1998), *Applied Regression Analysis,* 3rd ed., John Wiley & Sons, New York. A comprehensive book on regression written for statistically oriented readers.

Montgomery, D. C., Peck, E. A., and Vining, G. G. (2006), *Introduction to Linear Regression Analysis,* 4th ed., John Wiley & Sons, Hoboken, NJ. A comprehensive book on regression written for engineers and physical scientists.

Myers, R. H. (1990), *Classical and Modern Regression with Applications,* 2nd ed., PWS-Kent, Boston. Contains many examples with annotated SAS output. Very well written.

DESIGN OF EXPERIMENTS

Box, G. E. P., Hunter, W. G., and Hunter, J. S. (1978), *Statistics for Experimenters,* John Wiley & Sons, New York. An excellent introduction to the subject for those readers desiring a statistically oriented treatment. Contains many useful suggestions for data analysis.

Montgomery, D. C. (2005a), *Design and Analysis of Experiments,* 6th ed., John Wiley & Sons, Hoboken, NJ. Written at the same level as the Box, Hunter, and Hunter book, but focused on applications in engineering and science.

STATISTICAL QUALITY CONTROL AND RELATED METHODS

Duncan, A. J. (1974), *Quality Control and Industrial Statistics,* 4th ed., Richard D. Irwin, Homewood, IL. A classic book on the subject.

Grant, E. L., and Leavenworth, R. S. (1988), *Statistical Quality Control,* 6th ed., McGraw-Hill, New York. One of the first books on the subject; contains many good examples.

Montgomery, D. C. (2005b), *Introduction to Statistical Quality Control,* 5th ed., John Wiley & Sons, Hoboken, NJ.

A modern comprehensive treatment of the subject written at the same level as this book.

Western Electric Company (1956), *Statistical Quality Control Handbook,* Western Electric Company, Inc., Indianapolis, IN. An oldie but a goodie.

Appendix C
Answers
to Selected
Exercises

CHAPTER 2

Section 2-1

2-1. $\bar{x} = 56.09$, $s = 11.33$
2-3. $\bar{x} = 1288.43$, $s = 15.80$
2-5. $\bar{x} = 43.98$, $s = 12.29$
2-7. $\bar{x} = 810.514$, $s = 128.32$

Section 2-2

2-17.

	N	Median	Q1	Q3	5th	95th
Variable Cycles	70	1436.5	1097.8	1735.0	772.85	2113.5

2-19.

	N	Median	Q1	Q3	5th	95th
Variable Yield	90	89.25	86.10	93.125	83.055	96.58

Section 2-4

2-27. (a) $\bar{x} = 65.86$, $s = 12.16$
    (b) $Q_1 = 58.5$, $Q_3 = 75$
    (c) Median = 67.5
    (d) $\bar{x} = 66.86$, $s = 10.74$, $Q_1 = 60$, $Q_3 = 75$
2-29. (a) $\bar{x} = 2.415$, $s = 0.534$
2-31. (a) $\bar{x} = 83.11$, $s^2 = 50.55$, $s = 7.11$
    (b) $Q_1 = 79.5$, $Q_3 = 84.50$
    (d) $\bar{x} = 81$, $s = 3.46$, $Q_1 = 79.25$, $Q_3 = 83.75$
2-33. (a) $\bar{x} = 0.04939$, $s^2 = 0.00001568$
    (b) $Q_1 = 0.04738$, $Q_3 = 0.0513$
    (c) Median = 0.04975
    (e) $5^{th} = 0.03974$, $95^{th} = 0.057$

2-35. (a) $\bar{x} = 0.7481$, $s^2 = 0.00226$
    (b) $Q_1 = 0.7050$, $Q_3 = 0.7838$
    (c) Median = 0.742
    (e) $5^{th} = 0.5025$, $95^{th} = 0.821$
2-37. (a) High dose: $\bar{x} = 52.65$, $s^2 = 1490.32$
        Control: $\bar{x} = 382.7$, $s^2 = 175224.35$
    (b) High dose: $Q_1 = 21.70$, $Q_3 = 74.38$
        Control: $Q_1 = 101.9$, $Q_3 = 501.1$
    (c) High dose: Median = 45
        Control: Median = 215.4
    (e) High dose: $5^{th} = 13.125$
              $95^{th} = 133.67$
        Control: $5^{th} = 17.045$
              $95^{th} = 1460.23$

Section 2-6

2-45. (a) $X1$ has negative correlation with Y, $X2$ and $X3$ have positive correlation with Y
    (b) $r_{x_1} = -0.883$, $r_{x_2} = 0.585$, $r_{x_3} = 0.995$, agree with part (a)
2-47. (a) Negative (b) -0.852, -0.898
2-49. (a) Positive (b) 0.773

Supplemental Exercises

2-51. (a) $s^2 = 19.9$, $s = 4.46$ (b) $s^2 = 19.9$, $s = 4.46$
    (c) $s^2 = 19.9$, $s = 4.46$ (d) $s^2 = 1990$, $s = 44.6$
2-53. (c) $\bar{x} = 42.14$, $s = 4.41$
2-55. (a) $Range_1 = 4$, $Range_2 = 4$, same
    (b) $s_1 = 1.604$, $s_2 = 1.852$, $s_2 > s_1$
2-57. (b) $\bar{x} = 9.8$, $s = 3.611$

CHAPTER 3

Section 3-2

3-1. Continuous
3-3. Continuous
3-5. Discrete
3-7. Continuous
3-9. Continuous

Section 3-3

3-11. (a) Yes (b) 0.6 (c) 0.4 (d) 1
3-13. (a) 0.7 (b) 0.9 (c) 0.2 (d) 0.5
3-15. (a) 0.55 (b) 0.95 (c) 0.50
3-17. (a) 0.50 (b) 0.25 (c) 0.25 (d) 0.9

Section 3-4

3-19. (a) $k = 3/64, E(X) = 3, V(X) = 0.6$
(b) $k = 1/6, E(X) = 11/9, V(X) = 0.284$
(c) $k = 1, E(X) = 1, V(X) = 1$
(d) $k = 1, E(X) = 100.5, V(X) = 0.08333$
3-21. (a) 1 (b) 0.8647 (c) 0.8647 (d) 0.1353 (e) 9
3-23. (a) 0.7165 (b) 0.2031 (c) 0.6321 (d) 316.2
(e) $E(X) = 3000, V(X) = 9000000$
3-25. (a) 0.5 (b) 0.5 (c) 0.2 (d) 0.4
3-27. (b) $1 - x^{-2}$ (c) 2.0 (d) 0.96 (e) 0.0204
3-29. (a) 0.8 (b) 0.5
3-31. (a) 0.9 (b) 0.8 (c) 0.1 (d) 0.1
(f) $E(X) = 205, V(X) = 8.3333$
3-33. (a) 0.913 (b) $E(X) = 4.3101, V(X) = 51.4230$
(c) 12.9303

Section 3-5

3-35. (a) 0 (b) -3.09 (c) -1.18 (d) -1.11 (e) 1.75
3-37. (a) 0.97725 (b) 0.84134 (c) 0.68268
(d) 0.9973 (e) 0.47725 (f) 0.49865
3-39. (a) 0.99865 (b) 0.00023 (c) 0.47725
(d) 0.83513 (e) 0.69123
3-41. (a) 0.9938 (b) 0.1359 (c) 5835.51
3-43. (a) 0.0082 (b) 0.7211 (c) 0.5641
3-45. (a) 12.309 (b) 12.155
3-47. (a) 0.1587 (b) 90.0 (c) 0.9973
3-49. (a) 0.09012 (b) 0.501165 (c) 13.97
3-51. (a) 0.0668 (b) 0.8664 (c) 0.000214
3-53. (a) 0.9332 (b) 20952.2 (c) 145×10^{22}
3-55. (a) 0.03593 (b) 1.65 (c) 12.6965
3-57. $E(X) = 12000, V(X) = 3.61 \times 10^{10}$
3-59. (a) 0.5273 (b) 8862.3 (c) 0.00166
3-61. $E(X) = 2.5, V(X) = 1.7077$
3-63. (a) 120 (b) 1.32934 (c) 11.6317
3-65. $E(X) = 1.28, V(X) = 0.512$
3-67. $r = 3.24, \lambda = 0.72$

Section 3-7

3-79. (a) 0.433 (b) 0.409 (c) 0.316
(d) $E(X) = 3.319, V(X) = 3.7212$
3-81. (a) 4/7 (b) 3/7 (c) $E(X) = 11/7, V(X) = 26/49$

3-83. (a) 0.170 (b) 0.10 (c) 0.91
(d) $E(X) = 9.98, V(X) = 2.02$
3-85. (b) $E(X) = 2.5, V(X) = 2.05$ (c) 0.5 (d) 0.75

Section 3-8

3-89. (a) 0.0148 (b) 0.8684 (c) 0 (d) 0.1109
3-91. (a) 0.0015 (b) 0.9298 (c) 0 (d) 0.0686
3-93. 0.0043
3-95. (a) $n = 50, p = 0.1$ (b) 0.1117 (c) 0
3-97. (a) 0.9961 (b) 0.989
(c) $E(X) = 112.5, \sigma_X = 3.354$
3-99. (a) 0.13422 (b) 0.000001 (b) 0.301990
3-101. (a) 1 (b) 0.999997
(c) $E(X) = 12.244, \sigma = 1.504$

Section 3-9.1

3-103. (a) 0.7408 (b) 0.9997 (c) 0 (d) 0.0333
3-105. $E(X) = V(X) = 3.912$
3-107. (a) 0.0844 (b) 0.0103 (c) 0.0185 (d) 0.1251
3-109. (a) 4.54×10^{-5} (b) 0.6321
3-111. (a) 0.7261 (b) 0.0731
3-113. 0.2941
3-115. (a) 0.0076 (b) 0.1462
3-117. (a) 0.4566 (b) 0.047

Section 3-9.2

3-119. (a) 0.3679 (b) 0.0498 (c) 0.0183 (d) 14.9787
3-121. (a) 0.0821 (b) 0.5654 (c) 0.2246 (d) 27.63
3-123. (a) 0.1353 (b) 0.2707 (c) 5 (d) 0.0409
(e) 0.1353 (f) 0.1353
3-125. (a) 0.3679 (b) 0.3679 (c) 2

Section 3-10

3-127. (a) 0.010724 (b) 0.007760 (c) 0.107488
3-129. (a) 0.4471 (b) 0.428576
3-131. (a) 0.107488 (b) 0.440427
3-133. (a) $E(X) = 362, \sigma = 19.0168$ (b) 0.2555
(c) 392.7799
3-135. (a) 0.819754 (b) 0.930563 (c) 0.069437
(d) 0.694974

Section 3-11

3-137. (a) 0.2457 (b) 0.7641 (c) 0.5743 (d) 0.1848
3-139. (a) 0.372 (b) 0.1402 (c) 0.3437 (d) 0.5783
3-141. (a) 0.8404 (b) 0.4033 (c) 0
3-143. (a) $P(T_1) = 0.35, P(T_2) = 0.45, P(T_3) = 0.25,$
$P(T_4) = 0.10$ (b) 0.0175 (c) 0.0295
3-145. 0.8740
3-147. 0.988
3-149. 0.973675

Section 3-12

3-151. (a) 31 (b) 268
3-153. (a) 22 (b) 128 (c) 44 (d) 512
3-155. (a) 0.8770 (b) 0.2864 (c) 0.6826
3-157. (a) $E(T) = 3, \sigma_T = 0.141$ (b) 0.0169
3-159. (a) $E(D) = 6, \sigma_D = 0.1225$ (b) 0.2072

3-161. (a) 7 (b) 0.0041
3-163. $E(Y) = 800, V(Y) = 57600$
3-165. $E(P) = 160,000 \ V(P) = 16000000$
3-167. $E(G) = 0.07396, V(G) = 3.23 \times 10^{-7}$
3-169. $E(Y) = 24, V(Y) = 644$

Section 3-13

3-171. (a) Mean = 100, Variance = 81/16
 (b) 0.1870 (c) 0.0912 (d) 0.7753
3-173. (a) Mean = 20, Variance = 1/20
 (b) 0 (c) 0 (d) 1
3-175. (a) 0.0016 (b) 6
3-177. 0.4306
3-179. (a) 0.1762 (b) 0.8237 (c) 0.0005
3-181. (a) 0.0791 (b) 0.1038 (c) 0.1867

Supplemental Exercises

3-183. (a) 0.7769 (b) 0.7769 (c) 0.1733
 (d) 0 (e) 0.0498
3-185. (a) 0.6 (b) 0.8 (c) 0.7 (d) 3.9 (e) 3.09
3-187. Competitor has longer life
3-189. (a) 0.1298 (b) 0.8972 (c) 42.5
 (d) 1.51, 0.7356, 2.55
3-191. (a) Exponential, mean = 12 (b) 0.2865
 (c) 0.341 (d) 0.436
3-193. (a) Exponential, mean = 100 (b) 0.632
 (c) 0.1353 (d) 0.6065
3-195. (a) 0.0978 (b) 0.0006 (c) 0.00005
3-197. (a) 33.3 (b) 22.36
3-199. (a) Mean = 240, Variance = 0.42 (b) 0.0010
3-201. (a) 0.0084 (b) 0
3-203. (a) 0.0018 (b) 3.4×10^{-6}
3-205. (a) 0.919 (b) 0.402 (c) Machine 1 (d) 0.252
3-207. (a) No (b) No
3-209. 0.309
3-211. (a) $T = W + X + Y + Z$ (b) 73 (c) 106
 (d) 0.57534
3-213. (a) 0.9619 (b) 0.4305
3-215. (a) 6.92 (b) 0.77 (c) 0.2188
3-219. (c) 312.825
3-221. (b) 0.38 (c) 8.65
3-223. (a) 0.105650 (b) (172.16, 187.84)
3-225. (a) 0.2212 (b) 0.2865 (c) 0.2212

CHAPTER 4

Section 4-2

4-1. SE = 1.04, Variance = 9.7344

4-5. $\hat{\Theta}_1$ better
4-7. 0.5

Section 4-3

4-13. (a) 0.0129 (b) 0.0681 (c) 0.9319
4-15. (a) 13.687 (b) 0.0923 (c) 0.9077
4-17. (a) 0.0244 (b) 0.0122 (c) 0

4-19. (a) 0.057 (b) 0.0105 (c) 0.9895
4-21. (a) 0.0228 (b) 0.0004 (c) 0.9996
4-23. (a) 0.0574 (b) 0.265
4-25. 8.85, 9.16

Section 4-4

4-27. (a) 0.014286 (b) 0.126016 (c) 0.031556
 (d) 0.051176 (e) 0.802588
4-29. (a) $Z = 4, P\text{-value} = 0$ (b) 2-sided
 (c) (30.426, 31.974) (d) 0
4-31. (a) $Z_0 = 1.77 > 1.65$, reject H_0 (b) 0.04
 (c) 0 (d) 2 (e) $(4.003, \infty)$
4-33. (a) $P\text{-value} = 0.7188$, do not reject H_0
 (b) 5 (c) 0.680542 (d) (87.85, 93.11)
 (e) Do not reject H_0
4-35. (a) $P\text{-value} = 0.03673$, reject H_0 (b) 0.363169
 (c) 37 (d) $(1.994, \infty)$ (e) Do not reject H_0
4-37. (a) $Z_0 = -26.79 < -2.58$, reject H_0
 (b) $p\text{-value} = 0$, reject H_0 (c) (3237.53, 3273.31)
 (d) (3231.96, 3278.88)
4-39. 97

Section 4-5

4-41. (a) $0.01 < p\text{-value} < 0.025$
 (b) $0.001 < p\text{-value} < 0.0025$
 (c) $0.025 < p\text{-value} < 0.05$
 (d) $0.05 < p\text{-value} < 0.10$
4-43. (a) $0.005 < p\text{-value} < 0.01$
 (b) $0.025 < p\text{-value} < 0.05$
 (c) $0.0025 < p\text{-value} < 0.005$
 (d) $0.10 < p\text{-value} < 0.25$
4-45. (a) 11
 (b) StDev = 1.1639, 95%$L = (26.2853, \infty)$,
 T = 2.029
4-47. (a) $t_0 = 1.55 < 1.833$, do not reject H_0
 (b) No, $d = 0.3295$, power = 0.22
 (c) $(59732.78, \infty)$ (d) Do not reject H_0
4-49. (a) $t_0 = 2.14 > 1.761$, reject H_0
 (b) $(5522.3, \infty)$ (c) Reject H_0
4-51. (a) Normal (b) $t_0 = 0.8735 > -1.796$, do not reject
 H_0 (c) $(-\infty, 9.358)$ (d) do not reject H_0 (e) 60
4-53. (a) $t_0 = 3.58 > 1.711$, reject H_0 (b) $(4.030, \infty)$
4-55. (a) $t_0 = 30.625 > 2.201$, reject H_0
 (b) $t_0 = 30.625 > 1.711$, reject H_0
 (c) Yes, $d = 3.125$, power = 1 (d) (1.091, 1.105)
4-57. (a) $t_0 = 0.97 < 2.539$, do not reject H_0
 (b) Normal (c) $d = 0.42$, power = 0.3
 (d) $d = 0.52$, power = 0.9, $n = 50$ (e) $(23.326, \infty)$

Section 4-6

4-59. (a) $X_0^2 = 8.96 < 23.685$, do not reject H_0
 (b) $(0.00015, \infty)$ (c) Do not reject H_0
4-61. (a) $X_0^2 = 9.2112 < 16.919$, do not reject H_0
 (b) $0.10 < p\text{-value} < 0.50$
 (c) $(4899877.36, \infty)$ (d) Do not reject H_0

4-63. (a) $32.36 < X_0^2 = 55.88 < 71.42$, do not reject H_0
(b) $0.20 < p$-value < 1
(c) $(0.096, 0.2115)$ (d) Do not reject H_0

Section 4-7

4-65. (a) 1-sided   (b) Yes   (c) Sample $p = 0.69125$, $95\% L = (0.66445, \infty)$, p-value $= 0.014286$
4-67. (a) $Z_0 = 1.48 > 1.28$, reject H_0 (b) 0.5055
(c) 136   (d) $(0.254, \infty)$ (e) Reject H_0 (f) 1572
4-69. 666
4-71. 2401
4-73. (a) 0.8288   (b) 4397
4-75. (a) $Z_0 = 0.2917 > -1.645$, do not reject H_0
(b) 0.3859   (c) 0.4345   (d) 769
4-77. (a) 0.08535   (b) 0

Section 4-8

4-79. (a) $(54291.75, 68692.25)$ (b) $(52875.64, 70108.37)$
4-81. (a) $(4929.93, 6320.27)$ (b) $(4819.73, 6430.47)$
4-83. (a) $(7.617, 10.617)$ (b) $(6.760, 11.474)$

Section 4-10

4-85. (a) $X_0^2 = 2.094 < 9.49$, do not reject H_0 (b) 0.7185
4-87. (a) $X_0^2 = 5.79 < 9.49$, do not reject H_0 (b) 0.2154
4-89. (a) $X_0^2 = 10.39 < 7.81$, reject H_0 (b) 0.0155

Supplemental Exercises

4-95. (a) $0.05 < p$-value < 0.10
(b) $0.01 < p$-value < 0.025
(c) P-value < 0.005
4-97. (a) Normal   (b) $(16.99, \infty)$ (c) $(16.99, 33.25)$
(d) $(-\infty, 343.76)$ (e) $(28.23, 343.74)$
(f) $(15.81, 192.44)$ (g) mean: $(16.88, 33.14)$,
variance: $(28.23, 343.74)$
(i) $(-4.657, 54.897)$ (j) $(-13.191, 63.431)$
4-101. (a) 0.452   (b) 0.102   (c) 0.014
4-103. (a) $t_0 = 11.01 > 1.761$, reject H_0
(b) p-value < 0.0005 (c) $(590.95, \infty)$
(d) $(153.63, 712.74)$
4-105. (b) $0.05 < p$-value < 0.10, do not reject H_0
4-107. (a) $0.1 < p$-value < 0.5, do not reject H_0
(b) $0.025 < p$-value < 0.05, reject H_0
4-109. (a) $t_0 = 0.667 < 2.056$, do not reject H_0
(b) $0.5 < p$-value < 0.8 (c) $(246.84, 404.16)$
4-111. (a) $t_0 = 2.48 > 1.943$, reject H_0
(b) $d = 0.3125$, power $= 0.1$ (c) 100
(d) Reject H_0, $(303.2, \infty)$
4-113. Mean $= 19.514$, df $= 14$
4-115. (a) $X_0^2 = 1.75 > 1.24$, do not reject H_0
4-117. (a) $Z_0 = -1.15 > -2.33$, do not reject H_0,
P-value $= 0.1251$
4-119. (a) $(0.554, 0.720)$ (b) $(0.538, 0.734)$
4-121. (a) 0.0256   (b) 0.1314   (c) 0.8686

4-123. (a) $X_0^2 = 0.88971 < 9.49$, do not reject H_0 (b) 0.641
4-125. (a) SE $= 0.456825$, T $= 3.058$
(b) Do not reject H_0 (c) smaller
(d) $0.005 < p$-value < 0.01, reject H_0

CHAPTER 5

Section 5-2

5-1. (a) p-value $= 0.3222 > 0.05$, do not reject H_0
(b) 0.977   (c) $(-0.0098, 0.0298)$ (d) 12
5-3. (a) $Z_0 = -6.70 < -1.96$, reject H_0 (b) 0
(c) 0.2483   (d) $(-8.21, -4.49)$
5-5. (a) p-value $= 0.0036 < 0.05$, reject H_0 (b) 9
5-7. 8
5-9. $(-21.08, 7.72)$
5-11. (a) p-value $= 0.02872$, reject H_0

Section 5-3

5-13. (a) p-value $= 0.001 < 0.05$, reject H_0 (b) 2-sided
(c) Reject H_0 (d) Reject H_0 (e) -1.196
(f) p-value < 0.001
5-15. (a) p-value < 0.001, reject H_0 (b) $(14.93, 27.28)$
(c) $n > 8$
5-17. (a) p-value > 0.80, do not reject H_0
(b) $(-0.394, 0.494)$
5-19. (a) p-value < 0.0005, reject H_0 (b) $(1.065, \infty)$
(c) Yes
5-21. (a) $t_0 = -2.82 < -2.101$, reject H_0
(b) $0.010 < p$-value < 0.020
(c) $(-0.749, -0.111)$
5-23. (a) $t_0 = 3.14 > 1.345$, reject H_0
(b) $0.0025 < p$-value < 0.005 (c) $(0.045, 0.240)$
5-25. 38
5-27. p-value > 0.40, do not reject H_0
5-29. $(-15, \infty)$

Section 5-4

5-31. (a) $StDev_{x1} = 1.5432$, SE $Mean_{Diff} = 1.160$,
T $= -4.2534$, p-value $= 0.002 < 0.05$,
reject H_0
(b) 2-sided
(c) $(-8.7017, -1.1635)$
(d) $0.001 < p$-value < 0.0025
(e) P-value < 0.001
5-37. (a) $(-1.216, 2.55)$
5-39. (a) $0.01 < p$-value < 0.025, reject H_0
(b) $(-10.97, -0.011)$
5-41. $t_0 = -3.48 > 3.499$, do not reject H_0

Section 5-5

5-45. (a) 1.47   (b) 2.19   (c) 4.41   (d) 0.595
(e) 0.439   (f) 0.297
5-47. $0.025 < p -$ value < 0.05
5-49. $0.248 < f_0 = 3.34 < 4.03$, do not reject H_0

5-51. (a) $(0.2177, 8.92)$ (b) $(0.145, 13.40)$
(c) $(0.582, \infty)$

5-53. $0.265 < f_0 = 1.15 < 3.12$, do not reject H_0

5-55. $f_0 = 0.064 < 0.4386$, reject H_0

5-57. $(0.245, \infty)$

Section 5-6

5-59. (a) 1-sided (b) p-value $= 0.023 < 0.05$,
reject H_0 (c) reject H_0 (d) $(0.0024, 0.14618)$

5-61. (a) 0.819 (b) 383

5-63. (a) p-value $= 0.06142$, do not reject H_0
(b) 0.4592 (c) 0.90

5-65. $(-0.00565, \infty)$

Section 5-8

5-67. $\mathrm{DF}_{\mathrm{Factor}} = 4$, $\mathrm{SS}_{\mathrm{Factor}} = 987.71$, $\mathrm{MS}_{\mathrm{Error}} = 7.46$,
$F = 33.1$, p-value < 0.01

5-69. (a) p-value $= 0$, reject H_0

5-71. (a) p-value $= 0.559$, do not reject H_0

5-73. (a) $\mathrm{MS}_{\mathrm{Treatement}} = 252.64$, $\mathrm{DF}_{\mathrm{Block}} = 5$, $SS_{\mathrm{Block}} = 323.825$, $\mathrm{MS}_{\mathrm{Error}} = 8.4665$
$F = 7.65$, p-value$_{\mathrm{Treatment}} < 0.01$,
p-value$_{\mathrm{Block}} < 0.01$
(b) 6 (c) significance

5-75. (b) $s = 0.022$

Supplemental Exercises

5-81. (b) $f_0 = 0.609 > 0.459$, do not reject H_0

5-83. (a) $f_0 = 4 > 1.92$, reject H_0
(b) p-value < 0.01 (c) $(2.17, \infty)$

5-85. (a) $(-5.362, -0.638)$ (b) $(-6.340, 0.340)$

5-87. (a) $Z_0 = 6.55 > 1.96$, reject H_0
(b) $Z_0 = 6.55 > 2.57$, reject H_0

5-89. (a) $-1.96 < Z_0 = 0.88 < 1.96$, do not reject H_0
(b) $-1.65 < Z_0 = 0.88 < 1.65$, do not reject H_0
(d) $-1.96 < Z_0 = 1.25 < 1.96$, do not reject H_0
$-1.65 < Z_0 = 1.012 < 1.65$, do not reject H_0

5-91. (a) $Z_0 = -5.36 < -2.58$, reject H_0

5-93. 23

5-95. (b) $(-0.366, 0.264)$ (c) Yes

5-97. (a) $n = 4$ (b) Yes

5-99. (a) $\mathrm{SS}_{\mathrm{Error}} = 86.752$, $\mathrm{SS}_{\mathrm{Total}} = 181.881$,
$\mathrm{DF}_{\mathrm{Treatment}} = 4$, $\mathrm{DF}_{\mathrm{Error}} = 15$,
$\mathrm{MS}_{\mathrm{Treatment}} = 23.782$, $\mathrm{MS}_{\mathrm{Error}} = 5.783$
$F = 4.112$
(b) $f_0 = 4.112 > 3.06$, reject H_0

5-101. (b) p-value > 0.8, do not reject H_0 (c) $n = 22$

5-103. (b) 0.002

5-105. (b) p-value $= 0.228 > 0.05$, do not reject H_0

CHAPTER 6

Section 6-2

6-1. (a) $\hat{y} = 0.0249 + 0.129x$
(c) $SS_E = 0.000001370$, $\hat{\sigma}^2 = 0.000000342$

(d) $se(\hat{\beta}_1) = 0.007738$, $se(\hat{\beta}_0) = 0.001786$
(f) 98.6% (i) β_0: $(0.02, 0.03)$, β_1: $(0.107, 0.15)$
(k) $r = 0.993$, p-value $= 0$

6-3. (a) $\hat{y} = 0.393 + 0.00333x$
(c) $SS_E = 0.0007542$, $\hat{\sigma}^2 = 0.0000419$
(d) $se(\hat{\beta}_1) = 0.0005815$, $se(\hat{\beta}_0) = 0.04258$
(f) 64.5% (i) β_0: $(0.304, 0.483)$, β_1:
$(0.00211, 0.00455)$ (k) $r = 0.803$, p-value $= 0$

6-5. (a) $\hat{y} = 40.6 - 2.12x$
(c) $SS_E = 13.999$, $\hat{\sigma}^2 = 1.077$
(d) $se(\hat{\beta}_1) = 0.2313$, $se(\hat{\beta}_0) = 0.7509$ (f) 86.6%
(i) β_0: $(38.93, 41.18)$, β_1: $(-2.62, -1.62)$
(k) $r = -0.931$, p-value $= 0$

6-7. (a) 0.055137 (b) $(0.054460, 0.055813)$
(c) $(0.053376, 0.056897)$

6-9. (a) 36.095 (b) $(35.059, 37.131)$
(c) $(32.802, 39.388)$

6-11. (a) $\hat{y} = 32.049 - 2.77x$, $\hat{\sigma}^2 = 1.118$ (b) 14.045
(c) -1.385 (d) 3.61 (e) 1.574

6-13. (a) $T_x = 9.7141$, p-value$_x < 0.001$,
$\mathrm{R}^2_{\mathrm{Adjusted}} = 87.78\%$, $\mathrm{SS}_{\mathrm{Total}} = 4.1289$
$\mathrm{MS}_{\mathrm{Error}} = 0.0388$, $\mathrm{S} = 0.197$, $\mathrm{F} = 94.41$,
p-value$_{\mathrm{regression}} < 0.02$
(b) 0.0388 (c) significance (d) $(0.6444, 1.0171)$
(f) $\hat{y} = 0.6649 + 0.83075x$, 0.324
(g) 95% CI: $(1.787, 2.035)$, 95% PI: $(1.464, 2.358)$

Section 6-3

6-15. (a) $\hat{y} = 351 - 1.27x_1 - 0.154x_2$
(c) $\mathrm{SS}_E = 1950.4$, $\hat{\sigma}^2 = 650.1$ (d) $\mathrm{R}^2 = 86.2\%$,
$\mathrm{R}^2_{\mathrm{Adjusted}} = 77.0\%$ (f) $se(\hat{\beta}_0) = 74.75$,
$se(\hat{\beta}_1) = 1.169$, $se(\hat{\beta}_2) = 0.08953$
(h) β_0: $(113.17, 588.81)$, β_1: $(-4.99, 2.45)$,
β_2: $(-0.439, 0.131)$ (j) $VIF = 2.6$

6-17. (a) $\hat{y} = -103 + 0.605x_1 + 8.92x_2 + 1.44x_3$
$+ 0.014x_4$
(c) $\mathrm{SS}_E = 1699.0$, $\hat{\sigma}^2 = 242.7$
(d) $\mathrm{R}^2 = 74.5\%$, $\mathrm{R}^2_{\mathrm{Adjusted}} = 59.9\%$
(f) $se(\hat{\beta}_0) = 207.9$, $se(\hat{\beta}_1) = 0.3689$,
$se(\hat{\beta}_2) = 5.301$, $se(\hat{\beta}_3) = 2.392$,
$se(\hat{\beta}_4) = 0.7338$
(h) β_0: $(-594.38, 388.98)$, β_1: $(-0.267, 1.478)$,
β_2: $(-3.613, 21.461)$, β_3: $(-4.22, 7.094)$,
β_4: $(-1.722, 1.75)$ (k) $VIFs < 10$

6-19. (a) 149.9 (b) $(85.1, 214.7)$ (c) $(-12.5, 312.3)$

6-21. (a) 287.56 (b) $(263.77, 311.35)$
(c) $(243.69, 331.44)$

6-23. (a) $\hat{y} = 238.56 + 0.3339x_1 - 2.7167x_2$
(b) $\hat{\sigma}^2 = 1321$, $se(\hat{\beta}_0) = 45.23$,
$se(\hat{\beta}_1) = 0.6763$, $se(\hat{\beta}_2) = 0.6887$
(d) $(-37.0845, 160.1235)$ (e) $(18.9417, 104.0973)$

6-25. (a) $T_{x1} = 1.2859$, $T_{x2} = 13.8682$,
$0.20 < p\text{-value}_{x1} < 0.50$, $p\text{-value}_{x2} < 0.001$,
$R_2 = 88.50\%$, $SS_{Total} = 4.1289$, $DF_{Error} = 25$,
$MS_{Error} = 0.6933$, $S = 0.832646$,
$F = 96.182$, $p\text{-value}_{regression} < 0.02$
(b) $\hat{\sigma}^2 = 0.693$ (c) $F_0 = 96.18 > 3.39$, reject H_0
(d) β_1: $T_0 = 1.29 < 2.060$, not significant,
β_2: $T_0 = 13.87 > 2.060$, significant
(e) $(-0.4463, 1.9297)$ (f) $(7.7602, 10.4682)$

Section 6-4

6-27. (a) $\hat{y} = 643 + 11.4x_1 - 0.933x_2 - 0.0106x_1x_2$
$- 0.0272x_1^2 + 0.000471x_2^2$

Supplemental Exercises

6-37. (a) 0.12   (b) Points 17 and 18 are leverage points
6-39. (a) $F_0 = 1323.62 > 4.38$, reject H_0
6-41. (b) $\hat{y} = 0.47 + 20.6x$ (c) 21.038
(d) $\hat{y} = 101371$, $e = 1.6629$ (f) $\hat{y} = 21.031461x$
6-43. (b) $\hat{y} = 2625 - 37x$ (c) 1886.154
6-49. (a) $\hat{y} = -440 + 19.1x_1 + 68.1x_2$
(b) $\hat{\sigma}^2 = 55563$, $se(\hat{\beta}_0) = 94.20$, $se(\hat{\beta}_1) = 3.460$,
$se(\hat{\beta}_2) = 5.241$ (d) $(-299.8, 674.3)$
(e) $(91.9, 282.6)$

CHAPTER 7
Section 7-3

7-1. (a)

Term	Coeff	Se (coeff)
Material	9.313	7.730
Temperature	-33.938	7.730
Mat $\times$ Temp	4.687	7.730

(c) Only Temperature is significant
(d) Material: $(-12.294, 49.546)$
Temperature: $(-98.796, -36.956)$
Material $\times$ Temperature: $(-21.546, 40.294)$

7-3. (a)

Term	Coeff	Se (coeff)
PVP	7.8250	0.5744
Time	3.3875	0.5744
PVP $\times$ Time	-2.7125	0.5744

(c) All are significant
(d) PVP: $(13.352, 17.948)$
Time: $(4.477, 9.073)$
PVP $\times$ Time: $(-7.723, -3.127)$

7-5. (a)

Term	Coeff	Se (coeff)
Temp	-0.875	0.7181
%Copper	3.625	0.7181
Temp $\times$ Copper	1.375	0.7181

(c) %Copper is significant
(d) Temperature: $(-4.622, 1.122)$
%Copper: $(4.378, 10.122)$
Temperature $\times$ %Copper: $(-0.122, 5.622)$

7-7. (a)

Term	Coeff	Se (coeff)
Doping	-0.2425	0.07622
Anneal	1.0150	0.07622
Doping $\times$ Anneal	0.1675	0.07622

(c) Polysilicon doping and Anneal are significant
(d) Doping: $(-0.7898, -0.1802)$
Anneal: $(1.725, 2.34)$
Doping $\times$ Anneal: $(0.0302, 0.6398)$

7-9. (a) $MS_B = 0.80083$, $F_{interaction} = 50.7036$,
$P\text{-value}_A < 0.01$, $P\text{-value}_B < 0.01$
(b) All are significant
(c) $SE(\text{effect}) = 0.0866$
(d) $Coeff_A = 0.25835$, $SE(Coeff_A) = SE(Coeff_B)$
$= 0.04330$, $T_B = 5.965$, $P\text{-value}_A < 0.001$,
$P\text{-value}_{AB} < 0.001$
(e) $\hat{y} = 7.9750 + 0.8083x_A + 0.2583x_B$
$- 0.3083x_{AB}$
(f) 7.7333

Sections 7-4, 7-5, and 7-6

7-11. (a)

Term	Effect	Coeff
A	18.25	9.12
B	84.25	42.12
C	71.75	35.88
AB	-11.25	-5.63
AC	-119.25	-59.63
BC	-24.25	-12.13
ABC	-34.75	-17.38

(b) Tool life $= 413 + 9.1A + 42.1B + 35.9C$
$- 59.6AC$
7-13. Score $= 175 + 8.50A + 5.44C + 4.19D + 4.56AD$
7-15. (a) No terms are significant
(b) There is no appropriate model
7-17. (a) $\hat{\sigma}^2 = 8.2$ (b) $t = -9.20$, curvature is significant
7-19. Block 1: (1) ab ac bc; Block 2: a b c abc; none of the
factors or interactions appears to be significant using
only first replicate
7-21. Block 1: (1) acd bcd ab; Block 2: a b cd $abcd$; Block
3: d ac bc abd; Block 4: c ad bd abc; the factors A, C,
and D and the interaction AD and ACD are significant
7-23. ABC, CDE
7-25. (a) $SE(Coeff_{AB}) = 0.3903$, $T_C = 1.1209$,
$T_{AB} = 1.7615$, $T_{BC} = -0.08007$, $0.002 < P\text{-value}_A$
< 0.005, $0.05 < P\text{-value}_B < 0.01$, $0.1 < P\text{-value}_{AB}$
< 0.2, $P\text{-value}_{AC} > 0.8$, $0.1 < P\text{-value}_{ABC} < 0.2$
(b) $\hat{y} = 6.0625 + 1.6875x_1$ (c) 4.375

Section 7-7

7-27. (a) $I = ABCDE$
(b) $A = 1.4350$, $B = -1.4650$, $C = -0.2725$,
$D = 4.5450$, $E = -0.7025$, $AB = 1.15$,
$AC = -0.9125$, $AD = -1.23$, $AE = 0.4275$,
$BC = 0.2925$, $BD = 0.12$, $BE = 0.1625$

(c) color $= 2.77 + 0.718A - 0.732B + 2.27D$
  $+ 0.575AB - 0.615AD$

7-29. (a) $I = ACE = BDE = ABCD$

 (b) $A = -1.525, B = -5.175, C = 2.275,$
  $D = -0.675, E = 2.275, AB = 1.825,$
  $AD = -1.275$

 (c) Factor B is significant (using AB and AD for error)

7-31. Strength $= 3025 + 725A + 1825B + 875D$
  $+ 325E$

Section 7-8

7-37. (a) $\hat{y} = 499.26 + 85x_1 + 35x_2 - 71.67x_3$
  $+ 25.83x_1x_2$

 (b) $\hat{y} = 299 + 50.89x_1 + 75.78x_2 + 59.5x_3$
  $- 17.33x_1^2 - 34x_2^2 - 17.17x_3^2 + 13.33x_1x_2$
  $+ 26.83x_1x_3 - 17.92x_2x_3$

7-39. (a) Central composite design, not rotatable

 (b) Quadratic model not reasonable

 (c) Increase x_1

Section 7-9

7-41. (a) Interaction is significant

 (c) Paint type $= 1$, Drying time $= 25$ minutes

7-43. (a)

Source	F
Hardwood	7.55
Cook time	31.31
Freeness	19.71
Hardwood $\times$ cook time	2.89
Hardwood $\times$ freeness	2.94
Cook time $\times$ freeness	0.95
Hardwood $\times$ Cook time $\times$ freeness	0.94

 (b) Hardwood, cook time, freeness, and hardwood $\times$ freeness are significant

7-45. $SS_{Total} = 171.582, MS_B = 41.322, F_A = 19.17,$
  $F_{AB} = 1.24, P\text{-value}_B < 0.01, P\text{-value}_{AB} > 0.25$

Supplemental Exercises

7-47. (a)

Source	pH	Catalyst	pH $\times$ Catalyst
t_0	2.54	-0.05	5.02

 pH and pH $\times$ Catalyst are significant

7-49. (a)

Source	Level	Salt	Level $\times$ Salt
f_0	63.24	39.75	5.29

 Level, Salt, and Level $\times$ Salt are significant

 (b) Application level 1 increases flammability average

7-51. (a) $A = -2.74, B = -6.66, C = 3.49,$
  $AB = -8.71, AC = 7.04, BC = 11.46,$
  $ABC = -6.49$

 (b) $\hat{\sigma}_c^2 = 8.40$

 (c) Two-way interactions are significant (specifically BC interaction)

 (d) Delta line $= -11.8 - 1.37A - 3.33B$
  $+ 1.75C - 4.35AB + 3.52AC$

$+ 5.73BC$

7-53. (a)

Term	Effect	Term	Effect
V	15.75	FP	-6.00
F	-10.75	FG	19.25
P	-8.75	PG	-3.75
G	-25.00	VFP	1.25
VF	-8.00	VFG	-1.50
VP	3.00	VPG	0.50
VG	2.75		

 (b) V, F, P, G, VF, and FGP are possibly significant

 (c) V, G, and FPG are significant
  Surface Rough $= 101 + 7.88V - 5.37F$
  $- 4.38P - 12.5G - 6.25FPG$

7-55. Factors F, P, G, VF, and VP are significant. Use $G = VFP$ as a design generator.

7-61. Model: $y = 13.728 + 0.2966A - 0.4052B$
  $- 0.1240A^2 - 0.079B^2$
  Maximum viscosity is 14.425 using $A = 1.19$ and $B = -2.56$

CHAPTER 8
Section 8-4

8-1. (a) $\bar{x}$ chart: $UCL = 39.42, LCL = 28.48$
  R chart: $UCL = 13.77, LCL = 0$

 (b) $\bar{x}$ chart: $UCL = 39.34, LCL = 29.22$
  R chart: $UCL = 12.74, LCL = 0$

8-3. (a) $\bar{x}$ chart: $UCL = 18.20, LCL = 12.08$
  R chart: $UCL = 9.581, LCL = 0$

 (b) $\bar{x}$ chart: $UCL = 17.98, LCL = 12.31$
  R chart: $UCL = 8.885, LCL = 0$

8-5. (a) $\bar{x}$ chart: $UCL = 340.69, LCL = 271.87$
  R chart: $UCL = 107.71, LCL = 0$

 (b) $\hat{\mu} = \bar{\bar{x}} = 306.28, \hat{\sigma} = \dfrac{\bar{r}}{d_2} = \dfrac{47.2}{2.059} = 22.92$

8-7. (a) $\bar{x}$ chart: $UCL = 7.485, LCL = 5.613$
  R chart: $UCL = 2.922, LCL = 0$

 (b) $\bar{x}$ chart: $UCL = 7.369, LCL = 5.077$
  R chart: $UCL = 2.883, LCL = 0$

Section 8-5

8-9. (a) x chart: $UCL = 19.15, LCL = 12.83$
  MR chart: $UCL = 3.887, LCL = 0$

 (b) $\hat{\mu} = 15.99, \hat{\sigma} = 1.055$

8-11. (a) x chart: $UCL = 580.2, LCL = 380$
  MR chart: $UCL = 123, LCL = 0$
  Revised limits: x chart: $UCL = 582.3,$
    $LCL = 388.7$
  MR chart: $UCL = 118.9, LCL = 0$

 (b) $\hat{\mu} = 485.5, \hat{\sigma} = 32.26$

Section 8-6

8-13. (a) $C_P = 1.141, C_{pk} = 0.5932$ (b) 0.0376

8-15. $C_p = 0.5285$

8-17. Fallout is 0.0925, $C_P = 0.582, C_{pk} = 0.490$

8-19. $C_P = 0.248$, $C_{pk} = 0.138$
8-21. (388.72, 582.28)

Section 8-7

8-23. $UCL = 1.676$, $LCL = 0$
8-25. $UCL = 8.382$, $LCL = 0$

Section 8-8

8-27. (a) 0.2177   (b) 4.6
8-29. (a) 0.4203   (b) 2.38
8-31. (a) 0.3022   (b) 3.31
8-33. (a) 0.00413   (b) 242.13
8-35. (a) 0.06552   (b) 15.26

Section 8-9

8-37. (a) No significant difference
(b) $\sigma^2_{\text{Total}} = 0.0825$; $\sigma^2_{\text{gauge}} = 0.0825$; $\sigma^2_{\text{specimen}} = 0$
(c) 100%
8-39. (a) 2 replicates
(b) $\sigma^2_{\text{Total}} = 8242$; $\sigma^2_{\text{gauge}} = 574$; $\sigma^2_{\text{part}} = 7668$
(c) 26.39%

8-41. (a) No significant difference
(b) $\sigma^2_{\text{Repeat}} = 104.12$; $\sigma^2_{\text{Reprod}} = 1.44$
8-43. (a) $\sigma^2_{\text{Repeat}} = 3465$; $\sigma^2_{\text{Reprod}} = 20779$

Supplemental Exercises

8-45. (a) $\bar{x}$ chart: $UCL = 64.56$, $LCL = 63.71$
R chart: $UCL = 1.075$, $LCL = 0$
Revised limits: $\bar{x}$ chart: $UCL = 64.02$, $LCL = 63.98$
R chart: $UCL = 0.04623$, $LCL = 0$
(b) $\hat{\mu} = \bar{\bar{x}} = 64$, $\hat{\sigma} = 0.0104$
(c) $C_P = 0.641$ (d) $C_{pk} = 0.641$ (e) 0.025
(f) $ARL = 161$
8-47. (a) $UCL = 1.503$, $LCL = 0$
(b) Revised: $UCL = 1.302$, $LCL = 0$
8-49. (a) 0.0294   (b) 0.0003
8-51. (a) 1.9   (c) 6.25
8-53. (a) 0.01017   (b) 0.0101
(c) 0.098525, 0.0888
8-55. (a) 0.2476   (b) 0.5339
8-59. 0.00135

Index

Summary of Single-Sample Hypothesis Testing Procedures

Case	Null Hypothesis	Test Statistic	Alternative Hypothesis	P-Value	Criteria for Rejection, Fixed-Level Test	OC Curve Parameter	OC Curve Appendix A Chart V
1.	$H_0: \mu = \mu_0$ σ^2 known	$z_0 = \dfrac{\bar{x} - \mu_0}{\sigma/\sqrt{n}}$	$H_1: \mu \neq \mu_0$	$2[1 - \Phi(|z_0|)]$	$|z_0| > z_{\alpha/2}$	—	—
			$H_1: \mu > \mu_0$	$1 - \Phi(z_0)$	$z_0 > z_\alpha$	—	—
			$H_1: \mu < \mu_0$	$\Phi(z_0)$	$z_0 < -z_\alpha$	—	—
2.	$H_0: \mu = \mu_0$ σ^2 unknown	$t_0 = \dfrac{\bar{x} - \mu_0}{s/\sqrt{n}}$	$H_1: \mu \neq \mu_0$	Probability above t_0 plus probability below $-t_0$	$|t_0| > t_{\alpha/2, n-1}$	$d = |\mu - \mu_0|/\sigma$	a, b
			$H_1: \mu > \mu_0$	Probability above t_0	$t_0 > t_{\alpha, n-1}$	$d = (\mu - \mu_0)/\sigma$	c, d
			$H_1: \mu < \mu_0$	Probability below t_0	$t_0 < -t_{\alpha, n-1}$	$d = (\mu_0 - \mu)/\sigma$	c, d
3.	$H_0: \sigma^2 = \sigma_0^2$	$\chi_0^2 = \dfrac{(n-1)s^2}{\sigma_0^2}$	$H_1: \sigma^2 \neq \sigma_0^2$	2 (Probability beyond χ_0^2)	$\chi_0^2 > \chi_{\alpha/2, n-1}^2$ or $\chi_0^2 < \chi_{1-\alpha/2, n-1}^2$	—	—
			$H_1: \sigma^2 > \sigma_0^2$	Probability above χ_0^2	$\chi_0^2 > \chi_{\alpha, n-1}^2$	—	—
			$H_1: \sigma^2 < \sigma_0^2$	Probability below χ_0^2	$\chi_0^2 < \chi_{1-\alpha, n-1}^2$	—	—
4.	$H_0: p = p_0$	$z_0 = \dfrac{x - np_0}{\sqrt{np_0(1 - p_0)}}$	$H_1: p \neq p_0$	$2[1 - \Phi(|z_0|)]$	$|z_0| > z_{\alpha/2}$	—	—
			$H_1: p > p_0$	$1 - \Phi(z_0)$	$z_0 > z_\alpha$	—	—
			$H_1: p < p_0$	$\Phi(z_0)$	$z_0 < -z_\alpha$	—	—

Summary of Single-Sample Interval Estimation Procedures

Case	Problem Type	Point Estimate	Type of Interval	$100(1-\alpha)\%$ Confidence Interval
1.	Confidence interval on the mean μ, variance σ^2 known	$\bar{x}$	Two-sided	$\bar{x} - z_{\alpha/2}\sigma/\sqrt{n} \leq \mu \leq \bar{x} + z_{\alpha/2}\sigma/\sqrt{n}$
			One-sided lower	$\bar{x} - z_{\alpha}\sigma/\sqrt{n} \leq \mu$
			One-sided upper	$\mu \leq \bar{x} + z_{\alpha}\sigma/\sqrt{n}$
2.	Confidence interval on the mean μ of a normal distribution, variance σ^2 unknown	$\bar{x}$	Two-sided	$\bar{x} - t_{\alpha/2,n-1}s/\sqrt{n} \leq \mu \leq \bar{x} + t_{\alpha/2,n-1}s/\sqrt{n}$
			One-sided lower	$\bar{x} - t_{\alpha,n-1}s/\sqrt{n} \leq \mu$
			One-sided upper	$\mu \leq \bar{x} + t_{\alpha,n-1}s/\sqrt{n}$
3.	Confidence interval on the variance σ^2 of a normal distribution	s^2	Two-sided	$\dfrac{(n-1)s^2}{\chi^2_{\alpha/2,n-1}} \leq \sigma^2 \leq \dfrac{(n-1)s^2}{\chi^2_{1-\alpha/2,n-1}}$
			One-sided lower	$\dfrac{(n-1)s^2}{\chi^2_{\alpha,n-1}} \leq \sigma^2$
			One-sided upper	$\sigma^2 \leq \dfrac{(n-1)s^2}{\chi^2_{1-\alpha,n-1}}$
4.	Confidence interval on a proportion or parameter of a binomial distribution p	$\hat{p}$	Two-sided	$\hat{p} - z_{\alpha/2}\sqrt{\dfrac{\hat{p}(1-\hat{p})}{n}} \leq p \leq \hat{p} + z_{\alpha/2}\sqrt{\dfrac{\hat{p}(1-\hat{p})}{n}}$
			One-sided lower	$\hat{p} - z_{\alpha}\sqrt{\dfrac{\hat{p}(1-\hat{p})}{n}} \leq p$
			One-sided upper	$p \leq \hat{p} + z_{\alpha}\sqrt{\dfrac{\hat{p}(1-\hat{p})}{n}}$
5.	Prediction interval on a future observation from a normal distribution, variance unknown	$\bar{x}$	Two-sided	$\bar{x} - t_{\alpha/2,n-1}s\sqrt{1+\dfrac{1}{n}} \leq X_{n+1} \leq \bar{x} + t_{\alpha/2,n-1}s\sqrt{1+\dfrac{1}{n}}$
6.	Tolerance interval for capturing at least $\gamma\%$ of the values in a normal population with confidence level $100(1-\alpha)\%$.		Two-sided	$\bar{x} - ks, \; \bar{x} + ks$

Summary of Two-Sample Hypothesis Testing Procedures

Case	Null Hypothesis	Test Statistic	Alternative Hypothesis	P-Value	Criteria for Rejection, Fixed-Level Test	OC Curve Parameter	OC Curve Appendix A Chart IV
1.	$H_0: \mu_1 = \mu_2$ σ_1^2 and σ_2^2 known	$z_0 = \dfrac{\bar{x}_1 - \bar{x}_2}{\sqrt{\dfrac{\sigma_1^2}{n_1} + \dfrac{\sigma_2^2}{n_2}}}$	$H_1: \mu_1 \neq \mu_2$ $H_1: \mu_1 > \mu_2$ $H_1: \mu_1 < \mu_2$	$2[1 - \Phi(|z_0|)]$ $1 - \Phi(z_0)$ $\Phi(z_0)$	$|z_0| > z_{\alpha/2}$ $z_0 > z_\alpha$ $z_0 < -z_\alpha$	— — —	— — —
2.	$H_0: \mu_1 = \mu_2$ $\sigma_1^2 = \sigma_2^2$ unknown	$t_0 = \dfrac{\bar{x}_1 - \bar{x}_2}{s_p\sqrt{\dfrac{1}{n_1} + \dfrac{1}{n_2}}}$	$H_1: \mu_1 \neq \mu_2$ $H_1: \mu_1 > \mu_2$ $H_1: \mu_1 < \mu_2$	Probability above t_0 plus probability below t_0 Probability above t_0 Probability below t_0	$|t_0| > t_{\alpha/2, n_1 + n_2 - 2}$ $t_0 > t_{\alpha, n_1 + n_2 - 2}$ $t_0 < -t_{\alpha, n_1 + n_2 - 2}$	$d = |\mu - \mu_0|/2\sigma$ $d = (\mu - \mu_0)/2\sigma$ $d = (\mu_0 - \mu)/2\sigma$	a, b c, d c, d
3.	$H_0: \mu_1 = \mu_2$ $\sigma_1^2 \neq \sigma_2^2$ unknown	$t_0 = \dfrac{\bar{x}_1 - \bar{x}_2}{\sqrt{\dfrac{s_1^2}{n_1} + \dfrac{s_2^2}{n_2}}}$ $v = \dfrac{\left(\dfrac{s_1^2}{n_1} + \dfrac{s_2^2}{n_2}\right)^2}{\dfrac{(s_1^2/n_1)^2}{n_1 - 1} + \dfrac{(s_2^2/n_2)^2}{n_2 - 1}}$	$H_1: \mu_1 \neq \mu_2$ $H_1: \mu_1 > \mu_2$ $H_1: \mu_1 < \mu_2$	Probability above t_0 plus probability below t_0 Probability above t_0 Probability below t_0	$|t_0| > t_{\alpha/2, v}$ $t_0 > t_{\alpha, v}$ $t_0 < -t_{\alpha, v}$	— —	— — —
4.	Paired data $H_0: \mu_D = 0$	$t_0 = \dfrac{\bar{d}}{s_d/\sqrt{n}}$	$H_1: \mu_d \neq 0$ $H_1: \mu_d > 0$ $H_1: \mu_d < 0$	Probability above t_0 plus probability below t_0 Probability above t_0 Probability below t_0	$|t_0| > t_{\alpha/2, n-1}$ $t_0 > t_{\alpha, n-1}$ $t_0 < -t_{\alpha, n-1}$	— — —	— — —
5.	$H_0: \sigma_1^2 = \sigma_2^2$	$f_0 = s_1^2/s_2^2$	$H_1: \sigma_1^2 \neq \sigma_2^2$ $H_1: \sigma_1^2 > \sigma_2^2$ $H_1: \sigma_1^2 < \sigma_1^2$	2(Probability beyond f_0) Probability above f_0 Probability below f_0	$f_0 > f_{\alpha/2, n_1 - 1, n_2 - 1}$ or $f_0 < f_{1 - \alpha/2, n_1 - 1, n_2 - 1}$ $f_0 > f_{\alpha, n_1 - 1, n_2 - 1}$ $f_0 < f_{1 - \alpha, n_1 - 1, n_2 - 1}$	— — —	— — —
6.	$H_0: p_1 = p_2$	$z_0 = \dfrac{\hat{p}_1 - \hat{p}_2}{\sqrt{\hat{p}(1 - \hat{p})\left[\dfrac{1}{n_1} + \dfrac{1}{n_2}\right]}}$	$H_1: p_1 \neq p_2$ $H_1: p_1 > p_2$ $H_1: p_1 < p_2$	$2[1 - \Phi(|z_0|)]$ $1 - \Phi(z_0)$ $\Phi(z_0)$	$|z_0| > z_{\alpha/2}$ $z_0 > z_\alpha$ $z_0 < -z_\alpha$	— — —	— — —